C0-AMS-486

The Plant Cytoskeleton in Cell Differentiation and Development

Annual Plant Reviews

A series for researchers and postgraduates in the plant sciences. Each volume in this annual series will focus on a theme of topical importance and emphasis will be placed on rapid publication.

Titles in the series:

1. Arabidopsis
Edited by M. Anderson and J. Roberts

2. Biochemistry of Plant Secondary Metabolism
Edited by M. Wink

3. Functions of Plant Secondary Metabolites and their Exploitation in Biotechnology
Edited by M. Wink

4. Molecular Plant Pathology
Edited by M. Dickinson and J. Beynon

5. Vacuolar Compartments
Edited by D.G. Robinson and J.C. Rogers

6. Plant Reproduction
Edited by S.D. O'Neill and J.A. Roberts

7. Protein–Protein Interactions in Plant Biology
Edited by M.T. McManus, W.A. Laing and A.C. Allan

8. The Plant Cell Wall
Edited by J. Rose

9. The Golgi Apparatus and the Plant Secretory Pathway
Edited by D.G. Robinson

10. The Plant Cytoskeleton in Cell Differentiation and Development
Edited by P.J. Hussey

The Plant Cytoskeleton in Cell Differentiation and Development

Edited by

PATRICK J. HUSSEY
Chair of Plant Molecular Cell Biology
The Integrative Cell Biology Laboratory
School of Biological and Biomedical Sciences
University of Durham
UK

Editorial Offices:
Blackwell Publishing Ltd,
9600 Garsington Road, Oxford OX4 2DQ, UK
Tel: +44 (0)1865 776868
Blackwell Publishing Asia Pty Ltd,
550 Swanston Street, Carlton,
Victoria 3053, Australia
Tel: +61 (0)3 8359 1011

ISBN 1–84127–421–6
Originated as Sheffield Academic Press

Published in the USA and Canada (only) by
CRC Press LLC, 2000 Corporate Blvd., N.W.,
Boca Raton, FL 33431, USA
Orders from the USA and Canada (only) to
CRC Press LLC

USA and Canada only:
ISBN 0–8493–1981–1

First published 2004

Library of Congress
Cataloging-in-Publication Data:
A catalog record for this title is available from the Library of Congress

British Library Cataloguing-in-Publication Data:
A catalogue record for this title is available from the British Library

Set in 10/12 pt Times
by Integra Software Services Pvt. Ltd, Pondicherry, India
Printed and bound in Great Britain by
MPG Books Ltd, Bodmin, Cornwall

For further information on
Blackwell Publishing, visit our website:
www.blackwellpublishing.com

Contents

List of contributors

Dr Mark Beilstein	Department of Biology, University of Missouri-St Louis, 8001 Natural Bridge Road, St Louis, MO 63121
Dr Laszlo Bogre	School of Biological Sciences, Royal Holloway, University of London, Egham TW20 OEX, UK
Dr Luísa Camacho	Departamento de Biologia Vegetal, Faculdade de Ciências de Lisboa, Bloco C2, 1749-016 Lisboa, Portugal
Prof. Nicholas C. Carpita	Department of Botany and Plant Pathology, Purdue University, West Lafayette, IN47907, USA
Dr Jordi Chan	Plant Cytoskeleton Group, Department of Cell and Developmental Biology, John Innes Centre, Norwich NR4 7UH
Prof. Nam-Hai Chua	Laboratory of Plant Molecular Biology, Rockefeller University, 1230 York Avenue, New York, NY 10021, USA
Dr David A. Collings	Plant Cell Biology Group, Research School of Biological Sciences, The Australian National University
Dr John H. Doonan	Department of Cell Biology, John Innes Centre, Colney Lane, Norwich NR4 7UH
Dr Paula Duque	Laboratory of Plant Molecular Biology, Rockefeller University, 1230 York Avenue, New York, NY 10021, USA
Dr Claire Grierson	School of Biological Sciences, University of Bristol, Woodland Road, Clifton, Bristol BS8 1UG
Dr Franz Grolig	Fachbereich Biologie/Botanik, Philipps-Universität, Karl Von Frisch Str. 8 D-35032 Marburg, Germany
Prof. Patrick J. Hussey	School of Biological and Biomedical Sciences, University of Durham, South Road, Durham DH1 3LE
Dr Tijs Ketelaar	The Integrative Cell Biology Laboratory, School of Biological and Biomedical Sciences, University of Durham, South Road, Durham, DHI 3LE, UK

Prof. Clive Lloyd Plant Cytoskeleton Group, Department of Cell and Developmental Biology, John Innes Centre, Norwich NR4 7UH

Prof. Rui Malhó Dept. de Biologia Vegetal, FCL, Bloco C2, Campo Grande, 1749-016 Lisboa, Portugal

Dr Maureen C. McCann Department of Cell and Developmental Biology, John Innes Centre, Colney Lane, Norwich NR4 7UH

Dr Juan-Pablo Sánchez Laboratory of Plant Molecular Biology, Rockefeller University, 1230 York Avenue, New York, NY 10021, USA

Prof. Christopher J. Staiger Department of Biological Sciences, Purdue University, West Lafayette, IN 47907-1392, USA

Dr Keiko Sugimoto-Shirasu Department of Cell and Developmental Biology, John Innes Centre, Colney, Norwich, NR4 7UH, UK

Dr Dan Szymanski Department of Agronomy, Purdue University, 1150 Lilly Hall of Life Sciences, W. Lafayette IN 47907-1150, USA

Dr Geoffrey O. Wasteneys Plant Cell Biology Group, Research School of Biological Sciences, The Australian National University, GPO Box 475, Canberra ACT 2601, Australia

Dr Magdalena Weingarner Max-Planck-Institute of Molecular Plant Physiology, AM Mühlenberg 1, 14476 Golm, Germany

Preface

The plant cytoskeleton is a dynamic filamentous structure composed of actin and microtubule networks. Actin and microtubules in plants are no different structurally from their animal and fungal counterparts. However, the strategies of cell differentiation and development in plants require this network to respond appropriately to plant-specific developmental cues and to environmental factors. This book, written by biochemists, cell biologists and geneticists, views the cytoskeleton from different perspectives but, on the whole, as a network composed of structural and regulatory proteins controlled by internal and external stimuli that result in different aspects of cell differentiation.

Biochemical and genetic analyses, coupled with homology searches between the complete *Arabidopsis* and human genome sequences, have provided a fairly clear picture of the proteins that are common between the plant and animal cytoskeletons. From these analyses, it is apparent that plants do not have all the cytoskeletal proteins present in animals. Moreover, where plants have the animal homologues, the plant versions can have unique properties and can be utilised in different ways or through different pathways. Perhaps this is because plants have to respond to unique signals and, through evolution, have developed their own responsive strategies. Very few proteins have been found to be cytoskeletal-associated and at the same time plant-specific, but it is hard to believe that the immobile life of a plant cell has not evolved its own specific class of associated cytoskeletal factors.

The first part of this book describes the key molecules of the actin and microtubule networks (Chapters 1 and 2). These chapters are followed by chapters on basic functions that are attributed to the plant cytoskeleton: cell growth and expansion (Chapter 4), mitosis – including a comparison between plants and animals (Chapter 5), and factors that contribute to organelle movement and transport on the cytoskeleton (Chapter 7). Given the close association of the cytoskeleton with the external scaffolding of the plant cell, Chapter 6 considers the cell wall as the primary sensory panel for signal transduction to and from the cytoskeletal network. Moving on from these basic foundations, the last four chapters examine the role of a stimulus-responsive cytoskeletal network in cells commonly used as models for studying cell morphogenesis and signal transduction: root hairs (Chapter 7), pollen tubes (Chapter 8), trichomes (Chapter 9) and guard cells (Chapter 10).

I would like to take this opportunity to thank all the authors for their hard work in producing such excellent chapters and to thank the publishers. Together we have created an excellent volume of definitive reviews that will serve as a valuable resource for plant scientists.

Patrick J. Hussey

Part 1
The cytoskeleton: the machinery and key molecules

1 Microtubules and microtubule-associated proteins

Clive Lloyd, Jordi Chan and Patrick J. Hussey

1.1 Introduction

The asymmetrical and often sculptured shape of the plant cell is held by the cell wall – a highly complex structure involving the coordinated activity of hundreds of enzymes. In addition to the integration of numerous metabolic activities, there is a very strong spatial requirement to ensure that wall precursors are inserted in appropriate sites at appropriate times. This is where microtubules play an important part.

At different stages of the cell cycle, microtubules form four different scaffolds, and three of these are involved with wall formation and alignment. The cortical microtubule array is found beneath the plasma membrane during interphase. By directing precursors to the appropriate sites, these microtubules help to orchestrate the various assembly processes and thus help form an organized cell wall. Whereas the cytoplasmic microtubules of animal cells radiate to the periphery from the cell body, the interphase microtubules do not radiate from a fixed point but are dispersed over the cortex in a manner reflecting the diffuse growth of the body cells (as opposed to free-growing hairs, filament tubes that grow by tip growth). This characteristic organization is crucial in explaining the different ways in which plant and animal cells behave and how they attain their overall form. Plants are effectively pumped out into the environment by turgor pressures several times that inside a car tyre. Great strength is needed to contain such large forces and this is provided by the high tensile strength of cellulose microfibrils that wrap around the cell, fulfilling much the same purpose as the steel threads that reinforce a rubber tyre. The direction in which cell swelling occurs is controlled by the particular arrangement of microfibrils in the cell wall. However, it seems likely that controlled, directional growth will not occur without the overall coordinating influence of the cortical microtubules which are sandwiched in a thin cytoplasmic layer between the vacuole and the wall. While the cellulose microfibrils provide the high tensile strength necessary to resist the high internal pressure of the turgid cell, it appears that microtubules some how help to regulate the direction in which wall materials are deposited and, indirectly, the direction in which the cell expands. The parallelism between cortical microtubules and cellulose microfibrils may not be as tight as suggested by early studies but, as mentioned later, it does seem that without organized microtubules, cell wall formation eventually becomes disorganized, and directional growth perturbed. One long-held idea is that groups of parallel microtubules, which run transversely around elongating cells, provide tracks for the movement of the self-propelled cellulose synthases as they spin out the 30–50 chains of β1–4 linked glucosyl residues that make up each semi-crystalline cellulose microfibril. It must be emphasized, however, that this view of the wall-laying machine remains hypothetical (see Chapters 3 and 6). Reconstituting cellulose biosynthesis *in vitro* has proved largely elusive; demonstrating any kind of linkage (whether direct or indirect) between synthesizing complex and cytoskeleton has also proved difficult, with the consequence that the way we conceive the behaviour of this transmembrane, supramolecular assembly has hardly advanced at all. Examining the so-called microtubule–microfibril paradigm is not within the scope of this book, but in this chapter we will examine the rapid progress that has been made over the last few years in understanding the associated proteins that regulate the formation, the associative properties and the dynamic behaviour of plant microtubules. We shall also be examining the multiplicity of tubulin isotypes that make up the microtubule arrays, as well as mutant tubulins that affect the morphology of plants as well as properties such as herbicide resistance.

The second occasion that microtubules can influence the deposition of cell wall is during cytokinesis when the interdigitated double rings of microtubules constituting the phragmoplast guide secretory vesicles, not to the cortex, but to the new cell plate formed between the daughter nuclei. Another cortical array, the preprophase band, which looks like a bunched-up ring-like version of the interphase array, predicts the site where the phragmoplast will eventually join the mother cell wall. However, all cortical microtubules disappear upon entry into mitosis to be replaced by two kinds of bipolar microtubular structures involved in cell division. The mitotic spindle is subtly different from the animal spindle. For example, there are no centrioles at the spindle poles and the astral microtubules that radiate out of the back end of the spindle poles are less conspicuous (close inspection shows that there are such back end microtubules at anaphase). Also, the division plane is determined before mitosis by processes that align the preprophase band. Another bipolar microtubular structure, the phragmoplast, then guides secretory vesicles to the line between the two sets of sister chromatids, and the ring-like phragmoplast then grows out centrifugally to meet the mother wall at the site foretold by the transient placement of the preprophase band. These microtubule-based structures are therefore quite distinct from those found in animal cells; at least three of them (PPB, spindle, phragmoplast) are intimately involved with deposition of the cell wall.

Bioinformatic analysis of the various published eukaryotic genomes shows that although plants do not possess obvious homologues of the brain plant microtubule-associated proteins (MAPs), many of the MAPs discovered so far are shared by other eukaryotes (with some evidence of plant-specific sequences), so why study them in plants? The resounding answer is that molecules have more in common than the cells from which they come, and plant cells put these common building blocks to different uses. For example, the dispersed cortical microtubule array held together in parallel sheets underlying a carbohydrate-rich wall is different; biosynthesis of cellulose from sites within the plasma membrane itself is different from the secretion of pre-assembled collagen; the division plane is determined (and marked) prior to metaphase and the centrifugally growing phragmoplast, in which vesicles are guided by microtubules to a cell plate, is also unique. In the post-genomic phase it is important to decipher how protein machines work and in this respect the plant cytoskeleton offers exciting possibilities for studying protein complexes for which there are no other precedents. Microtubule-associated proteins (MAPs) are likely to provide the key to understanding the organization of the microtubule arrays and the processes they support. This review concentrates on the research over the last decade (much of it in the last few years) that provides a molecular basis for understanding the role of microtubule proteins in cell expansion and division. We begin with the distinctive features of plant tubulins.

1.2 Plant tubulin

The major microtubule protein is a heterodimer of α- and β-tubulin subunits. This heterodimer is dumbbell shaped and associates head to tail to form protofilaments. In general 13 protofilaments associate laterally in a staggered fashion to form the wall of a hollow tubule of external diameter 25 nm. Structurally, plant microtubules are no different from their animal/fungal counterparts. However, they seem to be more dynamic *in vivo*. Microinjection of fluorescent tubulin or expression of MAP4-GFP chimeras in plant cells, coupled with measuring fluorescence recovery after photobleaching (FRAP), has shown that cortical microtubules demonstrate dynamic instability (stochastic switching between microtubule growth and shrinkage) and are more dynamic in this respect than microtubules in many mammalian cells (Hush *et al.*, 1994; Yuan *et al.*, 1994; Marc *et al.*, 1998).

Plant microtubules also show different responses to anti-microtubule agents compared to mammalian microtubules. Plant microtubules are more resistant to breakdown by colchicine. Also, the dinitroaniline and the phosphorothioamidate anti-mitotic herbicides, which do not bind mammalian tubulin, have been shown to bind to purified plant tubulin and inhibit the *in vitro* assembly of plant microtubules. The dinitroanilines and phosphorothioamidate herbicides have been shown to have similar 3D geometry even though they have very different chemistry, indicating that they are likely to bind the same target site (Anthony & Hussey, 1999a). Extensive use of the dinitroaniline herbicides, coupled with their persistence in the soil, has had severe consequences in agriculture. Goosegrass is classed as one of the world's worst weeds, and dinitroaniline resistant biotypes are a persistent problem, particularly in the soybean and cotton fields of the southern states of North America. Two biotypes are now well characterized, a dinitroaniline resistant (R) biotype which has an ID 50 (the dose of herbicide for a 50% inhibition of growth) of 1.70 mg/l and an intermediate (I) biotype with an ID 50 of 0.18 mg/l. The sensitive (S) biotype for comparison has an ID 50 of 0.04 mg/l (Anthony *et al.*, 1998; Yamamoto *et al.*, 1998; Zeng & Baird, 1999). Both biotypes are cross-resistant to the phosphorothioamidate herbicides.

The cause of this resistance in both the R-biotype and the I-biotype (Anthony *et al.*, 1998; Yamamoto *et al.*, 1998) has been shown to be the result of a single amino acid change in an α-tubulin: Thr239 to Ile239 in the R-biotype and Met268 to Thr268 in the I-biotype. Both mutations have been shown to confer dinitroaniline resistance on transgenic maize calli (Anthony & Hussey, 1999b). Moreover, the combined R- and I-biotype mutations increase the herbicide tolerance of transgenic maize calli by a value close to the summation of the maximum herbicide tolerances of maize calli harbouring the single mutations. The threonine at 239 in a tubulin is positioned at the end of the long central helix close to the site of dimer–dimer interaction. The residue at 268, which is a methionine in plants and algae, and a proline in metazoans, is buried in the centre of the α-tubulin. Taken together, these data imply that each mutation is likely to exert its effect by a different mechanism. These mechanisms may involve increasing the stability of microtubules against the depolymerizing effects of the herbicide or changing the conformation of the α/β-tubulin dimer so that herbicide binding is less effective or a combination of the two.

Another mutation in plant tubulin that has a significant effect on the morphology of plant growth is the lefty mutant of *Arabidopsis thaliana* (Furutani *et al.*, 2000; Hashimoto, 2002; Thitamadee *et al.*, 2002). The lefty mutants are so-called because the various organs of the plant, including roots and shoots, twist to the left. The lefty mutation is in an α-tubulin gene which results in an amino acid change, a serine being replaced by a phenylalanine. The amino acid at position 180 is highly conserved (either a serine or an alanine) across taxonomically distant species and it is close to the interface between the α- and β-tubulin subunits and is in contact with the non-exchangeable GTP nucleotide. The function of the non-exchangeable GTP is unclear but it is possible that its presence favours the formation of the tubulin heterodimer in eukaryotic cells. In this context, mutations at or near this site may change the shape of the dimer which results in less stable microtubules (Thitamadee *et al.*, 2002). This change in shape of the dimer could also change the shape of the microtubules and/or affect their orientation. In the lefty mutants, the microtubules in the elongated epidermal cells are not transverse, but are offset at an angle to the long axis so that they form right-handed helices. It has long been believed that microtubules have some influence over the deposition of the cellulose microfibrils that constitute the cell wall and that the orientation of these microfibrils is parallel to the orientation of the microtubules (although there are exceptions). The cell wall maintains the shape of a cell that is set up by the underlying microtubules. Offsetting the angle of the microtubules to form right-handed helices may offset the orientation of the fibres that constitute the cell wall and result in a twist that is propagated along a cell file. It has been suggested that this could be the underlying cause of the lefty phenotype (Hussey, 2002; Lloyd & Chan, 2002).

1.3 Microtubule-associated proteins

Microtubule-associated proteins were first isolated from animal cells as proteins that co-purified with brain tubulin through rounds of temperature-induced assembly/disassembly. Proteins that were enriched were termed MAPs. This method depends upon the relatively high amount of neuronal tubulin in brain extracts which, in the warm, in the absence of calcium ions, and above the critical concentration for assembly, self-assemble and can be pelleted away from non-assembling proteins at relatively low speed centrifugation. For a long time MAPs could not be isolated from less specialized animal cells, and in plants progress was even slower. Another problem with this stringent definition of a MAP was that not all proteins now known to interact with microtubules could be identified by rounds of assembly/disassembly, for instance, the kinesins. The mechanochemical ATPase kinesin was only isolated because its affinity for microtubules was enhanced by blocking the releasing effect of ATP. Another problem is that few plant preparations have sufficient tubulin to allow self-assembly, although this can be overcome by the alternative use of taxol-stabilized neuronal microtubules as fishing rods. Many different classes of protein will show association with microtubules. Some of these will be classical structural MAPs, some will be enzymes and other interacting proteins that associate – perhaps less avidly – with microtubules *in vivo*, while others may be irrelevant basic proteins that bind the negatively charged tubulin polymers under *in vitro* conditions. The identity of a protein as a MAP therefore depends upon a cellular localization on microtubules in addition to the *in vitro* indications. But even the cellular assays are not without controversy for it is possible that proteins from a compartment never exposed to microtubules under normal physiological conditions are released during the preparatory steps for immunocytochemistry. In the following sections, MAPs have been categorized, for ease of discussion, into: structural MAPs that form filaments which cross-bridge microtubules; motor proteins that move microtubules and/or move cargo along them; proteins involved in microtubule nucleation; and a loose grouping of microtubule-interacting proteins, proteins that do not fall into any of these categories but nonetheless have a real claim for microtubule association.

1.3.1 Cross-bridging MAPs

Various plant cell fractions had been previously shown to bundle microtubules, but only over the last decade have specific proteins been isolated that are responsible for this activity. The focus upon bundling as an assay can be attributed to the fact that proteins responsible for maintaining the often regular inter-microtubule distance between cortical MTs are expected to collect MTs into bundles *in vitro*. Jiang *et al.* (1992) reported that a bundling activity could be isolated from evacuolated protoplasts (*miniprotoplasts*). Jiang and Sonobe (1993) then developed this method to purify MAP65. They stimulated the assembly of tobacco tubulin with taxol and then disassembled the endogenous MTs with cold Ca^{++} and high salt. After several rounds, they purified a quadruplet of protein bands responsible for the MT bundling activity. In the electron microscope, the MAP65 fraction was seen to form short (<10 nm) cross-bridges between the MTs. Chan *et al.* (1996) used taxol-stabilized brain MTs to isolate MAP65 from detergent-extracted cytoskeletons prepared from carrot suspension cells. This produced a triplet

(60, 62 and 68 kDa) of MAP65 bands on 1D SDS gels plus a couple of proteins less than 100 kDa. The isolated protein fraction stimulated sub-critical concentrations of tubulin to assemble into radiating arrays. Monospecific antibodies prepared against the three MAP65 bands stained all four MT arrays in carrot cells much as Jiang and Sonobe (1993) had found for tobacco. Later, Rutten *et al.* (1997) chromatographically purified the 60 kDa band. This did not stimulate the assembly of tubulin at sub-critical concentrations, but it did stimulate assembly at concentrations of tubulin that would just self-assemble. It also stabilized brain MTs against the depolymerizing effects of cold and calcium. MAP60 did not bundle MTs. To separate the MAP65 proteins from the less than 100 kDa proteins, Chan *et al.* (1999) separated the carrot MAP fraction by sucrose density gradient centrifugation. The purified mixture of 60, 62 and 68 kDa proteins did not stimulate the assembly of tubulin but it did form MT bundles. In the EM the bundles of MTs could be seen to be held together by 25–30 nm cross-bridges that often formed a herringbone/chevron type of pattern (Fig. 1.1). It seems unlikely that a ca. 60 kDa polypeptide can bind to two microtubules and span a 30 nm gap, and so it would seem that the herringbone bridges are composed of at least two polypeptide chains. These larger bridges are in the range of sizes seen for the fine filaments that span cortical MTs *in planta* (Lancelle *et al.*, 1986) and so it would seem that the MAP65 family is responsible for the parallel spacing of cortical MTs. However, since the MAP68 was virtually absent from the fraction, and MAP60 was previously found not to bundle microtubules (Rutten *et al.*, 1997), it would appear that MAP62 possesses the major bundling activity (although it is possible that MAP60 had been inactivated during chromatographic isolation). Microdensitometric computer analysis showed that the bridges are regularly spaced along the MT backbone, consistent with a symmetrical superlattice for 13 protofilament MTs. It was hypothesized that MAP65 could zipper-up adjacent dynamic microtubules upon the cortex.

Using the polyclonal anti-MAP65 serum raised to the biochemically purified tobacco MAP65, Smertenko *et al.* (2000) screened a cDNA library from tobacco BY-2 suspension cells and isolated NtMAP65-1a. Subsequently, NtMAP65-1b and NtMAP65-1c were isolated from a re-screen and shown to be 85% similar (Hussey *et al.*, 2002). An antiserum raised to recombinant NtMAP65-1a detected a subset of bands recognized by the more general anti-MAP65 antibody. Where the latter antiserum immuno-stained all four microtubule arrays, the more specific antiserum raised against NtMAP65-1a (Smertenko *et al.*, 2000) decorated only some of the cortical MTs, did not stain the MTs surrounding the prophase spindle, and – instead of giving blanket staining of the division structures – labelled only the region where the opposing sets of MTs overlapped in the anaphase spindle and in the cytokinetic phragmoplast.

Smertenko *et al.* (2000) showed that NtMAP65-1a increased microtubule polymerization and that it bound and bundled microtubules *in vitro*. However, in an experiment where cold treatment of tobacco BY-2 cells was used to depolymerize microtubules, and the repolymerization of microtubules monitored by incubation of the cells at 25°C, it was found that NtMAP65-1a appeared to decorate microtubules

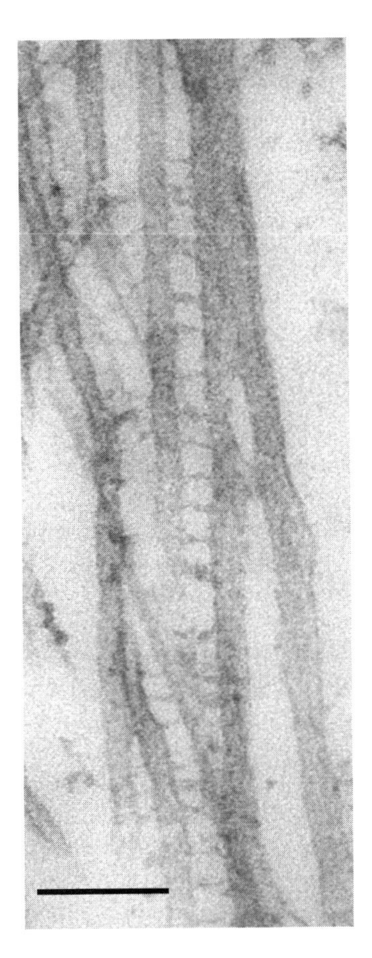

Fig. 1.1 MAP65 forms cross-bridges between microtubules. MAP65, biochemically isolated from carrot suspension cells, induces bundle formation when added to brain microtubules. The filamentous cross-bridges are 25–30 nm long and have a regular pattern along the microtubule backbone (see Chan *et al.*, 1999). Bar = 165 nm.

once they were polymerized. In studies on cytoskeletal organization during programmed cell death in *Picia abies* somatic embryogenesis, it was shown that MAP65 was present in the cytoplasm but not bound to cortical microtubules in a developmentally arrested cell line incapable of normal embryonic pattern formation, whereas MAP65 was bound to microtubules at the equivalent time point in normal embryogenic cell lines (Smertenko *et al.*, 2003). These data raised the possibility that the binding of the cortical microtubule cytoskeleton by MAP65 is important for embryo differentiation. In animals, MAP1b-stabilized microtubules were found to be essential for embryo development, as animals deficient in this MAP do not produce embryos and, moreover those with decreased amounts of

MAP1b showed severe embryo abnormalities (Edelmann *et al.*, 1996). Furthermore, Smertenko *et al.* (2003) showed that of the 65 kDa and 63 kDa MAP65 isoforms found in *Picia abies* embryos, the 63 kDa isoform was prevalent in the normal embryogenic cell line compared to the developmentally arrested cell line where the 65 kDa isoform was the major isoform. The observation that organized microtubules are bound by MAP65 in the normal line at all stages of embryo development analyzed, suggested that a critical quantity of 63 kDa isoform is essential for cell differentiation. Perhaps there is a similar necessity for the 63 kDa MAP in plant embrogenesis as MAP1b is essential for animal embryo development.

Nine MAP65 isoforms can be identified in the *Arabidopsis* genome database and, as with the multigene family of tubulins, this invites speculation on the function of the various family members. From findings on NtMAP65-1a it would appear that MAP65 isoforms may be targeted to specific parts of the various MT arrays, that some MAP65s may have a cell cycle-dependent pattern of expression, and that there may even be developmentally regulated patterns of expression. On 1D SDS gels of undifferentiated tobacco (Jiang & Sonobe, 1993) and carrot cell suspensions (Chan *et al.*, 1996), the MAP65 family appears as 3–4 bands. By examining different regions of azuki bean epicotyls, Sawano *et al.* (2000) report only one MAP65 band and that the amount of this was highest when cells were growing rapidly: it decreased when growth ceased and the proportion of cells exhibiting transverse microtubules also decreased. Although we do not know the number of MAP65 genes in the azuki bean genome, the single protein band (which could be one or several isoforms) is likely to represent a reduced pattern of expression. This could represent the expression of a specific isoform(s) as part of a developmental programme and/or – because the three division-related MT arrays are absent from the zone of cell elongation – it could be a cell cycle effect related to the entry of cells into interphase. Support for the latter idea comes from the report (Barroso *et al.*, 2000) that the 68 kDa component of the MAP65 triplet disappears when carrot suspensions stop dividing, undergo elongation and express only the cortical array. Subsequently, Chan *et al.* (2003) confirmed that as carrot suspension cells stop dividing and elongate, the 1D SDS MAP65 triplet reduces to a single band. Mass spectral analysis showed that this contained peptides identical to those from the 62 kDa band that was the major component of the biochemically purified MAP65 fraction demonstrated to form 25–30 nm cross-bridges between microtubule *in vitro* (Chan *et al.*, 1999). This protein, for which a cDNA was isolated, belongs to the most conserved sub-group of the MAP65 family. It would appear that this species of MAP65 is the sole (or at least, dominant) species responsible for the cross-bridging of microtubules in the cortical array; it would also appear that the various isoforms of MAP65 are subject to cell cycle-dependent patterns of expression.

Some animal MAPs form cross-bridges and prior to the molecular characterization of plant MAPs it was anticipated that the latter could turn out to contain relatives of the ubiquitous MAP4/MAP2/tau family which is characterized by several repeat motifs in the microtubule binding domain. Smertenko *et al.* (2000) have cloned MAP65 from tobacco BY-2 cells and found that it contains none of these repeats. However, it does show resemblance to PRC1, a microtubule bundling protein recently characterized in animal cells (Mollinari *et al.*, 2002), and to the anaphase spindle elongation protein, Ase1p, from yeast (Hussey *et al.*, 2002).

The *Arabidopsis* homologue of human TOGp and *Xenopus* MAP215 (XMAP215) has been identified genetically as MOR1 and GEM1 (Whittington *et al.*, 2001; Twell *et al.*, 2002). Whittington *et al.* (2001) showed that the extremely squat habit of the *Arabidopsis microtubule organization 1 (mor1)* mutant can be attributed to the fact that cells expand sideways instead of elongating. When placed at the restrictive temperature, microtubules of the cortical array shortened and lost alignment. However, when plants were put in the permissive temperature this disorder reverted within minutes to the normal orderly state of microtubule parallelism. The MOR1 mutations were found in the N-terminal heat repeat implying that the integrity of this heat repeat was essential for the establishment and maintenance of the interphase cortical array.

What is intriguing about the *mor1* mutants is that the mitotic arrays are not affected by these mutations and yet antibodies raised to the recombinant protein detect MOR1 in cortical and mitotic arrays (Twell *et al.*, 2002). In 1998, Park *et al.* characterized the *Arabidopsis gemini1-1 (gem1-1)* mutant and, more recently, a second allele *gem1-2* was identified (Twell *et al.*, 2002). The *gem1* mutants are homozygous lethal indicating that *GEM1* is an essential gene, and in the heterozygotes upto 50% of the pollen is aberrant as cytokinesis in pollen mitosis 1 (PM1) is defective. PM1 occurs when the microspore nucleus divides to form the larger vegetative cell and the smaller generative cell; the generative cell later goes on to divide in pollen mitosis II to form the two sperm cells. In the *gem1* mutants nuclear division in PM1 does occur but the cell plate is either absent or irregularly formed. Complementation and co-segregation analysis showed that *GEM1* is synonymous with *MOR1*. In heterozygotes of each of the two *gem1* mutants both wild-type and a 3′ truncated *GEM1* transcript was identified. The truncated transcripts lacked sequence that encode a C-terminal microtubule binding domain (but probably not the only one), indicating that the cytokinesis defect in the *gem1* mutants is the result of defective microtubule binding.

Some idea of the function of MOR1/GEM1 can be gained from what is known about its *Xenopus* relative, XMAP215, which stimulates microtubule polymerization (Vasquez *et al.*, 1994). Earlier studies had suggested that XMAP215 exerts its effect through microtubule stabilization (Kinoshita *et al.*, 2001; Popov *et al.* 2001; see Hussey & Hawkins, 2001 for discussion) and this effect was antagonized by the central motor kinesin, XKCM1 (Tournebize *et al.*, 2000). More recent studies, however, suggest that XMAP215 actually has a destabilizing effect on yeast (Van Breugel *et al.*, 2003) and *Xenopus* microtubules (Shirasu-Hiza *et al.*, 2003) and these findings need to be repeated for plants before the implications can be fully realised.

A homologue of MOR1/GEM1, has been isolated from telophase tobacco cells and named TMBP200 and shown to bundle microtubules together through cross-bridges (Yasuhara *et al.*, 2002). These cross-bridges were 10 nm; that is, some 15–20 nm shorter than those formed by MAP65 (Chan *et al.*, 1999). Although the cross-bridges between cortical microtubules seen in the rapid freeze study of Lancelle *et al.* (1986) were more consistent with the longer side-arms formed by carrot MAP65 *in vitro*, there is perhaps not enough information to decide whether MAP65 or MOR1/GEM1, or perhaps both, has a role in bringing cortical microtubules into parallel array.

Further possible hints to the function of MOR1/GEM1 come from recent work on XMAP215 by Popov *et al.* (2002). XMAP215 immobilized to beads induced the formation of microtubule asters in *Xenopus* egg extracts as well as in solutions of pure tubulin. Depletion of this protein impaired the ability of egg extracts to reconstitute the microtubule nucleation potential of salt-stripped centrosomes. An interesting conclusion was that XMAP215 not only stimulates microtubule nucleation but it anchors the minus ends of nascent microtubules. XMAP215 does not require γ-tubulin in order to nucleate microtubules from solutions of pure tubulin. It is too early to say how this translates to the effect of MOR1/GEM1 at the plant cortex, but it is possible that this protein may have a minus end role within the dispersed, multiple nucleation sites.

Using tubulin affinity chromatography to isolate carrot proteins, Durso and Cyr (1994) identified a 50 kDa protein as a homologue of elongation factor-1α. The protein induces taxol-stabilized brain microtubules to form bundles and this bundling effect was shown to be modulated by the addition of calcium and calmodulin. In fixed cells, immunofluorescence studies showed it to decorate all four microtubule arrays (Durso *et al.*, 1996). Some controversy has attached to EF-1α as an MAP, or, indeed, as an actin bundling protein (Gungabissoon *et al.*, 2001). While the criterion for an MAP – that it should be enhanced by rounds of tubulin de-/re-polymerization – has changed over recent years in favour of one of microtubule association, this relaxation carries with it perhaps a greater burden of proof that the protein actually associates with microtubules *in vivo*. The case for EF-1α seems to have been thrown open again by observations on transiently expressed GFP-EF-1α fusion proteins in fava bean leaf epidermal cells (Moore & Cyr, 2000). Under normal conditions, the signal produced by the fusion protein was indistinguishable from the general cytoplasmic signal of GFP alone. Microtubule-like staining could only be obtained by reducing the pH by adding organic acids to the medium. For the moment, the relevance to normal physiology of binding under conditions of cytoplasmic acidification is unknown.

Another protein with reported microtubule bundling activity is eukaryotic initiation factor eIF(iso)4F. This component of the protein translational machinery is composed of equimolar quantities of a p28 and a p86 subunit. Bokros *et al.* (1995) found that the p86 subunit binds and bundles microtubules *in vitro*. In cells it decorates cortical microtubules in coarse patches and it was suggested that this might reflect linkage of the protein synthesizing machinery and mRNAs to the cytoskeleton. The p86 subunit was also reported to anneal microtubules in an end-to-end fashion (Hugdahl *et al.*, 1995).

1.3.2 Proteins that link microtubules to the plasma membrane

The attachment of microtubules to the inner face of the plasma membrane is one of the unique features of the cortical array. Electron microscopy confirms that, in transverse section, cortical microtubules associate with the boundary membrane by short links (Lancelle *et al.*, 1986). It is not absolutely known if there are two populations of microtubules: one bound directly to the membrane and another bound indirectly to those microtubules, but it is likely that most cortical microtubules are attached to the plasma membrane to varying degrees. This radial form of linkage, together with the lateral inter-microtubule linkages, is a defining feature of the cortical array that largely accounts for the characteristic pattern of this cytoskeletal system. Although several studies have shown that microtubules remain stuck to substrate-attached plasma membrane disks torn out of burst protoplasts, none of these has identified the radial linkage. Marc *et al.* (1996) used a different approach. Starting with membranes from tobacco suspension cells, they isolated a detergent-soluble fraction and isolated proteins by tubulin affinity. Amongst these was a 90 kDa protein. Screening an *Arabidopsis* cDNA library with an antibody against this protein produced a partial clone encoding phospholipase D (Gardiner *et al.*, 2001). The antibody decorated cortical microtubules and when these were removed by calcium, the antibody labelled residual dots (which could be removed by detergent extraction). By immunofluorescence, the protein was shown to decorate all four microtubule arrays; not only were the cortical arrays of interphase and preprophase labelled, as well as the cell plate-depositing phragmoplast, but the mitotic spindle also. The protein could anchor microtubules at the cortex although its behaviour is more consistent that of a cytoplasmic peripheral protein whose targeting to the plasma membrane is regulated in a cell cycle manner, rather than an integral transmembrane protein.

1.3.3 Microtubule motor proteins

Kinesin was isolated first from squid axon. As its name implies, kinesin transports materials by walking along the microtubule backbone, using the energy of ATP hydrolysis to generate movement. This archetypal conventional kinesin is a tetramer comprised of two heavy chains and two light chains. Other similar movement proteins, showing variations in the basic structure, were later identified and these were called kinesin-like or kinesin-related proteins. The microtubule binding motor domain is quite well conserved across phylogeny. It can be found at the N-terminus or the C-terminus of the molecule, transporting cargo to the plus end or the minus end of the microtubule respectively. Kinesins are known in which the motor domain occurs in the central part, and some are thought to have a destabilizing effect on MT polymerization. Now that the *Arabidopsis* genome has been sequenced it is predicted that there are at least 61 kinesin-related proteins (Reddy & Day, 2001). Of the five sequenced eukaryotic genomes, *Arabidopsis* has the highest percentage of kinesins. A phylogenetic analysis of the *Arabidopsis* motor domain sequences reveals that it has seven of the nine recognized subfamilies of kinesins plus a high proportion (24/61) that do not fall into any existing category. The subfamilies involved in transport processes are underrepresented in *Arabidopsis*, which also does not contain the conventional heavy chain kinesins. There are no members of the KRP85/95 or Unc104/KIF1 subfamilies.

1.3.3.1 Kinesin-related proteins in cytokinesis

Of the 61 kinesins so far detected in the *Arabidopsis* genome, most of those that have been localized within cells have an association with the phragmoplast. The centrifugal deposition of a new cross wall is distinctly different from the contractile purse-string cytokinesis of animal cells and this disproportionate occurrence of kinesins in the microtubular structure that deposits the cross wall is perhaps a reflection of the importance of cell plate formation to the growth habit of plant cells. The first indication of phragmoplast-specific kinesins emerged from the work of Asada and Shibaoka (1994). Using isolated phragmoplasts from synchronized tobacco BY-2 cells, they isolated a fraction that supported the gliding of microtubules on a glass surface. This plus end-directed motor activity was found in a fraction containing a 120 kDa and a 125 kDa polypeptide. A cDNA for a 125 kDa protein was then isolated and characterized as a member of the bimC subgroup of the kinesin superfamily (Asada *et al.*, 1997). These proteins are known to form homotetramers in animals; with two microtubule-binding heads at each end, these bipolar proteins can cross-link anti-parallel microtubules. By walking towards the plus ends, these oligomers slide microtubules apart and are largely responsible for setting up the bipolarity of the spindle with plus ends overlapping at the centre. In BY-2 cells, the tobacco kinesin-related protein TKRP125 decorates cortical microtubules in G2, the preprophase band, the spindle (concentrated towards the mid-line) and the phragmoplast. Antibodies failed to decorate the line of overlap between the two half sets of phragmoplast microtubules – an unexpected finding since this is where plus end-directed motors would be expected to congregate. However, the ability of antibodies to the motor domain to inhibit translocation of microtubules within the phragmoplast does support a role for TKRP125 in cytokinesis. The bimC mutant phenotype, originally described by Enos and Morris (1990) consist of spindle poles that fail to separate, demonstrating its function in setting up spindle bipolarity: the work of Asada *et al.* (1997) suggests that the plant bimC, TKRP125, has a similar role in separating the two halves of the cytokinetic apparatus. In carrot, Barroso *et al.* (2000) biochemically isolated a 120 kDa protein that bound microtubules. This proved to be composed of two polypeptides (cf the 120 and 125 kDa proteins of Asada *et al.*, 1997). A partial genomic DNA cognate to one of the proteins was found to be homologous to the motor region of TKRP125 and was named DcKRP120-1. A full length cDNA was also isolated that coded for another bimC microtubule motor protein and was named DcKRP120-2. Unlike anti-TKRP125, antibodies to DcKRP120-2 labelled the mid-line of the phragmoplast, consistent with the plus end concentration of this protein (Fig. 1.2). If the oppositely directed phragmoplast microtubules can grow at their plus ends

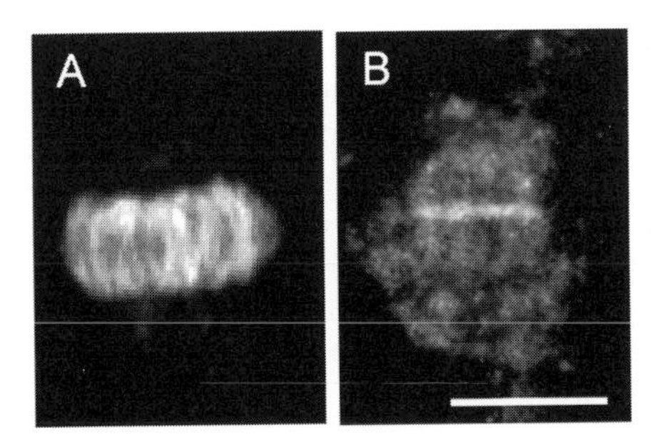

Fig. 1.2 The BIMC kinesin-related protein, DcKRP120-2, decorates the plus ends of phragmoplast microtubules. (A) shows the anti-tubulin staining of the phragmoplast microtubules. (B) shows that antibodies to the kinesin-related protein tend to decorate the mid-line of the cytokinetic phragmoplast – the line where the cell plate is deposited and where the plus ends of the two opposed sets of phragmoplast microtubules meet (see Barroso *et al.*, 2000).

(Asada *et al.*, 1991), this would cause the two half sets to inter-penetrate. However, this tendency of the two halves to grow into each other would be antagonized by the ability of bi-functional plus end-directed motors to slide the microtubules apart, maintaining a minimal line of overlap at the plus ends. This property of plus end bimC motors has been suggested by Barroso *et al.* (2000) to maintain a disc-like profile for the fusion of cell plate vesicles, as required for continuous centrifugal outgrowth (instead of the central aggregation) of the new cross wall. Both of the carrot bimC kinesin-related proteins are found in the cold-stable cytoskeleton fraction; they avidly bind microtubules and are suggested to have a role in the physical stabilization of mitotic and cytokinetic arrays.

Lee and Liu (2000) have reported another kinesin-related protein that concentrates in the mid-line of the phragmoplast. AtPAKRP1 (*Arabidopsis thaliana* phragmoplast-associated kinesin-related protein) does not belong to the bimC subgroup but resembles the ungrouped XKlp2 from *Xenopus*. Although its motility has not been demonstrated *in vitro* it is believed to be a plus end-directed motor. By immunofluorescence it has been shown to start to have a specific localization along interzonal microtubule bundles at late anaphase. Later it has a more restricted staining pattern at the plus ends of microtubules in the phragmoplast. Lee and Liu (2000) added anti-AtPAKRP antibodies or truncated fusion proteins to glycerinated BY-2 cells and found that the organization of the anti-parallel phragmoplast microtubules was lost, suggestive of a role for AtPAKRP in maintaining the integrity of the cytokinetic apparatus. It was further suggested that this kinesin-related protein could serve such a structural role once the bimC kinesins had slid the two half sets of microtubules apart. However, if tubulin is continually added to the plus ends of the phragmoplast microtubules then it is possible that bimC motors are continually needed to separate the two halves during its centrifugal outgrowth and not just during initial set-up.

In a following paper, Lee *et al.* (2001) described another kinesin-related protein, AtPAKRP2. This is an ungrouped N-terminal motor kinesin. It first appears in a concentrated form amongst the interzonal microtubules during late anaphase, when the sister chromatids are already well separated. At telophase, AtPAKRP2 appears very concentrated at the mid-line of the phragmoplast and it moved centrifugally with the outgrowing ring of microtubules. By comparison with AtPAKRP1, which neatly labelled the mid-line, AtPAKRP2 had a more punctate distribution around the equatorial plane. Since this distribution was disturbed by brefeldin A, which disrupts the Golgi apparatus, it was concluded that AtPAKRP2 associates with Golgi vesicles in the phragmoplast.

The microtubules of the phragmoplast appear to undergo treadmilling (Asada *et al.*, 1991) – that is, subunits flux through the tubule by addition of new subunits at the growing plus end and subtraction at the minus end. Stabilization of the microtubules with taxol prevents the expansion of the double ring of microtubules and so it would seem that turnover is an essential part of phragmoplast expansion (Yasuhara *et al.*, 1993). Phragmoplast-associated kinesins have also been reported by Nishihama *et al.* (2001) in their search for factors that activate NPK1 – a protein kinase kinase kinase that regulates the expansion of the cell plate at cytokinesis. The tobacco NPK activating kinase-like protein 1 (NACK1), which is a kinesin, was shown to bind to NPK1 and stimulate its kinase activity *in vitro*. NACK1 and a related protein NACK2 are similar to two kinesins (AtNACK1 and AtNACK2) in the *Arabidopsis* genome database and all four constitute a subfamily most closely related to human CENP-E. In animals CENP-E associates with centromeres but this localization could not be shown for plants and it was hypothesized that NACK1-related kinesins might participate in a process unique to plant cytokinesis.

Identical to AtNACK1 is the HINKEL kinesin-related protein of *Arabidopsis* (Strompen *et al.*, 2002). *Arabidopsis hinkel* (*hik*) mutants have multiple mitotic spindles and metaphase plates that, in the presence of incomplete cell walls, suggests failure to complete cytokinesis. Since the cytokinesis-specific syntaxin, KNOLLE, accumulates at the plane of cell division it would appear that HINKEL/AtNACK1 does not affect the delivery of vesicles. The defect was traced to the failure of microtubules to depolymerize from the centre of the expanding phramoplast, forming a solid microtubular disk instead of a hollow ring. The structure does, nevertheless, expand centrifugally but the presence of a solid bundle of microtubules, in the centre of what should be a hollow ring, seems to destabilize the nascent cell plate. In their study on *hinkel*, Strompen *et al.* (2002) saw only fully expanded phragmoplasts whereas Nishihama *et al.* (2001) saw different effects using dominant-negative constructs. Over-expression of truncated NACK1 lacking the motor domain produced cells that formed an initial cell plate but expanded no further, suggesting an arrest or delay of cytokinesis. Loss-of-function of NACK1 by virus-induced gene silencing also caused cytokinesis to fail, resulting in the formation of binucleate cells. Despite the discrepancies between the two studies they agree on the finding that this kinesin-related protein is essential for successful formation of the cell plate and its fusion with the mother wall.

In theory we can identify several classes of MAP that affect the structure and behaviour of the phragmoplast: (1) Proteins that hold the two half sets of microtubules together. Proteins like the bimC kinesins (TKRP125, DcKRP120-2) are required to cross-link the two half sets where they overlap and to slide them apart; AtPAKRP1 – although not a bimC – may also play a part in maintaining the fidelity of the line of overlap between the two half sets. (2) Proteins that affect the stability of the dynamic array. In a later section we will see that other classes of protein may stabilize the microtubule ends. At the moment we do not know exactly what function the C-terminal motors KATA/B/C have at the minus ends of the phragmoplast, whether they have a role in vesicle movement or in controlling the dynamics at the minus end (see next section; Mitsui *et al.*, 1993, 1994). (3) Proteins that move cargo to the mid-line. In addition to the structural MAPs that set up and maintain the essential bi-polarity of the cytokinetic apparatus there must be other movement proteins that guide secretory vesicles along the microtubules to the line of overlap where the cell plate develops. At the moment, only AtPAKRP2 is thought to function in this way but there must be others yet to be described.

1.3.3.2 Kinesin-related proteins in mitosis

The first plant kinesins to be identified were the *Arabidopsis kat* genes which code for an 89 kDa (KATA), an 82 kDa (KATB) and an 84 kDa (KATC) protein. Mitsui *et al.* (1993) used PCR primers from the conserved region of kinesin heavy chain to amplify five different clone types. cDNAs for *Arabidopsis katb* and *katc* were then isolated by Mitsui *et al.* (1994) and a truncated KATC protein shown to bind MTs and exhibit the characteristic MT-stimulated ATPase activity, showing that the putative kinesin gene coded a functionally competent kinesin motor protein. Antibodies raised against peptides encoded by the *KATA* gene were found to stain MT bundles; they also stained the mitotic apparatus of tobacco generative cells – by anaphase the staining was concentrated in the mid-zone and a complementary pattern of mid-line staining was seen in the early phragmoplast (Liu & Palevitz, 1996). In *Arabidopsis*, too, this antibody stained the spindle mid-zone during metaphase and anaphase (Liu *et al.*, 1996). By lysing the cells in the presence of taxol and ATP, the antigen moved to the spindle poles – consistent with this being a C-terminal motor which moves to the minus ends of microtubules. Marcus *et al.* (2002) confirmed through *in vitro* studies that the *Arabidopsis* protein, ATK1 (formerly known as KATA), does possess minus end-directed motility and that its movement is non-processive. The *atk1-1* mutant, of *Arabidopsis*, which has a defective *ATK1* gene, was found to have defective spindle organization and meiotic chromosome segregation (Chen *et al.*, 2002). Instead of the fusiform spindles of wild-type metaphase I cells, the mutant cells had broad, unfocused and multi-axial poles. ATK1/KATA – a minus end motor – would appear to play a crucial role in spindle morphogenesis by helping to draw together the minus ends. Consistent with this, microtubules in *Arabidopsis* cells mutant for ATK1 fail to accumulate at the spindle poles in prophase and have diminished ability to form bipolar spindles in prometaphase. These defects are nonetheless rectified by anaphase and so there would seem to be some error-correcting mechanism (Marcus *et al.*, 2003).

Using antibodies generated against truncated proteins expressed by KATB and KATC, Mitsui *et al.* (1996) detected a ca. 85 kDa kinesin-like protein that accumulated in M-phase, presumably with the spindle and phragmoplast. While conventional kinesins are typically thought of as transporting cargo along interphase microtubules, the KAT A,B,C kinesin-like proteins would seem to be involved in the construction and/or subsequent functioning of the two mirror image sets of microtubules that consecutively bring about nuclear and cytoplasmic separation. A fourth member of the kat family, KATD, appears to be structurally distinct in that it lacks the α-helical coiled-coil region typical of kinesin heavy chains (Tamura *et al.*, 1999), suggesting that it may work as a monomeric protein. KATD binds microtubules and has microtubule-stimulated ATPase activity. However, its motor domain occurs in the central region of the protein and further work is needed to determine whether it actually functions like some other members of this class, as a motor or in the alternative capacity as a microtubule destabilizer. Since it is expressed specifically in floral tissue it may have a role in processes unique to flowers.

The kinesin-like calmodulin binding protein, KCBP, first discovered in plants is a chimeric protein that both binds calmodulin and microtubules (Reddy *et al.*, 1996a,b). Bacterially expressed KCBP from *Arabidopsis* binds microtubules to a glass surface and causes them to glide in a manner consistent with a minus end-directed protein (Song *et al.*, 1997). However, in the presence of calcium and calmodulin, KCBP fails to bind microtubules to the glass surface. These dual functions provide for an interesting contribution to microtubule function *in vivo*. In the presence of calcium/calmodulin, KCBP has a substantially reduced affinity for microtubules (Deavours *et al.*, 1998). In synchronized cell cultures, the amount of KCBP was low during interphase but increased at M-phase. Immunolocalization showed that the protein occurs in preprophase bands, spindles and phragmoplasts (Bowser & Reddy, 1997) – a more detailed study then suggested how KCBP may function in division (Smirnova *et al.*, 1998). KCBP is first detected in the prophase spindle of *Haemanthus* endosperm, then labels the kinetochore fibres through metaphase. By mid-anaphase it redistributes to the spindle poles at a time when the barrel-shaped spindle becomes more convergent. Next, it labels the phragmoplast and cell plate. It would appear that KCBP has a role in maintaining the coherence of the kinetochore fibres, thus maintaining the integrity of the spindle poles in anaphase. The relocation of KCBP to the poles is consistent with its property as a minus end-directed microtubule motor and it is likely that, much as dynein is thought to function in animal spindles, KCBP draws MTs together as it travels towards the poles. In the absence of defined spindle pole bodies, minus end-directed proteins, capable of drawing microtubules together, are likely to have a particularly important role in maintaining the integrity of the poles and, hence, the binary nature of the spindle. Although it seems likely that other proteins will be discovered that maintain the integrity of the spindle poles, the converging activity of minus end motors like KCBP can also be inferred to be important for ensuring that the chromatin is split into two coherent lots. This may be particularly important in plants which show no significant spindle elongation during anaphase, and hence no concomitant distancing of the sister chromatids. The concentration of KCBP towards the cell plate (i.e., where the plus ends of the phragmoplast interdigitate) does not, however, appear to support a similar minus end role during cytokinesis and it is possible that converging activity is changed, perhaps even inactivated, as a result of cell cycle-dependent changes in calcium concentration. Such conclusions emerged from a subsequent study in which antibodies were microinjected into *Tradescantia* stamen hair cells in order to inactivate calmodulin, thereby activating KCBP (Vos *et al.*, 2000). It was suggested that KCBP is activated during nuclear envelope breakdown and anaphase, when it promotes the sliding and bundling of microtubules. Down-regulation of its activity during telophase would switch off the minus end bundling property and allow the parallel phragmoplast microtubules to be displaced outwards, instead of tying their minus ends into bunches. Vos *et al.* (2000) speculated that the closely associated tubular ER system regulates the concentration of calcium ions in microdomains during mitosis and cytokinesis and could therefore down-regulate the activity of KCBP. Phosphorylation by the KCBP-interacting protein kinase identified by Day *et al.* (2000) represents another possible regulatory mechanism.

Despite the apparently low amount of this protein during interphase, it has been shown to decorate the cortical microtubules of cotton fibres and so would seem to function around the microtubule cycle, not just at division (Preuss *et al.*, 2003). Bioinformatic analysis of the *Arabidopsis* genome sequence shows several proteins likely to be involved in mitosis. As Reddy and Day (2001) noted, *Arabidopsis* contains two internal motor kinesins that group with the MCSK/KIF2 subfamily known to be involved in chromosome movement. Also, three kinesins group with members of the kip3 subfamily also known to be involved in this nuclear function. Three *Arabidopsis* kinesins form a sub-branch off chromokinesin/KIF4 subfamily members that may have a function in chromosome movement and spindle positioning. Of the ungrouped kinesins in the *Arabidopsis* genome, seven cluster with the CENP-E subfamily. These proteins bind the kinetochore which, during mitosis, associates with microtubules to draw the chromatids to the spindle poles.

1.3.3.3 Kinesin-related proteins in interphase

Concerning interphase, few kinesins have so far been described to have a role during the expansion phase of growth. KCBP is also known to be the gene product of *ZWICHEL*, detected during an analysis of *Arabidopsis* trichome mutants. Although KCBP has not been demonstrated to be specifically associated with the interphase cytoskeleton, *zwi* mutants have a shorter stalk and fewer branches, indicating that it is likely to have some role during differentiation; however, apart from this, morphogenesis appears to be normal (Oppenheimer *et al.*, 1997).

A 90 kDa protein has been biochemically isolated from tobacco pollen tubes by Cai *et al.* (2000) and shown to have kinesin-like properties. It supports the motility of microtubules on glass and although it is not known if this is a plus- or a minus-ended motor, the protein is located with organelles that associate with cortical microtubules and so would seem to be involved in their movement *in vivo*.

Matsui *et al.* (2001) have described TBK5 which is highly expressed in interphase tobacco BY-2 cells. It has a predicted structure characterized by a central motor domain. Some other central motor kinesins, like *Xenopus* XKCM1 (Walczak *et al.*, 2002), seem not to function as movement proteins but to destabilize the tubule. There is no suggestion that TBK5 works in this way; it has a putative neck domain closely related to that of kinesins with C-terminal motor domains that move cargo to the minus ends. Functional studies are required to determine the kind of role this kinesin plays during interphase. The same authors report that they have isolated 11 cDNAs, named TBK1-11, from tobacco cells. TBK3 is closely related to kata of *Arabidopsis*, TBK6 and TKRP125 are related to the bimC subgroup, TBK10 is related to mouse KIF2 of the central motor subfamily, TBK1 and three others are related to *Arabidopsis* KATD and TBK 7 and 8 are related to *Xenopus* Klp2. TBK4 and 5 are phylogenetically distant from all other eukaryotic kinesins and appear to be novel subtypes specific to plants. Perhaps the best supported role for an interphase kinesin comes from the work of Zhong *et al.* (2002). They have shown that the *fragile fiber1* (*fra1*) mutant, which has shorter stems than wildtype, results from mutation of a kinesin-related protein. The phenotype was traced to a disorganization of the cellulose microfibrils in the secondary walls, but this was not accompanied by disorganization of the cortical microtubules. The mutated kinesin is most closely related to the KIF4 group of animal kinesins, implicated in vesicle transport and mitosis. Since FRA1 is located to the cortical cytoplasm and not to the nucleus (it does not have a nuclear localization signal) it would seem, by analogy with vertebrate KIF4 kinesins, to be involved in the plus end-directed movement of vesicles. However, the case for cortical microtubules in the transport of plant vesicles is not well developed and Zhong *et al.* (2002) did not find any evidence for changes in wall composition, which might have been expected if secretion were compromised. Instead, the disorganization of cellulose microfibrils suggests an organizational role. It was speculated that FRA1 could move membrane-bound cellulose synthases along cortical microtubules (the *monorail* hypothesis) or somehow stop synthases from escaping the channel formed between pairs of microtubules (the *guard rail* hypothesis). These ideas revive interest in Heath's (1974) hypotheses on microtubules and cellulose alignment, still unsubstantiated after more than a quarter century.

1.3.3.4 Dynein

Kinesin-related proteins can, depending on the location of the motor region within the molecule, move cargo to the plus or to the minus end of microtubules. Another microtubule motor, dynein, moves cargo exclusively to the minus ends of microtubules. The flowering plants do not possess flagella (and hence the flagellar form of dynein), which is responsible for sperm motility in lower plants. Such arguments have provided the theoretical basis for the absence of cytoplasmic dynein from higher plants. This conclusion was recently drawn, prematurely it would seem, from computer-based searches of higher plant expressed sequence tag (EST) databases and of the *Arabidopsis* genome (Lawrence *et al.*, 2001). Dynein is a multiprotein complex containing large heavy chains (>500 kDa) as well as several associated proteins. Moscatelli *et al.* (1995) isolated from tobacco pollen tubes, two ca. 400 kDa proteins that immunologically and enzymatically resemble the dynein heavy chains. This provisional identification received support from King (2002) who reported that the rice genome shotgun sequence does contain sequences corresponding to the highly conserved central region of β flagellar outer arm dynein heavy chain from the green alga *Chlamydomonas*. Since angiosperms do not contain cilia/flagella it is unlikely that plants use these proteins for the kinds of purposes to which they were put in the flagellar axoneme. Instead, it is likely that they have alternative cytoplasmic functions. In animals, where the minus ends of cytoplasmic microtubules are clustered around the centrosome in the cell body, the minus end-directed motor activity of dynein is responsible for maintaining the central location of the Golgi body. In higher plants, Golgi stacks are dispersed around the cytoplasm as multiple dictyosomes and so this pattern, consistent with the more generalized insertion of secretory vesicles into the diffusely growing cell cortex, does not apply. The cortical microtubules of plants are also not gathered together at their minus ends, they are essentially parallel and so there are few opportunities for dynein to draw together the minus ends of microtubules. One is in anaphase when the poles show several points of microtubule convergence, and another is during early interphase when microtubules radiate from the nuclear surface and, later, from dispersed microtubule nucleation sites at the cortex. In G2, microtubules radiate from the nuclear surface and contact the cortex; initially they radiate as a three-dimensional starburst but then reorganize into a division-plane-predicting sheet with the preprophase band at the cortex (Lambert & Lloyd, 1994). It will be interesting to see if these radial microtubules are reeled in by cortical dynein, as they appear to be in animal cells.

1.3.4 Proteins involved in microtubule nucleation and release: the formation of the cortical array

The concentration of tubulin in the cytoplasm is insufficient to allow promiscuous self-assembly at random sites around the cell. So-called typical animal cells, such as the fibroblasts used in tissue culture, present a pattern in which microtubules radiate from a fixed point adjacent to the nucleus in the centre of the cell. This radiating array is constructed from microtubules with their slow growing minus ends focused around the amorphous material surrounding the centriole pair. It is this material that nucleates, initiates, the polymerization process while the distal plus ends grow and shrink. In plants it is not possible to backtrack along a radial microtubule array to find the source of microtubule nucleating material because individual tubules are dispersed in parallel groups around the cortex. In the absence of markers for the nucleating material it is not known whether neighbouring microtubules are parallel or anti-parallel.

The nuclear surface is considered to be the most likely site for the initiation of microtubules. Following cytokinesis, at a time when the mother cell's cortical microtubule array is still depolymerized, microtubules radiate from the nuclear surface (Flanders *et al.*, 1990). However, this phase is transient since nucleus-associated arrays rarely persist beyond these early stages of outgrowth for microtubules, then appear upon the cell cortex and the nuclear surface seems to fall back into inactivity. In living interphase cells, labelled tubulin has not been seen to incorporate into, and be exported from, microtubules at the nuclear surface (Yuan *et al.*, 1994; Granger & Cyr, 2001).

Gamma tubulin, a member of the tubulin superfamily, is well established as a marker for the minus ends of microtubules. It forms a filament and according to different models, either forms a ring that acts as template for the addition of α-, β-tubulin dimers, or as a short rod against which other protofilaments grow. Antibodies to γ-tubulin label sites around the nucleus but also over the entire cortical array (Panteris *et al.*, 2000). This general pattern of cortical labelling could be a reflection of the staggered arrangement of microtubule ends within the cortical array (Hardham & Gunning, 1978) or it could mean that γ-tubulin has a different role in plants, acting as a general stabilizer not confined to minus ends. This scenario suggests that microtubule nucleating material is transported from the nuclear surface to the cortex. However, γ-tubulin – although found at the nuclear surface – is not detected in the expected discrete foci at the cortex but in a more dispersed form (Panteris *et al.*, 2000). The free ends of microtubules are staggered around the cell cortex and it is possible that a seemingly general immunofluorescence pattern is due to the overlapping of minus ends in a parallel array (Joshi & Palevitz, 1996). Alternatively, γ-tubulin may not be a component of higher plant microtubule nucleating material and could have a more general role in microtubule stabilization.

Another component of the γ-tubulin ring complex seems to show a more discrete cortical localization pattern than γ-tubulin itself. Erhardt *et al.* (2002) have reported that homologues of Spc98p are present in rice and *Arabidopsis*. This protein is also found at the nuclear surface in cells and where microtubules are nucleated on isolated tobacco nuclei. It has a more restricted distribution and does not co-localize γ-tubulin on cortical microtubules.

Research on katanin has provided a fresh look at the way that the cortical array is formed. This heterodimeric ATPase is found in animal centrosomes where it is thought to cut microtubules free from their minus end anchorage (McNally *et al.*, 1996). It has a 60 kDa catalytic subunit used for severing microtubules and an 80 kDa regulatory subunit. McClinton *et al.* (2001) identified a homologue of the p60 subunit of katanin, AtKSS, in *Arabidopsis* and, although no homologue of the 80 kDa regulatory subunit has been detected in the *Arabidopsis* genome, recombinant p60 is demonstrably active *in vitro* (Stoppin-Mellet *et al.*, 2002). The *Arabidopsis* katanin mutant, *fragile fibre* (*fra2*), shows the importance of katanin to plant growth. The cell walls of the mutant have reduced cellulose and this seems to provide the explanation for the mechanically weakened cell walls and reduced cell length resulting in a stunted growth habit (a 50% reduction in height) (Burk *et al.*, 2001). Since sequence comparison showed that FRA2 was similar to katanin it was renamed AtKTN1. In *fra2*, the mitotic and cytokinetic microtubules appear normal but the early stages of cortical array formation are perturbed. Just after exit from cytokinesis microtubule converging points remain at the nuclear envelope at the same time as microtubules appear at the cortex in radiating clusters. Since the percentage of cells in root tip squashes that contain microtubules radiating from the post-cytokinetic nucleus is very small, it appears that this stage is transient. Normally, this brief phase does not therefore persist once the cortical array is established. This can be interpreted as follows: microtubules are nucleated from material at the nuclear surface and as the microtubules contact the cortex this material is transported to the cell periphery. Microtubules have been seen bridging the post-cytokinetic nuclei to the cortex (e.g. Flanders *et al.*, 1990) and these could provide the conduit for the outflow of nucleating material from the perinuclear region to the plasma membrane, perhaps using plus end-directed microtubule motors. Alternatively, if no movement is involved, then dormant cortical nucleation sites could be reactivated. Once at the cortex, it is presumed that microtubules are then initiated at dispersed nucleation sites. Since microtubules are rarely seen gathered into foci they must be severed, enabling them to adopt a parallel arrangement. This interpretation is supported by observations on *fra2* in which conversion of these abnormally focused arrays to the evenly distributed parallel arrangement was delayed, and there was only limited expansion during interphase (Burk & Ye, 2002). This was also seen by Bichet *et al.* (2001) in the *Arabidopsis* mutant *botero 1*, which is allelic to *fra2*. In the wild type, cortical microtubules in cells near the division zone of the meristem were loosely organized but became more highly aligned in transverse arrays; the farther cells elongated away from the meristem. In *bot1/fra2* this crucial transition was not observed. This mutant shows more elegantly than any treatment with anti-microtubule drugs the importance of cortical microtubules for cell elongation and, in particular, the importance of microtubule severance in allowing the formation of parallel microtubule arrays necessary for proper cellulose distribution.

Wasteneys (2002) has proposed a model for the roles of katanin and microtubule nucleation in generating the cortical array. After being severed by katanin, the nucleating minus end of the cortical microtubule is suggested to be transported to the plus end of the larger severed fragment by a plus end-directed kinesin-related protein. Certainly, some mechanism like this is needed to account for the way in which the density of the array is maintained during quite often massive cell elongation. The severance of a microtubule from its nucleating material would leave a free plus-ended stub at the nucleation site to grow and, after subsequent rounds of severance, to grow again. It would be surprising if motors are not involved in the movement of cortical microtubules. Microtubule dynamics (dynamic instability, treadmilling) could aid in the dispersion of microtubules and their associated nucleation sites during cell expansion. As we saw in the section on structural MAPs, cross-linking proteins like MAP65 could then play a key role in uniting the individual microtubules into organized sheets.

1.3.5 Microtubule-interacting proteins

Proteins that help nucleate microtubules, cross-link them and move on them, form functionally recognizable groups of microtubule-interacting proteins. In this section we discuss various proteins that are not so easily categorized.

One of the least understood aspects of the plant cytoskeleton is the interaction between actin filaments and microtubules. Each of the four microtubule assemblies is known to contain parallel actin filaments (e.g. Traas *et al.*, 1987) and only the endoplasmic actin cables that support cytoplasmic streaming seem to be free of microtubules. One possibility is that non-cable actin gains its overall order by association with microtubules. When actin, depolymerized by microinjection of actin depolymerizing factor (ADF), is allowed to recover, it does not initially reconstitute the central actin cables but forms finer filaments that co-align with the transverse cortical microtubules, thereby supporting the idea that microtubules anchor and give form to the cortical actin filaments (Hussey *et al.*, 1998). One factor that may unite the two cytoskeletal systems is MAP190 (Igarishi *et al.*, 2000). This 190 kDa protein was isolated from tobacco BY-2 cells based on its ability to purify with microtubules through cycles of polymerization/depolymerization. It also binds filamentous actin. By immunofluorescence, MAP190 is detected in the nucleus but when the nuclear envelope breaks down it locates to the spindle and phragmoplast. Since it was not detected in the preprophase band or the cortical array, this protein may have a cell cycle-regulated role in cross-linking actin and microtubules only during M-phase. Hussey *et al.* (2002) noted that there is an endoplasmic reticulum-retention signal in the N-terminus and a calmodulin-like domain in the C-terminus. From this they speculated that microtubules, actin and ER in the phragmoplast might be cross-linked by MAP190 in a calcium-dependent manner. It will be recalled that Vos *et al.* (2000) made a similar suggestion for the different effects of the kinesin-like calmodulin-binding protein, KCBP: namely, that ER-controlled fluxes of calcium could switch off the microtubule-binding activity at the transition between mitosis and cytokinesis. Screening the *Arabidopsis* genome database revealed one homologue of tobacco MAP190.

Cells in the leaves of *tangled 1* (*tan1*) maize mutants divide abnormally (Smith *et al.*, 1996). The misalignment of the division planes is correlated with the failure of the outgrowing phragmoplast to be guided to the cortical site predicted by the preprophase band of microtubules. Smith *et al.* (2001) have shown that the product of the *TAN1* gene, the 41 kDa TAN1 protein, has characteristics of an MAP. It shows no strong homology with other groups of known MAPs but it does share a distant relationship with basic regions of the vertebrate adenomatous polyposis coli (APC) proteins known to bind microtubules. Using a blot overlay assay, brain microtubules were found to bind TAN1. Antibodies to TAN1 decorated microtubules in fixed cells. During interphase, the labelling was distributed through the cytoplasm as dots. Failure to label the cortical microtubule array is consistent with the expression of this protein only in actively dividing tissue. There is a rough correspondence of the dots with the preprophase band, spindle and phragmoplast, rather than a 1:1 co-alignment with microtubule bundles. From this pattern of labelling it appears that TAN1 does not evenly decorate the microtubule backbone or the polarized ends of division structures, such as the mid-line of the phragmoplast or the spindle poles.

Molecular chaperones are involved in the folding transport and storage of polypeptides. One of the plant chaperones, heat shock protein 90 from tobacco cells, has been shown to bind to nitrocellulose coated with tubulin dimers or microtubules. HSP90 also cofractionates with tubulin dimers that are isolated by affinity to the tubulin-binding herbicide, ethyl N-phenylcarbamate. The chaperone is also found in anti-tubulin immunoprecipitates (Freudenreich & Nick, 1998). In immunofluorescence studies, HSP90 was found to decorate the nuclear envelope and discontinuous foci along the radial microtubules that emanate from it (Petrasek *et al.*, 1998). Staining is also found in the preprophase band, not in the spindle, but is then seen at the edge of the phragmoplast. It was speculated that HSP90 could be involved in microtubule nucleation.

Ran, a highly conserved small GTP-binding protein of the Ras superfamily has been implicated in at least two major functions in animal cells: it directs the passage of proteins and RNA through nuclear pores during interphase, and it is involved in the polymerization of spindle microtubules. The nucleotide binding of Ran is regulated by accessory proteins including RanGAP. An indication that this system is effective in plants was provided by Pay *et al.* (2002). They showed that the yeast RanGAP mutant rna1 can be complemented with alfalfa and *Arabidopsis* cDNAs coding for putative plant RanGAPs. These investigators also expressed GFP fusion proteins of RanGAP and showed that the protein relocated from a patchy decoration of the interphase nuclear envelope to a more defined labelling of the anaphase spindle. This was followed by localization to the microtubules of the phragmoplast as well as the nuclear envelope that re-formed around the telophase nuclei. In extracts from mitotic cells, the RanGAP could be shown to co-assemble with tubulin stimulated to polymerize by addition of taxol; this could not be shown for interphase extracts, indicating cell cycle regulation of the tubulin–RanGAP interaction. It was hypothesized that RanGAP, in addition to its role in nuclear transport, could be involved in fusion of vesicles necessary for forming the cell plate and perhaps in masking microtubule nucleating sites at the nuclear envelope.

Import and export of proteins and ribonucleoproteins through the nuclear pore complex is mediated by a receptor-mediated process involving importins. Importin α is part of an $\alpha\beta$ heterodimer that binds to cytoplasmic proteins containing nuclear localization signals. Importin α co-localizes with cytoplasmic strands containing microtubules and microfilaments. When the actin filaments were depolymerized with cytochalasin B, importin α collected inside the nucleus – an effect not seen with microtubule depolymerizing drugs (Smith & Raikhel, 1998). Microtubules and microfilaments often co-localize in plants and depolymerization of one might affect the distribution of the other and the exact binding partner might be difficult to diagnose. However, *in vitro* spindown studies showed that importin α associates with either actin filaments or microtubules in the presence of a functional nuclear localization signal peptide. These studies suggest that importin α is capable of bi-functional linkage. Molecular dissection of the molecule should help reveal the microtubule binding domains.

The 70 kDa heat shock proteins help in the movement of proteins across membranes. Beet yellows closterovirus (BYV) encodes an Hsp70 homologue that acts as a viral movement protein and localizes to the plasmodesmatal pores that interconnect cells. Early work suggested that BYV performed as an MAP (Karasev *et al.*, 1992); more recent work suggests that the Hsp70 homologue helps form a tail domain associated with the virion capsid and is required for the intercellular movement of the virus (Alzhanova *et al.*, 2001). It is proposed that the Hsp70-like protein helps to move the virion through the plasmodesmatal pore much as Hsp70 itself is thought to translocate proteins into the endoplasmic reticulum.

Perhaps the best-studied viral movement protein is the 30 kDa protein from tobacco mosaic virus necessary for the symplastic spread of the virus through the plant. The movement protein facilitates viral transmission by increasing the size exclusion limit of the plasmodesmatal pore. Using GFP-movement protein fusions, Heinlein *et al.* (1995) demonstrated that the movement protein co-localized with cytoplasmic filaments, the decoration of which was disrupted with anti-microtubule but not anti-actin drugs. Boyko *et al.* (2000) then showed that a conserved sequence motif in tobamovirus MPs shares similarity with a region in tubulins proposed to mediate lateral contacts between the protofilaments forming the microtubule backbone. Point mutation of this region interfered with the ability of the movement protein to associate with microtubules *in vivo*. Movement proteins could therefore mimic tubulin–tubulin interfaces for association with microtubules.

1.4 Concluding remarks

The plant cytoskeleton field has made quantum leaps over recent years. A decade ago, when *The Cytoskeletal Basis of Plant Growth and Form* (Lloyd, 1991) was published, no plant MAPs had been fully characterized and we knew nothing about microtubule dynamics. That has all changed. Tagging of microtubules, first by microinjected fluorescent tubulin, then by transformation with green fluorescent protein chimeras, opened our eyes to the dynamicity of the plant cytoskeleton. The new generation of confocal microscopes will take us even closer to real time imaging; as fading becomes a less significant factor it will be possible to see how the arrays behave and reorganize at key transitions of the microtubule cycle. Genetic approaches have also begun to pay dividends. The number of mutants that turn out to have a cytoskeletal basis is growing and some (e.g., *MOR1/GEM1* and *FRA2* [katanin]) have added greatly to our understanding of microtubule behaviour. The list of MAPs grows apace. The major MAP2/4/tau family from vertebrates seems to be absent from plants but virtually all other plant MAPs turn out to share some sequence homology with known MAPs from yeast and animal cells. Publication of the *Arabidopsis* genome sequence, together with the availability of mutant collections, provides a great resource for studying how these conserved proteins function in plants. It is now quite feasible to obtain homologous genes, to couple them to fluorescent reporter genes in order to follow the distribution of the gene products, and to investigate functionality by gene truncations or silencing. This kind of analysis of universal microtubule proteins is likely to represent an important phase in plant cytoskeleton research. However, as mentioned in the introductory remarks, the message for plant biology lies not in the similarities but in the different ways that immobile plant cells use microtubule proteins for plant-specific processes.

References

Anthony, R.G., Waldin, T.R., Ray, J.A., Bright, S.W.J. & Hussey, P.J. (1998) Herbicide resistance caused by the spontaneous mutation of the cytoskeletal protein tubulin, *Nature*, **393**, 260–263.

Anthony, R.G. & Hussey, P.J. (1999a) Dinitroaniline herbicide resistance and the microtubule cytoskeleton, *Trends Plant Sci.*, **4**, 112–116.

Anthony, R.G. & Hussey, P.J. (1999b) Double mutation in *Eleusine indica* a tubulin increases the resistance of transgenic maize calli to dinitroaniline and phosphorothioamidate herbicides, *Plant J.*, **18**, 669–674.

Asada, T., Sonobe, S. & Shibaoka, H. (1991) Microtubule translocation in the cytokinetic apparatus of cultured tobacco cells, *Nature, Lond.*, **350**, 238–241.

Asada, T., Kuriyama, R. & Shibaoka, H. (1997) TKRP125, a kinesin-related protein involved in the centrosome-independent organization of the cytokinetic apparatus in tobacco By-2 cells, *J. Cell Sci.*, **110**, 179–189.

Barroso, C., Chan, J., Allan, V., Doonan, J., Hussey, P. & Lloyd, C.W. (2000) Two kinesin-related proteins associated with the cold stable cytoskeleton of carrot cells: characterization of a novel kinesin, DcKRP120-2, *Plant J.*, **24**, 869–968.

Bichet, A., Desnos, T., Turner, S., Grandjean, O. & Hofte, H. (2001) *BOTERO1* is required for normal orientation of cortical microtubules and anisotropic cell expansion in *Arabidopsis*, *Plant J.*, **25**, 137–148.

Bowser, J. & Reddy, A.S.N. (1997) Localization of a kinesin-like calmodulin-binding protein in dividing cells of *Arabidopsis* and tobacco, *Plant J.*, **12**, 1429–1437.

Boyko, V., Ferralli, J., Ashby, J., Schellenbaum, P. & Heinlein, M. (2000) Function of microtubules in intercellular transport of plant virus RNA, *Nat. Cell Biol.*, **2**, 826–832.

Burk, D.H., Liu, B., Zhong, R., Morrison, W.H. & Ye, Z.-H. (2001) A katanin-like protein regulates normal cell wall biosynthesis and cell elongation, *Plant Cell*, **13**, 807–827.

Burk, D.H. & Ye, Z.-H. (2002) Alteration of oriented deposition of cellulose microfibrils by mutation of a katanin-like microtubule severing protein, *Plant Cell*, **14**, 2145–2160.

Cai, G., Romagnioli, S., Moscatelli, A., *et al.* (2000) Identification and characterization of a novel microtubule-based motor associated with membranous organelles in tobacco pollen tubes, *Plant Cell*, **12**, 1719–1736.

Chan, J., Mao, G., Smertenko, A., *et al.* (2003) Identification of a MAP65 isoform involved in directional expansion of plant cells, *FEBS Lett.*, **534**, 161–163.

Chan, J., Rutten, T. & Lloyd, C.W. (1996) Isolation of microtubule-associated proteins from carrot cytoskeletons: a 120 kDa MAP decorates all four microtubule arrays and the nucleus, *Plant J.*, **10**, 251–259.

Chan, J., Jensen, C.G., Jensen, L.C.W., Bush, M. & Lloyd, C.W. (1999) The 65-kDa carrot microtubule-associated protein forms regularly arranged filamentous cross-bridges between microtubules, *Proc. Natl. Acad. Sci. USA*, **96**, 14931–14936.

Chen, C., Marcus, A., Li, W., *et al.* (2002) The *Arabidopsis* ATK1 gene is required for spindle morphogenesis in male meiosis, *Development*, **129**, 2401–2409.

Durso, N.A. & Cyr, R.J. (1994) A calmodulin-sensitive interaction between microtubules and a higher plant homolog of elongation factor-1 alpha, *Plant Cell*, **6**, 893–905.

Durso, N.A., Leslie, J.D. & Cyr, R.J. (1996) In situ immunocytochemical evidence that a homolog of protein translation factor EF-1a is associated with microtubules in carrot cells, *Protoplasma*, **190**, 141–150.

Edelmann, W., Zervas, M., Costello, P., *et al.* (1996) Neuronal abnormalities in microtubule associated protein 1B mutant mice, *Proc. Natl. Acad. Sci. USA*, **93**, 1270–1275.

Enos, A.P. & Morris, N.R. (1990) Mutation of a gene that encodes a kinesin-like protein blocks nuclear division in *A. nidulans*, *Cell*, **23**, 1019–1027.

Erhardt, M., Stoppin-Mellet, V., Campagne, S., *et al.* (2002) The plant Spc98p homologue colocalizes with γ-tubulin at microtubule nucleation sites and is required for microtubule nucleation, *J. Cell Sci.*, **115**, 2423–2431.

Flanders, D.J., Rawlins, D.J., Shaw, P.J. & Lloyd, C.W. (1990) Re-establishment of the interphase microtubule array in vacuolated plant cells, studied by confocal microscopy and 3-D imaging, *Development*, **110**, 897–904.

Furutani, I., Watanabe, Y., Prieto, R., *et al.* (2000) The SPIRAL genes are required for directional control of cell elongation in *Arabidopsis thaliana*, *Development*, **127**, 4443–4453.

Gardiner, J.C., Harper, J.D., Weerakoon, N.D., *et al.* (2001) A 90-kD phospholipase D from tobacco binds to microtubules and the plasma membrane, *Plant Cell*, **13**, 2143–2158.

Granger, C.L. & Cyr, R.J. (2001) Spatiotemporal relationships between growth and microtubule orientation as revealed in living root cells of *Arabidopsis thaliana* transformed with green-fluorescent-protein gene construct *GFP-MBD*, *Protoplasma*, **216**, 201–214.

Gungabissoon, R.A., Khan, S., Hussey, P.J. & Maciver, S.K. (2001) Interaction of elongation factor 1alpha from *Zea mays* (ZmEF-1alpha) with F-actin and interplay with the maize actin severing protein, ZmADF3, *Cell Motil. Cytoskeleton*, **49**, 104–111.

Hardham, A.R. & Gunning, B.E.S. (1978) Structure of cortical microtubule arrays in plant cells, *J. Cell Biol.*, **77**, 14–34.

Hashimoto, T. (2002) Molecular genetic analysis of left-right handedness in plants, *Phil. Trans. R. Soc. Lond. B*, **357**, 799–808.

Heath, I.B. (1974) A unified hypothesis for the role of membrane bound enzyme complexes and microtubules in plant cell wall synthesis, *J. Theor. Biol.*, **48**, 445–449.

Heinlein, M., Epel, B.L., Padgett, H.S. & Beachy, R.N. (1995) Interaction of tobamovirus movement proteins with the plant cytoskeleton, *Science*, **270**, 1983–1985.

Hush, J.M., Wadsworth, P., Callaham, D.A. & Hepler, P.K. (1994) Quantification of microtubule dynamics in living plant cells using fluorescence redistribution after photobleaching, *J. Cell Science*, **107**, 775–784.

Hussey, P.J. (2002) Microtubules do the twist, *Nature*, **417**, 128–129.

Hussey, P.J. & Hawkins, T.J. (2001) Plant microtubule-associated proteins: the HEAT is off in temperature-sensitive *mor1*, *Trends Plant Sci.*, **6**, 389–392.

Hussey, P.J., Yuan, M., Calder, M. & Lloyd, C.W. (1998) Microinjection of pollen-specific actin-depolymerizing factor, ZmADFI, reorientates F-actin strands in *Tradescantia* stamen hair cells, *Plant J.*, **14**, 353–357.

Hussey, P.J., Hawkins, T.J., Igarashi, I., Kaloriti, D. & Smertenko, A.P. (2000) The plant cytoskeleton: recent advances in the study of the plant microtubule-associated proteins MAP-65, MAP-190 and the Xenopus MAP215-like protein, MORI, *Plant Mol. Biol.*, **50**, 915–924.

Igarishi, H., Orii, H., Mori, H., Shimmen, T. & Sonobe, S. (2000) Isolation of a novel 190 kDa protein from tobacco By-2 cells: possible involvement in the interaction between actin filaments and microtubules, *Plant Cell Physiol.*, **41**, 920–931.

Jiang, C.-J. & Sonobe, S. (1993) Identification and preliminary characterization of a 65 kDa higher-plant microtubule-associated protein, *J. Cell Sci.*, **105**, 891–901.

Jiang, C.-J., Sonobe, S. & Shibaoka, H. (1992) Assembly of microtubules in a cytoplasmic extract of tobacco BY-2 miniprotoplasts in the absence of microtubule stabilizing agents, *Plant Cell Physiol.*, **33**, 497–501.

Joshi, H.C. & Palevitz, B.A. (1996) γ-Tubulin and microtubule organization in plants, *Trends Cell Biol.*, **6**, 41–44.

King, S.M. (2002) Dyneins motor on in plants, *Traffic*, **3**, 930–931.

Lambert, A.-M. & Lloyd, C.W. (1994) The higher plant microtubule cycle in *Microtubules* (eds J.S. Hyams & C.W. Lloyd), Alan R Liss, New York, pp. 327–341.

Lancelle, S.A., Callaham, D.A. & Hepler, P.K. (1986) A method for rapid freeze fixation of plant cells, *Protoplasma*, **131**, 153–165.

Lee, Y.-R.J. & Liu, B. (2000) Identification of a phragmoplast-associated kinesin-related protein in higher plants, *Curr. Biol.*, **10**, 797–800.

Lee, Y.-R.J., Giang, H.M. & Liu, B. (2001) A novel plant kinesin-related protein specifically associates with the phragmoplast organelles, *Plant Cell*, **13**, 2427–2439.

Liu, B., Cyr. R.J. & Palevitz, B.A. (1996) A kinesin-like protein, KatAp, in the cells of *Arabidopsis* and other plants, *Plant Cell*, **8**, 119–132.

Lloyd, C.W. (ed.) (1991) *The Cytoskeletal Basis of Plant Growth and Form*, Academic Press, London.

Lloyd, C.W. & Chan, J. (2002) Helical microtubule arrays and spiral growth, *Plant Cell*, **14**, 2319–2324.

Lloyd, C. & Hussey, P. (2001) Microtubule-associated proteins in plants – why we need a MAP, *Nat. Rev. Mol. Cell Biol. Rev.*, **2**, 40–47.

Marc, J., Sharkey, D.E., Durso, N.A., Zhang, M. & Cyr, R.J. (1996) Isolation of a 90-kD microtubule-associated protein from tobacco membranes, *Plant Cell*, **8**, 2127–2138.

Marc, J., Granger, C.L., Brincat, J., *et al.* (1998) A GFP-MAP4 reporter gene for visualising cortical microtubule rearrangements in living epidermal cells, *Plant Cell*, **10**, 1927–1939.

Marcus, A.I., Ambrose, J.C., Blickley, L., Hancock, W.O. & Cyr, R.J. (2002) *Arabidopsis* thaliana protein, ATK1, is a minus-end directed kinesin that exhibits non-processive movement, *Cell Motil. Cytoskeleton*, **52**, 144–50

Marcus, A.I., Li, W., Ma, H. & Cyr, R.J. (2003) A kinesin mutant with an atypical bipolar spindle undergoes normal mitosis, *Mol. Biol. Cell.*, **14**, 1717–1726.

McClinton, R.S., Chandler, J.S. & Callis, J. (2001) cDNA isolation, characterization, and protein intracellular localization of a katanin-like p60 subunit from *Arabidopsis thaliana*, *Protoplasma*, **216**, 181–190.

McNally, F.J., Okawa, K., Iwamatsu, A. & Vale, R.D. (1996) Katanin, the microtubule-severing ATPase, is concentrated at centrosomes, *J. Cell Sci.*, **109**, 561–567.

Mitsui, H., Yamaguchi-Shinozaki, K., Shinozaki, K., Nishikawa, K. & Takahashi, H. (1993) Identification of a gene family (*kat*) encoding kinesin-like proteins in *Arabidopsis thaliana* and the characterization of secondary structure of KatA, *Mol. Gen. Genet.*, **238**, 362–368.

Mollinari, C., Kleman, J.-P., Jiang, W., Hunter, T. & Margolis, R.L. (2002) PRC1 is a microtubule binding and bundling protein essential to maintain the mitotic spindle midzone, *J. Cell Biol.*, **157**, 1175–1186.

Moore, R.C. & Cyr, R.J. (2000) Association between elongation factor-1alpha and microtubules in vivo is domain dependent and conditional, *Cell Motil. Cytoskeleton*, **45**, 279–292.

Moscatelli, A., Del Casino, C., Lozzi, L., *et al.* (1995) High molecular weight polypeptides related to dynein heavy chains in *Nicotiana tabacum* pollen tubes, *J. Cell Sci.*, **108**, 1117–1125.

Nishihama, R., Ishikawa, M., Araki, S., Soyano, T., Asada, T. & Machida, Y. (2001) The NPK1 mitogen-activated protein kinase kinase kinase is a regulator of cell-plate formation in plant cytokinesis, *Genes Dev.*, **15**, 352–363.

Oppenheimer, D.G., Pollock, M.A., Vacik, J., *et al.* (1997) Essential role of a kinesin-like protein in *Arabidopsis* trichome morphogenesis, *Proc. Natl. Acad. Sci. USA*, **94**, 6261–6266.

Panteris, E., Apostolakos, P., Graf, R. & Galatis, B. (2000) Gamma-tubulin colocalizes with microtubule arrays and tubulin paracrystals in dividing vegetative cells of higher plants, *Protoplasma*, **210**, 179–187.

Park, S.K., Howden, R. & Twell, D. (1998) The *Arabidopsis thaliana* gametophytic mutation *gemini pollen 1* disrupts microspore polarity, division assymmetry and pollen cell fate, *Development*, **125**, 3789–3799.

Pay, A., Resch, K., Frohnmeyer, H., *et al.* (2002) Plant RanGAPs are localized at the nuclear envelope in interphase and associated with microtubules in mitotic cells, *Plant J.*, **30**, 699–709.

Popov, A.V., Pozniakovsky, A., Arnal, I., *et al.* (2001) XMAP215 regulates microtubule dynamics through two distinct domains, *EMBO J.*, **20**, 397–410.

Popov, A.V., Severin, F. & Karsenti, E. (2002) XMAP215 is required for the microtubule nucleating activity of centrosomes, *Curr. Biol.*, **12**, 1326–1330.

Preuss, M.L., Delmer, D. & Liu, B. (2003) The cotton kinesin-like calmodulin-binding protein associates with cortical microtubules in cotton fibers, *Plant Phys.*, **132**, 154–160.

Reddy, A.S. & Day, I.S. (2001) Kinesins in the *Arabidopsis* genome: a comparative analysis among eukaryotes, *BMC Genomics*, **2**(1), 2.

Reddy, A.S.N., Narasinkulu, S.B., Safadi, F. & Goloukin, M. (1996a) A plant kinesin heavy chain-like protein is a calmodulin binding protein, *Plant J.*, **10**, 9–21.

Reddy, A.S.N., Safadi, F., Narasinkulu, S.B., Golovkin, M. & Hu, X. (1996b) A novel plant calmodulin-binding protein with a kinesin heavy chain motor domain, *J. Biol. Chem.*, **271**, 7052–7060.

Rutten, T., Chan, J. & Lloyd, C.W. (1997) A 60 kDa plant microtubule-associated protein promotes the growth and stabilization of neurotubules in vitro, *Proc. Natl. Acad. Sci. USA*, **94**, 4469–4474.

Shirasu-Hiza, M., Coughlin, P. & Mitchison, T. (2003) Identification of XMAP215 as a microtubules-destabilizing factor in *Xenopus* egg extract by biochemical purification, *J. Cell Biol.*, **61**, 349–358.

Smertenko, A., Saleh, N., Igarashi, H., *et al.* (2000) A new class of microtubule-associated proteins in plants, *Nat. Cell Biol.*, **2**, 750–753.

Smertenko, A.P., Bozhkov, P.V., Filonova, L.H., von Arnold, S. & Hussey, P.J. (2003) Re-organisation of the cytoskeleton during developmental programmed cell death in *Picia abies* embryos, *Plant J.*, **33**, 813–824.

Smirnova, E., Reddy, A., Bowser, J. & Bajer, A. (1998) Minus end-directed kinesin-like motor protein, Kcbp, localizes to anaphase spindle poles in *Haemanthus* endosperm. *Cell Motil. Cytoskel.*, **41**, 271–280.

Smith, L.G., Gerttula, S.M., Han, S. & Levy, J. (2001) TANGLED1: a microtubule binding protein required for the spatial control of cytokinesis in maize, *J. Cell Biol.*, **152**, 231–236.

Song, H., Golovkin, M., Reddy, A.S. & Endow, S.A. (1997) *In Vitro* motility of ATKCBP, a calmodulin-binding kinesin protein of *Arabidopsis*. *Proc. Natl. Acad. Sci. USA*, **94**, 322–327.

Stoppin-Mellet, V., Gaillard, J. & Vantard, M. (2002) Functional evidence for in vitro microtubule severing by the plant katanin homolog, *Biochem. J.*, **365**, 337–372.

Strompen, G., El Kasmi, F., Richter, S., *et al.* (2002) The *Arabidopsis HINKEL* gene encodes a kinesin-related protein involved in cytokinesis and is expressed in a cell cycle-dependent manner, *Curr. Bio.*, **12**, 153–158.

Thitamadee, S., Tuchihara, K. & Hashimoto, T. (2002) Microtubule basis for left-handed helical growth in *Arabidopsis*, *Nature*, **417**, 193–196.

Tournebize, R., Popov, A., Kinoshita, K., *et al.* (2000) Control of microtubule dynamics by the antagonistic activities of XMAP215 and XKCM1 in *Xenopus* egg extracts, *Nature Cell Biol.*, **2**, 13–19.

Twell, D., Park, S.K., Hawkins, T.J., *et al.* (2002) MOR1/GEM1 plays an essential role in the plant-specific cytokinetic phragmoplast, *Nature Cell Biol.*, **4**, 711–714.

Van Breugel, M., Drechsel, D. & Hyman, A. (2003) Stu2p, the budding yeast member of the conserved Dis1/XMAP215 family of microtubule-associated protein is a plus end-binding microtubule destabilizer, *J. Cell Biol.*, **161**, 359–369.

Vasquez, R.J., Gard, D.L. & Cassimeris, L. (1994) XMAP from *Xenopus* eggs promotes rapid plus end assembly of microtubules and rapid microtubule polymer turnover, *J. Cell Biol.*, **127**, 985–993.

Vos, J.W., Safadi, F., Reddy, A.S.N. & Hepler, P.K. (2000) The kinesin-like calmodulin binding protein is differentially involved in cell division, *Plant Cell*, **12**, 979–990.

Walczak, C.E., Gan, E.C., Desai, A., Mitchison, T.J. & Kline-Smith, S.L. (2002) The microtubule-destabilising kinesin XKCM1 is required for chromosome positioning during spindle assembly, *Curr. Biol.*, **12**, 1885–1889.

Wasteneys, G.O. (2002) Microtubule organization in the green kingdom: chaos or self-order? *J. Cell Sci.*, **115**, 1345–1354.

Whittington, A.T., Vugrek, O., Wei, K.J., *et al.* (2001) MOR1 is essential for organizing cortical microtubules in plants, *Nature*, **411**, 610–613.

Yamamoto, E., Zeng, L. & Baird, W.V. (1998) α-Tubulin missense mutations correlate with antimicrotubule drug resistance in *Eleusine indica*, *Plant Cell*, **10**, 297–308.

Yasuhara, H., Sonobe, S. & Shibaoka, H. (1993) Effects of taxol on the development of the cell plate and of the phragmoplast in tobacco BY-2 cells, *Plant Cell Physiol.*, **34**, 21–29.

Yasuhara, H., Muraoka, M., Shogaki, H., Mori, H. & Sonobe, S. (2002) TBMP200, a microtubule bundling polypeptide isolated from telophase tobacco BY-2 cells is a MOR1 homologue, *Plant Cell Physiol.*, **43**, 595–603.

Yuan, M., Shaw, P.J., Warn, R.M. & Lloyd, C.W. (1994) Dynamic reorientation of cortical microtubules, from transverse to longitudinal in living plant cells, *Proc. Natl. Acad. Sci. USA*, **91**, 6050–6053.

Zeng, L. & Baird, W.V. (1999) Inheritance of tolerance to the anti-microtubule dinitroaniline herbicides in an intermediate biotype of goosegrass, *Am. J. Bot.*, **86**, 940–947.

Zhong, R., Burk, D.H., Morrison, W.H. & Ye, Z.-H. (2002) A kinesin-like protein is essential for oriented deposition of cellulose microfibrils and cell wall strength, *Plant Cell*, **14**, 3101–3117.

2 Actin and actin-modulating proteins

Christopher J. Staiger and Patrick J. Hussey

2.1 Introduction

Although cells of flowering plants do not crawl away from noxious stimuli or migrate towards sources of nutrient and developmental signals, it has been known for more than a century that they exhibit a multitude of fascinating and biologically relevant intracellular processes. Many of these, including the very prominent cytoplasmic streaming of certain cells, are among the fastest biological movements on the planet. At the heart of these movements is a dynamic network of polymeric elements, motor molecules and associated proteins called the cytoskeleton. The actin framework, consisting of actin filaments (F-actin), is perhaps the more dynamic of the two structures, but direct measurements of its turnover are lacking. Localization studies and the judicious use of actin inhibitors all point toward multiple intracellular functions for actin. Included among these roles are actin filaments serving as the molecular railroad tracks for myosin-based cytoplasmic streaming, allowing the translocation of specific organelles like Golgi bodies and peroxisomes, and anchoring chloroplasts at favorable positions in response to changing light conditions (see Chapter 5). The movement of smaller, less visible organelles like exo- and endocytotic vesicles can also be inferred to require actin filament function, but the molecular mechanisms underpinning all of these organelle movements and which if any class of motors are involved remain to be elucidated. At a whole cell level, the actin cytoskeleton cooperates with microtubules to mark the nascent cell division site and provides guidewires to direct the developing cell plate to the final site of fusion with the mother wall during cytokinesis. The stability and repositioning of transvacuolar cytoplasmic strands, as well as the position of the nucleus, hints that actin structures are responsible for much of the cytoplasmic architecture in vacuolated plant cells. Dramatic rearrangements of cytoskeletal polymers also signal the arrival of chemical and structural defenses to the site of attempted penetration by fungal pathogens and perhaps prevent the invasion by non-host species. And, although the shape of plant cells is largely dictated by the nature of the cell wall, through regulating the local deposition and composition of this extracellular matrix, the cytoskeleton is responsible for controlling cellular morphogenesis (see Chapters 3 and 6). Reverse-genetic evidence implicates both microtubules and actin filaments in controlling cellular elongation within different tissues of the plant cell body. An extreme example of cellular morphogenesis is epitomized by the tip growth of pollen tubes and root hairs (see Chapters 7 and 8), which requires a very specific arrangement of actin filaments to choreograph the arrival and fusion of secretory vesicles at the extreme apex of these cylindrical cells. Less well characterized, but equally remarkable, are the prominent, interdigitating lobes that form between many epidermal pavement cells. These too probably require a dynamic actin cytoskeleton, and may even share some basic features with tip growth phenomena.

Actin is an abundant cellular protein with a globular, bilobed shape and a molecular weight of 42 kDa. In cells it exists in two pools: a population of subunits (G-actin) and the filament network. The proportion of protein in each pool varies depending on the cell type and physiological status, but is typically something like 50:50. However, recent measurements of pollen grains and tubes show that a remarkably small amount (10–20%) of the total actin protein (100–200 μM) is present in filaments (Gibbon *et al.*, 1999; Snowman *et al.*, 2002). The polymer that is formed by actin polymerization can be described as a double-stranded string of pearls, with an average diameter of 7 nm, a right-handed helical twist, and 13–14 subunits per half turn. Polymerization is normally coupled to hydrolysis of ATP, which binds in a divalent cation- and nucleotide-binding site in the cleft between the two large domains. Filaments of actin show polarity with respect to subunit loss and addition, with the plus or barbed end being more dynamic than the pointed or minus end. Although actin polymerizes under physiological conditions in the test tube, its polymerization in cells is regulated by a multitude of accessory proteins and by post-translational modifications. In this chapter, we review what is known about the major regulators of actin assembly and dynamics in plant cells.

2.2 Actin

Actin was discovered nearly half a century ago and purified from skeletal muscle; its abundance and the ease of isolation have made rabbit muscle α-actin isoform the protein of choice for most biochemical studies. There are, however, obvious reasons for studying the physico-chemical properties of actin isolated from different tissues and other organisms. Indeed, actin isovariants from non-skeletal muscle tissues have distinct biochemical properties (Sheterline *et al.*, 1998), and actins from certain organisms show preferential binding to endogenous ABPs (Drubin *et al.*, 1988; Namba *et al.*, 1992; Kim *et al.*, 1996; Ono, 1999). Although numerous early reports of the isolation of plant actin can be found in the literature, most of these purifications had low yields and the protein often did not assemble without assistance from phalloidin. In 1992, Lungfei Yan and colleagues (Liu & Yen, 1992) described a multistep procedure for the isolation of milligram amounts of actin of maize pollen. Some years later, Staiger and colleagues took a different and more rapid approach using the same starting material (Ren *et al.*, 1997). Maize pollen actin can be purified to near homogeneity in a single chromatography step by formation of a heterogenous human profilin–plant actin complex that is then bound to poly-L-proline (PLP)-sepharose. The actin is recovered with a gentle, salt elution and, after dialysis, can be assembled to give electrophoretically pure actin.

Typical yields are 3–5 mg from 10 g of frozen maize pollen. A practical alternative to purification of the profilin–actin complex is DNase I chromatography (Schafer *et al.*, 1998), which has been used to isolate assembly-competent, native actin from zucchini hypocotyls (Hu *et al.*, 2000). In our experience, this latter approach suffers from the difficulty of recovering functional protein after the harsh conditions required to elute G-actin from the high-affinity DNase I complex. Analysis of the assembly properties for plant actins reveals very few obvious differences when compared with rabbit muscle actin (Yen *et al.*, 1995; Ren *et al.*, 1997). The only difference that has been quantified is a 10–20-fold greater intrinsic rate of nucleotide exchange for plant versus non-plant actin (Kovar *et al.*, 2001b); this may obviate the need for certain accessory proteins like profilin and ADF to modulate turnover. Several classes of plant ABP (profilin, ADF, myosin, fimbrin) have been tested for the ability to distinguish between plant and muscle actin isoforms with negative results. This is not to say that plant and muscle actin behave in identical fashion and that such differences will not be found in the future. One of the most compelling reasons to continue along this path is the observation that muscle actin appears to poison living plant cells, whereas pollen actin added by microinjection is perfectly fine (Hepler *et al.*, 1993; Ren *et al.*, 1997).

Cytoskeletal proteins in plants are typically encoded by large multigene families (Meagher & Williamson, 1994; Meagher *et al.*, 2000). Some plants, like tobacco and petunia contain dozens or hundreds of actin genes (Baird & Meagher, 1987; Thangavelu *et al.*, 1993), whereas *Arabidopsis* has just eight expressed actins that share 93–97% amino acid sequence identity (McDowell *et al.*, 1996b; McKinney & Meagher, 1998). One of the fundamental problems in plant biology is understanding the basis for the large gene families that encode cytoskeletal proteins. Meagher and coworkers (1999a) propose that elaboration of multiple isoforms allows for *isovariant dynamics* or 'the temporal and biochemical expansion of a biological system's responses as a result of the simultaneous expression and interaction of multiple isovariants of a protein'. This concept is not mutually exclusive with the idea that gene families are necessary for complex patterns of expression within the plant and during development. Indeed, the eight *Arabidopsis* actins can be grouped into two ancient classes and five subclasses that correspond with the pattern of expression in the plant. Meagher postulates an ancient duplication in cytoskeletal gene families, predating the divergence of monocots and dicots, that gives rise to reproductive and sporophytic isoforms (Meagher *et al.*, 1999b). As the genome of rice is completed and expression patterns for cytoskeletal proteins are uncovered, it will be interesting to see whether additional supporting data can be generated. Whether the reproductive and vegetative actin isovariants are functionally non-equivalent has not been addressed unambiguously. By analogy with classic genetic experiments with *Drosophila* sperm-specific β2-tubulin isoform (Hoyle & Raff, 1990), and *Drosophila* flight muscle-specific actin (Fyrberg *et al.*, 1998), the critical experiment that must be performed is to test whether a reproductive actin expressed behind a vegetative-actin promoter can complement the loss-of-function phenotype produced when a vegetative actin is mutated.

(The reciprocal experiment is also possible and is a valid test of the model.) Since mutant actin alleles with visible phenotypes are now available (see below) such approaches are immediately practical in *Arabidopsis*. Indeed, an elegant example of such a test is provided by Meagher and coworkers (Gilliland *et al.*, 2002); over-expression of a reproductive actin isovariant complements the root hair phenotype associated with ACT2 loss-of-function mutations.

Although actins are among the first genes to be identified in reverse-genetic screens for T-DNA insertion events (McKinney *et al.*, 1995), it was not until very recently that visible phenotypes have been reported. This lack of gross defects led many to assert that actin isovariants were functionally redundant. However, multigenerational studies indicate that mutant actin alleles tended to be lost from populations, suggesting that they are deleterious (Gilliland *et al.*, 1998). And quite recently, loss-of-function alleles for at least two actin isoforms with visible phenotypes affecting organ development or cell morphology have been reported. Root hairs are tip growing projections of specialized epidermal cells called trichoblasts; in *Arabidopsis*, root hairs express three different actin isovariants, ACT2, ACT7 and ACT8 (An *et al.*, 1996; McDowell *et al.*, 1996a). In a forward-genetic screen for deformed root hairs, the cloning of three *der1* alleles reveals that these encode ACT2 with single amino acid substitutions for two residues (A183V, R97H and R97C) (Ringli *et al.*, 2002). Mutant plants have defects that include altered site selection for initial bulge formation on trichoblasts, mispositioning of the tip growth machinery and perturbation of tip growth (see also Gilliland *et al.*, 2002). Many of these phenotypes resemble abnormalities produced with actin-disrupting drugs (Bibikova *et al.*, 1999; Miller *et al.*, 1999; Baluska *et al.*, 2000), confirming a role for actin in these processes. The wild-type *ACT2* cDNA, under the control of its native promoter or a root hair-specific promoter, complements the mutant phenotypes. Mutations in the *ACT7* gene cause more widespread defects to roots, reduced seed germination, as well as some perturbation to above-ground organs (Gilliland *et al.*, 2003). A reduction in ACT7 protein levels in homozygous mutant plants correlates with reduction in the length of the root axis and smaller cotyledons. Assuming that cell number is not affected, these results imply a role for actin in cell elongation. Similar conclusions can be drawn from the analysis of whole plants and seedlings treated with the F-actin depolymerizing agent, latrunculin B. Indeed, such pharmacological studies result in dwarfed plants with normal cell division patterns and quite normal looking organs that have reduced cell size (Baluska *et al.*, 2001a). How exactly actin contributes to cell expansion is not presently known, but the best guess would be that it acts through vesicle trafficking to and from the plasma membrane similar to its suggested role during tip growth (Mathur & Hülskamp, 2002; Smith, 2003).

The function of *Arabidopsis* actin isovariants has also been tested with over-expression and ectopic expression constructs (Kandasamy *et al.*, 2002). When a pollen-specific isovariant (ACT1) is expressed ectopically in the body of the plant, under the control of the vegetative *ACT2* promoter, detrimental effects are observed. The level of ectopic ACT1 protein expression correlates positively with abnormal organ morphology, reduced plant stature and delayed bolting or flowering. At the cellular level, ectopically expressed pollen ACT1 assembles into massive, highly bundled arrays in leaf cells from the dwarf plants. In contrast, overexpression of a vegetative isovariant (ACT2) under the control of the *ACT2* promoter has only modest consequences for plant growth, organ morphology and size, or actin organization. These data provide indirect evidence for the *isovariant dynamics* model, and are consistent with a mis-expressed actin not functioning properly in the foreign environment. Carrying this concept a step further, it is possible that pollen actin is detrimental to the sporophytic cell because it is unable to interact with the endogenous suite of vegetative actin-binding protein isoforms. Additional complicated experiments will be required to further test this hypothesis. Regardless, it is intriguing to note that both disruption of actin filaments and actin overexpression lead to inhibition of cell elongation. This observation begs for new, innovative approaches to dissect the molecular mechanism of actin's role during cell expansion.

Limited evidence for the regulation of plant actin by post-translational modifications is available. The pulvinus cells on the *sensitive plant* (*Mimosa pudica*) are remarkable for large changes in turgor pressure that result in closing of leaf pinnae and bending of the petioles in response to touch stimulation. Immunoblotting studies demonstrate that actin from the pulvini of *Mimosa* is highly phosphorylated on tyrosine residues, but this level drops precipitously during the touch response (Kameyama *et al.*, 2000). Importantly, a tyrosine kinase inhibitor (PAO) blocks the bending and actin dephosphorylation, suggesting that actin modification is necessary for the response. Earlier work by Fleurat-Lessard *et al.* (1988) showed that actin-disrupting drugs also block bending, but what the consequences of actin phosphorylation are on the biochemical properties of plant actin and what reorganization(s) occur in pulvini cells during stimulation are unknown.

A 63-kDa polypeptide in root nodules of *Phaseolus* is demonstrated to be a monoubiquitinylated actin isoform (Dantán-González *et al.*, 2001). This isovariant also accumulates in plants that are infected with fungal or bacterial pathogens, but is not induced with heat shock, wounding or osmotic stress. Although monoubiquitinylation appears to correlate with the increased resistance of actin to proteolytic degradation, the physiological consequences of this modification for actin polymerization have not been described.

2.3 Myosin

The molecular engine for the actin cytoskeleton is a mechanochemical enzyme, or motor protein, called myosin. Myosins belong to a superfamily of actin-activated ATPases that propel various cargo along actin tracks or that translocate actin filaments on other rather immobile objects like the plasma membrane, the endoplasmic reticulum or cytoskeletal elements. Like other motor molecules, the N-terminal actin-binding domain is highly conserved across kingdoms, whereas the C-terminal tail or cargo-binding domain is quite divergent. Originally, myosins were classified as Type I (unconventional) and Type II (conventional, muscle myosin). With the proliferation of genomic sequence information, we know today that myosins belong to at least 18 different classes, with three classes that are exclusive to plants and algae (Hodge & Cope, 2000). The plant and algal myosins belong to Class VIII, Class XI and Class XIII (Reichelt & Kendrick-Jones, 2000). The engine for cytoplasmic streaming from the green alga *Chara* is able to propel organelle movements during cytoplasmic streaming at rates up to 60 μm/s, a much higher rate than other myosins (Higashi-Fujime, 1991)! Cloning of the *Chara* myosin reveals that it is a Class XI myosin (Kashiyama *et al.*, 2000; Morimatsu *et al.*, 2000) as is a 175-kDa myosin that probably drives streaming in tobacco BY-2 cells (Tominaga *et al.*, 2003). The *Arabidopsis* genome contains 17 myosin-like sequences that fall into class VIII and XI (Reddy, 2001; Reddy & Day, 2001). Much work needs to be done to analyse mutant phenotypes associated with myosin loss-of-function mutations, to understand the cargo these myosins carry and to reveal the subcellular distribution of specific isoforms. Moreover, it will be important to examine the biochemical properties of these plant myosins, including the rates of translocation on filaments, regulation by phosphorylation and/or calcium, and the processivity and/or direction of travel. For further information on plant myosins, the reader is referred to several excellent recent reviews on this topic (Asada & Collings, 1997; Yamamoto *et al.*, 1999; Reichelt & Kendrick-Jones, 2000; Shimmen *et al.*, 2000; Yokota & Shimmen, 2000; Reddy, 2001; Reddy & Day, 2001).

2.4 Actin-binding proteins: overview

The dynamic nature of the cytoskeleton depends on the spatial distribution and the local activity of a complex mixture of actin-binding proteins (ABPs) present within all eukaryotic cells. In non-plant systems, 70 classes of ABPs have been characterized (Kreis & Vale, 1999). These are responsible for controlling the size and activity of the G-actin pool, the subcellular sites for filament nucleation, the rate of filament turnover and the organization of filaments into higher-order structures. ABPs can be grouped into several somewhat artificial categories, including:

- *monomer-binding proteins* that regulate polymerization and depolymerization
- *cross-linking and bundling proteins* that form orthogonal networks or bundles
- *capping proteins* that control the availability of filament ends for subunit addition
- *nucleation factors* that seed polymerization of new filaments, and
- *side-binding proteins* that stabilize filaments and/or regulate myosin binding.

Several of these classes have been identified in plants and detailed characterization is progressing rapidly (see below). Homology-based searches of the completed *Arabidopsis* genome indicate that other classes may be absent or poorly conserved in plants (Assaad, 2001; Hussey *et al.*, 2002). Notable absences include thymosin β4, WASp, coronin, α-actinin, spectrin, tropomyosin, filamin, fascin and SLA2p/Hip1. One intriguing possibility is that plants have duplicated and elaborated certain ancient classes of ABPs to fulfill multiple functions. For example, profilin isovariants could function both to regulate actin assembly (Pantaloni & Carlier, 1993; Kang *et al.*, 1999) and to buffer the monomer pool (a function attributed to thymosin β4 in metazoans). The proliferation of diverse fimbrin or villin isoforms may have obviated the need for α-actinins, filamins and spectrins. An alternative, and overlapping hypothesis, is that plants contain novel families of yet-to-be-discovered ABPs. Either way, the future of actin cytoskeletal research promises to be exciting.

ABPs also function at the crossroads between extracellular signals and rearrangements of the cytoskeleton (Nick, 1999; Staiger, 2000; Vantard & Blanchoin, 2003). Many, like profilin and ADF, are PtdIns(4,5)P_2-binding proteins or are regulated by calcium (Niggli, 2001). The monomeric GTP-binding protein, Rop, is also a convincing candidate for linking signals and calcium fluxes to changes in actin organization (Kost *et al.*, 1999; Fu *et al.*, 2001; Fu *et al.*, 2002). However, no complete cascade of intermediates linking extracellular stimulus to actin rearrangement exists for pollen or any other plant cell.

2.5 Monomer-binding proteins

2.5.1 *ADF/cofilin*

In general, members of the actin depolymerizing factor (ADF) or cofilin group of proteins bind both G-actin and F-actin and modulate actin filament dynamics. They do this by severing actin filaments, thereby providing more filament ends for polymerization, and by increasing the dissociation rate constant at the pointed end. Members of the ADF/cofilin group are regulated by pH, by specific phospholipids and by reversible phosphorylation. Moreover, ADF/cofilin activity can be enhanced by another protein called actin-interacting protein 1(AIP1) (see Section 2.9.3). Multiple isoforms of ADF exist in plants and animals and some have distinct properties (for reviews see Bamburg, 1999; Maciver & Hussey, 2002; for a comprehensive bibliography see 'The ADF/Cofilin Homepage' at http://www.bms.ed.ac.uk/research/smaciver/cofilin.htm). Here, we describe the major characteristics of plant ADFs.

A phylogenetic analysis of plant ADF versus animal, fungal and protistan ADF has been documented (Maciver & Hussey, 2002). What is surprising is the fact that plants have large numbers of ADF genes (12 in *Arabidopsis*, although how many are functional remains to be determined). Of those plants analysed, it has been shown that members of these gene families are expressed differentially (Lopez *et al.*, 1996; Mun *et al.*, 2000; Dong *et al.*, 2001a,b). There appears to be those that are expressed specifically in reproductive tissues (e.g. pollen) and those expressed either constitutively and/or in vegetative tissues only. Differences in the structure and biochemical activities between pollen and vegetative ADFs have been documented indicating that selective pressures have resulted in the evolution of specific ADF isotypes that perform subtly different roles in different tissues and organs of plants (Smertenko *et al.*, 2001). It is interesting to note that the *Arabidopsis* genome database does not contain the proteins that are structurally similar to ADFs that are present in animals, e.g. drebrin (Peitsch *et al.*, 1999), or in fungi, e.g. twinfilin (Lappalainen *et al.*, 1998). This may indicate that plants do not need these proteins or that within the large gene families of plant ADFs (the *Arabidopsis* genes fall into four divergent clades) there are isotypes that can perform the roles of these missing proteins.

The most biochemically well-characterized plant ADFs are an *Arabidopsis* ADF (Carlier *et al.*, 1997; Bowman *et al.*, 2000), a maize vegetatively expressed ADF (Gungabissoon *et al.*, 1998) and a lily pollen-expressed ADF (Allwood *et al.*, 2002). Moreover, Dong *et al.* (2001b) have shown by aberrant expression of ADFs in *Arabidopsis*, that ADF contributes to the control of cell and organ expansion. The plant ADFs have been shown to increase actin dynamics and they do so in a pH-dependent manner. At pH 6.0 these ADFs bind F-actin and at pH 9.0 they prefer G-actin. The crossover point, the point at which the amount of ADF bound to F- and G- actin is the same, for maize ADF3 (Gungabissoon *et al.*, 1998), is pH 7.7 which is higher than that for human ADF which is 7.3. This also reflects the fact that the resting pH of plant cells is more alkaline than in animal cells.

It has been shown that the activity of plant ADFs are inhibited by phosphatidylinositol 4,5-bisphosphate and to a lesser extent phosphatidylinositol 4-monophosphate. This inhibition is the result of direct binding of the phosphoinositides to the ADFs. Moreover, plant ADFs have been shown to inhibit the activity of plant phospholipase C. These data indicate that plant ADFs may be regulated by or be part of the phosphoinositide signal transduction pathway (Gungabissoon *et al.*, 1998).

Plant ADFs (with the exception of lily pollen ADF) can be phosphorylated on Ser6 (Smertenko *et al.*, 1998; Allwood *et al.*, 2001). This phosphorylation is by a calmodulin-domain like protein kinase and these kinases are specific to plants and protozoa. The *Arabidopsis* genome database does not contain the LIM kinases (Arber *et al.*, 1998) or the TESK kinases (Toshima *et al.*, 2001) known to phosphorylate metazoan ADF/cofilins. The phosphorylation of ADF can be mimicked by substituting the serine residue at position 6 for an aspartic or glutamic acid. This analogue of phosphorylated ADF has been shown not to bind G- or F-actin in cosedimentation assays and has been shown to have a much reduced effect on increasing the rate of actin dynamics using spectrofluorimetric assays (Smertenko *et al.*, 1998). These data indicate that some plant ADFs are likely to be regulated by reversible phosphorylation like their metazoan counterparts. However, as lily pollen ADF is not phosphorylated by plant cell extracts or by purified CDPK known to phosphorylate maize ADF3 (Allwood *et al.*, 2002), this would indicate that this ADF is more similar to those from yeast and *Dictyostelium discoideum* in this respect. The fact that CDPK phosphorylates plant ADFs also suggests that plant ADFs can be controlled indirectly by levels of Ca^{2+}, and many cellular processes that require a dynamic actin cytoskeleton are controlled by oscillations in Ca^{2+} levels, e.g. guard cell opening/closing (see Chapter 10). In metazoans, the LIM kinases are in turn regulated by Rho-GTPases (Yang *et al.*, 1998). Over-expression of Rho-GTPases in plant cells results in actin remodeling but how this occurs requires further study (Kost *et al.*, 1999; Fu *et al.*, 2001).

The lily pollen-specific ADF has characteristics different from the maize and *Arabidopsis* ADFs (Smertenko *et al.*, 2001; Allwood *et al.*, 2002). Although it binds G- and F-actin in a pH-dependent manner and is inhibited by phosphatidylinositol 4,5-bisphosphate and phosphatidylinositol 4-monophosphate, it has a weak effect on actin dynamics in severing and depolymerization assays. This F-actin binding capability but weak actin-depolymerizing activity *in vitro* has been used to explain why pollen ADF can be seen decorating F-actin in pollen grains. Together with an ADF from *Caenorhabditis elegans* (Ono, 2001), these are the only examples of ADF-decorating F-actin in cells apart from studies where ADF concentration has been artificially increased by microinjection or transfection (Moriyama *et al.*, 1996; Dong *et al.*, 2001a; Chen *et al.*, 2002). In pollen tubes, however, pollen ADF is no longer bound to F-actin and is mainly cytoplasmic. It has been proposed that the cytoplasmic localization of pollen ADF in the pollen tubes indicates that this protein is now available to actively turn over actin filaments (like the cytoplasmic localized maize ADF in root hairs (Jiang *et al.*, 1997). However, because of its low ADF activity other factors must contribute to increasing its activity, and/or perhaps other proteins must be involved in increasing actin dynamics to promote pollen tube extension. Plant floral-specific actin-interacting protein 1 (AIP1) has been shown to enhance the actin-depolymerizing activity of lily pollen ADF by approximately 60% compared to ADF (or AIP1) on its own. Moreover, pollen AIP1 and ADF have similar localizations in pollen tubes and pollen grains raising the possibility that these proteins could work as a couple to promote actin dynamics in pollen tubes.

2.5.2 *Profilin*

Profilin is a low molecular weight (12–15 kDa), ubiquitous, cytoplasmic protein. It is essential for the viability of fission yeast (Lu & Pollard, 2001), flies (Verheyen & Cooley, 1994) and mice (Witke *et al.*, 2001), whereas loss of profilin function severely impacts the growth of budding yeast (Haarer *et al.*, 1990) and *Dictyostelium* (Haugwitz *et al.*, 1994). Originally identified from calf spleen as a protein that forms a high affinity 1:1 complex with G-actin (Carlsson *et al.*, 1977), it is now understood that profilin can have complex effects on actin assembly. In the presence of capped ends, profilin acts like a simple sequestering protein by binding to G-actin and preventing its polymerization. However when filament ends are free, the profilin-actin complex can add to F-actin and allow assembly (Pantaloni & Carlier, 1993; Kang *et al.*, 1999). Which activity predominates is dictated by the size and activity of the G-actin pool, the presence of other actin-binding proteins and the stoichiometry of profilin to G-actin. In budding yeast and *Dictyostelium*, profilin has a sequestering function (Magdolen *et al.*, 1993; Haugwitz *et al.*, 1994). In contrast, loss-of-function mutants in *Drosophila* and *S. pombe* indicate that profilin is necessary to promote normal actin assembly (Cooley *et al.*, 1992; Balasubramanian *et al.*, 1994). Profilin function is likely to be cell-type specific and must, therefore, be examined on a case-by-case basis.

In addition to binding G-actin and actin-related proteins (Machesky *et al.*, 1994; Zhao *et al.*, 1998), profilin has a number of other protein and lipid ligands. For example, profilin's ability to bind contiguous stretches of proline (Tanaka & Shibata, 1985) is important for interactions with a number of proline-rich proteins (Wasserman, 1998; Tanaka, 2000; Holt & Koffer, 2001; Reinhard *et al.*, 2001; Deeks *et al.*, 2002). It is often inferred that the reduced viability associated with profilin mutants is due to failure of some actin-based function. However, evidence exists for the importance of non-actin functions of profilin. In yeasts, genetic data show that the ability to bind proline-rich sequences is essential (Ostrander *et al.*, 1999; Lu & Pollard, 2001). A site-directed mutant of profilin, that lacks poly-proline binding but is wild type for actin binding, does not complement the growth defect of *Dictyostelium* profilin mutants unless it is massively overexpressed (Lee *et al.*, 2000). Conversely, a mutant profilin that is defective for actin-binding, but does bind polyproline, can complement. In an *in vitro* assay, Cdc42-induced actin nucleation requires profilin with a high affinity for both actin and polyproline (Yang *et al.*, 2000). And, a mutated vertebrate profilin I that is unable to bind actin inhibits actin-based microspike formation in cultured cells, most likely by titrating out the endogenous profilin for binding to proline-rich signaling molecules which regulate actin polymerization (Suetsugu *et al.*, 1998, 1999). The picture emerging is that profilin-interacting proteins function both to target profilin to sites of actin assembly and to regulate profilin activity.

Proline-rich proteins that bind profilin belong to at least two major classes: VASP/Ena proteins and formin-homology (FH) proteins (Holt & Koffer, 2001; Reinhard *et al.*, 2001). The former are implicated in cAMP signaling, cell motility and axon outgrowth, but appear to be absent from the *Arabidopsis* genome. The FH-domain proteins are found in organisms from yeast to man and couple G-protein signaling to actin-based functions. Mutant phenotypes associated with FH proteins include defects in cell polarity and cytokinesis. Both classes of profilin-interacting protein contain a proline-rich motif (PRM1) with the consensus sequence XPPPPP, where X is G,A,L,S, I or V (Holt & Koffer, 2001). The stretch of prolines in PRM1 comprises 5–12 residues, and the motif is repeated 2–5 times.

In FH proteins, the proline-rich FH1 domain is always coupled to a conserved 130 amino acid FH2 domain (Wasserman, 1998). Recent evidence implicates FH2 as a microtubule-binding module (Ishizaki *et al.*, 2001; Palazzo *et al.*, 2001), and mammalian Diaphanous associates with the mitotic spindle and regulates micro-tubule organization (Kato *et al.*, 2000). Supported with evidence from yeast studies (Lee *et al.*, 1999; Feierback & Chang, 2001), FH proteins are thought to coordinate the microtubule and actin cytoskeleton during mitosis, cytokinesis and at the plasma membrane. *Arabidopsis* contains 21 *AtFH* genes, with predicted consecutive FH1 and FH2 modules, that can be grouped into two major families (Cvrcková, 2000; Deeks *et al.*, 2002). Type I AtFH proteins are unique among eukaryotic formins in containing putative signal peptides and/or predicted transmembrane motifs. The FH1 domain from one of these *Arabidopsis* formins, AtFH1, reportedly interacts with profilin in a yeast two-hybrid interaction (Banno & Chua, 2000), but no additional biochemical information exists presently. The Type II AtFH proteins are missing the membrane-targeting motifs. Within the two classes, there is enormous variation in the nature of the proline-rich sequences or FH1 domain, making theoretical predictions about their capacity for profilin-binding quite difficult. Both classes of AtFH protein are missing FH3 domains, implicated in subcellular localization, as well as the GTPase-binding domain, suggesting that their subcellular targeting and regulation may be unique to plants. Finally, it should be noted that certain formin proteins in yeast have been implicated in actin cable formation through the ability to associate with filament barbed ends and to nucleate actin polymerization (Evangelista *et al.*, 2002; Pruyne *et al.*, 2002; Sagot *et al.*, 2002a). This Arp2/3-independent mechanism for stimulation of actin polymerization requires the activity of profilin and its ability to facilitate actin assembly (Evangelista *et al.*, 2002; Sagot *et al.*, 2002b). Given that the conventional activators of Arp2/3 are missing or poorly conserved in *Arabidopsis* and that plant Arp2/3 has yet to be characterized biochemically (see below), attention should be given to this intriguing actin nucleation function of formins.

A role for profilin in vesicle trafficking and endocytosis has also been demonstrated. Vertebrate profilin has a high affinity for phosphoinositide lipids, and shows a marked preference for lipids phosphorylated at the D-3 position (Lu *et al.*, 1996). It also binds to PI-3-kinase and enhances lipid kinase activity *in vitro* (Singh *et al.*, 1996; Bhargavi *et al.*, 1998). D-3 phosphoinositides and PI-3-kinases play a major role in the trafficking of membranes through the endocytotic pathway (Cullen *et al.*, 2001; Gillooly *et al.*, 2001; Simonsen *et al.*, 2001; Vieira *et al.*, 2001). Indeed, profilin mutants in *Dictyostelium* show defects in endocytosis, macropinocytosis and phagocytosis (Haugwitz *et al.*, 1994; Lee *et al.*, 2000; Janssen & Schleicher, 2001; Janssen *et al.*, 2001), and genetic interactions between profilin and endocytotic components have been demonstrated (Temesvari *et al.*, 2000; Seastone *et al.*, 2001). Secretory vesicle formation in cell-free systems is inhibited by anti-profilin antibodies and stimulated by addition of profilin (Schmidt & Huttner, 1998; Dong *et al.*, 2000). Furthermore, purified Golgi and secretory vesicles contain abundant profilin (Dong *et al.*, 2000; Srinivasan *et al.*, 2001). In plant cells, PI-3-P is involved in trafficking of proteins between the trans-Golgi network and vacuoles (Kim *et al.*, 2001).

Profilin was first identified in plants as a birch pollen allergen (Valenta *et al.*, 1991) and as a pan-allergen in pollen and vegetable foods (Valenta *et al.*, 1992; Vallier *et al.*, 1992; vanRee *et al.*, 1992). Native proteins and cDNAs for profilin have now been isolated from a range of plant species, and the biochemical characterization of profilin isoforms is quite advanced (Staiger *et al.*, 1997; Gibbon & Staiger, 2000; Gibbon, 2001). Recombinant and native profilins bind to the three major profilin ligands: G-actin, polyproline and phosphoinositides. Intriguingly, phylogenetic analyses (Huang *et al.*, 1996; Meagher *et al.*, 1999b) and detailed characterization of multiple isoforms from maize (Gibbon *et al.*, 1998; Kovar *et al.*, 2000a) can be used to classify plant profilins into two distinct classes. The combination of biochemical properties held by different maize isoforms does not resemble the isoform variation among vertebrate profilins, raising the possibility that plant profilins have evolved unique functions. Although crystal structures for two plant profilins confirm that the overall fold of profilins from different kingdoms is strikingly similar (Fedorov *et al.*, 1997; Thorn *et al.*, 1997), mutagenesis studies and analysis of evolutionarily distant profilins emphasize the point that a few amino acid changes can dramatically alter the properties of profilin *in vitro*. In addition, a plant-specific patch that is spatially separated from the actin- and polyproline-binding sites (Fedorov *et al.*, 1997; Thorn *et al.*, 1997) offers a potential binding surface for unique interacting proteins.

Several reports indicate that profilin is distributed uniformly throughout the cytosol of pollen and pollen tubes (Grote *et al.*, 1993, 1995; Mittermann *et al.*, 1995). Indeed, work using GFP-profilin reporter constructs validates and extends these findings. Both classes of profilin are uniformly distributed in tobacco pollen tubes (Matsumoto, Franklin-Tong and Staiger, unpublished data). In contrast to these studies, profilin distributes in a polarized manner in root hairs from cress, maize and *Arabidopsis*, with a marked accumulation in the tip zone near the plasma membrane (Braun *et al.*, 1999; Baluska *et al.*, 2000). Profilin labeling is also found at the plasma membrane in developing pollen (von Witsch *et al.*, 1998) and associates with purified plasma membranes from *Arabidopsis* (Santoni *et al.*, 1998). Finally, profilin in maize root cells appears to shuttle between nucleus and cytoplasm depending on the stage of development and in response to inhibitors of signaling pathways (Baluska *et al.*, 2001b). Therefore, profilin distribution in plants depends on the particular isoform(s) expressed, and the cell type or developmental stage being examined.

Overexpression and loss-of-function studies reveal interesting roles for profilin. Increasing the concentration of profilin by microinjection of purified protein disrupts actin-based function in both stamen hair cells (Staiger *et al.*, 1994; Karakesisoglou *et al.*, 1996; Ren *et al.*, 1997; Valster *et al.*, 1997) and pollen tubes (Vidali *et al.*, 2001). The qualitative effects observed under these conditions are consistent with profilin acting as a sequestering protein (Gibbon & Staiger, 2000). A 20-fold increase in profilin protein levels in transgenic *Arabidopsis* plants results in longer roots and root hairs (Ramachandran *et al.*, 2000). Conversely, reduction in profilin levels with an antisense construct leads to a dwarf phenotype and hypocotyls that are 20–25% shorter than wild-type plants. Although profilin antisense plants are mimicked by treatments with latrunculin B (Baluska *et al.*, 2001a), it is not clear that the cell elongation phenotype results from actin-based defects. Surprisingly, plants with a T-DNA insertion in *PRF1* have phenotypes opposite to those of the antisense lines (McKinney *et al.*, 2001). The mutant seedlings of *prf1-1* plants have elongated hypocotyls. Compared with wild type, the *prf1-1* plants also have longer and increased numbers of root hairs. The differences between the two studies have not been explained. One possibility is that the antisense construct down-regulates multiple profilin isoforms, one of which has biochemical properties that are opposite to PRF1. An alternative explanation is that some physiological response of the plants to long-term down-regulation of profilin has occurred. Unfortunately, neither study presents evidence that actin organization is altered in mutant plants, so it is impossible to address whether profilin is acting as a sequestering protein. The possibility that cell elongation defects have nothing to do with profilin's ability to bind actin, but rather are related to vesicle trafficking, signal transduction or other yet to be discovered profilin functions must be considered.

The biochemical properties of plant profilins are best understood from studies of recombinant and native isoforms from maize. Before reconstituting pollen actin organization and dynamics *in vitro*, it is important to know how much profilin and actin exist in pollen and pollen tubes, and what their binding constants are *in vitro*. Several groups have used ELISA assays to determine the cellular concentrations of total actin and profilin pools from various pollen sources. In general, profilin and actin are abundant proteins and are present in equimolar amounts; however, the values range from 25 μM for lily pollen (Vidali & Hepler, 1997) to over 200 μM for poppy pollen (Snowman *et al.*, 2002). In maize pollen, the two proteins are present in equimolar amounts, 125 and 127 μM, respectively (Gibbon *et al.*, 1999). The differences in absolute values may reflect species-specific levels of these proteins or, more likely, methodological differences. Based on these values and on measurement of the affinity of native pollen profilin for pollen actin *in vitro*, it can be predicted that most of the actin pool would be in a profilin–actin complex and that there would be relatively little actin in polymer form (Kovar *et al.*, 2000a). Indeed, the data suggest that just 17.6 μM of the total actin pool would be present in actin filaments. In the first measurements of this kind for plant cells, Gibbon *et al.* (1999) determined F-actin levels directly and found that maize pollen contains just 11–14 μM actin in filament, or ~10% of the total pool. While this result is somewhat startling, to consider that pollen contains just 10% of its actin in polymer yet carries out such dramatic and dynamic functions, similar results have been obtained recently for poppy pollen (Snowman *et al.*, 2002). The simple model that can be derived from this molecular modeling and *in vitro* measurement is that profilin functions as an abundant, moderate affinity buffer of the G-actin pool in pollen (Fig. 2.1). This model predicts that increasing profilin concentration will lead to a reduction in F-actin levels and, conversely, loss of profilin function will result in increased F-actin. These predictions are testable with reverse-genetics and cell biological approaches. Another obvious approach is to measure the concentration of the profilin–actin complex directly. In the slime mold *Acanthamoeba*, the amount of P–A complex is consistent with profilin functioning as a buffer for the G-actin pool (Kaiser *et al.*, 1999).

Quantitative analyses for five recombinant maize profilin isoforms and native protein from pollen and endosperm provide biochemical evidence in support of the

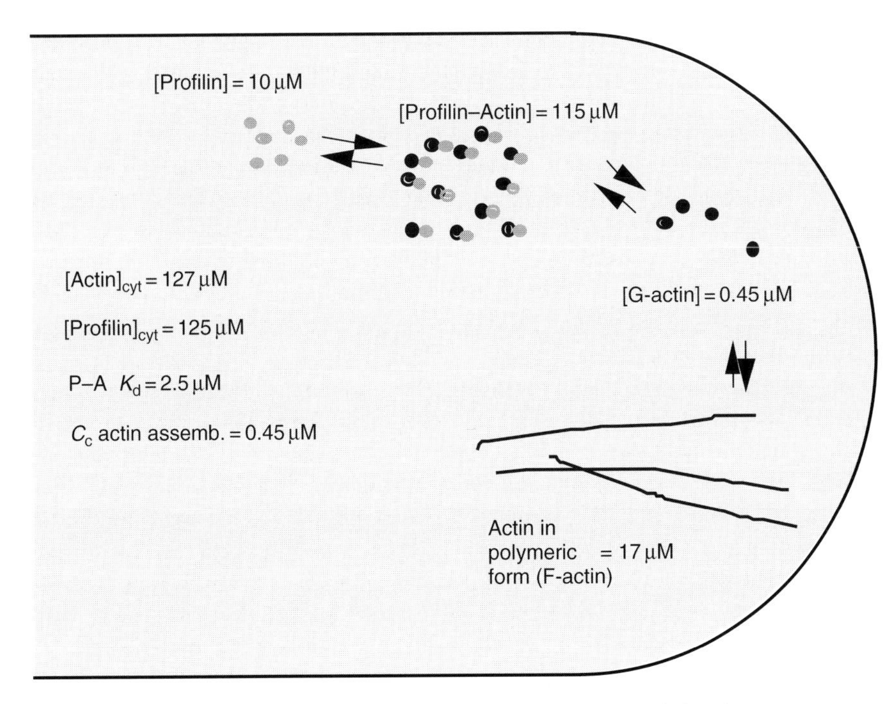

Fig. 2.1 A simple model for profilin function in pollen tubes. The cartoon is based on measurements of cellular concentrations for total profilin (127 μM) and total actin (125 μM) in maize pollen (Gibbon *et al.*, 1999). The apparent binding constant for native profilin binding to pollen actin (2.5 μM) and the critical concentration for maize pollen actin assembly (0.45 μM) at low calcium concentration (25 nM) were determined *in vitro* (Kovar *et al.*, 2000a). If we assume that actin assembles to the critical concentration and that profilin–actin (P–A) forms a 1:1 complex, then the bulk of monomer actin will be present in the P–A complex. This leaves very little actin available for polymerization, and the calculation such that 17 μM actin will be in polymer form. This is consistent with F-actin levels measured directly in maize pollen grains and pollen tubes, where values of 11 and 13 μM were determined, respectively (Gibbon *et al.*, 1999). The biochemical data are consistent with profilin functioning as the major sequestering protein in angiosperm pollen. A similar model is consistent with data obtained from poppy pollen (Snowman *et al.*, 2002). The model does not account for changes in profilin activity due to local calcium concentration, ignores differences in profilin isoforms, and assumes that filament barbed ends are largely blocked by yet to be identified capping factors.

isovariant dynamics model (Gibbon *et al.*, 1998; Kovar *et al.*, 2000a). These studies delineate two classes of functionally distinct profilin isoforms, and the presence of both classes in maize pollen necessitates a revision of the simple model shown in Fig. 2.1. Class II profilins have properties that are strikingly different from class I profilins. They have a higher affinity for poly-L-proline (PLP) and sequester more G-actin than do class I profilins. In contrast, class I profilins bind to PtdIns(4,5)P_2 more strongly than class II. These biochemical properties correlate with the ability of class II profilins to disrupt actin cytoplasmic architecture in live cells more rapidly than do class I profilins. We propose that the class I profilins are abundant, low affinity buffers of the G-actin pool and perform the major sequestering function according to the simple model of Fig. 2.1. On the other hand, the class II profilins are low abundance, high affinity actin-binding proteins that modulate actin assembly, perhaps at specific locations within the cell.

Like many ABPs, profilin is thought to function at the crossroads between intracellular signaling cascades and cytoskeletal reorganization. The interaction of plant profilin with phosphoinositide lipids has been demonstrated by two independent methods (Drøbak *et al.*, 1994; Kovar *et al.*, 2000a, 2001b), and the regulation of actin-sequestering activity by calcium is observed. The apparent K_d value for profilin–actin at physiologically relevant calcium levels decreases 2.8- to 6.2-fold as calcium increases (Kovar *et al.*, 2000a; Snowman *et al.*, 2002). The implication is that profilin will be a better sequestering protein at high calcium than it is at low calcium. This is relevant to the molecular mechanism of tip growth and for transducing signals from the extracellular environment. During pollen tube growth, an oscillatory, tip-focused calcium gradient correlates with pulses of growth (see Chapter 8). If profilin is a sensor for cellular calcium, then actin organization at the tip should oscillate in concert with calcium fluxes. Such a phenomenon has been observed by Yang and coworkers in elegant studies that examine the dynamics of GFP-mouse talin fusion protein in living pollen tubes (Fu *et al.*, 2001).

2.5.3 *Adenylyl cyclase-associated protein*

A third class of actin monomer-binding protein, the adenylyl cyclase-associated protein (CAP), comprises multifunctional proteins that are perhaps universal negative regulators of actin polymerization (Stevenson & Theurkauf, 2000). Originally identified in *S. cerevisiae* as a protein that has roles in signal transduction and cytoskeletal organization, an intriguing hint about its cellular role came from the demonstration that profilin overexpression could suppress defects associated with CAP/Srv2p mutations (Vojtek *et al.*, 1991). These multidomain proteins, which have been characterized from yeasts, fungi, animals and flies comprise an N-terminal region that binds to the catalytic subunit of adenylyl cyclase and to phosphoinositide lipids (Gottwald *et al.*, 1996). The middle domain has a couple of short tracts of proline-rich sequence that bind to SH3 proteins, like yeast ABP1p, and is important for subcellular localization (Freeman *et al.*, 1996). In *Dictyostelium*, it appears that the N-terminus is sufficient for subcellular targeting (Noegel *et al.*, 1999). The highly conserved C-terminus is generally believed to bind actin monomers and to prevent actin polymerization *in vitro* and *in vivo* (Gieselmann & Mann, 1992; Freeman *et al.*, 1995; Benlali *et al.*, 2000). Additionally, the C-terminal region contains a dimerization domain (Zelicof *et al.*, 1996). However, like profilin, the function of CAP may be more complex than simple actin sequestering. Recent studies show that the N-terminus of human CAP accelerates actin depolymerization and that the C-terminus can cooperate with cofilin to facilitate actin polymerization at filament barbed ends (Moriyama & Yahara, 2002). Surprisingly, the C-terminus also promotes nucleotide exchange on actin, much like vertebrate profilin. CAP is, therefore, qualitatively similar to profilin in its regulation by PtdIns(4,5)P_2, ability to enhance nucleotide exchange and complex effects on actin assembly. Indeed, microinjection of CAP1 from platelets into Swiss 3T3 cells can generate stress fiber-like actin filament bundles apparently by stimulating actin polymerization (Freeman & Field, 2000). Even more interesting activities can be expected, as human CAP associates with a high molecular weight actin–cofilin–AIP1 complex in cell extracts (Moriyama & Yahara, 2002).

Genes or cDNAs for CAPs have been identified from cotton (Kawai *et al.*, 1998) and *Arabidopsis* (Barrero *et al.*, 2002). The single *Arabidopsis* gene (AB014759, At4g34490) encodes a 476 residue, ~51-kDa polypeptide that shares ~30% amino acid sequence identity with orthologues from yeasts, flies and humans. Although the C-terminus seems to be better conserved than other regions of the molecule, recognizable RLE repeats are found in the N-terminal region and a poorly conserved proline-rich region separates N- from C-terminal domains. Transcripts for *AtCAP1* are found throughout vegetative portions of the plant and the protein is abundantly expressed in root tissue. To begin to understand the function of AtCAP1, overexpression studies were performed in yeast and plants. Overexpression of AtCAP1 rescues the actin-based mutant phenotype of yeast deficient for CAP. When overexpressed in transgenic *Arabidopsis* plants, behind an inducible promoter, a myriad of morphological abnormalities are observed, each consistent with perturbation of cytoskeletal function. These include reduction in organ size, decrease in cell numbers and defective cell elongation. Many of these abnormalities are phenocopied by treatment with latrunculin B (Baluska *et al.*, 2001a) or overexpression of ADF/cofilin (Dong *et al.*, 2001b), and can be used to infer that AtCAP causes actin depolymerization under these conditions. Indeed, when an excess of AtCAP1 is produced in transgenic tobacco BY-2 cells, actin filament abundance is diminished and severe cytoarchitecture defects are induced. Direct evidence for AtCAP binding to monomer actin is quite limited; however, a recombinant GST–AtCAP C-terminal region fusion protein can cosediment bovine actin and AtCAP coimmunoprecipitates with actin from BY-2 cell extracts. Future experiments should be directed towards understanding the affinity of AtCAP for monomer actin *in vitro*, the effect on actin polymerization, its subcellular localization, abundance and regulation. Moreover, it will be critical to examine whether loss-of-function alleles have actin-based phenotypes to corroborate the data from these over-expression approaches.

2.6 Cross-linking and bundling factors

2.6.1 Fimbrin

Fimbrins belong to a superfamily of actin filament cross-linking and bundling proteins that include α-actinin, spectrin, dystrophin and ABP120 (Correia &

Matsudaira, 1999). Fimbrin is a 68-kDa protein that was originally discovered as one of the major proteins that organizes the core actin bundle of intestinal brush border microvilli (Matsudaira & Burgess, 1979; Bretscher & Weber, 1980a). In addition to microvilli, fimbrin is concentrated in membrane ruffles, microspikes and cell adhesion sites. The 27-kDa, core actin-binding module of fimbrins, and other members of this superfamily, comprises a tandem pair of calponin-homology (CH) domains. Unlike other CH domain-containing ABPs, fimbrins have a pair of actin-binding domain (ABD) modules, allowing them to cross-link actin filaments as a monomer. Indeed, the bulk of the fimbrin molecule is composed of these two ABDs (Fig. 2.2). Most fimbrins have a moderately conserved N-terminal headpiece that contains two EF-hand calcium binding centers. This is likely to confer calcium-sensitivity to the actin-binding modules. Genetic studies in a variety of organisms indicate that fimbrin functions during cellular morphogenesis and polarity establishment, responses to osmotic stress, cytokinesis and endocytosis.

Plant fimbrins were discovered originally as a partial EST isolated from a cDNA library prepared from wheat tissue that had been subjected to aluminum stress (Cruz-Ortega *et al.*, 1997). Subsequently, with degenerate oligonucleotides and PCR, McCurdy and coworkers identified an *Arabidopsis* cDNA, *AtFIM1* (McCurdy & Kim, 1998). The *Arabidopsis* genome contains sequences for five AtFIM genes that encode predicted proteins ranging from 73.4 to 79.8 kDa and which are located on three different chromosomes (Table 2.1). All of the fimbrins contain an ~100 amino acid N-terminal region, but these have rather poorly conserved or

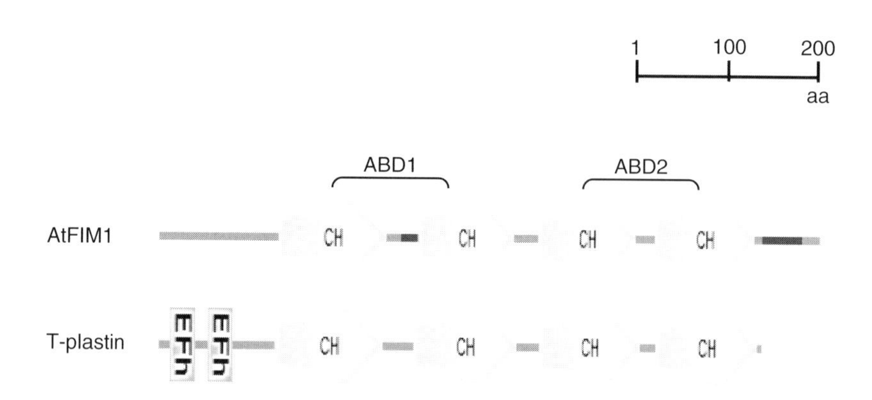

Fig. 2.2 Modular organization of fimbrins. Shown is the predicated domain structure for AtFIM1 (McCurdy & Kim, 1998; At4g26700) and vertebrate T-plastin (Genbank A34789). Members of the fimbrin/plastin family have a conserved, modular organization. This consists of an approx. 100 amino acid, N-terminal domain with two calcium-binding EF-hands. The C-terminal two-thirds of the protein comprises a pair of 27-kDa actin-binding modules (ABD1 and ABD2) that are constructed from a tandem pair of calponin-homology (CH) domains. Plant fimbrins, here typified by *Arabidopsis* fimbrin 1, often contain an extended C-terminal region of variable length and poor sequence conservation. Domain organization was predicted and figures generated with SMART (Schultz *et al.*, 1998; http//smart.embl-heidelberg.de/smart/).

Table 2.1 The *Arabidopsis* fimbrin family

Protein	Locus	Length (aa)	Mass (kDa)	pI	No. of EST	Full-length cDNA
AtFim1	At4g26700	687	76.9	6.13	4	√
AtFim2	At5g48460	654	73.7	8.52	6	√
AtFim3	At2g04750	652	73.4	6.00	0	
AtFim4	At5g55400	714	79.8	5.13	1	
AtFim5	At5g35700	687	76.8	5.13	3	√

missing EF-hands. The majority of the predicted AtFIM proteins comprises a tandem pair of 250–275 amino acid ABDs, each of which is constructed from two CH domains (Fig. 2.2). Interestingly, the *Arabidopsis* fimbrins also have a C-terminal region that is of variable length and is poorly conserved among the five family members. Phylogenetic analyses indicate that the five fimbrin proteins belong to three distinct classes: FIM3 and FIM5; FIM1 and FIM4; and FIM2. Because there are no other proteins in the *Arabidopsis* database with a tandem pair of CH domains, it seems probable that fimbrin family members have adopted the functions of α-actinin, spectrin and dystrophin found in other eukaryotes.

There is rather limited biochemical information on plant fimbrins to date, but these are likely to be roughly equivalent to proteins found in yeasts and vertebrates. Recombinant AtFIM1 binds with high affinity to both muscle and plant actin filaments and appears to have two biochemically-distinct binding sites that may correlate with the two different ABDs (Kovar *et al.*, 2000b). (Alternatively, docking of one ABD to the actin filament may alter the conformation and properties of the second ABD, as proposed by Volkmann *et al.*, 2001.) Unlike T-plastin, this interaction is calcium-independent. AtFIM1 crosslinks actin filaments into supramolecular structures but these are not typical of the bundles observed in the presence of most other fimbrins or 135-ABP (see below). As is the situation in yeast, plant fimbrin can stabilize actin arrays against depolymerization, both *in vitro* and *in vivo*. Specifically, AtFIM1 prevents profilin-mediated disassembly of plant actin filaments in the test tube and in living stamen hair cells. Fimbrins may also be useful reporters for examining actin organization and turnover in live plant cells; a fluorescent protein analog of AtFIM1 decorates a fine cortical actin array when microinjected into *Tradescantia* stamen hair cells (Kovar *et al.*, 2001a).

Major questions that remain to be answered include: do fimbrin isovariants perform distinct functions in the test tube and throughout the plant?; how is plant fimbrin activity regulated?; and is there cooperativity between fimbrin and other bundling or cross-linking factors to accomplish intracellular actin organization and function?

2.6.2 Villin and gelsolin-related proteins

Villins are a family of highly conserved cross-linking proteins that have been identified in diverse eukaryotic cells (Friederich & Louvard, 1999). Surprisingly, members of this family are absent from yeasts. Originally isolated as a major 95-kDa structural protein from the core actin filament bundle of intestinal epithelial cell microvilli (Matsudaira & Burgess, 1979; Bretscher & Weber, 1980b), villins cross-link F-actin in a calcium-independent manner. In the presence of micromolar calcium, villin can also sever actin filaments and cap filament barbed ends (Glenney *et al.*, 1980; Janmey & Matsudaira, 1988). Villins belong to a superfamily of ABPs that are constructed from 125 to 150 amino acid gelsolin-repeat domains (Mooseker, 1983; Weeds & Maciver, 1993). Villin and gelsolin contain six gelsolin repeats, whereas severin, fragmin and CapG contain three repeats (Fig. 2.3); the latter are calcium-dependent severing and/or capping proteins. A single gelsolin repeat has the capacity to bind monomeric actin (Way *et al.*, 1990; McLaughlin *et al.*, 1993), and each of the repeats of a multidomain protein confers different functions to the molecule (Pope *et al.*, 1991; Pollard *et al.*, 1994; Janmey *et al.*, 1998). The cross-linking activity of the villin subfamily correlates with the presence of a C-terminal 8.5-kDa headpiece domain that is also capable of binding to F-actin (Glenney *et al.*, 1981; Pope *et al.*, 1994).

Knock-out mice with a null allele for villin have normal intestinal microvilli, but are defective in calcium-stimulated fragmentation and destruction of the brush

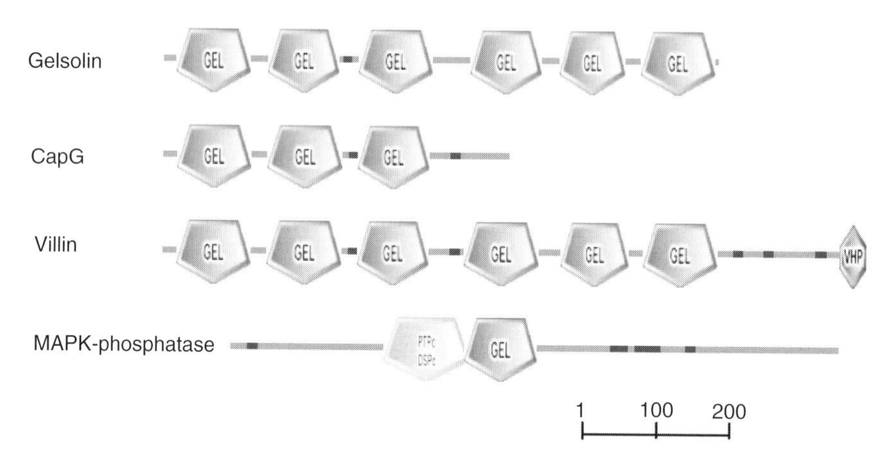

Fig. 2.3 Modular organization of villin/gelsolin family members in eukaryotic cells. Multidomain proteins comprising gelsolin-homology (GEL) and villin-headpiece (VHP) domains generate a variety of F-actin cross-linking, severing and capping factors (see text for details). Gelsolins contain six GEL domains, as exemplified by human adseverin (Q9Y6U3), and are able to bind F-actin, sever filaments and cap filament ends. Villins contain six GEL domains plus an additional F-actin binding module (VHP) and are capable of cross-linking actin filaments. Shown is the predicted domain structure for AtVLN1 (Klahre *et al.*, 2000; At2g29820). Proteins with three GEL domains, such as mammalian CapG, are barbed-end capping factors. Splice variants of AtVLN1 encode predicted CapG-like polypeptides (Klahre *et al.*, 2000; Huang & Staiger, unpublished). A single GEL domain is capable of binding F-actin, and an example of such a module linked to a MAPK phosphatase is present in maize, tomato and *Arabidopsis* (Ulm *et al.*, 2001; At3g55270). Domain organization was predicted and figures generated with SMART (Schultz *et al.*, 1998; http://smart.embl-heidelberg.de/smart/).

borders, leading to the conclusion that villin is essential for cytoskeletal reorganization in response to stimuli (Ferrary *et al.*, 1999). In contrast, *Drosophila* mutations in the *quail* gene lead to female sterility and have defects in actin bundle formation during oogenesis (Mahajan-Miklos & Cooley, 1994; Matova *et al.*, 1999). Although the core gelsolin-homology domains of QUAIL are highly conserved, the recombinant protein does not sever, cap or nucleate actin filaments *in vitro* (Matova *et al.*, 1999). This result emphasizes the need to make direct measurements of the actin-binding properties for villin family members, before assuming function *in vivo*.

Through elegant biochemical studies by Shimmen and colleagues, two bundling proteins from lily pollen, 135-ABP and 115-ABP, were purified (Yokota & Shimmen, 2000). Both proteins bind to vertebrate F-actin *in vitro* and form large filament bundles detectable by EM (Nakayasu *et al.*, 1998; Yokota *et al.*, 1998). Filament binding by 135-ABP is calcium-calmodulin dependent (Yokota *et al.*, 2000) and the protein associates with large actin bundles that support reverse-fountain streaming in pollen tubes (Yokota *et al.*, 1998; Vidali *et al.*, 1999) and root hairs (Tominaga *et al.*, 2000). Actin filaments within the bundles formed by 135-ABP *in vitro* have a uniform polarity (Yokota & Shimmen, 1999), which is like those observed in root hairs (Tominaga *et al.*, 2000), suggesting that 135-ABP is involved in actin cable formation or stabilization in tip-growing cells. In support of this hypothesis, the organization of actin bundles and streaming lanes is altered following microinjection of antibodies into living root hairs (Tominaga *et al.*, 2000). Moreover, 135-ABP antibody injections into *Arabidopsis* root hairs perturb the position of the nucleus with respect to the tip region, leading the authors to speculate that filament bundling is necessary for preventing nuclear migration into the extreme apex of growing hairs (Ketelaar *et al.*, 2002). Whether this bundling activity is required at the level of the large axial actin cables or in the zone of fine bundled actin just distal to the tip zone (Miller *et al.*, 1999) is not apparent, because actin organization was not observed directly following these treatments (Ketelaar *et al.*, 2002; see Chapter 7). Molecular cloning from a lily pollen cDNA library reveals that 135-ABP is most similar to villin-like proteins from *Arabidopsis*, lobster and mouse, and contains a six-repeat core domain and a reasonably well-conserved headpiece (Vidali *et al.*, 1999). However, a sequence identified by mutagenesis (KKEK) as being necessary for F-actin binding by headpiece (Friederich *et al.*, 1992, 1999) is not absolutely conserved. Whether these villin-like proteins function cooperatively with other classes of bundling factors, like the fimbrins for example, remains to be determined.

Arabidopsis contains five villin-like genes (*AtVLN1-5*, Klahre *et al.*, 2000; Huang & Staiger, unpublished data). *AtVLN2* and *AtVLN3* are most similar to 135-ABP, whereas *AtVLN1*, 4 and 5 are more divergent. One distinguishing feature of the plant villins, is the variable and extended linker located between the last gelsolin repeat and the villin headpiece (VHP). Although there is no biochemical or genetic evidence for the function of the AtVLNs, GFP-villin fusions associate with actin filaments *in vivo* (Klahre *et al.*, 2000). Specifically, a full-length

AtVLN3–GFP fusion protein, as well as headpiece constructs for AtVLN1, 2 and 3, decorates actin filaments in tobacco BY-2 cells and in various *Arabidopsis* tissues. These results confirm that the headpiece domains, which lack several conserved residues, are functional actin-binding modules. These studies indicate the molecular basis for filament bundling, with a functional F-actin binding module in the VHP and a predicted one as part of the gelsolin repeat modules. Alternatively, given that native 135-ABP migrates as a putative homodimer (Yokota *et al.*, 1998), the VHP may be sufficient for bundling actin filaments. We predict that the AtVLNs are involved in formation of major actin bundles that support cytoplasmic streaming in a variety of cell types. Moreover, the hypothesis that plant villins are involved in cell expansion or elongation (Klahre *et al.*, 2000) can now be tested directly using reverse genetics.

Although the calcium-regulated, severing activity of villin (or gelsolin) is postulated to act at the center of signal-mediated cytoskeletal rearrangements in root hairs and pollen tubes (Kohno & Shimmen, 1988; Cárdenas *et al.*, 1998; Geitmann *et al.*, 2000), these biochemical properties have yet to be demonstrated directly. Indeed, as for *Drosophila* QUAIL (see above), the full-length plant villins may lack severing, capping or nucleating activity. It will also be interesting to determine whether splice variants, which resemble gelsolin (i.e. containing six gelsolin repeats but missing the headpiece), for the five VLNs exist in *Arabidopsis* or other plants.

A single 150 amino acid gelsolin-homology domain is sufficient for F-actin binding. A query of the SMART database reveals genes in maize and *Arabidopsis* with a single gelsolin repeat. The full-length *Arabidopsis* gene product is characterized as a MAP kinase phosphatase (AtMKP1) that is activated in response to cellular damage (Ulm *et al.*, 2001). The AtMKP1 gelsolin-like domain shares ~35% identity with mouse villin and gelsolin. If verified, interactions with the actin cytoskeleton could function to target this phosphatase to specific subcellular locales.

Finally, the VHP itself is a conserved, ubiquitous 8.5-kDa actin-binding module that is present in several other cytoskeleton-associated proteins (Vardar *et al.*, 1999). For example, headpiece occurs as part of protovillin (Hofmann *et al.*, 1993) as well as linked to a talin domain in a morphogenesis protein from *Dictyostelium* (Tsujioka *et al.*, 1999). It is also associated with a coronin-like domain in an *Entamoeba* ABP (Willhoeft & Tannich, 1999), with dematin/band 4.9 from erythrocytes and other mammalian tissues (Rana *et al.*, 1993; Azim *et al.*, 1995), and with a LIM domain in Limatin (Kim *et al.*, 1997). The headpieces for AtVLN1, AtVLN2 and AtVLN3 interact with F-actin in plant cells, but there is no evidence for direct association with filaments *in vitro*. The *Arabidopsis* VHPs are ~45% identical and 65% similar to chicken VHP, but each isoform contains only a subset of the residues implicated in binding to actin (Doering & Matsudaira, 1996; Vardar *et al.*, 1999). Searches of the *Arabidopsis* database with chicken and plant sequences reveals other candidate VHP-containing genes, which are potential candidates for novel plant ABPs.

2.6.3 115-ABP

During the purification of pollen myosin and lily villin (135-ABP), another F-actin binding factor was identified (Yokota & Shimmen, 1994; Yokota *et al.*, 1998). This protein, when purified to homogeneity, binds and bundles actin filaments *in vitro* (Nakayasu *et al.*, 1998). Although it has not been identified at the molecular level, 115-ABP is unlikely to be a proteolytic fragment of lily villin or α-actinin based on antibody cross-reactivity. Moreover, its activity appears to be independent of ATP and calcium, the former observation making it unlikely that this is an unconventional myosin. Unfortunately, the preparation of antibodies against 115-ABP has been difficult, making assignment of function and localization in plant cells impossible (Yokota & Shimmen, 2000). However, it is postulated that this factor somehow cooperates with lily villin to achieve the formation of actin bundles in pollen tubes (Yokota & Shimmen, 2000).

2.6.4 eEF-1α

The eukaryotic elongation factor 1α (eEF-1α) is a multifunctional protein that may link protein synthesis to the cytoskeleton (Edmonds *et al.*, 1999). A 50-kDa polypeptide that binds and bundles F-actin in a pH-dependent manner was purified from *Dictyostelium* (Dharmawardhane *et al.*, 1991). The cloning and sequencing of its corresponding cDNA revealed identity with eEF-1α, a GTP-binding protein that functions to bind aminoacyl tRNA to the A-site on the ribosome. Its actin-binding, as well as protein translation, properties appear to be highly conserved across kingdoms. Plant eEF-1α also binds and bundles microtubules, as well as binding CaM and a PI-4-kinase (Durso & Cyr, 1994; Yang & Boss, 1994). A plant GFP-eEF-1α fusion protein labels the microtubule cytoskeleton predominantly (Moore & Cyr, 2000), but in other studies and cell types eEF-1α is localized along actin filaments on protein storage bodies (Clore *et al.*, 1996) or on actin cables that support cytoplasmic streaming (Collings *et al.*, 1994).

The *Dictyostelium* protein binds both G- and F-actin with K_d values of 0.06–10 μm, but is dependent on ionic strength and pH *in vitro*. Binding is saturated at 1.5:1 and the filaments are bundled into square-packed arrays that may exclude hexagonal cross-linkers like fimbrin and villin. eEF-1α reduces rates of polymerization and depolymerization and can increase the level of polymer at steady-state. Initial evidence for plant eEF-1α binding to actin comes from analysis of the purified PI-4-kinase activator (PIK-A49), which is able to co-polymerize with chicken muscle F-actin and co-sediments at low centrifugal forces, indicative of bundling activity (Yang *et al.*, 1993). There is also limited evidence for co-purification with a plasma membrane-associated, cytoskeletal fraction from carrot suspension cells (Yang & Boss, 1994). A maize endosperm eEF-1α appears to bundle F-actin networks *in vitro* (Sun *et al.*, 1997). Also maize eEF-1α has been shown to bind and bundle actin filaments *in vitro* and the bundling activity is enhanced by ADF, suggesting that eEF-1α is capable of cooperating with other ABPs (Gungabissoon *et al.*, 2001). Whether such activities occur *in vivo*, and what the precise role of the interaction of plant eEF-1α with the actin cytoskeleton *in vivo* is, remains to be established.

Another component of the protein synthesis machinery in plants, the EF1βγ complex, may link translation to the cytoskeleton. EF1βγ mediates exchange of GDP for GTP on EF-1α, thereby activating the latter. The complex isolated from wheat germ binds with high affinity to monomeric actin ($K_d = 0.17\,\mu M$) and stimulates the initial rate of actin polymerization *in vitro* (Furukawa *et al.*, 2001). This is a dramatically different effect from EF-1α and points to the diversity of interactions between protein synthesis machinery and the cytoskeleton.

2.6.5 Spectrin

A two-dimensional lattice of cytoskeletal polymers, the membrane skeleton, underlies the plasma membrane in many cell types and functions to regulate membrane stability and elasticity (Morrow, 1999). In vertebrate erythrocytes, the major polymeric elements are heterodimers or heterotetramers of α/β-spectrin, which are 220/240-kDa polypeptides, respectively. Spectrin is an elongated, rod-like molecule, comprised mostly of 17–20 repeats of an 106-amino acid motif. It associates with F-actin, ankyrin and band 4.1 to form a planar network that stabilizes the plasma membrane, but can also be found in association with the Golgi apparatus. The F-actin binding site is at the N-terminus of the β-subunit, and is homologous with the tandem CH domain, actin-binding module of fimbrins, α-actinin, dystrophin and ABP-120. This multifunctional protein also associates with phosphoinositides, calcium, calmodulin and is regulated by phosphorylation.

The occurence of spectrin-like molecules, and a spectrin-based membrane skeleton, in higher plants is rather ambiguous. Numerous studies use antibodies against vertebrate or avian spectrins to detect polypeptides on Western blots and/or to immunostain plant cells. Although there are examples of cross-reactivity with an appropriate size polypeptide (Michaud *et al.*, 1991; Wang & Yen, 1991; de Ruijter & Emons, 1993; Faraday & Spanswick, 1993), often lower molecular weight (<100 kDa) proteins are detected (Sikorski *et al.*, 1993; Bisikirska & Sikorski, 1997; Li *et al.*, 1999). This may be attributed to the difficulties associated with protein isolation from plant extracts and concomitant proteolysis. Spectrin-like antigens are sometimes localized near the plasma membrane (Michaud *et al.*, 1991; de Ruijter & Emons, 1993), perhaps in putative signaling foci (Reuzeau *et al.*, 1997), but can also be detected on or near intracellular organelles (de Ruijter & Emons, 1993; Li *et al.*, 1999). In root hairs of *Vicia faba*, a spectrin-like antigen accumulates at the apical tip region of actively growing hairs (de Ruijter *et al.*, 1998). No cDNA for a spectrin-like protein has been reported, however, and scrutiny of the *Arabidopsis* genome suggests that this family of ABP is not present (Hussey *et al.*, 2002; Staiger, unpublished). Indeed the CH domains, which are used in a plethora of signaling and ABPs from yeasts, flies and vertebrates, are found only in the fimbrins (see above) and in the tails of an interesting subclass of kinesin-related proteins from *Arabidopsis* (Korenbaum & Rivero, 2002). Consistent with this, we are also skeptical about the occurence of α-actinin and filamin proteins in *Arabidopsis*; these should also contain the conserved tandem CH-domain ABD. Perhaps the need for a membrane skeleton is obviated by the presence of a cell wall or is usurped by a predominantly actin- and microtubule-based membrane skeleton in plants. Alternatively, the function of spectrin (and α-actinin) may have been accommodated by other families of cross-linking protein (e.g. the fimbrins and villins) during evolution.

2.7 Capping factors

2.7.1 Capping protein (CP)

Capping protein (CP: also CapZ or Cap32/34) was first identified in *Acanthamoeba* (Isenberg *et al.*, 1980), and is well characterized in lower eukaryotes, yeast, flies and vertebrates (Cooper *et al.*, 1999; Cooper & Schafer, 2000). This ubiquitous ABP consists of a heterodimer of α and β subunits, each of which has a M_r of 28–36 kDa. Neither subunit alone has biochemical activity, but the two recombinant polypeptides can be combined *in vitro* (Haus *et al.*, 1991) to yield an activity that caps the fast-growing (barbed) ends of actin filaments with high affinity. CP stabilizes F-actin by preventing the loss and addition of subunits from filament ends, but it also facilitates filament nucleation. Binding is calcium independent but can be regulated by PtdIns(4,5)P_2, potentially linking signaling cascades to actin rearrangement. During vertebrate muscle development CapZ plays a critical role in the assembly of actin filaments that comprise the sarcomere (Schafer *et al.*, 1995). In *Dictyostelium*, loss-of-function mutants result in defects in F-actin levels and reduced cell motility (Hug *et al.*, 1995). Extensive analysis of CP function in yeast shows that it stabilizes F-actin against depolymerization and regulates actin organization (Amatruda *et al.*, 1990, 1992; Karpova *et al.*, 1995). Yeast CP is also involved in regulation of cell shape, polarity and localized cell wall deposition. Surprisingly, CP is not essential for yeast cells, but is lethal in the multicellular fly (Hopmann *et al.*, 1996) with death occurring during early larval development. Heterozygous mutant flies are viable, but show defects in actin assembly and bundle formation during bristle development. Further emphasizing its central role in basic cellular processes, CP is one of three ABPs necessary to reconstitute actin comet-tail formation and intracellular motility *in vitro* (Loisel *et al.*, 1999).

Plant CP has not been described biochemically, but BLAST searches of the *Arabidopsis* genome with yeast CP reveal two strong candidates for CP α/β subunits. AtCPA (At3g05520) is ~39 kDa and shares 29–35% amino acid sequence identity with α-subunits from amoeba, yeasts and vertebrates. The 31-kDa AtCPB (At1g71790) is more highly conserved, sharing more than 50% identity with *Dictyostelium* and vertebrate β subunits. Both proteins appear to be encoded by single genes, but there may be splice variant isoforms for the α-subunit based on cDNA sequence and EST information. The few measurements of actin levels in plant cells indicate that actin is predominantly in monomeric form (Gibbon *et al.*, 1999; Snowman *et al.*, 2000). This suggests that regulating the size and activity of the subunit pool will be critical, functions that are currently assigned to profilin and ADF/cofilin (Staiger *et al.*, 1997; Gibbon & Staiger, 2000; Kovar & Staiger, 2000). However, the amount of profilin and its affinity for G-actin cannot account for all of the monomer pool. It seems probable that the majority of filament ends in interphase plant cells are capped and that this is a conserved mechanism for regulating actin dynamics (Staiger *et al.*, 1997). We predict that AtCP is a major regulator of F-actin assembly and dynamics in many different cells and tissues.

2.7.2 CapG

Proteins comprising three gelsolin-homology domains (Fig. 2.3), like CapG from vertebrates (Yu *et al.*, 1990; Sun *et al.*, 1995), severin from *Dictyostelium* (Brown *et al.*, 1982) and fragmin from *Physarum* (Ampe & Vandekerckhove, 1987) are F-actin capping and/or severing factors (Noegel, 1999; Friederich & Louvard, 1999). These ABPs are often regulated by calcium, phosphorylation or phosphoinositides, and function as monomers. Although separate genes for such proteins do not exist in *Arabidopsis*, the possibility for splice variants of villin-like transcripts seems quite likely. Indeed, based on a premature stop codon in the EST H5D9T7 for *AtVLN1* and similar RT-PCR products, Klahre and colleagues suggested the presence of an *AtVLN1* splice variant (Klahre *et al.*, 2000). We have cloned one such full-length cDNA from a root cDNA library and expressed the recombinant protein for functional studies (Huang & Staiger, unpublished data). It encodes a predicted polypeptide of 446 amino acids that shares 100% identity to the N-terminal three gelsolin repeats of AtVLN1. The transcript is distinguished by altered splicing of intron 9 which adds 13 nucleotides and results in a frameshift beginning at aa393. Hence, the 50.4 kDa (pI of 5.6) AtCapG protein has a unique C-terminal tail of 54 amino acids which is not present in AtVLN1. This CapG-like protein from *Arabidopsis* may provide a fundamental link between environmental stimuli and actin rearrangements.

There is also compelling biochemical evidence for a calcium-dependent severing protein from *Mimosa pudica* that is related to the gelsolin/fragmin proteins (Yamashiro *et al.*, 2001). Because of the high affinity of monomer actin for DNase I, chromatography on DNase I-Sepharose has been used to isolate a variety of G- and F-actin binding proteins from different organisms. When extracts of *Mimosa* plants are applied to DNase I-Sepharose in the presence of calcium, and eluted with EGTA, two major polypeptides with M_r of 42 kDa and 90 kDa are eluted. When the EGTA is removed, the eluate increases the amount of actin in the supernatant by sedimentation analysis and generates short actin filaments as assessed by EM. Moreover, the eluate is able to facilitate actin polymerization in the presence of 0.2 mM calcium and decreases the lag period for assembly. All of these properties are diagnostic for a severing protein. The 42- kDa protein, which is shown to be distinct from actin by immunoblotting and based on its mobility in 2D SDS-PAGE, has limited amino acid sequence similarity (one of four peptide sequences aligned) with *Arabidopsis* villins and 135-ABP. However, until each protein is purified to homogeneity, attributing function to either is impractical. Unfortunately, it appears that the 90-kDa from *Mimosa* is rather unstable during purification.

2.7.3 Others

A 75-kDa polypeptide with a pI of 5.8 from *Impatiens* fruit pods was purified to homogeneity (Pal & Biswas, 1994). The protein exists as a homotetramer with M_r of 300 kDa and is a glycoprotein. When examined for potential interactions with cytoskeletal polymers, the 75-kDa protein failed to bind microtubules from plant or animal sources, but did interact with actin filaments. In seeded actin assembly reactions, substoichiometric amounts of 75 kDa markedly reduced the rate and extent of polymerization. Inhibition of actin assembly was also observed by viscometry assay, but preformed actin filaments were not depolymerized in the presence of 75 kDa. The increased number of short actin filaments in the EM and the ability to cosediment with actin polymers are all consistent with an F-actin capping activity. Unfortunately, primary sequence information or a specific antibody are not yet available for this intriguing protein.

2.8 Nucleation complexes

2.8.1 Arp2/3

A ubiquitous, multiprotein complex has garnered enormous attention as the key regulator of actin nucleation in eukaryotic cells (Higgs & Pollard, 2001; Welch & Mullins, 2002; Vantard & Blanchoin, 2003). The Arp2/3 complex is named for its two major subunits, the actin-related proteins (Arp) 2 & 3, and was identified originally from *Acanthamoeba* extracts by affinity chromatography on profilin-Sepharose (Machesky *et al.*, 1994). The biochemical properties of the complex are well defined and include the ability to seed filament assembly, cap filament pointed ends and bind to the sides of existing actin filaments. The combination of filament nucleation and side-binding allows the Arp2/3 complex to assemble dendritically-branched networks of filaments *in vitro* and *in vivo*. The Arp2/3 complex lies at the heart of a remarkable form of actin-based intracellular motility, comet-tail formation. Originally characterized as the machinery that drives the food-borne pathogen *Listeria* through the cytoplasm of infected host cells, this form of motility is now known to facilitate movement of a wide range of pathogenic microbes, as well as endogenous endosomal vesicles. Comet-tail formation can be reconstituted *in vitro* with just a few components that include actin, the Arp2/3 complex, capping protein and ADF/cofilin (Loisel *et al.*, 1999).

The basal activity of the Arp2/3 complex is rather low, however, and multiple different activators of the nucleation complex are being discovered (Welch & Mullins, 2002). Perhaps best understood is the Wiskott-Aldrich Syndrome protein (WASp) and relatives, which contain an acidic carboxy terminus that binds Arp2/3 complex and links G-protein signaling to actin polymerization. Other activators include cortactin, myosin I, CARMIL, Pan1p/Eps15 and ABP1. *Arabidopsis* contains genes for Arp2 and Arp3 (Klahre & Chua, 1999; McKinney *et al.*, 2002), as well as the other five subunits (our unpublished data). There is no doubt that plants will use Arp2/3 to regulate formation of actin arrays; however, no data on complex location, functional properties or activators exist at present. Regarding activators, it is of interest to note that the known regulators of Arp2/3 are missing or poorly conserved in *Arabidopsis*. Moreover, there is no direct evidence for comet-tail motility or branched, dendritic networks in plant cells. These networks will have to be resolved at the level of the electron microscope, perhaps by adaptation of the methods from Svitkina and Borisy (1999) or, better still, by cryoelectron tomography (Medalia *et al.*, 2002). One potential regulator of Arp2/3 is the product of the *BRK1* gene of maize that encodes a conserved 8-kDa protein (Frank & Smith, 2002). The mammalian ortholog HSPC300 is implicated as part of an activation complex (Eden *et al.*, 2002). However, the molecular details of this mechanism has yet to be thoroughly elucidated.

2.9 Other F-actin binding proteins

2.9.1 SuSy

Sucrose synthase (SuSy) is a predominantly cytosolic enzyme that catalyzes a reversible reaction that converts fructose and UDP-glucose into sucrose and UDP. In sink tissues, for example expanding maize leaves (Nguyen-Quoc *et al.*, 1990), SuSy is involved in sucrose utilization providing UDP-glucose for cell wall deposition and starch synthesis (Chourey *et al.*, 1998). There is also indirect evidence that SuSy can utilize hydrolysis of sucrose to channel UDP-glucose directly to the plasma membrane cellulase synthase in cotton fibers (Amor *et al.*, 1995). Consistent with this activity, a population of membrane-associated SuSy has been reported by several groups (Carlson & Chourey, 1996; Winter *et al.*, 1997), and membrane partitioning may be regulated by protein phosphorylation (Winter *et al.*, 1997). In addition, a population of SuSy from elongating maize leaves or pulvini is present in the detergent-insoluble fraction from microsomal membranes, which is a crude estimation of cytoskeletal interaction (Winter *et al.*, 1998). SuSy from maize is a homotetramer of ~92 kDa subunits that binds with high affinity to F-actin *in vitro* and saturates at a stoichiometry of 1:20. Apparently, it is mostly the non-phosphorylated, cytosolic form of SuSy that binds actin filaments. Both maize and *Arabidopsis* contain two genes for SuSy proteins (Martin *et al.*, 1993; Chourey *et al.*, 1998), but whether both isoforms are capable of binding to actin is not known.

2.9.2 *ABP/MAP190*

Actin filaments and microtubules codistribute in many plant cytoskeletal arrays and it is a commonly held view that cooperation between the two polymer systems is necessary for fundamental cellular functions (Collings & Allen, 2000). The most logical candidates for microtubule–actin-filament cross-linkers are the structural microtubule-associated proteins (MAPs), like MAP-2 and tau (Olmsted, 1986). However, these are apparently not present in the *Arabidopsis* genome, perhaps being replaced by plant-specific MAPs. A strong candidate for such a linker molecule, at least in dividing plant cells, was identified recently (Igarashi *et al.*, 2000). A 190-kDa MAP partially purified from tobacco BY-2 cells bundles microtubules and cross-links actin filaments *in vitro*. Immunolocalization shows that 190-kDa ABP/MAP distributes with the mitotic spindle and phragmoplast, but not along interphase cytoskeletal arrays. The tobacco cDNA encodes a protein with a calmodulin-like EF-hand sequence but no recognizable actin- or microtubule-binding motifs. Unfortunately, the native 190-kDa was not purified to homogeneity, nor were the biochemical properties of a recombinant molecule examined. The authors do, however, note the presence of a sequence in *Arabidopsis* that is 41% identical to the tobacco protein (Hussey *et al.*, 2002). We predict that this putative ABP is a cross-linker of microtubules and actin filaments that plays a central role in cell division and cytokinesis in *Arabidopsis*. The biochemical interaction with actin filaments and its localization in dividing cells remains to be determined.

2.9.3 *AIP1*

Actin-interacting protein, AIP1, is a highly conserved, ubiquitous, WD-repeat protein. WD repeats are protein-folding domains that usually occur in multiple copies within a single polypeptide. Generally, WD-repeat proteins bind to additional proteins and function as part of multiprotein complexes. AIP1 was identified originally by a yeast two-hybrid screen with yeast actin as the bait (Amberg *et al.*, 1995). The function of AIP1 in cells is likely to involve interactions with ADF/cofilin (Konzok *et al.*, 1999; Rodal *et al.*, 1999), because AIP1 enhances the ability of ADF to destabilize actin filaments and facilitates depolymerization (Iida & Yahara, 1999; Okada *et al.*, 1999; Rodal *et al.*, 1999). Mutational analysis further assigns a role for AIP1 in actin polymer dynamics and organization *in vivo*, especially during fundamental cellular processes like motility, endocytosis and phagocytosis (Konzok *et al.*, 1999).

Arabidopsis contains two WD-repeat proteins (AtAIP1-1, AtAIP1-2), that share ~30% amino acid sequence identity with yeast AIP1 (Allwood *et al.*, 2002). The two products are 67% identical with each other and are expressed in a tissue-specific manner, with *AIP1-1* transcript found only in floral tissues and *AIP1-2* transcript present everywhere in the plant except floral tissue. The recombinant AtAIP1-1 protein cooperates with lily pollen ADF to enhance actin-depolymerizing activity *in vitro*, and perhaps binds weakly to filaments on its own.

Cytologically, pollen AIP1 is found along major actin bundles and aggregates in ungerminated *Narcissus* pollen grains, virtually identical to ADF localization. Furthermore, also like ADF, AIP1 localization becomes largely cytosolic upon germination. Because the lily pollen ADF is not regulated by phosphorylation and is a somewhat poorer regulator of actin dynamics when compared with vegetative isoforms, regulation by AIP1 may play a substantial role in modulating the cellular location and activity of ADF in pollen. As pointed out above, AIP1 is found in a multiprotein complex with CAP, ADF/cofilin and actin in vertebrate cells (Moriyama & Yahara, 2002). The molecular components are present in pollen for such an actin regulatory complex, but its presence has not yet been verified.

2.9.4 Annexin

Annexins are a large family of calcium-dependent, phospholipid-binding proteins (Gerke & Moss, 1997; Kaetzel & Dedman, 1999), that are involved in vesicle motility, exocytosis and endocytosis. The core of each annexin protein is built from 4 repeats of a 70-amino acid motif, with each of these endonexin folds containing a Ca^{2+}- and phospholipid-binding domain. The annexins probably function as homodimers or heterodimers of 32–37 kDa subunits. Many annexin isoforms have been shown to bind F-actin *in vitro* and *in vivo*, and some are able to bundle actin filaments. The C-terminal 15 amino acids of annexin II are sufficient for actin binding, but other studies implicate the N-terminus in actin interactions. The significance of binding to F-actin may relate to the recent demonstration that annexin is essential for actin comet-tail motility of macropinosomes (Merrifield *et al.*, 2001).

Plant annexins have been purified and cDNAs cloned from many higher plants and ferns (Blackbourn *et al.*, 1992; Clark & Roux, 1995; Clark *et al.*, 1995; Delmer & Potikha, 1997). Generally, they are Ca^{2+}-dependent phospholipid-binding proteins, and associate with Golgi-derived secretory vesicles or the vacuolar membrane. However, both biochemical and crystal structure data suggest that the properties of plant annexins are quite different from those of animal annexins (Hofmann *et al.*, 2000), including poor conservation of all but one endonexin fold and the presence of endogenous ATPase or GTPase activity (McClung *et al.*, 1994; Calvert *et al.*, 1996; Lim *et al.*, 1998). Moreover, cotton fiber annexin associates with purified plasma membranes and correlates with decreased β-1,3-glucan synthase activity (Andrawis *et al.*, 1993) and wheat annexins have potential calcium channel activity (Breton *et al.*, 2000). Tomato p34 and p35 (Calvert *et al.*, 1996), and zucchini hypocotyl annexin (Hu *et al.*, 2000), are able to bind F-actin, whereas maize coleoptile and green pepper annexins do not apparently interact with actin (Blackbourn *et al.*, 1992; Hoshino *et al.*, 1995). Although annexins accumulate in the apex of growing pollen tubes (Blackbourn *et al.*, 1992) and fern rhizoid cells (Clark *et al.*, 1995), an association with actin filaments has not been demonstrated cytologically. *Arabidopsis* contains seven annexin (*ANNAT*) genes that display organ- and cell type-specific gene expression (Clark *et al.*, 2001). ANNAT1, 2, 6 and 7 share the highest level of sequence identity with the C-terminal 15 amino acids of human and tomato annexins, however, none of the ANNATs have been characterized for actin binding.

2.9.5 Gephyrin/AtCNX1

Gephyrin is a peripheral membrane-associated protein, originally identified in vertebrate post-synaptic cells, that is essential for the localization and clustering of inhibitory neurotransmitter receptors. Cloning of the gephyrin gene demonstrated a similarity to enzymes from bacteria, flies and plants that are involved in biosynthesis of molybdenum cofactor (Moco) (Ramming *et al.*, 2000). The vertebrate protein binds to microtubules with high affinity (Kirsch *et al.*, 1991), and the yeast protein has been shown to bind profilin (Mammoto *et al.*, 1998), perhaps linking their functions to the cytoskeleton. The *Arabidopsis CNX1* gene encodes a Moco biosynthetic enzyme; however, the two conserved domains are inverted with respect to the domain structure of rat gephyrin (Schwarz *et al.*, 2000). The G-domain located at the C-terminus of the 85-kDa polypeptide binds to molybdopterin, whereas the N-terminal E-domain is essential for activating molybdenum. The proline-rich tract positioned in the middle of the rat protein is missing from *Arabidopsis* CNX1. A particularly careful quantitative analysis of cosedimentation data reveals that full-length CNX1 binds with reasonable affinity (2 μM) to actin filaments from rabbit muscle. Deletion analysis shows that the interaction is mediated by the E-domain. It remains to be demonstrated that CNX associates with actin filaments within cells, but the possibility for anchoring the Moco-synthetic machinery to the cytoskeleton remains an intriguing possibility.

2.9.6 AtSH3P

Clathrin-coated vesicles (CCV) are hallmarks for the trafficking pathway that transports materials recovered from the plasma membrane by endocytosis, as well as for the targeting of NTPP-containing vacuolar proteins moving from the trans-Golgi network to the prevacuolar compartment (Holstein, 2002). In yeast and vertebrate cells, growing evidence implicates both actin and ABPs in CCV formation, function and endocytosis (Qualmann *et al.*, 2000; Jeng & Welch, 2001; Lanzetti *et al.*, 2001). Many of the relevant ABPs, like Hip1R/Sla2p, ABP1p and Rvs167 are absent or poorly conserved in plants (see above). SH3-domain containing proteins may be involved in CCV formation, fission or uncoating. This domain is capable of binding to proline-rich proteins with a PXXP motif, and is often present in multidomain proteins that interact with the actin cytoskeleton. *Arabidopsis* contains several such proteins and one of these, AtSH3P1, colocalizes with clathrin and interacts with an auxillin-like protein (Lam *et al.*, 2001). Ultrastructural evidence is provided for localization along actin filaments in pollen grains and preliminary data are provided for binding of recombinant AtSH3P1 (± SH3 domain) to actin *in vitro*. Detailed studies to measure the affinity of AtSH3P1 for plant actin filaments need to be performed; however, the plant protein does complement the endocytosis and actin deficiencies associated with the yeast *rvs167* mutant. It shall be interesting to watch future progress in the analysis of this potential linker between endocytosis (or vacuolar targeting) and the actin cytoskeleton.

2.9.7 Caldesmon

Preliminary biochemical studies have uncovered possible caldesmon-like proteins in plants (Krauze *et al.*, 1998), yet their presence in the *Arabidopsis* database is ambiguous (Hussey *et al.*, 2002). Caldesmon (CaD), an F-actin side-binding protein, is best characterized from vertebrate smooth muscle and non-muscle cells (Huber *et al.*, 1999). It is generally held to be a regulator of actomyosin interactions, inhibiting myosin ATPase when bound to actin filaments, and is itself negatively regulated by Ca-CaM and protein phosphorylation. In addition to myosin binding, vertebrate CaD also binds tropomyosin, troponin C and S100 proteins. The overall domain structure comprises four modules; most importantly, a myosin-binding module is located at the N-terminus and CaM and actin-binding sites are found at the C-terminus. A 107-kDa polypeptide from *Ornithogalum virens* pollen tubes cross-reacts with antibodies raised against the N-terminal region of chicken gizzard CaD (Krauze *et al.*, 1998). A limited biochemical analysis with crude extracts from pollen suggests that the putative CaD cosediments with muscle actin filaments and binds bovine brain CaM in the presence of calcium. Sequence identification is required to verify its molecular relationship to vertebrate and lower eukaryotic CaDs, and electrophoretically pure protein must be re-examined for binding to the major ligands with careful attention to appropriate controls.

2.9.8 Tropomyosin

Tropomyosins (Tm) in a wide variety of organisms are found as heterodimers of two 30–40 kDa polypeptides (Smillie, 1999). These thin, extended molecules bind along the length of actin filaments, with a single Tm heterodimer spanning 6–7 actin monomers in the filament. In muscle cells, Tm regulates access of myosin to the thin filament in a calcium-dependent manner. Stabilization of F-actin is another known function of Tm family members, with convincing evidence for an antagonist effect on ADF/cofilin-mediated actin dynamics shown in elegant studies on *C. elegans* (Ono & Ono, 2002). In yeast, mutational analysis demonstrates a role in stabilization and formation of F-actin cables, but not cortical patches (Liu & Bretscher, 1989; Pruyne *et al.*, 1998). The fission yeast Tm is an essential gene and functions during cytokinesis (Balasubramanian *et al.*, 1992). Turkina and colleagues have attempted to purify plant Tm by taking advantage of its predicted heat-stable property (Turkina & Akatova, 1994; Turkina *et al.*, 1995). Extracts from cow parsnip phloem and wheat callus cells were boiled and subjected to further fractionation. The resulting protein mixtures were cosedimented with muscle actin filaments and bound peptides analysed by SDS-PAGE. Both plants yielded a series of binding partners with M_r of 30–38 kDa. However, the polypeptides were not sequenced or shown to be immunologically related to Tm. Another protein of 86 kDa, also characterized by heat-stability and F-actin binding in these same studies, was predicted to be a plant caldesmon. Several groups have indicated that the *Arabidopsis* genome database contains Tm-like sequences (Assaad, 2001); however, we remain skeptical of such assertions. Often, a putative actin-binding sequence comprising little more than a short stretch of residues (DAIKKL) is used for these searches. Until cDNAs encoding proteins of appropriate molecular weight, and with similar domain organization to non-plant Tm, are cloned and the recombinant proteins evaluated for F-actin interaction *in vitro*, it is best to remain cautious.

2.9.9 Vinculin

Focal contacts and adherens junctions represent points of cell–cell and cell–ECM adhesion in vertebrate cells. One of the major proteins on the cytoplasmic face of focal contacts is vinculin, a 117-kDa protein discovered originally from chicken gizzard (Burridge & Chrzanowska-Wodnicka, 1996; Geiger, 1999). Mutational analysis confirms a role for vinculin as a junctional adapter protein that is necessary for the construction or maintenance of cellular adhesions, and probably links integral membrane proteins to the actin cytoskeleton. High affinity binding of the C-terminus of vinculin by F-actin is masked through an intramolecular, self-association with the N-terminal head region. This autoinhibition is alleviated when vinculin binds to PtdIns(4,5)P_2, and also reveals a talin-binding site at the N-terminus. Other binding partners for this plasma membrane–cytoskeletal adapter protein include paxillin, α-actinin, VASP and tensin. In plants, cross-reactivity with human β-catenin or human vinculin antibodies detects 115–120-kDa polypeptides in extracts from maize root and coleoptile cells, as well as several lower molecular weight species (Endlé *et al.*, 1998; Baluska *et al.*, 1999). The staining pattern in root cells is quite variable; in certain cell types punctate labeling is observed on the plasma membrane, but more typically is associated with cytoplasmic organelles (Baluska *et al.*, 1999). In dividing tobacco BY-2 cells, a vinculin-like epitope is quite prominent at the cell plate, a raft of Golgi-derived vesicle fusion products delivered by the phragmoplast. Analysis of the *Arabidopsis* genome does not, however, reveal any convincing candidates for plant vinculin (Hussey *et al.*, 2002). Thus, models for vinculin-like molecules supporting cell–cell adhesions or coordinating actin organization during cytokinesis require further experimental support.

2.9.10 LIM proteins

Several proteins with a pair of conserved zinc-finger modules belong to a group of multidomain proteins that are implicated in development, signaling, transcription and cytoskeletal organization. The LIM domain is a conserved, cysteine-rich module that contains two zinc-binding fingers and which functions as a protein-binding module (Khurana *et al.*, 2002b). Members of the CRP/group 2 family are primarily associated with the actin cytoskeleton. Vertebrate CRP proteins appear to associate with actin indirectly through binding α-actinin, β-spectrin or zyxin. By contrast, two *Dictyostelium* LIM proteins (LIMC and LIMD) bind to F-actin directly *in vitro* and associate with the cortical cytoskeleton (Khurana *et al.*, 2002a). These low molecular weight (~20 kDa) proteins comprising two or one LIM domains, respectively, are suggested to provide rigidity to the cell cortex. Sunflower, tobacco and *Arabidopsis* appear to contain at least three LIM proteins, each with two LIM domains (Eliasson *et al.*, 2000). Like other cytoskeleton-associated proteins, these can be grouped into reproductive and sporophytic isoforms. A pollen-specific isoform from sunflower (HaPLIM1) is present on abundant organelles or vesicles and accumulates at the F-actin rich germination sites; a role during pollen germination has been proposed (Baltz *et al.*, 1992, 1999). A sporophytic isoform HaWLIM1 from sunflower is implicated in cytokinesis and localizes to the phragmoplast (Mundel *et al.*, 2000). In neither case, however, has evidence for association with the actin cytoskeleton directly or indirectly been provided.

2.10 Concluding remarks

Structurally, plant F-actin is no different from its animal counterpart. For the most part the biochemistry of plant and animal actins are also the same. There is high sequence conservation between the plant and animal actins as well as between some of the principal ABPs. In general, the conserved ABPs are represented by relatively large gene families. In many cases, the divergence within these gene families is not as great as between like proteins in animal cells. For example, although there is a large villin gene family there appears to be no scinderin, severin, fragmin as in animal cells, and also no gelsolin-like activity has yet been demonstrated experimentally. This phenomenon of lack of divergence between like proteins is perhaps best exemplified by the profilin-binding proteins. The only conserved PLP profilin-binding protein in plants is formin – there appears to be no WASp, VASP, WIP or WAVE. Within the majority of the plant ABP sequences there does appear to be plant-specific regions; so perhaps in the evolution of plants, characters beneficial to plant ABPs that enable them to respond to their unique intracellular and extracellular signals have been established and maintained. In such cases members of the large villin, formin, ADF or profilin gene families for example, may be subtly tailored to suit the needs of plants whilst also covering some of the important functions of proteins such as WASp, thymosin β, or drebrin, that are missing from databases. In this background of database searching, it is apparent that plants do not contain many of the actin-binding or -modulating proteins present in animals. It is not too difficult for us to imagine then, that the plant actin cytoskeleton has exploited other ways of regulating its dynamics during plant cell morphogenesis. Therefore, elucidating the subtle characteristics of members of the large plant ABP gene families has to be one goal for the future. In addition, genetic and biochemical approaches to finding plant-specific ABPs is the other goal and there is some scope, as described in this chapter, that such studies will be identifying novel plant ABPs in the near future.

Acknowledgements

We are grateful for the advice, discussions and input of numerous faculty, post-docs and students in our labs during the last several years. In particular, CJS thanks the Purdue Motility Group (http://www.bio.purdue.edu/pmg/) for providing stimulating interactions. Thanks to Ms Lisa Gao (Purdue University) for constructing Figs 2.2 and 2.3 and Table 2.1, and to Dr Shanjin Huang (Purdue) for providing unpublished data. This work was funded by grant support from the US Department of Agriculture (02-35304-12412), US Department of Energy, Energy Bioscience Division (DE-FG02-99ER20337A01) and the US National Science Foundation (0130576-MCB), to CJS. PJH is supported by the Biotechnology and Biological Sciences Research Council, UK, and the Research Training Network of the European Union.

References

Allwood, E.G., Smertenko, A.P. & Hussey, P.J. (2001) Phosphorylation of plant actin-depolymerising factor by calmodulin-like domain protein kinase, *FEBS Lett.*, **499**, 97–100.

Allwood, E.G., Anthony, R.G., Smertenko, A.P., *et al.* (2002) Regulation of the pollen-specific actin-depolymerizing factor LlADF1, *Plant Cell*, **14**, 2915–2927.

Amatruda, J.F., Cannon, J.F., Tatchell, K., Hug, C. & Cooper, J.A. (1990) Disruption of the actin cytoskeleton in yeast capping protein mutants, *Nature*, **344**, 352–354.

Amatruda, J.F., Gattermeir, D.J., Karpova, T.S. & Cooper, J.A. (1992) Effects of null mutations and overexpression of capping protein on morphogenesis, actin distribution and polarized secretion in yeast, *J. Cell Biol.*, **119**, 1151–1162.

Amberg, D.C., Basart, E. & Botstein, D. (1995) Defining protein interactions with yeast actin in vivo, *Struct. Biol.*, **2**, 28–34.

Amor, Y., Haigler, C.H., Johnson, S., Wainscott, M. & Delmer, D. (1995) A membrane-associated form of sucrose synthase and its potential role in synthesis of cellulose and callose in plants, *Proc. Natl. Acad. Sci. USA*, **92**, 9353–9357.

Ampe, C. & Vandekerckhove, J. (1987) The F-actin capping proteins of *Physarum polycephalum*: cap42(a) is very similar, if not identical, to fragmin and is structurally and functionally very homologous to gelsolin; cap42(b) is *Physarum* actin, *EMBO J.*, **6**, 4149–4157.

An, Y.-Q., McDowell, J.M., Huang, S.R., McKinney, E.C., Chambliss, S. & Meagher, R.B. (1996) Strong, constitutive cxpression of the *Arabidopsis* ACT2/ACT8 actin subclass in vegetative tissues, *Plant J.*, **10**, 107–121.

Andrawis, A., Solomon, M. & Delmer, D.P. (1993) Cotton fiber annexins: a potential role in the regulation of callose synthase, *Plant J.*, **3**, 763–772.

Arber, S., Barbayannis, F.A., Hanser, H., *et al.* (1998) Regulation of actin dynamics through phosphorylation of cofilin by LIM-kinase, *Nature*, **393**, 805–809.

Asada, T. & Collings, D. (1997) Molecular motors in higher plants, *Trends Plant Sci.*, **2**, 29–37.

Assaad, F.F. (2001) Of weeds and men: what genomes teach us about plant cell biology, *Curr. Opin. Plant Biol.*, **4**, 478–487.

Azim, A.C., Knoll, J.H.M., Beggs, A.H. & Chishti, A.H. (1995) Isoform cloning, actin binding, and chromosomal location of human erythroid dematin, a member of the villin superfamily, *J. Biol. Chem.*, **270**, 17407–17413.

Baird, W.V. & Meagher, R.B. (1987) A complex gene superfamily encodes actin in petunia, *EMBO J.*, **6**, 3223–3231.

Balasubramanian, M.K., Helfman, D.M. & Hemmingsen, S.M. (1992) A new tropomyosin essential for cytokinesis in the fission yeast *S. pombe*, *Nature*, **360**, 84–87.

Balasubramanian, M.K., Hirani, B.R., Burke, J.D. & Gould, K.L. (1994) The *Schizosaccharomyces pombe cdc3*[+] gene encodes a profilin essential for cytokinesis, *J. Cell Biol.*, **125**, 1289–1301.

Baltz, R., Domon, C., Pillay, D.T.N. & Steinmetz, A. (1992) Characterization of a pollen-specific cDNA from sunflower encoding a zinc finger protein, *Plant J.*, **2**, 713–721.

Baltz, R., Schmit, A.-C., Kohnen, M., Hentges, F. & Steinmetz, A. (1999) Differential localization of the LIM domain protein PLIM-1 in microspores and mature pollen drains from sunflower, *Sex. Plant Reprod.*, **12**, 60–65.

Baluska, F., Samaj, J. & Volkmann, D. (1999) Proteins reacting with cadherin and catenin antibodies are present in maize showing tissue-, domain-, and development-specific associations with endoplasmic-reticulum membranes and actin microfilaments in root cells, *Protoplasma*, **206**, 174–187.

Baluska, F., Salaj, J., Mathur, J., *et al.* (2000) Root hair formation: F-actin-dependent tip growth is initiated by local assembly of profilin-supported F-actin meshworks accumulated within expansin-enriched bulges, *Develop. Biol.*, **227**, 618–632.

Baluska, F., Jasik, J., Edelmann, H.G., Salajová, T. & Volkmann, D. (2001a) Latrunculin B-induced plant dwarfism: plant cell elongation is F-actin-dependent, *Develop. Biol.*, **231**, 113–124.

Baluska, F., von Witsch, M., Peters, M., Hlavacka, A. & Volkmann, D. (2001b) Mastoparan alters subcellular distribution of profilin and remodels F-actin cytoskeleton in cells of maize root apices, *Plant Cell Physiol.*, **42**, 912–922.

Bamburg, J.R. (1999) Proteins of the ADF/cofilin family: essential regulators of actin dynamics, *Annu. Rev. Cell Develop. Biol.*, **15**, 185–230.

Banno, H. & Chua, N.-H. (2000) Characterization of the *Arabidopsis* formin-like protein AFH1 and its interacting protein, *Plant Cell Physiol.*, **41**, 617–626.

Barrero, R.A., Umeda, M., Yamamura, S. & Uchimiya, H. (2002) *Arabidopsis* CAP regulates the actin cytoskeleton necessary for plant cell elongation and division, *Plant Cell*, **14**, 149–163.

Benlali, A., Draskovic, I., Hazelett, D.J. & Treisman, J.E. (2000) *act up* controls actin polymerization to alter cell shape and restrict hedgehog signaling in the *Drosophila* eye disc, *Cell*, **101**, 271–281.

Bhargavi, V., Chari, V.B. & Singh, S.S. (1998) Phosphatidylinositol 3-kinase binds to profilin through the p85a subunit and regulates cytoskeletal assembly, *Biochem. Mol. Biol. Int.*, **46**, 241–248.

Bibikova, T.N., Blancaflor, E.B. & Gilroy, S. (1999) Microtubules regulate tip growth and orientation in root hairs of *Arabidopsis thaliana*, *Plant J.*, **17**, 657–665.

Bisikirska, B. & Sikorski, A.F. (1997) Some properties of spectrin-like proteins from *Pisum sativum*, *Z. Naturforsch.*, **52**, 180–186.

Blackbourn, H.D., Barker, P.J., Huskisson, N.S. & Battey, N.H. (1992) Properties and partial protein sequences of plant annexins, *Plant Physiol.*, **99**, 864–871.

Bowman, G.D., Nodelman, I.M., Hong, Y., Chua, N.-H., Lindberg, U. & Schutt, C.E. (2000) A comparative structural analysis of the ADF/cofilin family, *Proteins: Struct. Func. Gen.*, **41**, 374–384.

Braun, M., Baluska, F., von Witsch, M. & Menzel, D. (1999) Redistribution of actin, profilin and phosphatidylinositol-4,5-bisphosphate in growing and maturing root hairs, *Planta*, **209**, 435–443.

Breton, G., Vazquez-Tello, A., Danyluk, J. & Sarhan, F. (2000) Two novel intrinsic annexins accumulate in wheat membranes in response to low temperature, *Plant Cell Physiol.*, **41**, 177–184.

Bretscher, A. & Weber, K. (1980a) Fimbrin, a new microfilament-associated protein present in microvilli and other cell surface structures, *J. Cell Biol.*, **86**, 335–340.

Bretscher, A. & Weber, K. (1980b) Villin is a major protein of the microvillus cytoskeleton which binds both G and F actin in a calcium-dependent manner, *Cell*, **20**, 839–847.

Brown, S.S., Yamamoto, K. & Spudich, J.A. (1982) A 40,000-Dalton protein from *Dictyostelium discoideum* affects assembly properties of actin in a Ca^{2+}-dependent manner, *J. Cell Biol.*, **93**, 205–210.

Burridge, K. & Chrzanowska-Wodnicka, M. (1996) Focal adhesions, contractility, and signaling, *Annu. Rev. Cell Develop. Biol.*, **12**, 463–518.

Calvert, C.M., Gant, S.J. & Bowles, D.J. (1996) Tomato annexins p34 and p35 bind to F-actin and display nucleotide phosphodiesterase activity inhibited by phospholipid binding, *Plant Cell*, **8**, 333–342.

Cárdenas, L., Vidali, L., Domínguez, J., *et al.* (1998) Rearrangement of actin microfilaments in plant root hairs responding to *Rhizobium etli* nodulation signals, *Plant Physiol.*, **116**, 871–877.

Carlier, M.-F., Laurent, V., Santolini, J., *et al.* (1997) Actin depolymerizing factor (ADF/cofilin) enhances the rate of filament turnover: implication in actin-based motility, *J. Cell Biol.*, **136**, 1307–1322.

Carlson, S.J. & Chourey, P.S. (1996) Evidence for plasma membrane-associated forms of sucrose synthase in maize, *Mol. Gen. Genet.*, **252**, 303–310.

Carlsson, L., Nyström, L.E., Sundkvist, I., Markey, F. & Lindberg, U. (1977) Actin polymerizability is influenced by profilin, a low molecular weight protein in non-muscle cells, *J. Mol. Biol.*, **115**, 465–483.

Chen, C.Y., Wong, E.I., Vidali, L., *et al.* (2002) The regulation of actin organization by actin-depolymerizing factor in elongating pollen tubes, *Plant Cell*, **14**, 2175–2190.

Chourey, P.S., Taliercio, E.W., Carlson, S.J. & Ruan, Y.L. (1998) Genetic evidence that the two isozymes of sucrose synthase present in developing maize endosperm are critical, one for cell wall integrity and the other for starch biosynthesis, *Mol. Gen. Genet.*, **259**, 88–96.

Clark, G.B. & Roux, S.J. (1995) Annexins of plant cells, *Plant Physiol.*, **109**, 1133–1139.

Clark, G.B., Sessions, A., Eastburn, D.J. & Roux, S.J. (2001) Differential expression of members of the annexin multigene family in *Arabidopsis*, *Plant Physiol.*, **126**, 1072–1084.

Clark, G.B., Turnwald, S., Tirlapur, U.K., *et al.* (1995) Polar distribution of annexin-like proteins during phytochrome-mediated initiation and growth of rhizoids in the ferns *Dryopteris* and *Anemia*, *Planta*, **197**, 376–384.

Clore, A.M., Dannenhoffer, J.M. & Larkins, B.A. (1996) EF-1α is associated with a cytoskeletal network surrounding protein bodies in maize endosperm cells, *Plant Cell*, **8**, 2003–2014.

Collings, D.A., Wasteneys, G.O., Miyazaki, M. & Williamson, R.E. (1994) Elongation factor 1α is a component of the subcortical actin bundles of characean algae, *Cell Biol. Int.*, **18**, 1019–1024.

Collings, D.A. & Allen, N.S. (2000) Cortical actin interacts with the plasma membrane and microtubules, in *Actin: A Dynamic Framework for Multiple Plant Cell Functions* (eds C.J. Staiger, F. Baluska., D. Volkmann & P.W. Barlow), Kluwer Academic Publishers, Dordrecht, pp. 145–163.

Cooley, L., Verheyen, E. & Ayers, K. (1992) *chickadee* encodes a profilin required for intercellular cytoplasm transport during Drosophila oogenesis, *Cell*, **69**, 173–184.

Cooper, J.A., Hart, M.C., Karpova, T.S. & Schafer, D.A. (1999) Capping protein, in *Guidebook to the Cytoskeletal and Motor Proteins*, 2nd edn (eds T. Kreis & R. Vale), Oxford University Press, New York, pp. 62–64.

Cooper, J.A. & Schafer, D.A. (2000) Control of actin assembly and disassembly at filament ends, *Curr. Opin. Cell Biol.*, **12**, 97–103.

Correia, I. & Matsudaira, P. (1999) Fimbrin, in *Guidebook to the Cytoskeletal and Motor Proteins*, 2nd edn (eds T. Kreis & R. Vale), Oxford University Press, New York, pp. 97–99.

Cruz-Ortega, R., Cushman, J.C. & Ownby, J.D. (1997) cDNA clones encoding 1,3-ß-glucanase and a fimbrin-like cytoskeletal protein are induced by Al toxicity in wheat roots, *Plant Physiol.*, **114**, 1453–1460.

Cullen, P.J., Cozier, G.E., Banting, G. & Mellor, H. (2001) Modular phosphoinositide-binding domains – their role in signalling and membrane trafficking, *Curr. Biol.*, **11**, R882–R893.

Cvrcková, F. (2000) Are plant formins integral membrane proteins?, *Genome Biol.*, **1**, 001.001–001.007.

Dantán-González, E., Rosenstein, Y., Quinto, C. & Sánchez, F. (2001) Actin monoubiquitylation is induced in plants in response to pathogens and symbionts, *Molec. Plant Micr. Int.*, **14**, 1267–1273.

de Ruijter, N. & Emons, A.M. (1993) Immunodetection of spectrin antigens in plant cells, *Cell Biol. Int. Rep.*, **17**, 169–182.

de Ruijter, N.C.A., Rook, M.B., Bisseling, T. & Emons, A.M.C. (1998) Lipochito-oligosaccharides re-initiate root hair tip growth in *Vicia sativa* with high calcium and spectrin-like antigen at the tip, *Plant J.*, **13**, 341–350.

Deeks, M.J., Hussey, P.J. & Davies, B. (2002) Formins: intermediates in signal transduction cascades that affect cytoskeletal reorganization, *Trends Plant Sci.*, **7**, 492–498.

Delmer, D.P. & Potikha, T.S. (1997) Structures and functions of annexins in plants, *Cell. Mol. Life Sci.*, **53**, 546–553.

Dharmawardhane, S., Demma, M., Yang, F. & Condeelis, J. (1991) Compartmentalization and actin binding properties of ABP-50: the elongation factor-1 alpha of *Dictyostelium*, *Cell Motil. Cytoskeleton*, **20**, 279–288.

Doering, D.S. & Matsudaira, P. (1996) Cysteine scanning mutagenesis at 40 of 76 positions in villin headpiece maps the F-actin binding site and structural features of the domain, *Biochemistry*, **35**, 12677–12685.

Dong, J., Radau, B., Otto, A., Müller, E.-C., Lindschau, C. & Westermann, P. (2000) Profilin I attached to the Golgi is required for the formation of constitutive transport vesicles at the trans-Golgi network, *Biochim. Biophys. Acta*, **2**, 253–260.

Dong, C.-H., Kost, B., Xia, G. & Chua, N.-H. (2001a) Molecular identification and characterization of the *Arabidopsis* AtADF1, AtADF5 and AtADF6 genes, *Plant Mol. Biol.*, **45**, 517–527.

Dong, C.-H., Xia, G.-X., Hong, Y., Ramachandran, S., Kost, B. & Chua, N.-H. (2001b) ADF proteins are involved in the control of flowering and regulate F-actin organization, cell expansion, and organ growth in *Arabidopsis*, *Plant Cell*, **13**, 1333–1346.

Drøbak, B.K., Watkins, P.A.C., Valenta, R., Dove, S.K., Lloyd, C.W. & Staiger, C.J. (1994) Inhibition of plant plasma membrane phosphoinositide phospholipase C by the actin-binding protein, profilin, *Plant J.*, **6**, 389–400.

Drubin, D.G., Miller, K.G. & Botstein, D. (1988) Yeast actin-binding proteins: Evidence for a role in morphogenesis, *J. Cell Biol.*, **107**, 2551–2561.

Durso, N.A. & Cyr, R.J. (1994) A calmodulin-sensitive interaction between microtubules and a higher plant homolog of elongation factor-1α, *Plant Cell*, **6**, 893–905.

Eden, S., Rohatgi, R., Podtelejnikov, A.V., Mann, M. & Kirschner, M.W. (2002) Mechanism of regulation of WAVE1-induced actin nucleation by Rac1 and Nck, *Nature*, **418**, 790–793.

Edmonds, B.T., Liu, G. & Condeelis, J. (1999) Elongation factor 1A (eEF1A), In *Guidebook to the Cytoskeletal and Motor Proteins*, 2nd edn (eds T. Kreis & R. Vale), Oxford University Press, New York, pp. 84–87.

Eliasson, Å., Gass, N., Mundel, C., *et al.* (2000) Molecular and expression analysis of a LIM protein gene family from flowering plants, *Mol. Gen. Genet.*, **264**, 257–267.

Endlé, M.-C., Stoppin, V., Lambert, A.-M. & Schmit, A.-C. (1998) The growing cell plate of higher plants is a site of both actin assembly and vinculin-like antigen recruitment, *Eur. J. Cell Biol.*, **77**, 10–18.

Evangelista, M., Pruyne, D., Amberg, D.C., Boone, C. & Bretscher, A. (2002) Formins direct Arp2/3-independent actin filament assembly to polarize cell growth in yeast, *Nature Cell Biol.*, **4**, 32–41.

Faraday, C.D. & Spanswick, R.M. (1993) Evidence for a membrane skeleton in higher plants. A spectrin-like polypeptide co-isolates with rice root plasma membranes, *FEBS Lett.*, **318**, 313–316.

Fedorov, A.A., Ball, T., Mahoney, N.M., Valenta, R. & Almo, S.C. (1997) The molecular basis for allergen cross-reactivity: crystal structure and IgE-epitope mapping of birch pollen profilin, *Structure*, **5**, 33–45.

Feierback, B. & Chang, F. (2001) Roles of the fission yeast formin for3p in cell polarity, actin cable formation and symmetric cell division, *Curr. Biol.*, **11**, 1656–1665.

Ferrary, E., Cohen-Tannoudji, M., Pehau-Arnaudet, G., *et al.* (1999) In vivo, villin is required for Ca^{2+}-dependent F-actin disruption in intestinal brush borders, *J. Cell Biol.*, **146**, 819–829.

Fleurat-Lessard, P., Roblin, G., Bonmort, J. & Besse, C. (1988) Effects of colchicine, vinblastine, cytochalasin B and phalloidin on the seismonastic movement of *Mimosa pudica* leaf and on motor cell ultrastructure, *J. Exp. Bot.*, **39**, 209–221.

Frank, M.J. & Smith, L.G. (2002) A small, novel protein highly conserved in plants and animals promotes the polarized growth and division of maize leaf epidermal cells, *Curr. Biol.*, **12**, 849–853.

Freeman, N.L., Chen, Z., Horenstein, J., Weber, A. & Field, J. (1995) An actin monomer-binding activity localizes to the carboxyl terminal half of the *Saccharomyces cerevisiae* cyclase-associated protein, *J. Biol. Chem.*, **270**, 5680–5685.

Freeman, N.L., Lila, T., Mintzer, K.A., *et al.* (1996) A conserved proline-rich region of the *Saccharomyces cerevisiae* cyclase-associated protein binds SH3 domains and modulates cytoskeletal localization, *Mol. Cell Biol.*, **16**, 548–556.

Freeman, N.L. & Field, J. (2000) Mammalian homolog of the yeast cyclase associated protein, CAP/Srv2p, regulates actin filament assembly, *Cell Motil. Cytoskeleton*, **45**, 106–120.

Friederich, E. & Louvard, D. (1999) Villin, in *Guidebook to the Cytoskeletal and Motor Proteins* (eds T. Kries & R. Vale), Oxford University Press, New York.

Friederich, E., Vancompernolle, K., Huet, C., *et al.* (1992) An actin-binding site containing a conserved motif of charged amino acid residues is essential for the morphogenetic effect of villin, *Cell*, **70**, 81–92.

Friederich, E., Vancompernolle, K., Louvard, D. & Vandekerckhove, J. (1999) Villin function in the organization of the actin cytoskeleton. Correlation of *in vivo* effects to its biochemical activities in vitro, *J. Biol. Chem.*, **274**, 26751–26760.

Fu, Y., Wu, G. & Yang, Z. (2001) Rop GTPase-dependent dynamics of tip-localized F-actin controls tip growth in pollen tubes, *J. Cell Biol.*, **152**, 1019–1032.

Fu, Y., Li, H. & Yang, Z. (2002) The ROP2 GTPase controls the formation of cortical fine F-actin and the early phase of directional cell expansion during *Arabidopsis* organogenesis, *Plant Cell*, **14**, 777–794.

Furukawa, R., Jinks, T.M., Tishgarten, T., Mazzawi, M., Morris, D.R. & Fechheimer, M. (2001) Elongation factor 1β is an actin-binding protein, *Biochim. Biophys. Acta*, **1527**, 130–140.

Fyrberg, E.A., Fyrberg, C.C., Biggs, J.R., Saville, D., Beall, C.J. & Ketchum, A. (1998) Functional nonequivalence of *Drosophila* actin isoforms, *Biochem. Genet.*, **36**, 271–287.

Geiger, B. (1999) Vinculin, in *Guidebook to the Cytoskeletal and Motor Proteins* (eds T. Kreis & R. Vale), Oxford University Press, New York, pp. 172–175.

Geitmann, A., Snowman, B.N., Emons, A.M.C. & Franklin-Tong, V.E. (2000) Alterations in the actin cytoskeleton of pollen tubes are induced by the self-incompatibility reaction in *Papaver rhoeas*, *Plant Cell*, **12**, 1239–1251.

Gerke, V. & Moss, S.E. (1997) Annexins and membrane dynamics, *Biochim. Biophys. Acta*, **1357**, 129–154.

Gibbon, B.C. (2001) Actin monomer-binding proteins and the regulation of actin dynamics in plants, *J. Plant Growth Reg.*, **20**, 103–112.

Gibbon, B.C., Kovar, D.R. & Staiger, C.J. (1999) Latrunculin B has different effects on pollen germination and tube growth, *Plant Cell*, **11**, 2349–2363.

Gibbon, B.C. & Staiger, C.J. (2000) Profilin, in *Actin: A Dynamic Framework for Multiple Plant Cell Functions* (eds C.J. Staiger, F. Baluska, D. Volkmann & P. Barlow), Kluwer Academic Publisher, Dordrecht, The Netherlands, pp. 45–65.

Gibbon, B.C., Zonia, L.E., Kovar, D.R., Hussey, P.J. & Staiger, C.J. (1998) Pollen profilin function depends on interaction with proline-rich motifs, *Plant Cell*, **10**, 981–994.

Gieselmann, R. & Mann, K. (1992) ASP-56, a new actin sequestering protein from pig platelets with homology to CAP, an adenylate cyclase-associated protein from yeast, *FEBS Lett.*, **298**, 149–153.

Gilliland, L.U., Kandasamy, M.K., Pawloski, L.C. & Meagher, R.B. (2002) Both vegetative and reproductive actin isovariants complement the stunted root hair phenotype of the *Arabidopsis act2-1* mutation, *Plant Physiol.*, **130**, 2199–2209.

Gilliland, L.U., McKinney, E.C., Asmussen, M.A. & Meagher, R.B. (1998) Detection of deleterious genotypes in multigenerational studies. I. Disruptions in individual *Arabidopsis* actin genes, *Genetics*, **149**, 717–725.

Gilliland, L.U., Pawloski, L.C., Kandasamy, M.K. & Meagher, R.B. (2003) *Arabidopsis* actin gene ACT7 plays an essential role in germination and root growth, *Plant J.*, **33**, 319–328.

Gillooly, D.J., Simonsen, A. & Stenmark, H. (2001) Phosphoinositides and phagocytosis, *J. Cell Biol.*, **155**, 15–17.

Glenney, J.R.J., Bretscher, A. & Weber, K. (1980) Calcium control of the intestinal microvillus cytoskeleton: its implications for the regulation of microfilament organizations, *Proc. Natl. Acad. Sci. USA*, **77**, 6458–6462.

Glenney, J.R.J., Geisler, N., Kaulfus, P. & Weber, K. (1981) Demonstration of at least two different actin-binding sites in villin, a calcium-regulated modulator of F-actin organization, *J. Biol. Chem.*, **256**, 8156–8161.

Gottwald, U., Brokamp, R., Karakesisoglou, I., Schleicher, M. & Noegel, A.A. (1996) Identification of a cyclase-associated protein (CAP) homologue in *Dicytostelium discoideum* and characterization of its interaction with actin, *Mol. Biol. Cell*, **7**, 261–272.

Grote, M., Swoboda, I., Meagher, R.B. & Valenta, R. (1995) Localization of profilin- and actin-like immunoreactivity in in vitro-germinated tobacco pollen tubes by electron microscopy after special water-free fixation techniques, *Sex. Plant Reprod.*, **8**, 180–186.

Grote, M., Vrtala, S. & Valenta, R. (1993) Monitoring of two allergens, Bet vI and profilin, in dry and rehydrated birch pollen by immunogold electron microscopy and immunoblotting, *J. Histochem. Cytochem.*, **41**, 745–750.

Gungabissoon, R.A., Jiang, C.-J., Drøbak, B.K., Maciver, S.K. & Hussey, P.J. (1998) Interaction of maize actin-depolymerising factor with actin and phosphoinositides and its inhibition of plant phospholipase C, *Plant J.*, **16**, 689–696.

Gungabissoon, R.A., Khan, S., Hussey, P.J. & Maciver, S.K. (2001) The interaction of maize elongation factor 1α (ZmEF-1α) with F-actin and interplay with the maize actin severing protein, ZmADF3. *Cell Motil. Cytoskel.*, **49**(2), 104–111.

Haarer, B.K., Lillie, S.H., Adams, A.E.M., Magdolen, V., Bandlow, W. & Brown, S.S. (1990) Purification of profilin from *Saccharomyces cerevisiae* and analysis of profilin-deficient cells, *J. Cell Biol.*, **110**, 105–114.

Haugwitz, M., Noegel, A.A., Karakesisoglou, J. & Schleicher, M. (1994) *Dictyostelium amoeba* that lack G-actin-sequestering profilins show defects in F-actin content, cytokinesis, and development, *Cell*, **79**, 303–314.

Haus, U., Hartmann, H., Trommler, P., Noegel, A.A. & Schleicher, M. (1991) F-actin capping by Cap32/34 requires heterodimeric conformation and can be inhibited with PIP2, *Biochem. Biophys. Res. Comm.*, **181**, 833–839.

Hepler, P.K., Cleary, A.L., Gunning, B.E.S., Wadsworth, P., Wasteneys, G.O. & Zhang, D.H. (1993) Cytoskeletal dynamics in living plant cells, *Cell Biol. Int.*, **17**, 127–142.

Higashi-Fujime, S. (1991) Reconstitution of active movement in vitro based on the actin-myosin interaction, *Int. Rev. Cytol.*, **125**, 95–138.

Higgs, H.N. & Pollard, T.D. (2001) Regulation of actin filament network formation through Arp2/3 complex: activation by a diverse array of proteins, *Annu. Rev. Biochem.*, **70**, 649–676.

Hodge, T. & Cope, J. (2000) A myosin family tree, *J. Cell Sci.*, **113**, 3353–3354.

Hofmann, A., Noegel, A.A., Bomblies, L., Lottspeich, F. & Schleicher, M. (1993) The 100 kDa F-actin capping protein of *Dictyostelium amoeba* is a villin prototype ('protovillin'), *FEBS Lett.*, **328**, 71–76.

Hofmann, A., Proust, J., Dorowski, A., Schantz, R. & Huber, R. (2000) Annexin 24 from *Capsicum annuum* X-ray structure and biochemical characterization, *J. Biol. Chem.*, **275**, 8072–80782.

Holstein, S.E.H. (2002) Clathrin and plant endocytosis, *Traffic*, **2002**, 614–620.

Holt, M.R. & Koffer, A. (2001) Cell motility: proline-rich proteins promote protrusions, *Trends Cell Biol.*, **11**, 38–46.

Hopmann, R., Cooper, J.A. & Miller, K.G. (1996) Actin organization, bristle morphology, and viability are affected by actin capping protein mutations in *Drosophila*, *J. Cell Biol.*, **133**, 1293–1305.

Hoshino, T., Mizutani, A., Chida, M., Hidaka, H. & Mizutani, J. (1995) Plant annexin form homodimer during Ca^{2+}-dependent liposome aggregation, *Biochem. Mol. Biol. Int.*, **35**, 749–755.

Hoyle, H.D. & Raff, E.C. (1990) Two Drosophila beta tubulin isoforms are not functionally equivalent, *J. Cell Biol.*, **111**, 1009–1026.

Hu, S.Q., Brady, S.R., Kovar, D.R., *et al.* (2000) Identification of plant actin-binding proteins by F-actin affinity chromatography, *Plant J.*, **24**, 127–137.

Huang, S.R., McDowell, J.M., Weise, M.J. & Meagher, R.B. (1996) The *Arabidopsis* profilin gene family – evidence for an ancient split between constitutive and pollen-specific profilin genes, *Plant Physiol.*, **111**, 115–126.

Huber, P.A.J., Marston, S.B., Hodgkinson, J.L. & Wang, C.-L.A. (1999) Caldesmon, in *Guidebook to the Cytoskeletal and Motor Proteins*, 2nd edn (eds T. Kreis & R. Vale), Oxford University Press, New York, pp. 52–66.

Hug, C., Jay, P.Y., Reddy, I., *et al.* (1995) Capping protein levels influence actin assembly and cell motility in *Dictyostelium*, *Cell*, **81**, 591–600.

Hussey, P.J., Allwood, E.G. & Smertenko, A.P. (2002) Actin-binding proteins in the *Arabidopsis* genome database: properties of functionally distinct plant actin-depolymerizing factors/cofilins, *Phil. Trans. R. Soc. Lond. B.*, **357**, 791–798.

Hussey, P.J., Hawkins, T.J., Igarashi, I., Kaloriti, D. & Smertenko, A.P. (2002) The plant cytoskeleton: recent advances in the study of the plant microtubule-associated proteins MAP-65, MAP-190 and the Xenopus MAP215-like protein, MOR1, *Plant Mol. Biol.*, **50**, 915–924.

Igarashi, H., Orii, H., Mori, H., Shimmen, T. & Sonobe, S. (2000) Isolation of a novel 190 kDa protein from tobacco BY-2 cells: possible involvement in the interaction between actin filaments and microtubules, *Plant Cell Physiol.*, **41**, 920–931.

Iida, K. & Yahara, I. (1999) Cooperation of two actin-binding proteins, cofilin and Aip1, in Saccharomyces cerevisiae, *Genes Cells*, **4**, 21–32.

Isenberg, G., Aebi, U. & Pollard, T.D. (1980) An actin-binding protein from *Acanthamoeba* regulates actin filament polymerization and interactions, *Nature*, **288**, 455–459.

Ishizaki, T., Morishima, Y., Okamoto, M., Furuyashiki, T., Kato, T. & Narumiya, S. (2001) Coordination of microtubules and the actin cytoskeleton by the Rho effector mDia1, *Nature Cell Biol.*, **3**, 8–14.

Janmey, P.A. & Matsudaira, P.T. (1988) Functional comparison of villin and gelsolin. Effects of Ca^{2+}, KCl, and polyphosphoinositides, *J. Biol. Chem.*, **263**, 16738–16743.

Janmey, P.A., Stossel, T.P. & Allen, P.G. (1998) Deconstructing gelsolin: identifying sites that mimic or alter binding to actin and phosphoinositides, *Chem. Biol.*, **5**, R81–R85.

Janssen, K.P., Rost, R., Eichinger, L. & Schleicher, M. (2001) Characterization of CD36/LIMPII homologues in *Dictyostelium discoideum*, *J. Biol. Chem.*, **276**, 38899–38910.

Janssen, K.P. & Schleicher, M. (2001) *Dictyostelium discoideum*: a genetic model system for the study of professional phagocytes – profilin, phosphoinositides and the *lmp* gene family in *Dictyostelium*, *Biochim. Biophys. Acta*, **1525**, 228–233.

Jeng, R.L. & Welch, M.D. (2001) Actin and endocytosis – no longer the weakest link, *Curr. Biol.*, **11**, R691–R694.

Jiang, C.-J., Weeds, A.G. & Hussey, P.J. (1997) The maize actin-depolymerizing factor, ZmADF3, redistributes to the growing tip of elongating root hairs and can be induced to translocate into the nucleus with actin, *Plant J.*, **12**, 1035–1043.

Kaetzel, M.A. & Dedman, J.R. (1999) Annexins, in *Guidebook to the Cytoskeletal and Motor Proteins*, 2nd edn (eds T. Kreis & R. Vale), Oxford University Press, New York, pp. 38–42.

Kaiser, D.A., Vinson, V.K., Murphy, D.B. & Pollard, T.D. (1999) Profilin is predominantly associated with monomeric actin in *Acanthamoeba*, *J. Cell Sci.*, **112**, 3779–3790.

Kameyama, K., Kishi, Y., Yoshimura, M., Kanzawa, N., Sameshima, M. & Tsuchiya, T. (2000) Tyrosine phosphorylation in plant bending – puckering in a ticklish plant is controlled by dephosphorylation of its actin, *Nature*, **407**, 37.

Kandasamy, M.K., McKinney, E.C. & Meagher, R.B. (2002) Functional nonequivalency of actin isovariants in *Arabidopsis*, *Mol. Biol. Cell*, **13**, 251–261.

Kang, F., Purich, D.L. & Southwick, F.S. (1999) Profilin promotes barbed-end actin filament assembly without lowering the critical concentration, *J. Biol. Chem.*, **274**, 36963–36972.

Karakesisoglou, I., Schleicher, M., Gibbon, B.C. & Staiger, C.J. (1996) Plant profilins rescue the aberrant phenotype of profilin-deficient *Dictyostelium* cells, *Cell Motil. Cytoskeleton*, **34**, 36–47.

Karpova, T.S., Tatchell, K. & Cooper, J.A. (1995) Actin filaments in yeast are unstable in the absence of capping protein or fimbrin, *J. Cell Biol.*, **131**, 1483–1493.

Kashiyama, T., Kimura, N., Mimura, T. & Yamamoto, K. (2000) Cloning and characterization of a myosin from characean alga, the fastest motor protein in the world, *J. Biochem.*, **127**, 1065–1070.

Kato, T., Watanabe, N., Morishima, Y., Fujita, A., Ishizaki, T. & Narumiya, S. (2000) Localization of a mammalian homolog of diaphanous, mDia1, to the mitotic spindle in HeLa cells, *J. Cell Sci.*, **114**, 775–784.

Kawai, M., Aotsuka, S. & Uchimiya, H. (1998) Isolation of a cotton CAP gene: a homologue of adenylyl cyclase-associated protein highly expressed during fiber elongation, *Plant Cell Physiol.*, **39**, 1380–1383.

Ketelaar, T., Faivre-Moskalenko, C., Esseling, J.J., *et al.* (2002) Positioning of nuclei in *Arabidopsis* root hairs: an actin-regulated process of tip growth, *Plant Cell*, **14**, 2941–2955.

Khurana, B., Khurana, T., Khaire, N. & Noegel, A.A. (2002a) Functions of LIM proteins in cell polarity and chemotactic motility, *EMBO J.*, **21**, 5331–5342.

Khurana, T., Khurana, B. & Noegel, A.A. (2002b) LIM proteins: association with the actin cytoskeleton, *Protoplasma*, **219**, 1–12.

Kim, A.C., Peters, L.L., Knoll, J.H.M., *et al.* (1997) Limatin (LIMAB1), an actin-binding LIM protein, maps to mouse chromosome 19 and human chromosome 10q25, a region frequently deleted in human cancers, *Genomics*, **46**, 291–293.

Kim, D.H., Eu, Y.-J., Yoo, C.M., *et al.* (2001) Trafficking of phosphatidylinositol 3-phosphate from the *trans*-Golgi network to the lumen of the central vacuole in plant cells, *Plant Cell*, **13**, 287–301.

Kim, E., Miller, C.J. & Reisler, E. (1996) Polymerization and *in vitro* motility properties of yeast actin: a comparison with rabbit skeletal α-actin, *Biochemistry*, **35**, 16566–16572.

Kirsch, J., Langosch, D., Prior, P., Littauer, U.Z., Schmitt, B. & Betz, H. (1991) The 93-kDa glycine receptor-associated protein binds to tubulin, *J. Biol. Chem.*, **266**, 22242–22245.

Klahre, U. & Chua, N.-H. (1999) The *Arabidopsis ACTIN-RELATED PROTEIN 2 (AtARP2)* promoter directs expression in xylem precursor cells and pollen, *Plant Mol. Biol.*, **41**, 65–73.

Klahre, U., Friederich, E., Kost, B., Louvard, D. & Chua, N.-H. (2000) Villin-like actin-binding proteins are expressed ubiquitously in *Arabidopsis*, *Plant Physiol.*, **122**, 35–47.

Kohno, T. & Shimmen, T. (1988) Mechanism of Ca^{2+} inhibition of cytoplasmic streaming in lily pollen tubes, *J. Cell Sci.*, **91**, 501–509.

Konzok, A., Weber, I., Simmeth, E., Hacker, U., Maniak, M. & Müller-Taubenberger, A. (1999) DAip1, a *Dictyostelium* homologue of the yeast actin-interacting protein 1, is involved in endocytosis, cytokinesis, and motility, *J. Cell Biol.*, **146**, 453–464.

Korenbaum, E. & Rivero, F. (2002) Calponin homology domains at a glance, *J. Cell Sci.*, **115**, 3543–3545.

Kost, B., Lemichez, E., Spielhofer, P., *et al.* (1999) Rac homologues and compartmentalized phosphatidylinositol 4,5-bisphosphate act in a common pathway to regulate polar pollen tube growth, *J. Cell Biol.*, **145**, 317–330.

Kovar, D.R. & Staiger, C.J. (2000) Actin depolymerizing factor, in *Actin: A Dynamic Framework for Multiple Plant Cell Functions* (eds C.J. Staiger, F. Baluska, D. Volkmann & P. Barlow), Kluwer Academic Publishers, Dordrecht, The Netherlands, pp. 67–85.

Kovar, D.R., Drøbak, B.K. & Staiger, C.J. (2000a) Maize profilin isoforms are functionally distinct, *Plant Cell*, **12**, 583–598.

Kovar, D.R., Staiger, C.J., Weaver, E.A. & McCurdy, D.W. (2000b) AtFim1 is an actin filament crosslinking protein from *Arabidopsis thaliana*, *Plant J.*, **24**, 625–636.

Kovar, D.R., Gibbon, B.C., McCurdy, D.W. & Staiger & C.J. (2001a) Fluorescently labeled fimbrin decorates a dynamic actin filament network in live plant cells, *Planta*, **213**, 390–395.

Kovar, D.R., Yang, P., Sale, W.S., Drøbak, B.K. & Staiger, C.J. (2001b) *Chlamydomonas reinhardtii* produces a profilin with unique biochemical properties, *J. Cell Sci.*, **114**, 4293–4305.

Krauze, K., Makuch, R., Stepka, M. & Dabrowska, R. (1998) The first caldesmon-like protein in higher plants, *Biochem. Biophys. Res. Comm.*, **247**, 576–579.

Kreis, T. & Vale, R. (1999) *Guidebook to the Cytoskeletal and Motor Proteins*, 2nd edn, Oxford University Press, New York.

Lam, B.C.-H., Sage, T.L., Bianchi, F. & Blumwald, E. (2001) Role of SH3 domain-containing proteins in clathrin-mediated vesicle trafficking in *Arabidopsis*, *Plant Cell*, **13**, 2499–2512.

Lanzetti, L., Di Fiore, P.P. & Scita, G. (2001) Pathways linking endocytosis and actin cytoskeleton in mammalian cells, *Exp. Cell Res.*, **271**, 45–56.

Lappalainen, P., Kessels, M.M., Cope, M.J.T.V. & Drubin, D.G. (1998) The ADF homology (ADF-H) domain: a highly exploited actin-binding module, *Mol. Biol. Cell*, **9**, 1951–1959.

Lee, L., Klee, S.K., Evangelista, M., Boone, C. & Pellman, D. (1999) Control of mitotic spindle position by the *Saccharomyces cerevisiae* formin Bni1p, *J. Cell Biol.*, **144**, 947–961.

Lee, S.S., Karakesisoglou, I., Noegel, A.A., Rieger, D. & Schleicher, M. (2000) Dissection of functional domains by expression of point-mutated profilins in *Dictyostelium* mutants, *Eur. J. Cell Biol.*, **79**, 92–103.

Li, Y., Yan, L., Zee, S. Y. & Huang, B.Q. (1999) Membrane skeleton spectrin in pollen and pollen tube, *Chin. Sci. Bull.*, **44**, 930–933.

Lim, E.K., Roberts, M.R. & Bowles, D.J. (1998) Biochemical characterization of tomato annexin p35. Independence of calcium binding and phosphatase activities, *J. Biol. Chem.*, **273**, 34920–34925.

Liu, H. & Bretscher, A. (1989) Disruption of the single tropomyosin gene in yeast results in the disappearance of actin cables from the cytoskeleton, *Cell*, **57**, 233–242.

Liu, X. & Yen, L.-F. (1992) Purification and characterization of actin from maize pollen, *Plant Physiol.*, **99**, 1151–1155.

Loisel, T.P., Boujemaa, R., Pantaloni, D. & Carlier, M.-F. (1999) Reconstitution of actin-based motility of *Listeria* and *Shigella* using pure proteins, *Nature*, **401**, 613–616.

Lopez, I., Anthony, R.G., Maciver, S.K., *et al.* (1996) Pollen specific expression of maize genes encoding actin depolymerizing factor-like proteins, *Proc. Natl. Acad. Sci. USA*, **93**, 7415–7420.

Lu, J. & Pollard, T.D. (2001) Profilin binding to poly-L-proline and actin monomers along with ability to catalyze actin nucleotide exchange is required for viability of fission yeast, *Mol. Biol. Cell*, **12**, 1161–1175.

Lu, P.-J., Shieh, W.-R., Rhee, S.G., Yin, H.L. & Chen, C.-S. (1996) Lipid products of phosphoinositide 3-kinase bind human profilin with high affinity, *Biochemistry*, **35**, 14027–14034.

Machesky, L.M., Atkinson, S.J., Ampe, C., Vandekerckhove, J. & Pollard, T.D. (1994) Purification of a cortical complex containing two unconventional actins from *Acanthamoeba* by affinity chromatography on profilin – agarose, *J. Cell Biol.*, **127**, 107–115.

Maciver, S.K. & Hussey, P.J. (2002) The ADF/cofilin family: actin-remodeling proteins, *Genome Biol.*, **3**, 3007.3001–3007.3012.

Magdolen, V., Drubin, D.G., Mages, G. & Bandlow, W. (1993) High levels of profilin suppress the lethality caused by overproduction of actin in yeast cells, *FEBS Lett.*, **316**, 41–47.

Mahajan-Miklos, S. & Cooley, L. (1994) The villin-like protein encoded by the Drosophila quail gene is required for actin bundle assembly during oogenesis, *Cell*, **78**, 291–301.

Mammoto, A., Sasaki, T., Asakura, T., *et al.* (1998) Interactions of drebrin and gephyrin with profilin, *Biochem. Biophys. Res. Comm.*, **243**, 86–89.

Martin, T., Frommer, W.B., Salanoubat, M. & Willmitzer, L. (1993) Expression of an *Arabidopsis* sucrose synthase gene indicates a role in metabolization of sucrose both during phloem loading and in sink organs, *Plant J.*, **4**, 367–377.

Mathur, J. & Hülskamp, M. (2002) Microtubules and microfilaments in cell morphogenesis in higher plants, *Curr. Biol.*, **12**, R669–R676.

Matova, N., Mahajan-Miklos, S., Mooseker, M.S. & Cooley, L. (1999) *Drosophila* Quail, a villin-related protein, bundles actin filaments in apoptotic nurse cells, *Development*, **126**, 5645–5657.

Matsudaira, P.T. & Burgess, D.R. (1979) Identification and organization of the components in the isolated microvillus cytoskeleton, *J. Cell Biol.*, **83**, 667–673.

McClung, A.D., Carroll, A.D. & Battey, N.H. (1994) Identification and characterization of ATPase activity associated with maize (*Zea mays*) annexins, *Biochem. J.*, **303**, 709–712.

McCurdy, D.W. & Kim, M. (1998) Molecular cloning of a novel fimbrin-like cDNA from *Arabidopsis thaliana*, *Plant Mol. Biol.*, **36**, 23–31.

McDowell, J.M., An, Y.-Q., Huang, S., McKinney, E.C. & Meagher, R.B. (1996a) The *Arabidopsis* ACT7 actin gene is expressed in rapidly developing tissues and responds to several external stimuli, *Plant Physiol.*, **111**, 699–711.

McDowell, J.M., Huang, S., McKinney, E.C., An, Y.-Q. & Meagher, R.B. (1996b) Structure and evolution of the actin gene family in *Arabidopsis thaliana*, *Genetics*, **142**, 587–602.

McKinney, E.C., Ali, N., Traut, A., *et al.* (1995) Sequence-based identification of T-DNA insertion mutations in *Arabidopsis*: actin mutants *act2-1* and *act4-1*, *Plant J.*, **8**, 613–622.

McKinney, E.C., Kandasamy, M.K. & Meagher, R.B. (2001) Small changes in the regulation of one *Arabidopsis* profilin isovariant, PRF1, alter seedling development, *Plant Cell*, **13**, 1179–1191.

McKinney, E.C., Kandasamy, M.K. & Meagher, R.B. (2002) *Arabidopsis* contains ancient classes of differentially expressed actin-related protein genes, *Plant Physiol.*, **128**, 997–1007.

McKinney, E.C. & Meagher, R.B. (1998) Members of the *Arabidopsis* actin gene family are widely dispersed in the genome, *Genetics*, **149**, 663–675.

McLaughlin, P.J., Gooch, J.T., Mannherz, H.-G. & Weeds, A.G. (1993) Structure of gelsolin segment 1-actin complex and the mechanism of filament severing, *Nature*, **364**, 685–692.

Meagher, R.B., McKinney, E.C. & Kandasamy, M.K. (1999a) Isovariant dynamics expand and buffer the responses of complex systems: the diverse plant actin gene family, *Plant Cell*, **11**, 995–1005.

Meagher, R.B., McKinney, E.C. & Kandasamy, M.K. (2000) The significance of diversity in the plant actin gene family, in *Actin: A dynamic framework for multiple plant cell functions* (eds C.J. Staiger, F. Baluska, D. Volkmann, & P.W. Barlow), Kluwer Academic Publishers, Dordrecht, The Netherlands, pp. 3–27.

Meagher, R.B., McKinney, E.C. & Vitale, A.V. (1999b) The evolution of new structures – clues from plant cytoskeletal genes, *Trends Genet.*, **15**, 278–284.

Meagher, R.B. & Williamson, R.E. (1994) The plant cytoskeleton, in *Arabidopsis* (eds E. Meyerowitz & C. Somerville), Cold Spring Harbor Press, Cold Spring Harbor, NY, pp. 1049–1084.

Medalia, O., Weber, I., Frangakis, A.S., Nicastro, D., Gerisch, G. & Baumeister, W. (2002) Macromolecular architecture in eukaryotic cells visualized by cryoelectron tomography, *Science*, **298**, 1209–1213.

Merrifield, C.J., Rescher, U., Almers, W., *et al.* (2001) Annexin 2 has an essential role in actin-based macropinocytic rocketing, *Curr. Biol.*, **11**, 1136–1141.

Michaud, D., Guillet, G., Rogers, P.A. & Charest, P.M. (1991) Identification of a 220 kDa membrane-associated plant cell protein immunologically related to human β-spectrin, *FEBS Lett.*, **294**, 77–80.

Miller, D.D., de Ruijter, N.C.A., Bisseling, T. & Emons, A.M.C. (1999) The role of actin in root hair morphogenesis: studies with lipochito-oligosaccharide as a growth stimulator and cytochalasin as an actin perturbing drug, *Plant J.*, **17**, 141–154.

Mittermann, I., Swoboda, I., Pierson, E., *et al.* (1995) Molecular cloning and characterization of profilin from tobacco (*Nicotiana tabacum*): increased profilin expression during pollen maturation, *Plant Mol. Biol.*, **27**, 137–146.

Moore, R.C. & Cyr, R.J. (2000) Association between elongation factor-1α and microtubules in vivo is domain dependent and conditional, *Cell Motil. Cytoskeleton*, **45**, 279–292.

Mooseker, M.S. (1983) Actin binding proteins of the brush border, *Cell*, **35**, 11–13.

Morimatsu, M., Nakamura, A., Sumiyoshi, H., *et al.* (2000) The molecular structure of the fastest myosin from green algae, *Chara*, *Biochem. Biophys. Res. Comm.*, **270**, 147–152.

Moriyama, K. & Yahara, I. (2002) Human CAP1 is a key factor in the recycling of cofilin and actin for rapid actin turnover, *J. Cell Sci.*, **115**, 1591–1601.

Moriyama, K., Iida, K. & Yahara, I. (1996) Phosphorylation of Ser-3 of cofilin regulates its essential function on actin, *Genes cell*, **1**, 73–86.

Morrow, J.S. (1999) Spectrins, in *Guidebook to the Cytoskeletal and Motor Proteins*, 2nd edn (eds T. Kreis & R. Vale), Oxford University Press, New York, pp. 138–141.

Mun, J.H., Yu, H.J., Lee, H.S., *et al.* (2000) Two closely related cDNAs encoding actin-depolymerizing factors of petunia are mainly expressed in vegetative tissues, *Gene*, **257**, 167–176.

Mundel, C., Baltz, R., Eliasson, Å., *et al.* (2000) A LIM-domain protein from sunflower is localized to the cytoplasm and/or nucleus in a wide variety of tissues and is associated with the phragmoplast in dividing cells, *Plant Mol. Biol.*, **42**, 291–302.

Nakayasu, T., Yokota, E. & Shimmen, T. (1998) Purification of an actin-binding protein composed of 115-kDa polypeptide from pollen tubes of lily, *Biochem. Biophys. Res. Comm.*, **249**, 61–65.

Namba, Y., Ito, M., Zu, Y., Shigesada, K. & Maruyama, K. (1992) Human T cell L-plastin bundles actin filaments in a calcium-dependent manner, *J. Biochem.*, **112**, 503–507.

Nguyen-Quoc, B., Krivitzky, M. & Huber, S.C.A.L. (1990) Sucrose synthase in developing maize leaves. Regulation of activity by protein level during the import to export transition, *Plant Physiol.*, **94**, 516–523.

Nick, P. (1999) Signals, motors, morphogenesis – the cytoskeleton in plant development, *Plant Biol.*, **1**, 169–179.

Niggli, V. (2001) Structural properties of lipid-binding sites in cytoskeletal proteins, *Trends Biochem. Sci.*, **26**, 604–611.

Noegel, A.A. (1999) Severin, in *Guidebook to the Cytoskeletal and Motor Proteins*, 2nd edn (eds T. Kreis & R. Vale), Oxford University Press, New York, pp. 125–127.

Noegel, A.A., Rivero, F., Albrecht, R., *et al.* (1999) Assessing the role of the ASP56/CAP homologue of *Dictyostelium discoideum* and the requirements for subcellular localization, *J. Cell Sci.*, **112**, 3195–3203.

Okada, K., Obinata, T. & Abe, H. (1999) XAIP1: a *Xenopus* homologue of yeast actin interacting protein 1 (AIP1), which induces disassembly of actin filaments cooperatively with ADF cofilin family proteins, *J. Cell Sci.*, **112**, 1553–1565.

Olmsted, J.B. (1986) Microtubule-associated proteins, *Ann. Rev. Cell Biol.*, **2**, 421–458.

Ono, S. (1999) Purification and biochemical characterization of actin from *Caenorhabditis elegans*: its difference from rabbit muscle actin in the interaction with nematode ADF/cofilin, *Cell Motil. Cytoskeleton*, **43**, 128–136.

Ono, S. (2001) The *Caenorhabditis elegans unc-78* gene encodes a homologue of actin interacting protein 1 required for organised assembly of muscle actin filaments, *J. Cell Biol.*, **152**, 1313–1319.

Ono, S. & Ono, K. (2002) Tropomyosin inhibits ADF/cofilin-dependent actin filament dynamics, *J. Cell Biol.*, **156**, 1065–1076.

Ostrander, D.B., Ernst, E.G., Lavoie, T.B. & Gorman, J.A. (1999) Polyproline binding is an essential function of human profilin in yeast, *Eur. J. Biochem.*, **262**, 26–35.

Moriyama, K., Iida, K. & Yaharra, I. (1996) Phosphorylation of ser-3 of cofilin regulates its essential function on actin, *Genes Cells*, **1**, 73–86.

Pal, M. & Biswas, S. (1994) A novel protein accumulated during maturation of the pods of the plant *Impatiens balsmina*, *Mol. Cell. Biochem.*, **130**, 111–120.

Palazzo, A.F., Cook, T.A., Alberts, A.S. & Gundersen, G.G. (2001) mDia mediates Rho-regulated formation and orientation of stable microtubules, *Nature Cell Biol.*, **3**, 723–729.

Pantaloni, D. & Carlier, M.-F. (1993) How profilin promotes actin filament assembly in the presence of thymosin ß4, *Cell*, **75**, 1007–1014.

Peitsch, W.K., Grund, C., Kuhn, C., *et al.* (1999) Drebrin is a widespread actin-associating protein enriched at junctional plaques, defining a specific microfilament anchorage system in polar epithelial cells, *Eur. J. Cell Biol.*, **78**, 767–778.

Pollard, T.D., Almo, S., Quirk, S., Vinson, V. & Lattman, E.E. (1994) Structure of actin binding proteins: Insights about function at atomic resolution, *Annu. Rev. Cell Biol.*, **10**, 207–249.

Pope, B., Way, M., Matsudaira, P.T. & Weeds, A. (1994) Characterisation of the F-Actin binding domains of villin: classification of F-Actin binding proteins into two groups according to their binding sites on actin, *FEBS Lett.*, **338**, 58–62.

Pope, B., Way, M. & Weeds, A.G. (1991) Two of the three actin-binding domains of gelsolin bind to the same subdomain of actin, *FEBS Lett.*, **280**, 70–74.

Pruyne, D., Evangelista, M., Yang, C.S., *et al.* (2002) Role of formins in actin assembly: nucleation and barbed-end association, *Science*, **297**, 612–615.

Pruyne, D.W., Schott, D.H. & Bretscher, A. (1998) Tropomyosin-containing actin cables direct the Myo2p-dependent polarized delivery of secretory vesicles in budding yeast, *J. Cell Biol.*, **143**, 1931–1945.

Qualmann, B., Kessels, M.M. & Kelly, R.B. (2000) Molecular links between endocytosis and the actin cytoskeleton, *J. Cell Biol.*, **150**, F111–F116.

Ramachandran, S., Christensen, H.E.M., Ishimaru, Y., *et al.* (2000) Profilin plays a role in cell elongation, cell shape maintenance, and flowering in *Arabidopsis*, *Plant Physiol.*, **124**, 1637–1647.

Ramming, M., Kins, S., Werner, N., Hermann, A., Betz, H. & Kirsch, J. (2000) Diversity and phylogeny of gephyrin: tissue-specific splice variants, gene structure, and sequence similarities to molybdenum cofactor-synthesizing and cytoskeleton-associated proteins, *Proc. Natl. Acad. Sci USA*, **97**, 10266–10271.

Rana, A.P., Ruff, P., Maalouf, G.J., Speicher, D.W. & Chishti, A.H. (1993) Cloning of human erythroid dematin reveals another member of the villin family, *Proc. Natl. Acad. Sci. USA*, **90**, 6651–6655.

Reddy, A.S.N. & Day, I.S. (2001) Analysis of the myosins encoded in the recently completed *Arabidopsis thaliana* genome sequence, *Genome Biol.*, **2**, R0024.0001–0024.0017.

Reddy, A.S.N. (2001) Molecular motors and their functions in plants, *Int. Rev. Cytol.*, 97–178.

Reichelt, S. & Kendrick-Jones, J. (2000) Myosins, in *Actin: A Dynamic Framework for Multiple Plant Cell Functions* (eds C.J. Staiger, F. Baluska, D. Volkmann & P. Barlow), Kluwer Academic Publishers, Dordrecht, pp. 29–44.

Reinhard, M., Jarchau, T. & Walter, U. (2001) Actin-based motility: stop and go with Ena/VASP proteins, *Trends Biochem. Sci.*, **26**, 243–249.

Ren, H., Gibbon, B.C., Ashworth, S.L., Sherman, D.M., Yuan, M. & Staiger, C.J. (1997) Actin purified from maize pollen functions in living plant cells, *Plant Cell*, **9**, 1445–1457.

Reuzeau, C., Doolittle, K.W., McNally, J.G. & Pickard, B.G. (1997) Covisualization in living onion cells of putative integrin, putative spectrin, actin, putative intermediate filaments, and other proteins at the cell membrane and in an endomembrane sheath, *Protoplasma*, **199**, 173–197.

Ringli, C., Baumberger, N., Diet, A., Frey, B. & Keller, B. (2002) ACTIN2 is essential for bulge site selection and tip growth during root hair development of *Arabidopsis*, *Plant Physiol.*, **129**, 1464–1472.

Rodal, A.A., Tetreault, J.W., Lappalainen, P., Drubin, D.G. & Amberg, D.C. (1999) Aip1p interacts with cofilin to disassemble actin filaments, *J. Cell Biol.*, **145**, 1251–1264.

Sagot, I., Klee, S.K. & Pellman, D. (2002a) Yeast formins regulate cell polarity by controlling the assembly of actin cables, *Nature Cell Biol.*, **4**, 42–50.

Sagot, I., Rodal, A.A., Moseley, J., Goode, B.L. & Pellman, D. (2002b) An actin nucleation mechanism mediated by Bni1 and profilin, *Nature Cell Biol.*, **4**, 626–631.

Santoni, V., Rouquie, D., Doumas, P., *et al.* (1998) Use of a proteome strategy for tagging proteins present at the plasma membrane, *Plant J.*, **16**, 633–641.

Schafer, D.A., Hug, C. & Cooper, J.A. (1995) Inhibition of CapZ during myofibrillogenesis alters assembly of actin filaments, *J. Cell Biol.*, **128**, 61–70.

Schafer, D.A., Jennings, P.B. & Cooper, J.A. (1998) Rapid and efficient purification of actin from non-muscle sources, *Cell Motil. Cytoskeleton*, **39**, 166–171.

Schmidt, A. & Huttner, W.B. (1998) Biogenesis of synaptic-like microvesicles in perforated PC12 cells, *Methods*, **16**, 160–169.

Schwarz, G., Schulze, J., Bittner, F., *et al.* (2000) The molybdenum cofactor biosynthetic protein Cnx1 complements molybdate-repairable mutants, transfers molybdenum to the metal binding pterin, and is associated with the cytoskeleton, *Plant Cell*, **12**, 2455–2471.

Schultz, J., Milpetz, F., Bork, P. & Ponting, C.P. (1998) SMART, a simple modular architecture research tool: identification of signaling domains, *Proc. Natl. Acad. Sci. USA*, **95**, 5857–5864.

Seastone, D.J., Harris, E., Temesvari, L.A., Bear, J.E., Saxe, C.L. & Cardelli, J. (2001) The WASp-like protein Scar regulates macropinocytosis, phagocytosis and endosomal membrane flow in *Dictyostelium*, *J. Cell Sci.*, **114**, 2673–2683.

Sheterline, P., Clayton, J. & Sparrow, J.C. (1998) Actin, *Protein Profile*, **4**, 1–272.

Shimmen, T., Ridge, R.W., Lambiris, I., Plazinski, J., Yokota, E. & Williamson, R.E. (2000) Plant myosins, *Protoplasma*, **214**, 1–10.

Sikorski, A.F., Swat, W., Brzesinska, M., Wroblewski, Z. & Bisikirska, B. (1993) A protein cross-reacting with anti-spectrin antibodies is present in higher plant cells, *Z. Naturforsch.*, **48c**, 580–583.

Simonsen, A., Wurmser, A.E., Emr, S.D. & Stenmark, H. (2001) The role of phosphoinositides in membrane transport, *Curr. Opin. Cell Biol.*, **13**, 485–492.

Singh, S.S., Chauhan, A., Murakami, N. & Chauhan, V.P.S. (1996) Profilin and gelsolin stimulate phosphatidylinositol 3-kinase activity, *Biochemistry*, **35**, 16544–16549.

Smertenko, A.P., Allwood, E.G., Khan, S., *et al.* (2001) Interaction of pollen-specific actin-depolymerizing factor with actin, *Plant J.*, **25**, 203–212.

Smertenko, A.P., Jiang, C.-J., Simmons, N.J., Weeds, A.G., Davies, D.R. & Hussey, P.J. (1998) Ser6 in the maize actin-depolymerizing factor, ZmADF3, is phosphorylated by a calcium-stimulated protein kinase and is essential for the control of functional activity, *Plant J.*, **14**, 187–194.

Smillie, L.B. (1999) Tropomyosins, in *Guidebook to the Cytoskeletal and Motor Proteins*, 2nd edn (eds T. Kreis & R. Vale), Oxford University Press, New York, pp. 159–164.

Smith, L.G. (2003) Cytoskeletal control of plant cell shape: getting the fine points. *Curr. Opin. Plant. Biol.*, **6**, 63–73.

Snowman, B.N., Geitmann, A., Emons, A.M.C. & Franklin-Tong, V.E. (2000) Actin rearrangements in pollen tubes are stimulated by the self-incompatibility (SI) response in *Papaver rhoeas* L., in *Actin: A Dynamic Framework for Multiple Plant Cell Functions* (eds C.J. Staiger, F. Baluska, D. Volkmann & P. Barlow) Kluwer Academic Publishers, Dordrecht, The Netherlands, pp. 347–360.

Snowman, B.N., Kovar, D.R., Shevchenko, G., Franklin-Tong, V.E. & Staiger, C.J. (2002) Signal-mediated depolymerization of actin in pollen during the self-incompatibility response, *Plant Cell*, **14**, 2613–2626.

Srinivasan, S., Traini, M., Herbert, B., *et al.* (2001) Proteomic analysis of a developmentally regulated secretory vesicle, *Proteomics*, **1**, 1119–1127.

Staiger, C.J. (2000) Signaling to the actin cytoskeleton in plants, *Annu. Rev. Plant Physiol. Plant Mol. Biol.*, **51**, 257–288.

Staiger, C.J., Gibbon, B.C., Kovar, D.R. & Zonia, L.E. (1997) Profilin and actin depolymerizing factor: modulators of actin organization in plants, *Trends Plant Sci.*, **2**, 275–281.

Staiger, C.J., Yuan, M., Valenta, R., Shaw, P.J., Warn, R.M. & Lloyd, C.W. (1994) Microinjected profilin affects cytoplasmic streaming in plant cells by rapidly depolymerizing actin microfilaments, *Curr. Biol.*, **4**, 215–219.

Stevenson, V.A. & Theurkauf, W.E. (2000) Putting a CAP on actin polymerization, *Curr. Biol.*, **10**, R695–R697.

Suetsugu, S., Miki, H. & Takenawa, T. (1998) The essential role of profilin in the assembly of actin for microspike formation, *EMBO J.*, **17**, 6516–6526.

Suetsugu, S., Miki, H. & Takenawa, T. (1999) Distinct roles of profilin in cell morphological changes: microspikes, membrane ruffles, stress fibers, and cytokinesis, *FEBS Lett.*, **457**, 470–474.

Sun, H.-Q., Kwiatkowska, K., Wooten, D.C. & Yin, H.L. (1995) Effects of CapG overexpression on agonist-induced motility and second messenger generation, *J. Cell Biol.*, **129**, 147–156.

Sun, Y.J., Carneiro, N., Clore, A.M., Moro, G.L., Habben, J.E. & Larkins, B.A. (1997) Characterization of maize elongation factor 1A and its relationship to protein quality in the endosperm, *Plant Physiol.*, **115**, 1101–1107.

Svitkina, T.M. & Borisy, G.G. (1999) Arp2/3 complex and actin depolymerizing factor cofilin in dendritic organization and treadmilling of actin filament array in lamellipodia, *J. Cell Biol.*, **145**, 1009–1026.

Tanaka, K. (2000) Formin family proteins in cytoskeletal control, *Biochem. Biophys. Res. Comm.*, **267**, 479–481.

Tanaka, M. & Shibata, H. (1985) Poly (L-proline)-binding proteins from chick embryos are a profilin and a profilactin, *Eur. J. Biochem.*, **151**, 291–297.

Temesvari, L., Zhang, L., Fodera, B., Janssen, K.-P., Schleicher, M. & Cardelli, J.A. (2000) Inactivation of *lmpA*, encoding a LIMPII-related endosomal protein, suppresses the internalization and endosomal trafficking defects in profilin-null mutants, *Mol. Biol. Cell*, **11**, 2019–2031.

Thangavelu, M., Belostotsky, D., Bevan, M., Flavell, R.B., Rogers, H.J. & Lonsdale, D.M. (1993) Partial characterization of the *Nicotiana tabacum* actin gene family: evidence for pollen specific expression of one of the gene family members, *Mol. Gen. Genet.*, **240**, 290–295.

Thorn, K.S., Christensen, H.E.M., Shigeta, R.J., *et al.* (1997) The crystal structure of a major allergen from plants, *Structure*, **5**, 19–32.

Tominaga, M., Yokota, E., Vidali, L., Sonobe, S., Hepler, P.K. & Shimmen, T. (2000) The role of plant villin in the organization of the actin cytoskeleton, cytoplasmic streaming and the architecture of the transvacuolar strand in root hair cells of *Hydrocharis*, *Planta*, **210**, 836–843.

Tominaga, M., Kojima, H., Yokota, E., *et al.* (2003) Higher plant myosin XI moves processively on actin with 35nm steps at high velocity, *EMBO J.*, **22**, 1263–1272.

Toshima, J., Toshima, J.Y., Amano, T., Yang, N., Narumiya, S. & Mizuno, K. (2001) Cofilin phosphorylation by protein kinase testicular protein kinase 1 and its role in integrin-mediated actin reorganization and focal adhesion formation, *Mol. Biol. Cell*, **12**, 1131–1145.

Tsujioka, M., Machesky, L.M., Cole, S.L., Yahata, K. & Inouye, K. (1999) A unique talin homologue with a villin headpiece-like domain is required for multicellular morphogenesis in *Dictyostelium*, *Curr. Biol.*, **9**, 389–392.

Turkina, M.V. & Akatova, L.Z. (1994) Heat-stable actin-binding proteins from the phloem of *Heracleum sosnowskyi*, *Russ. J. Plant Physiol*, **41**(3), 367–373.

Turkina, M.V., Kulikova, A.L., Koppel, L.A., Akatova, L.Z. & Butenko, R.G. (1995) Actin and heat-stable actin binding proteins in wheat callus culture, *Russ. J. Plant Physiol.*, **42**, 303–309.

Ulm, R., Revenkova, E., di Sansebastiano, G.-P., Bechtold, N. & Paszkowski, J. (2001) Mitogen-activated protein kinase phosphatase is required for genotoxic stress relief in *Arabidopsis*, *Genes Develop.*, **15**, 699–709.

Valenta, R., Duchene, M., Ebner, C., *et al.* (1992) Profilins constitute a novel family of functional plant pan-allergens, *J. Exp. Med.*, **175**, 377–385.

Valenta, R., Duchêne, M., Pettenburger, K., *et al.* (1991) Identification of profilin as a novel pollen allergen; IgE autoreactivity in sensitized individuals, *Science*, **253**, 557–560.

Vallier, P., Dechamp, C., Valenta, R., Vial, O. & Deviller, P. (1992) Purification and characterization of an allergen from celery immunochemically related to an allergen present in several other plant species. Identification as profilin, *Clinical Exp. Allergy*, **22**, 774–782.

Valster, A.H., Pierson, E. S., Valenta, R., Hepler, P.K. & Emons, A.M.C. (1997) Probing the plant actin cytoskeleton during cytokinesis and interphase by profilin microinjection, *Plant Cell*, **9**, 1815–1824.

vanRee, R., Voitenko, V., vanLeeuwen, W. A. & Aalberse, R.C. (1992) Profilin is a cross-reactive allergen in pollen and vegetable foods, *Int. Arch. Allergy Immunol.*, **98**, 97–104.

Vantard, M. & Blanchoin, L. (2003) Actin polymerization processes in plant cells, *Curr. Opin. Plant Biol.*, **5**, 502–506.

Vardar, D., Buckley, D.A., Frank, B.S. & McKnight, C.J. (1999) NMR structure of an F-actin-binding 'headpiece' motif from villin, *J. Mol. Biol.*, **294**, 1299–1310.

Verheyen, E.M. & Cooley, L. (1994) Profilin mutations disrupt multiple actin-dependent processes during *Drosophila* development, *Development*, **120**, 717–728.

Vidali, L. & Hepler, P.K. (1997) Characterization and localization of profilin in pollen grains and tubes of *Lilium longiflorum*, *Cell Motil. Cytoskeleton*, **36**, 323–338.

Vidali, L., McKenna, S.T. & Hepler, P.K. (2001) Actin polymerization is essential for pollen tube growth, *Mol. Biol. Cell*, **12**, 2534–2545.

Vidali, L., Yokota, E., Cheung, A.Y., Shimmen, T. & Hepler, P.K. (1999) The 135 kDa actin-bundling protein from *Lilium longiflorum* pollen is the plant homologue of villin, *Protoplasma*, **209**, 283–291.

Vieira, O.V., Botelho, R.J., Rameh, L., *et al.* (2001) Distinct roles of class I and class III phosphatidylinositol 3-kinases in phagosome formation and maturation, *J. Cell Biol.*, **155**, 19–25.

Vojtek, A., Haarer, B., Field, J., *et al.* (1991) Evidence for a functional link between profilin and CAP in the yeast *S. cerevisiae*, *Cell*, **66**, 497–505.

Volkmann, N., DeRosier, D., Matsudaira, P. & Hanein, D. (2001) An atomic model of actin filaments cross-linked by fimbrin and its implications for bundle assembly and function, *J. Cell Biol.*, **153**, 947–956.

von Witsch, M., Baluska, F., Staiger, C.J. & Volkmann, D. (1998) Profilin is associated with the plasma membrane in microspores and pollen, *Eur. J. Cell Biol.*, **77**, 303–312.

Wang, Y.-D. & Yen, L.-F. (1991) Immunochemical identification of spectrins on the plasma membrane of leaf cells of *Vicia faba*, *Chi. Sci. Bull.*, **36**, 862–866.

Wasserman, S. (1998) FH proteins as cytoskeletal organizers, *Trends Cell Biol.*, **8**, 111–115.

Way, M., Pope, B., Gooch, J., Hawkins, M. & Weeds, A.G. (1990) Identification of a region in segment 1 of gelsolin critical for actin binding, *EMBO J.*, **9**, 4103–4109.

Weeds, A. & Maciver, S. (1993) F-actin capping proteins, *Curr. Opin. Cell Biol.*, **5**, 63–69.

Welch, M.D. & Mullins, R.D. (2002) Cellular control of actin nucleation, *Annu. Rev. Cell Dev. Biol.*, **18**, 247–288.

Willhoeft, U. & Tannich, E. (1999) The electrophoretic karyotype of *Entamoeba histolytica*, *Mol. Biochem. Parasitol.*, **99**, 41–53.

Winter, H., Huber, J.L. & Huber, S.C. (1997) Membrane association of sucrose synthase: changes during the graviresponse and possible control by protein phosphorylation, *FEBS Lett.*, **420**, 151–155.

Winter, H., Huber, J.L. & Huber, S.C. (1998) Identification of sucrose synthase as an actin-binding protein, *FEBS Lett.*, **430**, 205–208.

Witke, W., Sutherland, J.D., Sharpe, A., Arai, M. & Kwiatkowski, D.J. (2001) Profilin I is essential for cell survival and cell division in early mouse development, *Proc. Natl. Acad. Sci. USA*, **98**, 3832–3836.

Yamamoto, K., Hamada, S. & Kashiyama, T. (1999) Myosins from plants, *Cell. Mol. Life Sci.*, **56**, 227–232.

Yamashiro, S., Kameyama, K., Kanzawa, N., Tamiya, T., Mabuchi, I. & Tsuchiya, T. (2001) The gelsolin/fragmin family protein identified in the higher plant *Mimosa pudica*, *J. Biochem.*, **130**, 243–249.

Yang, C., Huang, M., DeBiasio, J., *et al.* (2000) Profilin enhances Cdc42-induced nucleation of actin polymerization, *J. Cell Biol.*, **150**, 1001–1012.

Yang, N. & Higuchi, O., Ohashi, K., *et al.* (1998) Cofilin phosphorylation by LIM-kinase 1 and its role in Rac-mediated actin reorganization, *Nature*, **393**, 809–812.

Yang, W. & Boss, W.F. (1994) Regulation of phosphatidylinositol 4-kinase by the protein activator PIK-A49, *J. Biol. Chem.*, **269**, 3852–3857.

Yang, W., Burkhart, W., Cavallius, J., Merrick, W.C. & Boss, W.F. (1993) Purification and characterization of a phosphatidylinositol 4-kinase activator in carrot cells, *J. Biol. Chem.*, **268**, 392–398.

Yen, L.-F., Liu, X. & Cai, S. (1995) Polymerization of actin from maize pollen, *Plant Physiol.*, **107**, 73–76.

Yokota, E., Muto, S. & Shimmen, T. (2000) Calcium-calmodulin suppresses the filamentous actin-binding activity of 135-kilodalton actin-bundling protein isolated from lily pollen tubes, *Plant Physiol.*, **123**, 645–654.

Yokota, E. & Shimmen, T. (1994) Isolation and characterization of plant myosin from pollen tubes of lily, *Protoplasma*, **177**, 153–162.

Yokota, E. & Shimmen, T. (1999) The 135-kDa actin-bundling protein from lily pollen tubes arranges F-actin into bundles with uniform polarity, *Planta*, **209**, 264–266.

Yokota, E. & Shimmen, T. (2000) Characterization of native actin-binding proteins from pollen: Myosin and the actin-bundling proteins, 135-ABP and 115-ABP, in *Actin: A Dynamic Framework for Multiple Plant Cell Functions* (eds C.J. Staiger, F. Baluska, D. Volkmann & P. Barlow), Kluwer Academic Publishers, Dordrecht, The Netherlands, pp. 103–118.

Yokota, E., Takahara, K.-I. & Shimmen, T. (1998) Actin-bundling protein isolated from pollen tubes of lily, *Plant Physiol.*, **116**, 1421–1429.

Yu, F.X., Johnston, P.A., Sudhof, T.C. & Yin, H.L. (1990) gCap39, a calcium ion- and polyphosphoinositide-regulated actin capping protein, *Science*, **250**, 1413–1415.

Zelicof, A., Protopopov, V., David, D., Lin, X.Y., Lustgarten, V. & Gerst, J.E. (1996) Two separate functions are encoded by the carboxyl terminal domains of the yeast cyclase-associated protein and its mammalian homologs – dimerization and actin binding, *J. Biol. Chem.*, **271**, 18243–18252.

Zhao, K., Wang, W., Rando, O.J., *et al.* (1998) Rapid and phosphoinositol-dependent binding of the SWI/SNF-like BAF complex to chromatin after T lymphocyte receptor signaling, *Cell*, **95**, 625–636.

Part 2

Fundamental cytoskeletal activities

3 Expanding beyond the great divide: the cytoskeleton and axial growth

Geoffrey O. Wasteneys and David A. Collings

3.1 Introduction

Cell expansion is the single most important process in plant morphogenesis. After all, cell division makes no direct contribution to organ size and relatively little contribution to organ shape. Cell division in the absence of cell expansion, as in endosperm tissue (Olsen, 2001), results simply in cells becoming more numerous with no net gain in organ size. Conversely, there are examples of remarkably normal organogenesis when the division process is perturbed or absent altogether. The alga *Acetabularia acetabulum* can form exquisitely complex shapes as single cells (Mandoli, 1998). Even in higher plants there is evidence that organogenesis can proceed despite serious cell division defects. Mutants defective in cell division plane alignment, including *fass* and *ton* of *Arabidopsis* (Traas *et al.*, 1995) and *tangled* of maize (Smith *et al.*, 2001), still develop organs in the right places even if they are somewhat deformed. These examples highlight the importance of directional cell expansion for plant morphogenesis.

One success story of higher plant evolution is the production of inconspicuous meristems and organ primordia. Meristems harbour many small and cytoplasmically dense cells that retain the potential to give rise to a wide variety of cell types. The progeny of these stem cells produce tissue precursors, which then follow a precise series of divisions to give rise to distinct tissue patterns that identify organ primordia. As the distance and number of partitions from the meristem centre increases, the basic patterns of organ primordia become established. Cells eventually undergo a transition, showing a reduced tendency to divide, and sometimes pass through a post-mitotic isotropic growth phase. But at a critical threshold, usually triggered by an abiotic signal, cells undergo massive increases in volume accompanied by vacuole formation. This inflation generally leads to terminal differentiation, and often precludes any capacity for further partitioning into new cells. Moreover, most cells grow predominantly or even exclusively in one direction, an anisotropic process we refer to as axial growth.

In this chapter, we investigate the cytoskeleton-mediated processes that control axial growth. Axial growth is distinct from polar growth, which strictly involves highly localized exocytosis. Tip-growing root hairs and pollen tubes expand by polar growth (see Chapters 7 and 8). While the normal cytokinetic process is also highly polarized, it plays an important role in setting up cells for axial growth. In this chapter, we therefore explore the polar secretion that occurs during cytokinesis, and discuss how this process helps to define the properties of cells that will undergo anisotropic expansion. We then explore the relationship between the cytoskeleton and axial cell expansion during interphase and post-mitotic expansion. These two phases are commonly lumped together as interphase but we stress the importance, in a developmental context, to distinguish these two events.

3.2 Division planes and the establishment of axiality

We begin by considering the cytoskeleton's function during cytokinesis, the subject of several recent reviews (microtubules: Otegui & Staehelin, 2000; Hepler *et al.*, 2002; microfilaments: Schmit, 2000; Hepler *et al.*, 2002). Cell plate construction during cytokinesis partitions the contents of post-mitotic cells through the activity of the substantially cytoskeletal phragmoplast. Unlike the primary cell wall, cell plates are not subjected to turgor pressure-driven strain deformation. Studying cell plate and cross wall construction thus provides a unique insight into building a wall from scratch, a process that does not have an existing template of wall material to build on and one that does not face immediate rearrangement of these materials by mechanical stresses. Cell plate formation is also of considerable relevance to the axial expansion that follows because the division plane is usually perpendicular to the subsequent axis of expansion, even when cells are partitioned along their long axis (Green, 1984).

3.2.1 Cell plate formation and expansion

Constructing the typical central-forming cell plate begins with fusion of secretory vesicles between the separated nuclei and is followed by centrifugal plate expansion towards the parent cell wall. Preservation by high pressure freezing for transmission electron microscopy shows plate formation progressing through several distinct stages, including a tubulo-vesicular network, a smooth tubular network, and a fenestrated cell plate (Samuels *et al.*, 1995; Bednarek & Falbel, 2002). During plate expansion, these stages can be seen in a radial gradient in individual cells.

In the tubulo-vesicular stage of cell plate formation, phragmoplast microtubules associate with the plate's leading edge. This is where the majority of vesicle fusion as well as the majority of clathrin-coated vesicles are found (Samuels *et al.*, 1995). Otegui *et al.* (2001) calculated that endocytosis removes 75% of the membrane added to the cell plate during *Arabidopsis* endosperm syncytial divisions. Both vesicle delivery and clathrin-coated vesicle recovery continue at low levels until the fenestrated cell plate stage (Samuels *et al.*, 1995), but microtubules do not associate with this region of the cell plate. Whether the trafficking mechanism and vesicle content differ from that during formation of the tubulo-vesicular network is unknown.

Wall formation is first detected at the smooth tubular network stage with the deposition of callose, which remains abundant in the new cross wall until some time after cytokinesis (Samuels *et al.*, 1995; Hong *et al.*, 2001a). A gene for the catalytic subunit of callose synthase, *CALS1*, has been cloned from *Arabidopsis*, and the encoded protein distributes in the cell plate when expressed as a GFP fusion protein in tobacco (Hong *et al.*, 2001a). *CALS1* interacts with the cell plate-specific dynamin, phragmoplastin and a UDP-glucose transferase. This entire complex might be regulated by the small GTPase ROP1 (Hong *et al.*, 2001b). Significant amounts of cellulose are synthesized at the fenestrated cell plate stage, before the establishment of any cortical microtubule array. Cellulose synthesis is required for the completion of cell plate formation, as its inhibition by the cellulose synthase inhibitor dichlorobenzonitrile results in wavy or incomplete cell plates (Vaughn *et al.*, 1996). The *korrigan* mutant of *Arabidopsis* shows a similar incomplete cytokinetic phenotype. The *KOR* gene encodes an endo-1,4-β-glucanase that is located at the cell plate (Zuo *et al.*, 2000) and is required for normal cellulose synthesis. A temperature-sensitive *KOR* allele, *rsw2*, has decreased amounts of cellulose at its restrictive temperature (Lane *et al.*, 2001). These observations suggest that the cellulose in the cell plate contributes to the stiffening process required for completion of partitioning. However, regulatory factors controlling sequential callose and cellulose deposition have yet to be identified.

3.2.2 Phragmoplast microtubule and microfilament organization

Hepler *et al.* (2002) define formation and expansion as two distinct stages of phragmoplast development. The phragmoplast forms late in anaphase, initially as a poorly defined mass of microfilaments and microtubules between the daughter nuclei, but soon develops into a barrel-like structure with well-defined polarity. Microtubules develop into two arrays of opposite polarity that interdigitate at the division plane, with their fast-growing or plus ends towards the centre (Asada *et al.*, 1997; Wasteneys, 2002). Microfilaments also develop into two parallel arrays, interdigitated at the division plane and with the barbed-ends central (Kakimoto & Shibaoka, 1988). The site of intersection of both the microtubules and microfilaments is the site of active fusion of Golgi-derived secretory vesicles. The phragmoplast expands outward until it reaches the approximate diameter of the two adjacent nuclei (Hepler *et al.*, 2002). The second stage of phragmoplast growth is its outward expansion from the perinuclear region towards the parent cell wall. During this expansion, microtubules and microfilaments at the centre of the phragmoplast depolymerize, so that the phragmoplast assumes a ring-like or toroidal appearance, with the developing and expanding cell plate at its centre (Wasteneys, 2002).

Subtle differences in the organization of microtubules and microfilaments in the phragmoplast suggest that mechanisms similar to those in force in other highly polarized exocytotic events may also operate during cell plate formation. In rapidly growing pollen tubes, root hairs, and lobe-forming epidermal cells, the cytoplasm and growth are highly polarized. Microtubules are typically excluded whereas fine networks of microfilaments are closer to but not in direct contact with the sites of

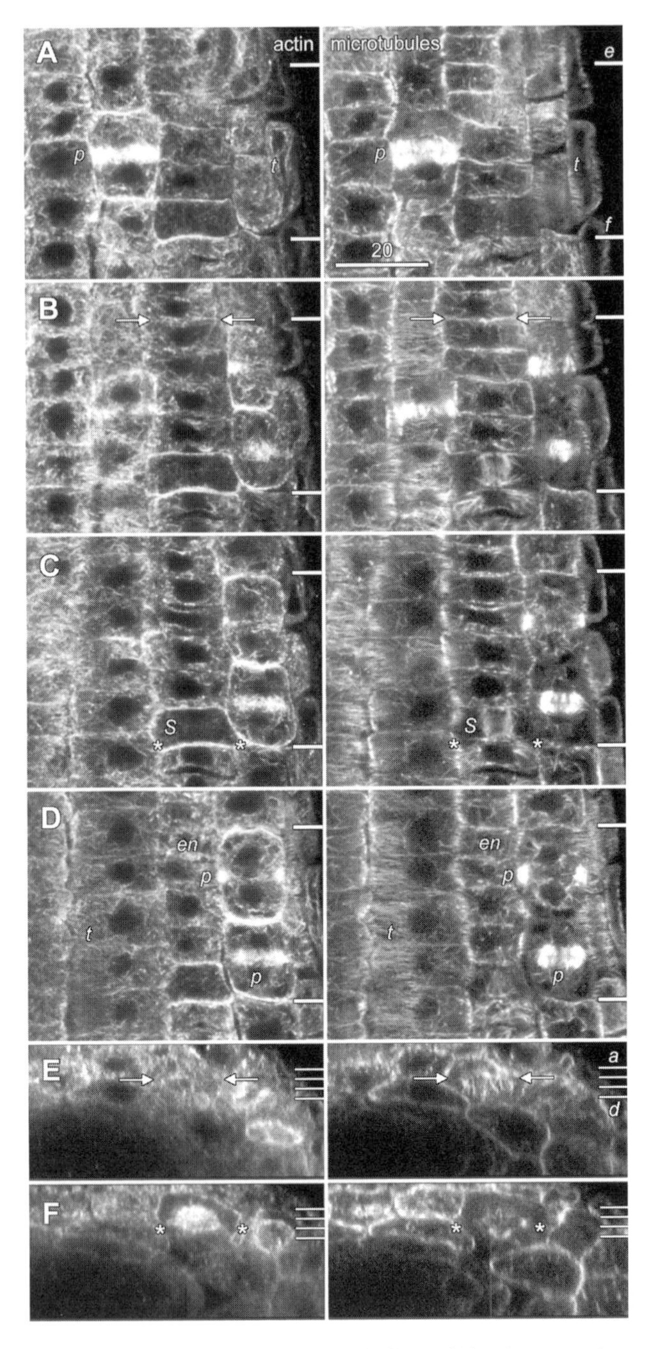

Fig. 3.1 Microfilament and microtubule organization in the *Arabidopsis* root meristem. *Arabidopsis* roots immunolabelled for tubulin and actin (Collings & Wasteneys, unpublished data) were optically sectioned by confocal microscopy. A–D Optical sections separated by 2.10 μm show extensive cytoskeletal

active secretion (reviewed by Wasteneys & Galway, 2003). Hepler *et al.* (2002) state that microtubules and microfilaments 'show a similar dynamic rearrangement during cytokinesis'. Just how similar is a difficult but important question to address. Some studies demonstrate that rearrangements of microtubules and microfilaments, albeit similar, are not identical. Electron micrographs show that phragmoplast microtubules are more highly ordered than are phragmoplast microfilaments (Kakimoto & Shibaoka, 1988). Various studies also indicate that early phragmoplast microfilaments do not extend as far from the equatorial plane as microtubules (Kakimoto & Shibaoka, 1988; Liu & Palevitz, 1992; Zhang *et al.*, 1993; Collings *et al.*, in press) (Fig. 3.1A–D). Moreover, as the phragmoplast matures, depolymerization of microtubules may precede the loss of microfilaments from the centre of the phragmoplast. This generates a microtubule annulus that surrounds either a ring of microfilaments (Zhang *et al.*, 1993) or a microtubule annulus that is not as broad as the microfilament annulus (Collings *et al.*, in press).

3.2.3 Motor proteins during phragmoplast formation and expansion

Phragmoplast microtubules and microfilaments are both oriented appropriately to act as tracks for vesicle transport to the cell plate. Identifying phragmoplast motor proteins is therefore a pursuit of several research groups. There are many motors to choose from. Higher plants have a wide variety of motor proteins including microfilament-associated myosins and microtubule-associated kinesins. In the *Arabidopsis* genome alone, there are some 17 different myosins belonging to two different plant-specific myosin subfamilies (4 class VIII myosins and 13 class XI myosins; Berg *et al.*, 2000). *Arabidopsis* also has at least 61 different kinesins, which, according to biochemical studies and sequence analysis, represent examples from most of the numerous subfamilies, including both minus and plus end-directed kinesins (Lawrence *et al.*, 2002; Reddy & Day, 2002). Motor proteins are likely to serve two major functions during cytokinesis. By converting energy into directed movement, they are probably active in transporting vesicles to and from the cell plate. They can also act as structural proteins by binding to the cytoskeleton in either

---

Fig. 3.1 (continued)
preservation. Interphase cells in the epidermis and cortex contain random endoplasmic microfilaments and microtubules (*en*), and transverse cortical microfilaments and microtubules (*t*). Dividing cells contain microfilaments and microtubules in phragmoplasts, where microfilaments form a narrower band than the microtubules (*p*), but only microtubules in the spindle (*s*). Dividing cells also have an increased density of microfilaments adjacent to the entire plasma membrane, except in the zone of actin depletion at the division plane. (E), (F) Computer-generated vertical sections, showing radial organization of the cytoskeleton and its pattern on end walls, were calculated for the locations in the optical sections marked with short bars (indicated as *e* and *f*). The planes of the four optical sections are also marked in the vertical sections with bars (indicated as *a* and *d*). An epidermal cell undergoing mitosis (between asterisks) shows an extensive microfilament array but lacks microtubules on the end wall, while an interphase epidermal cell has microfilaments and randomly oriented microtubules at the ends of the cell (between arrows). The scale bar in (A) represents 20 μm in all images.

an inactivated state or through cytoskeletal binding domains outside of the motor regions (see Wasteneys, 2002). For example, the tail regions of six *Arabidopsis* kinesins contain the calponin-homology domain that can mediate actin binding (Reddy & Day, 2002) while the tail of the kinesin-like calmodulin-binding protein, KCBP, contains a further microtubule-binding domain (Narasimhulu & Reddy, 1998).

3.2.3.1 Vesicle transport in the phragmoplast could be kinesin-based

In expanding and differentiating cells of higher plants, the transport of organelles and vesicles seems to depend solely on actomyosin. With the exception of as yet unproven roles in vesicle transport in growing pollen tubes (Cai *et al.*, 2000) and in the movement or positioning of nuclei and chloroplasts (Sato *et al.*, 2001), microtubules apparently do not contribute to vesicle and organelle movements. In the phragmoplast, the opposite situation may occur, though unequivocal evidence is still lacking.

Phragmoplast microfilaments are oriented in the correct configuration for myosin-based vesicle delivery, with their barbed-ends towards the division plane. Several observations, however, suggest that myosin does not participate in cell plate construction. Treatments that disrupt actin do not prevent cell plate formation although plates may become misaligned (see Section 3.2.3.3). Immunolabelling studies with antibodies specific for either myosin VIII (Reichelt *et al.*, 1999) or myosin XI (Liu *et al.*, 2001) do not label the developing phragmoplast. This is, however, in contrast to earlier studies using antibodies raised against a 170 kDa plant myosin (presumably a myosin XI), which labelled punctate vesicle-like structures within the phragmoplast (Lin *et al.*, 1994). It remains possible that the configuration of myosins associated with phragmoplast vesicles obscures epitopes such that access of antibodies raised to be specific to myosins VIII and/or XI is prevented. Myosin does, however, play at least one role in organelle transport during cell division. In certain monocots, notably onion, peroxisomes aggregate at the division plane in late anaphase, and remain associated with the inner edge of the microtubule phragmoplast through the later stages of cytokinesis. Aggregation is actomyosin-dependent, as cytochalasin, latrunculin and BDM all generate cytokinetic cells with randomized peroxisomes (Collings *et al.*, in press). Peroxisome aggregation has not, however, been observed at *Arabidopsis* cell plates so it may reflect subtle differences between cell wall formation in different plant taxa.

Evidence for microtubule-specific motor activity in cell plate formation is stronger. Orientation of phragmoplast microtubules is consistent with vesicle delivery by plus end-directed motors, and an *Arabidopsis* kinesin, AtPAKRP2, that was isolated from phragmoplasts fractionates along with endomembranes and vesicles. Furthermore, immunolabelling for this kinesin gives a punctate pattern that concentrates at the division plane, but whose localization is disrupted by brefeldin treatments, consistent with localization on secretory vesicles (Lee *et al.*, 2001). Unfortunately, this is the only direct evidence for a role of molecular motors in the delivery of vesicles to the phragmoplast. The association of vesicles with microtubules in syncytial cell phragmoplasts of *Arabidopsis* endosperm, mediated by kinked, rod-like structures that resemble kinesins when viewed in high resolution electron micrographs (Otegui *et al.*, 2001), has yet to be corroborated by either biochemical or immunological data.

3.2.3.2 Structural MAPs and kinesins function in phragmoplast formation and expansion

Two structural microtubule-associated proteins (MAPs) associate with and probably have critical roles in organizing the phragmoplast. Antibodies to tobacco MAP65 proteins (NtMAP65-1) label the zone of overlapping phragmoplast microtubules (Smertenko *et al.*, 2000), suggesting that these proteins act as cross-linkers of anti-parallel microtubules. Recent work also demonstrates that the 217 kDa MOR1/GEM1 protein associates with phragmoplast microtubules (Whittington *et al.*, 2001; Twell *et al.*, 2002; see Chapter 1). Whereas cell division continues in the *mor1-1* and *mor1-2* temperature-sensitive alleles at restrictive temperature, the stronger *gemini* alleles of *MOR1* in *Arabidopsis* are homozygous-lethal and generate gametophytic defects (Twell *et al.*, 2002). Cell plates do not form properly in *gem1* microspores, suggesting MOR1/GEM1 is essential for normal phragmoplast formation and function (Twell *et al.*, 2002). A tobacco homologue of MOR1, named TBMP200, was purified from phragmoplast-enriched extracts and has been shown to cross-link microtubules *in vitro* (Yasuhara *et al.*, 2002). MOR1/GEM1's distribution in the phragmoplast is still not clear. Antibodies generated to a C-terminal expression fragment localize to the cell plate zone (Twell *et al.*, 2002) whereas antibodies generated to an N-terminal peptide localize along the entire length of phragmoplast microtubules (Wasteneys, unpublished data).

The combination of opposing forces created by plus end- and minus end-directed motors allows the formation and stabilization of the anti-parallel microtubule arrays of the phragmoplast (Liu & Lee, 2001). The 125 kDa tobacco kinesin-related protein, TKRP125, is a bimC kinesin. It was isolated from phragmoplasts of synchronized tobacco BY-2 cells, and localizes to the central overlapping region. Antibodies against TKRP125 prevent the sliding expansion of the interdigitated microtubule arrays in *in vitro* assays (Asada *et al.*, 1997). TKRP125 homologues occur in all plants examined so far including *Arabidopsis* (Liu & Lee, 2001; Lawrence *et al.*, 2002; Reddy & Day, 2002). Spatial and functional data for TKRP125 are consistent with the activities of other bimC family members. These kinesins generate an outward force between opposing arrays of microtubules in the overlapping region of animal spindles and form homotetramers with motor domains at the opposing ends (Lawrence *et al.*, 2002). Another plus end-directed motor found at the phragmoplast midplane in *Arabidopsis*, AtPAKRP1, may also contribute to phragmoplast stability (Lee & Liu, 2000; Lee *et al.*, 2001).

Minus end-directed kinesins from the C-terminal motor subfamily act in the opposite direction to bimC kinesins to help stabilize the antiparallel array. Such kinesins, including katA, katB and katC in *Arabidopsis*, localize to the phragmoplast but have not been characterised biochemically (Mitsui *et al.*, 1993; Liu *et al.*, 1996). The kinesin-like calmodulin-binding protein (KCBP) also associates with the phragmoplast (Bowser & Reddy, 1997). This minus end-directed, C-terminal motor domain kinesin was isolated as a calmodulin-binding protein (Reddy *et al.*, 1996), and is found in all higher plants examined so far but not in other kingdoms. Motor activity in KCBP is regulated by calcium/calmodulin, so that in the presence of calcium, the motor is inactive. Antibodies raised against the calmodulin-binding domain of KCBP prevent calmodulin-binding, giving a constitutively active motor even in the presence of calcium. Significantly, microinjection of these antibodies into *Tradescantia virginiana* stamen hair cells results in aberrant cell division, suggesting that KCBP activity is downregulated during telophase (Vos *et al.*, 2000). KCBP was also identified in screens of *Arabidopsis* trichome branching mutants, and plays a role in the control of branching (Oppenheimer *et al.*, 1997), but this role may be unrelated to its functioning in cell division.

3.2.3.3 Expansion of the phragmoplast and cell plate requires both kinesins and myosins

Recent analysis has demonstrated that kinesins are also required for phragmoplast expansion. The seedling-lethal *hinkel* mutant of *Arabidopsis* has defective cytokinesis that often results in multinucleate cells. *HIK* encodes a member of a plant-specific subfamily of kinesins with an N-terminal motor, and is presumed to be plus end-directed. The *HIK* gene's expression is cell-cycle dependent, and HIK is found only at the phragmoplast midplane. In the *hik* mutant, the central microtubules of the second stage phragmoplast fail to depolymerize and secretory vesicle localization occurs normally. It is therefore assumed that the HIK kinesin is required for microtubule turnover rather than vesicle transport (Strompen *et al.*, 2002). The NACK1 protein of tobacco is a HIK homologue and also associates with the cell plate. Binding of NACK1 to NPK1, a MAP kinase kinase kinase, is required for the correct expansion of the phragmoplast and cell plate. Over-expression of a truncated NACK1 that contains the NPK1-binding domain but lacks the motor domain, prevents microtubule depolymerization and gives a similar cytokinesis-defective phenotype (Nishihama *et al.*, 2001, 2002) as does virus-induced silencing of NACK1 in tobacco, or T-DNA knockouts of the NACK1 homologue (HIK) in *Arabidopsis* (Nishihama *et al.*, 2002).

Actomyosin activity is essential for cell plate expansion. Treatments that impair microfilament organization, such as cytochalasins (Palevitz, 1987), or injections of the actin monomer-binding protein profilin (Valster *et al.*, 1997), disrupt the formation and correct alignment of the cell plate, as do treatments with the myosin antagonists BDM and ML-7 (Molchan *et al.*, 2002). Microfilaments extend between the phragmoplast and plasma membrane (Endlé *et al.*, 1998; Molchan *et al.*, 2002; Collings *et al.*, in press). Thus, one of the functions of microfilaments during cytokinesis seems to be to direct the expansion of the phragmoplast, and hence the cell plate, to the sites on the parent plasma membrane that are marked by the zone of microfilament depletion (Schmit, 2000).

3.2.4 Cytoskeletal mutants defective in cytokinesis

Several cytokinesis-defective mutants have been isolated. The four *PILZ* genes of *Arabidopsis*, along with *KIESEL*, are required for the folding of α- and β-tubulin and the formation of tubulin dimers. Loss of function alleles of these genes cannot produce microtubules and, not surprisingly, are blocked in cell division and are embryo-lethal (Steinborn *et al.*, 2002). Laurie Smith's group has also characterized plant cytokinesis mutants from maize, including *tangled*, *discordia*, *pangloss* and *brick*, that all show similar phenotypes to those induced by actin disruption (Gallagher & Smith, 2000). *Tangled* has defective longitudinal divisions, suggesting that the normal mechanism for specifying specialized division planes does not operate properly. The TAN protein has microtubule-binding properties and is a homologue of the adenomatous polyposis coli (APC) tumour-suppressing protein (Smith *et al.*, 2001). APC proteins are activators of a guanine nucleotide exchange factor that in turn activates a small GTPase. The discovery of TAN shows that small GTPases may play roles in specifying division planes in plant cells.

3.3 Setting up for axial growth: distinguishing lateral and end walls

For axial growth, end walls and lateral walls must have distinct physiological and mechanical properties. End walls expand relatively little or sometimes not at all, while lateral walls generally extend to become many times their original length. *Arabidopsis* root cells have perfect anisotropy, with expansion limited to the lateral walls, and no appreciable expansion of the end walls. In this sense, the last cross walls to form by transverse divisions never become primary walls and default immediately to secondary wall status (by definition, secondary walls are not extensible). End walls are therefore fundamentally different to the lateral walls of the same cells, and are likely to differ considerably in composition, both of polysaccharide material and wall proteins that include loosening enzymes. Differences in the organization and dynamic properties of the cytoskeleton reflect these differences.

3.3.1 The cytoskeleton at end walls of elongating cells

Most studies of cytoskeletal organization during cell elongation have focused on lateral walls, where the majority of growth occurs. End walls have been largely ignored, partly because imaging these is problematic. One extensive analysis by electron microscopy compared microtubule patterns in radially expanding and non-expanding end walls of *Azolla pinnata* root primordia. Microtubules were oriented transverse to the major axis of expansion at end walls but were radially oriented at non-expanding end walls (Busby & Gunning, 1983). We examined microtubules and microfilament distribution in elongating characean algal internodal cells by confocal microscopy and observed a sharp shift from abundant transverse cortical microtubules at the lateral walls to sparse and randomly oriented microtubules at the end walls (Wasteneys *et al.*, 1996). We also found that subcortical actin bundles extended around the ends of these giant cells but were unable to detect cortical microfilaments as observed along the lateral walls. In higher plants, however, microfilaments appear to be abundant at the end walls of expanding cells. Using sectioning analysis of *Arabidopsis* and maize roots, Baluška and coworkers reported that the apical and basal ends of root cells accumulate microfilaments, and that in dwarf mutants and latrunculin-treated plants, this actin array is not found (Baluška *et al.*, 2001a,b). Moreover, antibodies specific to myosin VIII were reported to concentrate at the plasma membrane of newly formed transverse cell walls of maize and *Arabidopsis* roots (Reichelt *et al.*, 1999; Šamaj *et al.*, 2000), suggesting a role in anchoring microfilaments rather than a motile function. Tissue sectioning, however, is not optimal for determining the overall pattern of cytoskeletal organization in such tissues and can only reveal specific patterns on cell faces that are tangential to the section plane.

We have recently carried out similar analysis of cytoskeletal distribution in intact *Arabidopsis* roots using double immunolocalization and confocal reconstructions (Collings & Wasteneys, unpublished data). In both the meristem (Fig. 3.1) and the distal elongation zone (Fig. 3.2), the cortical arrays of both microtubules and microfilaments are transversely aligned adjacent to the side walls. This is clearly visible in epidermal (Fig. 3.1A; *t*) as well as cortex cells (Fig. 3.1D; *t*). Interphase cells contain extensive endoplasmic microfilaments and microtubules, but once cells reach the elongation zone, endoplasmic microtubules (but not microfilament bundles) are no longer detected (Fig. 3.2). Using confocal microscopy, we can reconstruct the organization of the cytoskeleton along the end walls and there, microtubules appear to have a random orientation pattern (Fig. 3.1E; arrows).

3.4 Establishing axial growth

It is too simplistic to state that the cell cycle stops when terminal elongation growth begins. Not all tissues stop dividing a set distance from the meristem, though all must maintain the same rate of elongation (Section 3.6). In roots, the pericycle tissue layer continues to divide throughout the elongation zone (Tobias Baskin & Brian Gunning, personal communications), perhaps reflecting how these cells, which later give rise to lateral roots, remain small and cytoplasmically dense. In many expanding cells, ploidy levels increase by DNA endoreduplication, a mechanism that may enable cell enlargement. Presumably, division could also achieve this but endoreduplication may be optimal at sites distal to the meristem, where diploidy does not need to be maintained. Ethylene and gibberellins control endoreduplication in hypocotyls (Gendreau *et al.*, 1999) but recent evidence also links brassinosteroids and possibly abscisic acid to this process via the topoisomerase VI complex (Sugimoto-Shirazu *et al.*, 2002; Yin *et al.*, 2002). The formation of large central vacuoles is also a process that enables cells to achieve massive size,

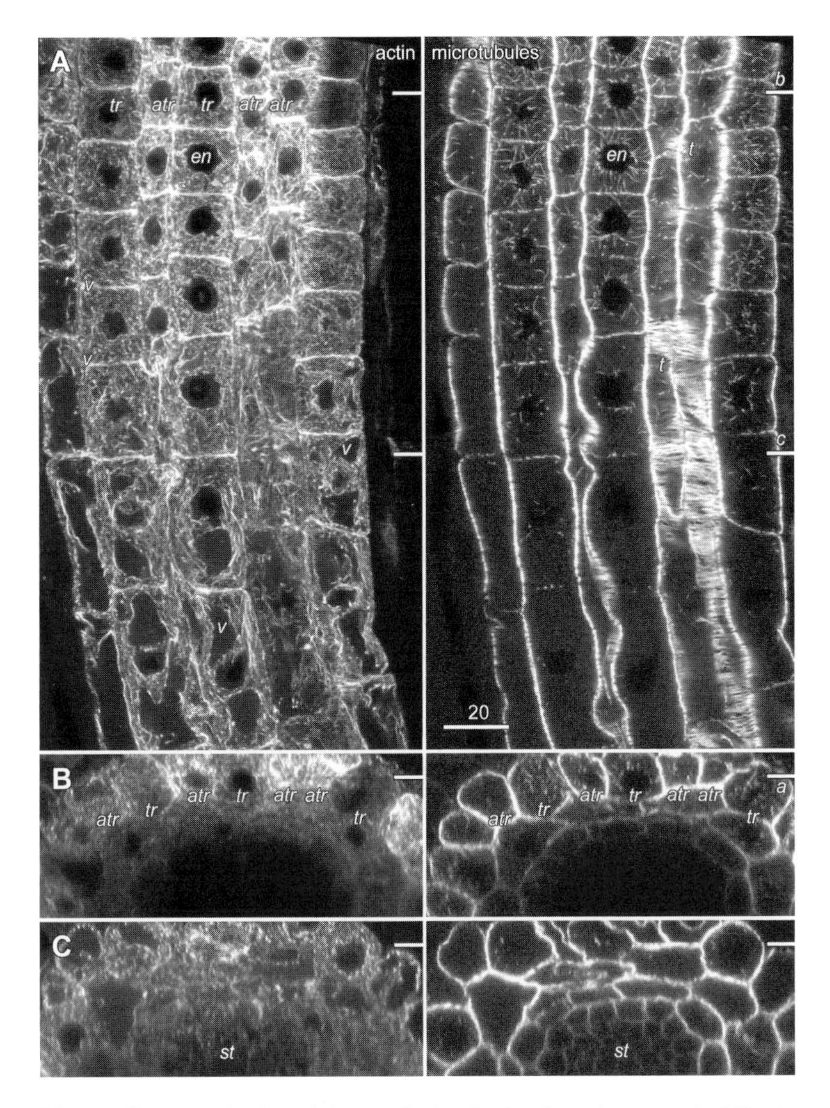

Fig. 3.2 Microfilament and microtubule organization in the elongation zone. *Arabidopsis* roots immunolabelled for tubulin and actin were optically sectioned by confocal microscopy. (A) An optical section shows that changes occur in the cytoskeleton as cells begin to elongate. Although cortical microtubules remain transverse (*t*), endoplasmic microtubules (*en*) disappear. Endoplasmic microfilaments are retained, but trichoblast cell files (*tr*) consistently contain fewer microfilaments than atrichoblast files (*atr*) especially in the distal parts of the elongation zone. Rapid cell elongation also coincides with the development of vacuoles (*v*). (B), (C) Computer-generated transverse sections show the radial organization of the cytoskeleton. They confirm the differences in microfilament organization in trichoblasts and atrichoblasts, and confirm the loss of endoplasmic microtubules between the distal elongation zone (B) and elongation zone (C). They also show the marked increase in antibody penetration in the elongation zone including into the central stele of the root (*st*). The scale bar in (A) represents 20 μm in all images.

since these compartments, which are essential for regulating turgor, occupy most of the volume. The efficient mixing of the thin layer of cytoplasm at the cell periphery depends on the myosin motor activity along microfilament networks (Wasteneys, 2002).

3.4.1 A transverse cortical microtubule array is essential for axial growth

Establishing a transverse cortical microtubule array during the onset of rapid expansion is a critical feature of axial growth. This is a process of both dispersal and alignment that is probably mediated by a combination of mechanisms including microtubule nucleation, assembly, severing, cross-linking, stabilization and selective disassembly. Candidate proteins and protein complexes that fulfil these activities are rapidly being discovered. One recent model proposes that severing of nucleating templates from the slow-growing ends of microtubules, and motor-dependent movement of these templates along microtubules is one way to generate a dispersed microtubule array throughout the cortex (Wasteneys, 2002). This dispersal process begins in early G1-phase of the cell cycle as a perinuclear array. The requirement for severing in the process is inferred from mutational analysis of the katanin p60 subunit homologue AtKSS (McClinton *et al.*, 2001), whose ATP-dependent severing activity has recently been demonstrated (Stoppin-Mellet *et al.*, 2002). Mutants with defective katanin p60 fail to produce a transverse cortical microtubule array during the onset of elongation, resulting in loss of growth anisotropy (Bichet *et al.*, 2001; Burk *et al.*, 2001), problems with cell fate specification (Webb *et al.*, 2002) and wall defects (Burk *et al.*, 2001; Burk & Ye, 2002). Just how these mutations affect the dynamic properties and ability of microtubules to function properly remains to be determined. Nucleating complexes are likely to include the recently identified Spc98p homologue (Erhardt *et al.*, 2002) and γ-tubulins, whose mutational analysis in plant cells is anticipated.

Selective stabilization of microtubules at the plasma membrane requires the activity of MAPs. The high molecular mass MOR1 protein, a DIS1-TOGp-XMAP215 homologue, is the strongest contender for this role. Two mutant alleles with single point mutations, *mor1-1* and *mor1-2*, generate temperature-dependent disorganization of the cortical arrays, left-handed twisting of organs (Section 3.6.2) and loss of growth anisotropy (Whittington *et al.*, 2001). MOR1 may act as a key component of a protein complex, with multiple binding partners. Mechanisms by which MOR1 stabilizes cortical microtubules have been proposed (Wasteneys, 2002), including the obvious cross-linking of microtubules to the plasma membrane. In xenopus egg extracts it has been shown that MOR1's homologue, XMAP215, stabilizes microtubules by opposing the activity of the kinesin XKCM1 (Tournebize *et al.*, 2000; Hussey & Hawkins, 2001). Whether MOR1 interacts with homologues of XKCM1 in plant cells is one avenue to explore in understanding the balanced turnover of microtubules that is required to maintain a fully functional cortical microtubule array.

3.4.2 *Microtubules and their relationship with cellulose microfibrils and xyloglucans*

Over the years there has been considerable acceptance, despite only limited evidence, for the cellulose synthase constraint model. The model proposes that cortical microtubules, through their interaction with the plasma membrane, constrain the movement of cellulose synthase complexes so as to control the orientation of cellulose microfibrils. Some recent studies have raised doubts about this model. Fisher and Cyr (1998) perturbed microtubule organization in tobacco BY-2 culture cells using drugs that reduced cellulose synthesis and Sugimoto *et al.* (2000) showed that cellulose microfibrils in *Arabidopsis* roots are more uniformly aligned than microtubules, as cells enter elongation. The *Arabidopsis fra2* mutant has disordered microtubules and also has somewhat altered cellulose microfibril patterns. Nevertheless, the microfibril texture is not as disturbed as would be predicted from the grossly disturbed microtubule organization. Furthermore, the *fra2* mutants have significantly depleted cellulose levels, a condition that alone generates disordered microfibrils in dichlorobenzonitrile-treated seedlings (Sugimoto *et al.*, 2001), and the *rsw1-1* (Sugimoto *et al.*, 2001) and *kobito* (Pagant *et al.*, 2002) mutants.

New data demonstrate that disturbance or removal of microtubules causes radial swelling without altering the parallel orientation of cellulose microfibrils (Sugimoto *et al.*, 2003). Himmelspach *et al.* (2003) have extended these studies, showing that cellulose microfibrils can re-establish parallel transverse order during microtubule disruption in the *mor1* mutant, even when a template of transverse microfibrils is destroyed by dichlorobenzonitrile treatment. Together, these studies in *Arabidopsis* roots suggest that: (1) cellulose microfibril orientation is largely self-ordered; (2) transverse orientation of both cortical microtubules and cellulose microfibrils is essential for axial growth; and (3) cortical microtubules do not directly regulate the movement of cellulose synthase complexes.

In exploring the cytoskeleton's role during axial growth, most emphasis has been placed on the relationship between cortical microtubules and cellulose microfibril alignment in lateral walls (Baskin, 2001). Yet it is now clear that this is not the only process responsible for maintaining anisotropy. The *rsw4* and *rsw7* mutants have substantially normal microtubule and cellulose microfibril organization and yet undergo radial swelling under restrictive conditions (Wiedemeier *et al.*, 2002). Relatively little focus has been given to how the cytoskeleton might modulate secretion of other wall components or the activity of wall loosening enzymes. Recent work, however, links microtubule organization and xyloglucan metabolism, though microtubule activity seems to be more responsive than regulatory. Takeda *et al.* (2002) demonstrated that xyloglucan application to pea stems can suppress cell elongation and stimulate a shift in microtubule orientation from transverse to longitudinal, while application of an oligosaccharide derivative of xyloglucan stimulates elongation. Some results of this study are notable from the perspective of microtubule organization. Like the findings of gravitropic bending studies (Section 3.6.1), neither microtubule reorientation from transverse to longitudinal nor its maintenance in a transverse direction controlled the growth response. Taxol stabilization of transverse microtubule orientation did not affect the ability of the xyloglucan application to reduce growth, while longitudinal microtubule reorientation generated by the kinase inhibitor 6-dimethylaminopurine did not prevent growth stimulation by the oligosaccharide derivative. Nevertheless, taxol treatments enhanced the elongation stimulated by oligosaccharide application, and 6-dimethylaminopurine-treatments enhanced oligosaccharide-induced growth inhibition. These results support the concept that cellulose orientation is largely self-ordered and dependent on the correct levels of polymer synthesis, while microtubule coalignment with microfibrils contributes to, but is not sufficient to generate axial growth. The results also support the concept that microtubule orientation depends on cues from the cell wall (Fisher & Cyr, 1998; Wasteneys, 2000; Baskin, 2001).

3.4.3 Does the cytoskeleton regulate wall polysaccharide and protein composition?

Axial growth requires that different processes act simultaneously on end and lateral walls. At the end walls, loosening needs to be greatly restricted or prevented altogether. At lateral walls, loosening activity must be upregulated relative to that at the end walls and yet restricted by the mechanical properties of the wall to expand in only one direction. Constraining expansion of end walls while maximizing loosening of lateral walls requires differential synthesis and secretion of matrix polysaccharides as well as wall-specific enzymes and other proteins. Current models of wall extensibility consider the interactions between xyloglucans and cellulose microfibrils to be the major load-bearing mechanism, though the importance of pectic polysaccharides is also recognized (Cosgrove, 2001). The ratio of xyloglucan to cellulose, and perhaps also the amount and activity of xyloglucan endotransglycosylase, are critical determinants of differential growth on end and lateral walls, and between growing and non-growing cells. We envisage that wall enzymes, including xyloglucan endotransglycosylases and endoglucanases, but also proteins like expansins are likely to differ in abundance and/or activity at the lateral and end walls. This can now be tested by immunocytochemistry. Arabinogalactan proteins (AGPs) are also likely to play important roles in elongation growth (van Hengel & Roberts, 2002). Interestingly, some AGPs influence microtubule organization. This has been shown for AGPs in the case of the *reb1* mutant, in which reduced AGP levels in trichoblasts lead to bulging cells with disturbed microtubule organization (Andème-Onzighi *et al.*, 2002).

3.4.4 Hormones, cytoskeleton and wall extensibility

As demonstrated by Baskin *et al.* (1999), environmental conditions dictate the degree of anisotropy. In response to abiotic cues, hormones reinforce or reduce anisotropy. Ethylene is frequently associated with altering the expansion axis of diffusely expanding cells. It may act to alter microtubule and microfibril orientation and change the growth direction from longitudinal to radial (Lang *et al.*, 1982) but it can also stimulate elongation, as has been shown in light-grown *Arabidopsis* hypocotyls (Smalle *et al.*, 1997) or stems of submerged rice (van der Straeten *et al.*, 2001). Gibberellins promote both anisotropic expansion and transverse cortical microtubules in aerial tissues. In barley, Wenzel *et al.* (2000) demonstrated that gibberellin-deficient dwarfs undergo considerable radial swelling during the early phase of cell expansion when microtubules are in fact still transversely aligned, supporting the idea that gibberellins act independently on wall loosening and microtubule orientation. Auxin has a similar major role on wall loosening by inducing matrix polysaccharide hydrolysis. As summarized by Cosgrove (2001) it also stimulates wall synthesis, increases wall plasticity and activates genes for wall enzymes and proton ATPases. The involvement of the cytoskeleton in polar auxin transport is considered in detail in Section 3.5.

3.4.5 How does the actin cytoskeleton contribute to cell elongation?

How are microfilaments organized during the crucial phase of cell development when rapid cell elongation begins? Bundles of microfilaments are located in the subcortex and in the transvacuolar strands where they generate cytoplasmic streaming, but a fine network of microfilaments is also found at the plasma membrane, parallel to the cortical microtubules (for example, Collings *et al.*, 1998; Blancaflor, 2000; Collings & Allen, 2000). The presence of this network in diffusely expanding cells is consistent with the actin cytoskeleton's presence near sites of rapid exocytosis during tip growth and wound wall formation (Wasteneys & Galway, 2003). The actin cytoskeleton does not undergo significant modification as elongation commences or as cells undergo rapid vacuolation. By comparison, the endoplasmic microtubules rapidly disappear (Fig. 3.2). Microfilaments are consistently more abundant in atrichoblasts than in trichoblasts (Fig. 3.2A), a result confirmed in computer-generated reconstructions (Fig. 3.2B) but opposite to that reported previously (Baluška *et al.*, 2001b).

Disruption of the actin cytoskeleton in *Arabidopsis* and maize with low concentrations of cytochalasin or latrunculin reduces cell elongation (Baskin & Bivens, 1995; Baluška *et al.*, 2001b). Actin disruption also generates a swollen root phenotype (Baskin & Bivens, 1995; Blancaflor, 2000) as does the myosin inhibitor 2,3-butanedione 2-monoxime (BDM) (Baskin & Bivens, 1995). This phenotype is similar to the effects of drug-induced microtubule depolymerization, the inhibition of exocytosis with brefeldin or monensin, or the inhibition of cellulose synthesis. The phenotype is not, however, induced by a large range of other metabolic or growth inhibitors (Baskin & Bivens, 1995). Furthermore, the transgenic over- and under-expression of certain actin-binding proteins in *Arabidopsis*, including profilin, actin depolymerization factor (ADF) and cyclase-associated protein, generate morphological changes that sometimes extend beyond simply altered growth rates and suggest effects on microtubule organization and anisotropic cell elongation

(Ramachandran *et al.*, 2000; Dong *et al.*, 2001; Barrero *et al.*, 2002). Over-expressing profilin also generated other morphological phenotypes including apparent left-handed stem twisting, and a reduction in the lobing of epidermal pavement cells (Ramachandran *et al.*, 2000), both features of mutants that show microtubule disruption (Whittington *et al.*, 2001; Thitamadee *et al.*, 2002).

How then do microfilaments contribute to cell elongation? One possible mechanism is through interactions with cortical microtubules. Numerous drug studies indicate that such interactions occur (Collings & Allen, 2000), although it is uncertain whether these interactions are direct or whether the coordination between the two systems is via an indirect pathway. Actomyosin-dependent streaming is needed to efficiently mix the thin layer of cytoplasm of vacuolated cells and to transport transcription and translation products to all parts of the expanding periphery. Furthermore, auxin transport depends on the activity of microfilament networks (Section 3.5.1) and not microtubules (Hasenstein *et al.*, 1999). But as microtubule organization is also one of the targets of auxin signalling (Takasue & Shibaoka, 1999), and as auxin promotes cell wall loosening (Taguchi *et al.* 1999), any changes in auxin levels by disruption to the actin cytoskeleton will have downstream consequences on microtubules and microtubule-dependent anisotropy.

3.5 Polar auxin transport and its regulation by the actin cytoskeleton

What factors define asymmetry in organs and individual cells, and how is this determined? Plant hormones help to define, reinforce and/or perturb cell polarity and axiality. Their effects on growth through the cytoskeleton were extensively reviewed several years ago (Shibaoka, 1994), but recent years have seen considerable advances in understanding how the cytoskeleton and plant hormones interact at a molecular level. Auxin is the best example of this, and in this section, we discuss how the cytoskeleton is involved in auxin transport and how it responds to auxin gradients.

3.5.1 Auxin transport and the chemiosmotic theory

The chemiosmotic theory of polar auxin transport (Raven, 1975) established a theoretical framework that accounts for the apical to basal movement through plants of indoleacetic acid (IAA), the most common biologically occurring auxin (Fig. 3.3A). The relative acidity of the cell wall (pH 5) causes weak acids such as auxin to protonate, forming a non-charged species that can diffuse across the plasma membrane into the cytoplasm where it then deprotonates to give a charged species that cannot re-cross the plasma membrane. Movement of auxin out of the cell is limited to an efflux carrier that is restricted solely to the basal plasma membrane, resulting in the polarised auxin transport downwards through the stem and roots (Fig. 3.3B). This theory has been validated experimentally in recent years. Eight auxin efflux carriers, encoded by members of the *PIN* gene family, have been identified in *Arabidopsis*, initially through the cloning of genes from mutants whose auxin-dependent development and/or gravitropic responses are compromised (Chen *et al.*, 1998; Gälweiler *et al.*, 1998; Luschnig *et al.*, 1998; Müller *et al.*, 1998; Friml & Palme, 2002). Localizations of these efflux carriers match the chemiosmotic theory's predictions (Fig. 3.3B). AtPIN1 is found at the lower or basal plasma membrane in shoots (relative to the shoot apical meristem), and at the lower or apical plasma membrane (towards the root apex) within the central stele of the root (Gälweiler *et al.*, 1998). AtPIN2 distributes to the basal plasma membrane of epidermal and cortical cells of the root tip (Müller *et al.*, 1998). Within the root tip, AtPIN4 may contribute to the increased auxin levels found in the quiescent zone cells (Friml *et al.*, 2002a), while AtPIN3 distributes evenly throughout the plasma membrane of root cap columella cells but redistributes to the lower side of the root after gravistimulation (Friml *et al.*, 2002b). These patterns are consistent with the paths that auxin takes through the plant. Interestingly, the auxin influx carrier AUX1 has an asymmetric distribution to the basal plasma membrane of cells within the protophloem of the root tip, again consistent with auxin movements (Swarup *et al.*, 2001) (Fig. 3.3B).

How do plants maintain these asymmetries? Separate streams of evidence indicate that the actin cytoskeleton, through its function in vesicle trafficking, helps to regulate the polarized location of auxin efflux carriers at the plasma membrane. Auxin efflux carriers undergo rapid cycling between the plasma membrane and a perinuclear endomembrane compartment (Fig. 3.3C,D). Brefeldin A, by inhibiting vesicle trafficking and endocytosis, causes a reversible build-up of PIN proteins and AUX1 within the perinuclear compartment (Steinmann *et al.*, 1999; Geldner *et al.*, 2001; Grebe *et al.*, 2002). Cycloheximide treatments that block protein synthesis do not prevent the build-up of AtPIN1, demonstrating that it cycles between the perinuclear compartment and the plasma membrane (Geldner *et al.*, 2001). Disruption of microfilaments with either cytochalasin or latrunculin has little effect on AtPIN1 localization, but cytochalasin pre-treatment prior to brefeldin A prevents AtPIN1 build-up in the perinuclear compartment, while cytochalasin addition after brefeldin A's removal prevents redirection of AtPIN1 to the plasma membrane (Geldner *et al.*, 2001). Similar results were also obtained for AtPIN3 (Friml *et al.*, 2002b). These results show that an intact and functional microfilament network is essential for both delivery to and recovery of PIN proteins from the plasma membrane (Fig. 3.3D). This functioning likely includes both myosin-dependent movement of vesicles along microfilament bundles, but may also indicate further roles for microfilaments in exocytosis and/or endocytosis.

Gloria Muday and colleagues have theorised that proteins regulating the auxin efflux carrier bind to the actin cytoskeleton (Fig. 3.3C; reviewed in Muday, 2000; Muday & DeLong, 2001). Naphthylphthalamic acid (NPA), an auxin transport inhibitor that inhibits auxin-dependent growth and gravitropism (Rashotte *et al.*, 2000), functions through high-affinity coupling to the NPA-binding protein that regulates the activity of auxin efflux carriers at the plasma membrane. If microfilaments are disrupted, NPA-binding activity recovered from plasma membrane preparations is reduced (Butler *et al.*, 1998). NPA-binding activity can also be recovered from F-actin affinity columns, and although the protein(s) responsible could not be identified, the results are consistent with the NPA-binding protein being an actin-binding protein (Hu *et al.*, 2000). Recent developments have, however, complicated the interpretation of the NPA-binding protein being a link between auxin efflux carriers and the actin cytoskeleton. NPA and other auxin transport inhibitors have a complicated mode of action that includes the brefeldin A-like suppression of efflux carrier cycling between the plasma membrane and perinuclear compartment (Geldner *et al.*, 2001), although these effects occur at significantly higher concentrations (about 200 μM) than required to block gravi-

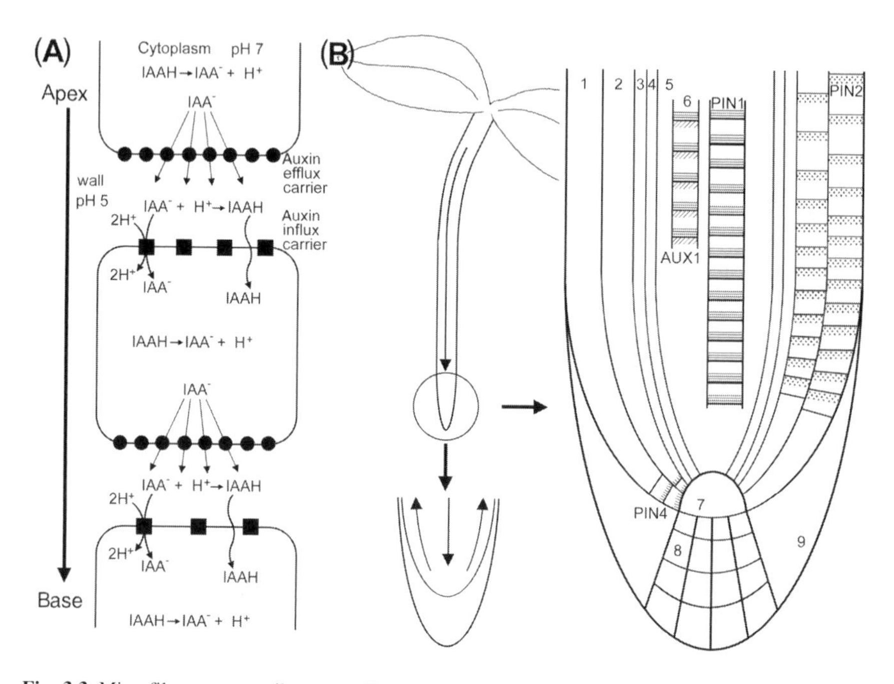

Fig. 3.3 Microfilaments contribute to cell elongation by maintaining polar auxin transport. (A) The chemiosmotic theory explains polar auxin transport through the asymmetric distribution of an auxin efflux carrier. Auxins are weak acids that occur in the membrane-permeant protonated form (IAAH) in the acidic cell wall, and which accumulate in the cytoplasm because they deprotonate to give a membrane-impermeant ion (IAA^-). The asymmetric distribution of auxin efflux carriers (circles) to the basal plasma membrane generates basipetal movement of auxin. This process is aided by the localization to the apical plasma membrane of auxin influx carriers (squares), which use the trans-membrane proton gradient to speed auxin uptake. (B) In *Arabidopsis* seedlings, basipetal auxin transport occurs through the central stele of the hypocotyl to the root tip, from where auxin circulates back through the cortical and epidermal cells. Auxin efflux carriers from the PIN family of proteins and the auxin influx carrier AUX1, whose locations match the predictions of the chemiosmotic theory, generate this pattern. In roots, AtPIN1 localises to the lower (tipward or apical) plasma membrane of vascular cells, while AtPIN2 localises to the upper (basal) plasma membrane of epidermal and cortex cells. AtPIN3 occurs in root cap cells, while AtPIN4 localises asymmetrically to the apical plasma membrane of cells adjacent to the quiescent centre. AUX1 also distributes asymmetrically to the basal ends of protophloem cells, the opposite end to AtPIN1. Cell types are: 1 = epidermis; 2 = cortex; 3 = endodermis; 4 = pericycle; 5 = vascular tissue, which includes protophloem (6); 7 = quiescent zone and cell file initials; 8 = columella; and 9 = lateral root cap.

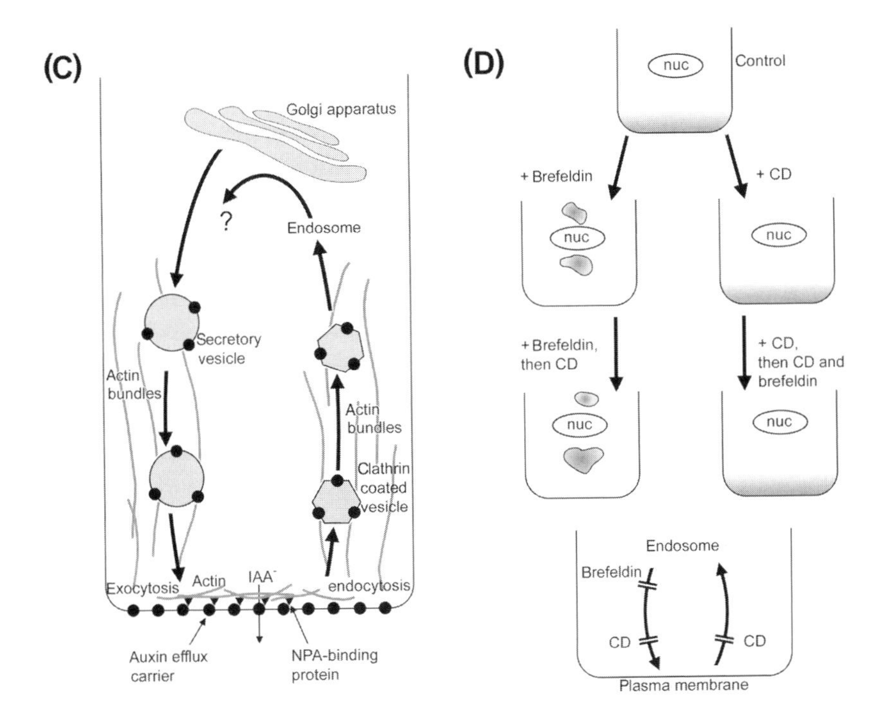

Fig. 3.3 (continued) (C), (D) Auxin efflux carriers such as AtPIN1 (circles) undergo rapid actin-dependent cycling between the plasma membrane and an endocytotic compartment, here labelled as an endosome, and may be regulated at the plasma membrane by the NPA-binding protein (triangles) and microfilaments (C). In control root cells (D), AtPIN1 (shading) distributes to the apical plasma membrane. Brefeldin A inhibits exocytosis and causes a rapid accumulation of AtPIN1 into a perinuclear endocytotic compartment, but microfilament disruption with cytochalasin (CD) has little effect on AtPIN1 localization. Serial brefeldin and cytochalasin treatments, however, demonstrate that microfilaments are required for cycling, and show that brefeldin blocks exocytosis, but microfilament disruption with cytochalasin D inhibits both exocytosis and endocytosis. This inhibition may involve both the prevention of vesicle movement along actin bundles, and/or a direct involvement in exocytosis and endocytosis.

tropic bending (about 1 μM; Rashotte *et al.*, 2000). Moreover, several *Arabidopsis* proteins involved in auxin transport, including aminopeptidases (Muday & Murphy, 2002; Murphy *et al.*, 2002), bind NPA with low affinity. These proteins have not been shown to bind to the actin cytoskeleton, and high-affinity NPA-binding proteins that interact with the actin cytoskeleton have not been identified (Muday & Murphy, 2002).

3.5.2 Important questions concerning auxin transport and the actin cytoskeleton

Why do the PIN proteins undergo rapid cycling? In their recent review of polar auxin transport, Friml and Palme (2002) suggested possible reasons for rapid cycling. Perhaps cycling allows for rapid changes in auxin efflux carrier location to be achieved or perhaps there is some sensing role to the cycling, with efflux carriers acting as membrane receptors. Alternatively, when vesicles containing PIN efflux carriers move along the actin cytoskeleton, they are also actively transporting auxin through the cell as a specific cargo. We view this third possibility as unlikely because the majority of auxin efflux carrier proteins such as AtPIN1 are found in polarized locations at the plasma membrane rather than associated with vesicles (Gälweiler *et al.*, 1998). Moreover, the acidification of secretory vesicles, which occurs before secretion in animal cells (Miesenböck *et al.*, 1998), and presumably also in plants cells, would limit auxin build-up because protonated auxin could diffuse back into the cell. Instead, we suggest that the rapid cycling of auxin efflux carriers may also contribute to their asymmetric distribution. Were targeted secretion to correctly deliver efflux carriers to one part of the plasma membrane, the rapid recovery of proteins from throughout the plasma membrane would limit diffusion of the efflux carriers through the cell. This would also eliminate the need for the efflux carriers to be tethered into the correct location through a cytoskeleton-based fixing mechanism such as the NPA-binding protein.

How are the PIN proteins targeted asymmetrically to the plasma membrane? A comparison of AtPIN1 and AUX1 location in the protophloem of roots shows that they are found at opposite ends of the cell (Gälweiler *et al.*, 1998; Swarup *et al.*, 2001) (Fig. 3.3B), confirming that cells must have multiple methods of ensuring different types of asymmetric exocytosis. How differences between the contents of exocytotic vesicles at end- and side-walls are generated and how the polar distribution of auxin efflux carriers re-establish after cell division are also unresolved questions. When a cell that has asymmetrically localized auxin efflux carriers divides, the daughter cells need to re-establish this polarization to ensure continued polar auxin transport. Initial evidence shows that AtPIN1 colocalizes with the KNOLLE protein in the cell plate, presumably without there being any distribution asymmetry (Geldner *et al.*, 2001). How and how quickly the polarized pattern re-establishes itself has not been determined.

3.5.3 Small GTPases may be a key to the shuttling of auxin efflux carriers

ROP (Rho of Plants) proteins are small GTPases that act as molecular switches. In tobacco suspension cells, the *Arabidopsis* ROP4-GFP fusion protein localises to the cell plate and cross wall. Immunolocalization suggests that ROP4 (and probably other closely related ROPs such as ROP6) is preferentially distributed at the apical and basal ends of elongating cells in *Arabidopsis* root meristems (Molendijk *et al.*, 2001). In the same study, expression of a constitutively active GFP-AtROP4 construct under an inducible promoter caused isotropic swelling of epidermal cells of hypocotyls, cotyledons and roots (Molendijk *et al.*, 2001). The identification of an end-wall associated ROP is consistent with a general ROP association with sites of active vesicle secretion, as well as the formation of fine actin networks, such as those reported at the ends of cells (Baluška *et al.*, 2001b). These ROPs could regulate the shuttling of auxin efflux carriers between their endomembrane compartment and the plasma membrane at the ends of cells (Section 3.5.1), either directly or by remodelling the actin cytoskeleton.

3.5.4 Auxin and gene expression

Auxin elicits many and diverse effects in plants. As discussed below (Section 3.6.1), auxin's involvement in gravitropism includes separate but tightly integrated effects on microtubule orientation and cell elongation. However, auxin also causes the rapid induction and repression of many different genes through multiple pathways (Guilfoyle *et al.*, 1998; Leyser, 2002) that may be modulated, at least in part, through ROP GTPases (Li *et al.*, 2001; Tao *et al.*, 2002). Some of the auxin-induced changes in gene expression directly affect the cytoskeleton. The *Arabidopsis ACT7* gene encodes an actin that is preferentially expressed in rapidly elongating tissue. *ACT7* expression responds to hormones, including auxin, abscisic acid and cytokinins. Exogenous auxin reduces *ACT7* expression along with reducing root elongation (McDowell *et al.*, 1996) but increases *ACT7* expression during auxin-dependent callus formation (Kandasamy *et al.*, 2001). Significantly, *act7-1*, a T-DNA insertional mutant, was slow to form callus in response to external auxin, suggesting a role for this actin isoform in callus formation (Kandasamy *et al.*, 2001).

3.6 Bending and twisting – the consequences of differential growth

For plant organs to grow in one direction, the different tissue layers must all elongate at the same rate. Any major differences in elongation rates between tissue layers result in buckling and rupture, with devastating consequences. Subtle differences in growth rates between cells in adjacent tissue layers can, however, alter the direction of organ growth. Gravitropic and phototropic bending responses all involve coordinated differential flank growth, in which there is evidence for cytoskeletal involvement. Uncoordinated differences in growth can lead to the helical arrangement of cell files or the twisting of whole organs. Microtubules may have a regulatory role in twisting. Mutational, transgenic and drug-dependent perturbation of microtubule organization can generate either left- or right-handed twisting of both axial and lateral organs (Table 3.1). We present, in turn, evidence for cytoskeletal involvement in bending and twisting.

3.6.1 Tropic bending responses

Tropic responses to light (phototropism), gravity (gravitropism) and other stimuli are adaptive growth responses in which the differential expansion of cells on opposite sides of organs, and not cell division, generate bending. The cytoskeleton's involvement in the perception of a gravity stimulus through amyloplast sedimentation

Table 3.1 Genotypes and treatments generating organ-twisting phenotypes in *Arabidopsis thaliana*

Genotype or treatment	Cortical microtubules	Protein targeted	Tissues targeted	Organ handedness	Ref.
wild type	transverse			ecotype-specific	1
propyzamide (low concs)	right-handed helix in epidermis	tubulin		left	2, 6
taxol (low concs)	stabilized	tubulin		left	2, 6
spr1	disordered in root endodermis and cortex; left-handed helical in root epidermis	unknown	axial organs; isotropic expansion of endodermis and cortex, elongation of epidermal cells not affected	right (axial organs)	2, 6
spr2	not determined	unknown		right (lateral organs)	2, 6
mor1-1, mor1-2 (29°C)	short and disordered	217 kDa MAP	all organs	left (loss of anisotropy after 24 h)	3, 4
lefty1	right-handed helix in root epidermis	α-tubulin 4	all organs	left	5, 6
lefty2	right-handed helix in root epidermis	α-tubulin 6	all organs	left	5, 6
35S::GFP-TUA6	transverse	overexpression of *Arabidopsis* α-tubulin 6	petioles, petals	right	6
35S::MBD-GFP	transverse	overexpression of mouse MAP4 microtubule-binding domain-GFP		right	6

References: 1, Migliaccio *et al.* (2000); 2, Furutani *et al.* (2000); 3, Whittington *et al.* (2001); 4, Wasteneys (2002); 5, Thitamadee *et al.* (2002); 6, Hashimoto (2002).

and the generation of the auxin asymmetry that results in asymmetric growth have been extensively debated, and recently reviewed (Blancaflor, 2002). The contradictory nature of many of the observations derives, in part, from the variety of tissues studied (e.g. roots, hypocotyls, coleoptiles and various types of stems). These may show different adaptations of an underlying gravitropic process. Using the bending response as a measure of signal perception also confounds analysis of the events in perception.

In this section, we consider the possible roles of the cytoskeleton in gravitropic bending responses. These responses require cell elongation, so microtubules should play a critical role. The relationship between microtubules and gravitropic bending has been extensively studied in maize coleoptiles, roots and stems. In gravistimulated coleoptiles and roots, there is a consistent change in microtubule orientation from transverse to longitudinal in the epidermal cells whose growth is inhibited. In upward bending coleoptiles these are the cells on the upper flank where auxin levels are reduced (Nick *et al.*, 1990; Himmelspach *et al.*, 1999) whereas in downward bending roots this occurs on the lower flank where auxin accumulates (Blancaflor & Hasenstein, 1995). This microtubule reorientation in coleoptiles is not coupled to growth and even occurs when the coleoptiles are prevented from growing and bending by gluing them to glass slides (Himmelspach & Nick, 2001). Similarly, Takesue and Shibaoka (1999) showed that auxin-induced longitudinal to transverse microtubule reorientation can occur when growth is suppressed by anaerobic conditions. In downward bending maize roots, microtubule reorientation does not begin until after bending has commenced and can be mimicked by applying auxin (Blancaflor & Hasenstein, 1995). These observations demonstrate that the reorientation of microtubules to the longitudinal direction has no intrinsic function in the bending response. Indeed, transverse to longitudinal reorientation of microtubules is a normal consequence of growth cessation (Liang *et al.*, 1996; Sugimoto *et al.*, 2000).

Reinforcement of transverse microtubules in the growth-stimulated cells on the outer side of the bend is potentially of greater significance to gravitropic bending but is likely to be a requirement rather than a regulatory mechanism. Indeed, microtubule depolymerization does not inhibit gravitropic bending of roots (Baluška *et al.*, 1996; Hasenstein *et al.*, 1999), though the time required for bending is unlikely to alter the mechanical properties of walls of the expanding cells. These experiments serve to demonstrate that the key to a bending response is the stimulation of growth on one flank and suppression on the other regardless of whether the stimulated growth is isotropic or anisotropic. The opposite responses of coleoptiles and roots to auxin (growth stimulation and maintenance of transverse microtubules in coleoptiles; growth inhibition and loss of transverse microtubules in roots) are intriguing.

The maize stem pulvinus, like the coleoptile, generates upward bending responses, with stimulation of growth on the lower flank by auxin accumulation (Long *et al.*, 2002). This system, however, is fundamentally different from gravi-responding roots and coleoptiles in that the pulvinal cells, which occur at nodes along the stem, remain in a quiescent state until gravistimulated. Once the stem is induced to bend upward by horizontal placement, the microtubules remain transverse in pulvinal cells on both the upper, non-elongating and lower, rapidly elongating flanks (Collings *et al.*, 1998). Here, the key to microtubules remaining transverse is the need to retain versatility. The pulvinus enables maize stems to recover vertical growth when flattened by wind, crop circle devotees (or alien spaceships) (Levengood & Talbott, 1999) but also allows for more moderate adjustments to stem attitude requiring sequential bending in more than one direction from a single node.

Several genes involved in gravity perception, signal transmission and response have also been identified through mutational strategies, and the cytoskeleton's roles in these processes dissected. As discussed in Section 3.5.1, the *PIN* family of auxin efflux carriers was first identified in agravitropic mutants in *Arabidopsis* hypocotyls and roots (Chen *et al.*, 1998; Gälweiler *et al.*, 1998; Luschnig *et al.*, 1998; Müller *et al.*, 1998), as was the auxin influx carrier AUX1 (Bennett *et al.*, 1996), and actin plays a fundamental role in the distribution of these proteins. The gene affected in the *altered response to gravity* (*arg1*) mutant of *Arabidopsis* encodes a DNA-J-like protein, with a coiled-coil region that may interact with the cytoskeleton (Sedbrook *et al.*, 1999) although this has yet to be shown conclusively.

Another gravity-response mutant with links to the cytoskeleton is the *Yin-Yang* mutant of rice. The coleoptiles of *Yin-Yang* are ten times more sensitive to cytochalasin D than wild-type coleoptiles but this sensitivity is auxin-dependent. When exposed to auxin, the actin cytoskeleton of the mutant and its gravitropic response resemble those of the cytochalasin-D-treated wild type. Despite its clear actin-based phenotype, *Yin-Yang* was isolated on the basis of its cell elongation being resistant to microtubule inhibitors, suggesting that its altered actin organization influences microtubule-dependent processes (Wang & Nick, 1998). The gene affected in *Yin-Yang* has not yet been identified.

3.6.2 Twisting

Twisting of organs is a feature of circumnutation and thigmotropic responses such as twining. Microtubule disruption can generate twisting in axial and lateral organs. That different perturbations to the microtubule cytoskeleton cause twisting that is consistently either left- or right-handed (Table 3.1), indicates the presence of some basic underlying mechanism. Can we understand the involvement of microtubules in cell expansion by studying how microtubule disruption influences twisting?

The most likely cause of twisting is in fact the differences in expansion between the outer and inner tissue layers. In the *spr1* mutant, Furutani *et al.* (2000) demonstrated that while the epidermal cells maintain normal anisotropic expansion, the cortex and endodermal cells grow isotropically. This more rapid elongation in the epidermis introduces physical stresses to which the epidermis can respond in two possible ways:

1. It can buckle, leading to tissue damage and probably seedling death.
2. If the difference in growth rates is small, the epidermal files can change their growth direction, resulting in twisting.

Although a detailed assessment has not been carried out in the *spr1* mutant, it seems likely that the twisting results from the third type of response.

All mutations and treatments that generate twisting also reduce anisotropic expansion to some extent, although this in itself is not enough to cause twisting. For example, taxol and propyzamide cause twisting but at higher concentrations produce radial swelling (Furutani *et al.*, 2000), while at its restrictive temperature, the temperature-sensitive *mor1-1* mutant gradually loses anisotropic expansion (Whittington *et al.*, 2001). The *lefty1* and *lefty2* mutants have thicker roots than wild-type, and *lefty1lefty2* (Thitamadee *et al.*, 2002) and *spr1spr2* (Furutani *et al.*, 2000) double mutants have strong radial swelling phenotypes. Many radial swelling mutants do not, however, show any twisting growth. Mutants with reduced cellulose levels, or plants in which microfilaments have been disrupted, show radial swelling but not twisting (although *Arabidopsis* plants over-expressing profilin show radially swollen hypocotyls that may show twisting – see Fig. 3.3 in Ramachandran *et al.*, 2000). Mutations affecting the microtubule severing protein katanin p60 also do not generate twisting. While this seems at odds with the strong correlation between twisting and microtubule defects, cellulose levels are significantly depressed in the *fra2* mutant (Burk *et al.*, 2001), unlike the left-twisting *mor1-1* mutant (Sugimoto *et al.*, 2003). Thus, mutations and treatments affecting microtubules, but not cellulose synthesis, can generate twisting.

What determines the direction of twisting? Hashimoto and coworkers observed that microtubule orientation is right handed in the left-twisting *lefty* mutants, and left-handed in the right-twisting *spr1* mutant, and have hypothesized that it is the handedness of cortical microtubules in the epidermis that determines the handedness of twisting (Furutani *et al.*, 2000; Hashimoto, 2002; Thitamadee *et al.*, 2002). We note, however, that if the orientation of the root is taken into account, the cortical microtubules are almost perpendicular to the gravity vector in these mutants, and although there is no evidence to suggest that microtubules sense gravity, this possibility should not be discounted. Furthermore, the model of Hashimoto and coworkers fails to explain the strong left-twisting of *mor1* mutants in which microtubules become disordered with no preferred handedness, and it also cannot explain the strong right-handed twisting caused by the transgenic expression of the MBD-GFP and GFP-TUA6 fusion proteins.

We suggest that various radial gradients generate at least some of the twisting phenotypes, acting in concert with an inherent torsional handedness of the whole organ. This torsion could be generated by the staggered helical division patterns that follow the periclinal divisions of the lateral root cap/epidermis initial cells. Defects in these divisions generate strong twisting in the *tornado* mutants (Cnops *et al.*, 2000). This inherent torsion might also explain why microtubules consistently reorient from transverse to longitudinal via a right-handed helix as root cells of maize and *Arabidopsis* (Liang *et al.*, 1996; Sugimoto *et al.*, 2000) cease elongation. Inward gradients, or treatments that first reduce anisotropic growth in the outer tissues, will generate left-handed twisting of cell files. This is consistent with *mor1's* temperature-dependent phenotype (a temperature gradient) and drug treatments such as propyzamide and taxol that diffuse inward from the epidermal layer (Furutani *et al.*, 2000). We would also predict that in the *lefty* mutants, expression of the mutated α-tubulin 4 and α-tubulin 6 would be greater in the epidermis than in the inner tissues, although this has not yet been determined. Conversely, right-handed twisting phenotypes may be generated by outward flowing gradients. This is clearly the case for the *spr1* mutant, with endodermis and cortex cells broader and shorter than the epidermal cells (Furutani *et al.*, 2000). Clearly, however, more detailed analyses of microtubule handedness in twisting organs other than roots are also required, because agar-based seedling culture introduces considerable artefacts. Growth on hard agar surfaces means that root twisting will cause skewing of roots in one direction across the plate, whereas similarly affected soil-grown roots would grow in a helical direction like a corkscrew (Buer *et al.*, 2000).

3.7 Conclusions and future perspectives

Axial growth is an inherent feature of plant function and essential at all stages of development, beginning with the zygotic embryo and continuing through until all reproductive organs are complete. Anisotropic expansion enables plants to produce organs of unlimited complexity, with features that help them compete for light, bend against gravity or towards light, minimize wind damage, twine up other plants, penetrate seemingly impenetrable soils and even optimize pollination strategies. We have examined the role microtubules and microfilaments play in axial growth, from the onset of cell plate formation in telophase, through the establishment of axial growth and on to more complex processes of differential expansion. Recent investigations assisted by the genome projects have uncovered a plethora of new factors, whose functions will begin to become clear as thoughtful investigations are carried out. Unfortunately, picking through the morass of confusing results will undoubtedly continue for some time.

Acknowledgements

Geoffrey Wasteneys and David Collings are supported by ARC Discovery Project Grants DP0208872 and DP0208806 respectively. We thank Moira Galway, Tobias Baskin and Takashi Hashimoto for helpful discussions.

References

Andème-Onzighi, C., Sivaguru, M., Judy-March, J., Baskin, T.I. & Driouich, A. (2002) The *reb1-1* mutation of *Arabidopsis* alters the morphology of trichoblasts, the expression of arabinogalactan-proteins and the organization of cortical microtubules, *Planta*, **215**, 949–958.

Asada, T., Kuriyama, R. & Shibaoka, H. (1997) TKRP125, a kinesin-related protein involved in the centrosome-independent organization of the cytokinetic apparatus in tobacco BY-2 cells, *J. Cell Sci.*, **110**, 179–189.

Baluška, F., Busti, E., Dolfini, S., Gavazzi, C. & Volkmann, D. (2001a) *Lilliputian* mutant of maize lacks cell elongation and shows defects in organization of actin cytoskeleton, *Dev. Biol.*, **236**, 478–491.

Baluška, F., Hauskrecht, M., Barlow, P.W. & Sievers, A. (1996) Gravitropism of the primary root of maize: a complex pattern of differential cellular growth in the cortex independent of the microtubular cytoskeleton, *Planta*, **198**, 310–318.

Baluška, F., Jasik, J., Edelmann, H.G., Salajová, T. & Volkmann, D. (2001b) Latrunculin B-induced plant dwarfism: plant cell elongation is F-actin-dependent, *Dev. Biol.*, **231**, 113–124.

Barrero, R.A., Umeda, M., Yamamura, S. & Uchimiya, H. (2002) Arabidopsis CAP regulates the actin cytoskeleton necessary for plant cell elongation and division, *Plant Cell*, **14**, 149–163.

Baskin, T.I. (2001) On the alignment of cellulose microfibrils by cortical microtubules: a review and a model, *Protoplasma*, **215**, 150–171.

Baskin, T.I. & Bivens, N.J. (1995) Stimulation of radial expansion in Arabidopsis roots by inhibitors of actomyosin and vesicle secretion but not by various inhibitors of metabolism, *Planta*, **197**, 514–521.

Baskin, T.I., Meeks, H.T.H.M., Liang, B.M. & Sharp, R.E. (1999) Regulation of growth anisotropy in well-watered and water-stressed maize roots. II. Role of cortical microtubules and cellulose microfibrils, *Plant Physiol.*, **118**, 681–692.

Bednarek, S.Y. & Falbel, T.G. (2002) Membrane trafficking during plant cytokinesis, *Traffic*, **3**, 621–629.

Bennett, M.J., Marchant, A., Green, H.G., *et al.* (1996) Arabidopsis AUX1 gene: a permease-like regulator of root gravitropism, *Science*, **273**, 948–950.

Berg, J.S., Powell, B.C. & Cheney, R.E. (2000) A millennial myosin census, *Mol. Biol. Cell*, **12**, 780–794.

Bichet, A., Desnos, T., Turner, S., Grandjean, O. & Höfte, H. (2001) *BOTERO1* is required for normal orientation of cortical microtubules and anisotropic cell expansion in *Arabidopsis*, *Plant J.*, **25**, 137–148.

Blancaflor, E.B. (2000) Cortical actin filaments potentially interact with cortical microtubules in regulating polarity of cell expansion in primary roots of maize (*Zea mays* L.), *J. Plant Growth Regul.*, **19**, 406–414.

Blancaflor, E.B. (2002) The cytoskeleton and gravitropism in higher plants, *J. Plant Growth Regul.*, **21**, 120–136.

Blancaflor, E.B. & Hasenstein, K.H. (1995) Time course and auxin sensitivity of cortical microtubule reorientation in maize roots, *Protoplasma*, **185**, 72–82.

Bowser, J. & Reddy, A.S.N. (1997) Localization of a kinesin-like calmodulin-binding protein in dividing cells of arabidopsis and tobacco, *Plant J.*, **12**, 1429–1437.

Buer, C.S., Masle, J. & Wasteneys, G.O. (2000) Growth conditions modulate root-wave phenotypes in Arabidopsis, *Plant Cell Physiol.*, **41**, 1164–1170.

Burk, D.H., Liu, B., Morrison, W.H. & Ye, Z.-H. (2001) A katanin-like protein regulates normal cell wall biosynthesis and cell elongation, *Plant Cell*, **13**, 807–827.

Burk, D.H. & Ye, Z.H. (2002) Alteration of oriented deposition of cellulose microfibrils by mutation of a katanin-like microtubule-severing protein, *Plant Cell*, **14**, 2145–2160.

Busby, C.H. & Gunning, B.E.S. (1983) Orientation of microtubules against transverse cell walls in roots of *Azolla pinnata* R. Br., *Protoplasma*, **116**, 78–85.

Butler, J.H., Hu, S., Brady, S., Dixon, M.W. & Muday, G.K. (1998) *In vitro* and *in vivo* evidence for actin association of the naphthylphthalamic acid-binding protein from zucchini hypocotyls, *Plant J.*, **13**, 291–301.

Cai, G., Romagnioli, S., Moscatelli, A., *et al.* (2000) Identification and characterization of a novel microtubule-based motor associated with membranous organelles in tobacco pollen tubes, *Plant Cell*, **12**, 1719–1736.

Chen, R., Hilson, P., Sedbrook, J., Rosen, E., Caspar, T. & Masson, P.H. (1998) The *Arabidopsis thaliana AGRAVITROPIC 1* gene encodes a component of the polar-auxin-transport efflux carrier, *Proc. Natl. Acad. Sci. USA*, **95**, 15112–15117.

Collings, D.A. & Allen, N.S. (2000) Cortical actin interacts with the plasma membrane and microtubules, in *Actin: A Dynamic Framework for Multiple Plant Cell Functions* (eds C.J. Staiger, F. Baluška, D. Volkmann & P.W. Barlow), Kluwer Academic, Dordrecht, pp. 145–163.

Collings, D.A., Vaughn, K.C. & Harper, J.D.I. The association of peroxisomes with the developing cell plate in dividing onion root cells depends on actin microfilaments and myosin, *Planta*, in press. [Published on-line, DOI 10.1007/s00425-003-1096-2.]

Collings, D.A., Winter, H., Wyatt, S.E. & Allen, N.S. (1998) Growth dynamics and cytoskeletal organization during stem maturation and gravity-induced stem bending in *Zea mays* L., *Planta*, **207**, 246–258.

Cosgrove, D.J. (2001) Wall structure and wall loosening. A look backwards and forwards, *Plant Physiol.*, **125**, 131–134.

Cnops, G., Wang, X., Linstead, P., van Montagu, M., van Lijsebettens, M. & Dolan, L. (2000) *TORNADO1* and *TORNADO2* are required for the specification of radial and circumferential pattern of the *Arabidopsis* root, *Development*, **17**, 3385–3394.

Dong, C.-H., Xia, G.-X., Hong, Y., Ramachandran, S., Kost, B. & Chua, N.-H. (2001) ADF proteins are involved in the control of flowering and regulate F-actin organization, cell expansion, and organ growth in arabidopsis, *Plant Cell*, **13**, 1333–1346.

Endlé, M.-C., Stoppin, V., Lambert, A.-M. & Schmit, A.-C. (1998) The growing cell plate of higher plants is a site of both actin assembly and vinculin-like antigen recruitment, *Eur. J. Cell Biol.*, **77**, 10–18.

Erhardt, M., Stoppin-Mellet, V., Campagne, S., *et al.* (2002) The plant Spc98p homologue colocalizes with γ-tubulin at microtubule nucleation sites and is required for microtubule nucleation, *J. Cell Sci.*, **115**, 2423–2431.

Fisher, D.D. & Cyr, R.J. (1998) Extending the microtubule/microfibril paradigm – cellulose synthesis is required for normal cortical microtubule alignment in elongating cells, *Plant Physiol.*, **116**, 1043–1051.

Friml, J., Benaková, E., Blilou, I., *et al.* (2002a) AtPIN4 mediates sink-driven auxin gradients and root patterning in *Arabidopsis*, *Cell*, **108**, 661–673.

Friml, J. & Palme, K. (2002) Polar auxin transport – old questions and new concepts?, *Plant Mol. Biol.*, **49**, 273–284.

Friml, J., Winiewska, J., Benaková, E., Mendgen, K. & Palme, K. (2002b) Lateral relocation of auxin efflux regulator PIN3 mediates tropism in *Arabidopsis*, *Nature*, **415**, 806–809.

Furutani, I., Watanabe, Y., Prieto, R., *et al.* (2000) The *SPIRAL* genes are required for directional control of cell elongation in *Arabidopsis thaliana*, *Development*, **127**, 4443–4453.

Gallagher, K. & Smith, L.G. (2000) Roles for polarity and nuclear determinants in specifying daughter cell fates after an asymmetric cell division in the maize leaf, *Curr. Biol.*, **10**, 1229–1232.

Gälweiler, L., Guan, C., Müller, A., *et al.* (1998) Regulation of polar auxin transport by AtPIN1 in *Arabidopsis* vascular tissue, *Science*, **282**, 2226–2230.

Geldner, N., Friml, J., Stierhof, Y.-D., Jürgens, G. & Palme, K. (2001) Auxin transport inhibitors block PIN1 cycling and vesicle trafficking, *Nature*, **413**, 425–428.

Gendreau, E., Orbovic, V., Höfte, H. & Traas, J. (1999) Gibberellin and ethylene control endoreduplication levels in the *Arabidopsis thaliana* hypocotyl, *Planta*, **209**, 513–516.

Grebe, M., Friml, J., Swarup, R., *et al.* (2002) Cell polarity signaling in *Arabidopsis* involves a BFA-sensitive auxin influx, *Curr. Biol.*, **12**, 329–334.

Green, P.B. (1984) Shifts in plant cell axiality: histogenetic influences on cellulose orientation in the succulent *Graptopetalum*, *Dev. Biol.*, **103**, 18–27.

Guilfoyle, T., Hagen, G., Ulmasov, T. & Murfett, J. (1998) How does auxin turn on genes?, *Plant Physiol.*, **118**, 341–347.

Hasenstein, K.H., Blancaflor, E.B. & Lee, J.S. (1999) The microtubule cytoskeleton does not integrate auxin transport and gravitropism in maize roots, *Physiol. Plant*, **105**, 729–738.

Hashimoto, T. (2002) Molecular genetic analysis of left-right handedness in plants, *Phil. Trans. R. Soc. Lond. B*, **357**, 799–808.

Hepler, P.K., Valster, A., Molchan, T. & Vos, J.W. (2002) Roles for kinesin and myosin during cytokinesis, *Phil. Trans. R. Soc. Lond. B*, **357**, 761–766.

Himmelspach, R. & Nick, P. (2001) Gravitropic microtubule reorientation can be uncoupled from growth, *Planta*, **212**, 184–189.

Himmelspach, R., Williamson, R.E. & Wasteneys, G.O. (2003) Cellulose microfibril alignment recovers from DCB-induced disruption despite microtubule disorganization, *Plant J.* (in press).

Himmelspach, R., Wymer, C.L., Lloyd, C.W. & Nick, P. (1999) Gravity-induced reorientation of cortical microtubules observed *in vivo*, *Plant J.*, **18**, 449–4453.

Hong, Z., Delauney, A.J. & Verma, D.P.S. (2001a) A cell plate-specific callose synthase and its interaction with phragmoplastin, *Plant Cell*, **13**, 755–768.

Hong, Z., Zhang, Z., Olson, J.M. & Verma, D.P.S. (2001b) A novel UDP-glucose transferase is part of the callose synthase complex and interacts with phragmoplastin at the forming cell plate, *Plant Cell*, **13**, 769–779.

Hu, S., Brady, S.R., Kovar, D.R., *et al.* (2000) Identification of plant actin-binding proteins by F-actin affinity chromatography, *Plant J.*, **24**, 127–137.

Hussey, P.J. & Hawkins, T.J. (2001) Plant microtubule-associated proteins: the HEAT is off in temperature-sensitive *mor1*, *Trends Plant Sci.*, **6**, 389–392.

Kakimoto, T. & Shibaoka, H. (1988) Cytoskeletal ultrastructure of phragmoplast-nuclei complexes isolated from cultured tobacco cells, *Protoplasma* (Suppl. 2), 95–103.

Kandasamy, M.K., Gilliland, L.U., McKinney, E.C. & Meagher, R.B. (2001) One plant actin isovariant, ACT7, is induced by auxin and required for normal callus formation, *Plant Cell*, **13**, 1541–1554.

Lane, D.R., Wiedemeier, A., Peng, L., *et al.* (2001) Temperature-sensitive alleles of *RSW2* link the KORRIGAN endo-1,4-β-glucanase to cellulose synthesis and cytokinesis in arabidopsis, *Plant Physiol.*, **126**, 278–288.

Lang, J.M., Eisinger, W.R. & Green, P.B. (1982) Effects of ethylene on the orientation of microtubules and cellulose microfibrils in pea epicotyl cells with polylamellate cell walls, *Protoplasma*, **110**, 5–14.

Lawrence, C.J., Malmberg, R.L., Muszynski, M.G. & Dawe, R.K. (2002) Maximum likelihood methods reveal conservation of function among closely related kinesin families, *J. Mol. Evol.*, **54**, 42–53.

Lee, Y.-R.J., Giang, H.M. & Liu, B. (2001) A novel plant kinesin-related protein specifically associates with the phragmoplast organelles, *Plant Cell*, **13**, 2427–2439.

Lee, Y.-R.J. & Liu, B. (2000) Identification of a phragmoplast-associated kinesin-related protein in higher plants, *Curr. Biol.*, **10**, 797–800.

Levengood, W.C. & Talbott, N.P. (1999) Dispersion of energies in worldwide crop formations, *Physiol. Plant.*, **105**, 615–624.

Leyser, O. (2002) Molecular genetics of auxin signaling, *Ann. Rev. Plant Biol.*, **53**, 377–398.

Li, H., Shen, J.-J., Zheng, Z.-L., Lin, Y. & Yang, Z. (2001) The Rop GTPase switch controls multiple developmental processes in Arabidopsis, *Plant Physiol.*, **126**, 670–684.

Liang, B.M., Dennings, A.M., Sharp, R.E. & Baskin, T.I. (1996) Consistent handedness of microtubule helical arrays in maize and Arabidopsis primary roots, *Protoplasma*, **190**, 8–15.

Lin, Q., Jablonsky, P.P., Elliott, J. & Williamson, R.E. (1994) A 170 kDa polypeptide from mung bean shares multiple epitopes with rabbit skeletal myosin and binds ADP-agarose, *Cell Biol. Int.*, **18**, 1035–1047.

Liu, B., Cyr, R.J. & Palevitz, B.A. (1996) A kinesin-like protein, KatAp, in the cells of arabidopsis and other plants, *Plant Cell*, **8**, 119–132.

Liu, B. & Lee, Y.R.J. (2001) Kinesin-related proteins in plant cytokinesis, *J. Plant Growth Regul.*, **20**, 141–150.

Liu, B. & Palevitz, B.A. (1992) Organization of cortical microfilaments in dividing root cells, *Cell Motil. Cytoskel.*, **23**, 252–264.

Liu, L., Zhou, J. & Pesacreta, T.C. (2001) Maize myosins: diversity, localization and function, *Cell Motil. Cytoskel.*, **48**, 130–148.

Long, J.C., Zhao, W., Rashotte, A.M., Muday, G.K. & Huber, S.C. (2002) Gravity-stimulated changes in auxin and invertase gene expression in maize pulvinal cells, *Plant Physiol.*, **128**, 591–602.

Luschnig, C., Gaxiola, R.A., Grisafi, P. & Fink, G.R. (1998) EIR1, a root-specific protein involved in auxin-transport, is required for gravitropism in *Arabidopsis thaliana*, *Genes Dev.*, **12**, 2175–2187.

Mandoli, D.F. (1998) Elaboration of body plan and phase change during development of *Acetabularia*: How is the complex architecture of a giant unicell built?, *Ann. Rev. Plant Physiol. Plant Mol. Biol.*, **49**, 173–198.

McClinton, R.S., Chandler, J.S. & Callis, J. (2001) cDNA isolation, characterization, and protein intracellular localization of a katanin-like p60 subunit from *Arabidopsis thaliana*, *Protoplasma*, **216**, 181–190.

McDowell, J.M., An, Y.-Q., Huang, S., McKinney, E.C. & Meagher, R.B. (1996) The arabidopsis *ACT7* gene is expressed in rapidly developing tissues and responds to several external stimuli, *Plant Physiol.*, **111**, 699–711.

Migliaccio, F., Piconese, S. & Trolelli, G. (2000) The right-handed slanting of *Arabidopsis thaliana* roots is due to the combined effects of positive gravitropism, circumnutation and thigomtropism, *J. Gravitational Physiol.*, **7**, 1–6.

Miesenböck, G., de Angelis, D.A. & Rothman, J.E. (1998) Visualizing secretion and synaptic transmission with pH-sensitive green fluorescent proteins, *Nature*, **394**, 192–195.

Mitsui, H., Yamaguchi-Shinozaki, K., Shinozaki, K., Nishikawa, K. & Takahashi, H. (1993) Identification of a gene family (*kat*) encoding kinesin-like proteins in *Arabidopsis thaliana* and the characterization of secondary structure of KatA, *Mol. Gen. Genet.*, **238**, 362–368.

Molchan, T.M., Valster, A.H. & Hepler, P.K. (2002) Actomyosin promotes cell plate alignment and late lateral expansion in *Tradescantia* stamen hair cells, *Planta*, **214**, 683–693.

Molendijk, A.J., Bischoff, F., Rajendrakumar, C.S.V., *et al.* (2001) *Arabidopsis thaliana* Rop GTPases are localized to tips of root hairs and control polar growth, *EMBO J.*, **20**, 2779–2788.

Muday, G.K. (2000) Maintenance of asymmetric cellular localization of an auxin transport protein through interaction with the actin cytoskeleton, *J. Plant Growth Regul.*, **19**, 385–396.

Muday, G.K. & DeLong, A. (2001) Polar auxin transport: controlling where and how much, *Trends Plant Sci.*, **6**, 535–542.

Muday, G.K. & Murphy, A.S. (2002) An emerging model of auxin transport regulation, *Plant Cell*, **14**, 293–299.

Murphy, A.S., Hoogner, K.R., Peer, W.A. & Taiz, L. (2002) Identification, purification, and molecular cloning of N-1-naphthylphthalamic acid-binding plasma membrane-associated aminopeptidases from arabidopsis, *Plant Physiol.*, **128**, 935–950.

Müller, A., Guan, C., Gälweiler, L., *et al.* (1998) *AtPIN2* defines a locus of *Arabidopsis* for root gravitropism control, *EMBO J.*, **17**, 6903–6911.

Narasimhulu, S.B. & Reddy, A.S.N. (1998) Characterization of microtubule binding domains in the arabidopsis kinesin-like calmodulin binding protein, *Plant Cell*, **10**, 957–965.

Nick, P., Bergfeld, R., Schafer, E. & Schopfer, P. (1990) Unilateral reorientation of microtubules at the outer epidermal wall during photo- and gravitropic curvature of maize coleoptiles and sunflower hypocotyls, *Planta*, **181**, 162–168.

Nishihama, R., Ishikawa, M., Araki, S., Soyano, T., Asada, T. & Machida, Y. (2001) The NPK1 mitogen-activated protein kinase kinase kinase is a regulator of cell-plate formation in plant cytokinesis, *Genes Dev.*, **15**, 352–361.

Nishihama, R., Soyano, T., Ishikawa, M., *et al.* (2002) Expansion of the cell plate in plant cytokinesis requires a kinesin-like protein/MAPKKK complex, *Cell*, **109**, 87–99.

Olsen, O.A. (2001) Endosperm development: cellularization and cell fate, *Ann. Rev. Plant Physiol. Plant Mol. Biol.*, **52**, 233–267.

Oppenheimer, D.G., Pollock, M.A., Vacik, J., *et al.* (1997) Essential role of a kinesin-like protein in *Arabidopsis* trichome morphogenesis, *Proc. Natl. Acad. Sci. USA*, **94**, 6261–6266.

Otegui, M. & Staehelin, L.A. (2000) Cytokinesis in flowering plants: more than one way to divide a cell, *Curr. Op. Plant Biol.*, **3**, 493–502.

Otegui, M.S., Mastronarde, D.N., Kang, B.-H., Bednarek, S.Y. & Staehelin, L.A. (2001) Three-dimensional analysis of syncitial-type cell plates during endosperm cellularization visualized by high resolution electron tomography, *Plant Cell*, **13**, 2033–2051.

Pagant, S., Bichet, A., Sugimoto, K., *et al.* (2002) *KOBITO1* encodes a novel plasma membrane protein necessary for normal synthesis of cellulose during cell expansion in arabidopsis, *Plant Cell*, **14**, 2001–2013.

Palevitz, B.A. (1987) Accumulation of F-actin during cytokinesis in *Allium*. Correlation with microtubule distribution and the effects of drugs, *Protoplasma*, **141**, 24–32.

Ramachandran, S., Christensen, H.E.M., Ishimaru, Y., *et al.* (2000) Profilin plays a role in cell elongation, cell shape maintenance and flowering in arabidopsis, *Plant Physiol.*, **124**, 1637–1647.

Rashotte, A.M., Brady, S.R., Reed, R.C., Ante, S.J. & Muday, G.K. (2000) Basipetal auxin transport is required for gravitropism in roots of arabidopsis, *Plant Physiol.*, **122**, 481–490.

Raven, J.A. (1975) Transport of indoleacetic acid in plant cells in relation to pH and electrical potential gradients, and its significance for polar IAA transport, *New Phytol.*, **74**, 163–172.

Reddy, A.S.N. & Day, I.S. (2002) Kinesins in the Arabidopsis genome: a comparative analysis among eukaryotes, *BMC Genomics*, **2**, 2.

Reddy, A.S.N., Safadi, F., Narasimhulu, S.B., Golovkin, M. & Hu, X. (1996) A novel plant calmodulin-binding protein with a kinesin heavy chain motor domain, *J. Biol. Chem.*, **271**, 7052–7060.

Reichelt, S., Knight, A.E., Hodge, T.P., *et al.* (1999) Characterization of the unconventional myosin VIII in plant cells and its localization at the post-cytokinetic cell wall, *Plant J.*, **19**, 555–567.

Samuels, A.L., Giddings Jr., T.H. & Staehelin, L.A. (1995) Cytokinesis in tobacco BY-2 and root tip cells: new model of cell plate formation in higher plant cells, *J. Cell Biol.*, **130**, 1345–1347.

Sato, Y., Wada, M. & Kadota, A. (2001) Choice of tracks, microtubules and/or actin filaments for chloroplast photo-movement is differentially controlled by phytochrome and a blue light receptor, *J. Cell Sci.*, **114**, 269–279.

Schmit, A.-C. (2000) Actin during mitosis and cytokinesis, in *Actin: A Dynamic Framework for Multiple Plant Cell Functions* (eds C.J. Staiger, F. Baluška, D. Volkmann & P.W. Barlow) Kluwer Academic, Dordrecht, pp. 437–456.

Sedbrook, J.C., Chen, R. & Masson, P.H. (1999) *ARG1* (Altered Response to Gravity) encodes a DNA-J-like protein that potentially interacts with the cytoskeleton, *Proc. Natl. Acad. Sci. USA*, **96**, 1140–1145.

Shibaoka, H. (1994) Plant hormone-induced changes in the orientation of cortical microtubules: alterations in the cross-linking between microtubules and the plasma membrane, *Ann. Rev. Plant Physiol. Plant Mol. Biol.*, **45**, 527–544.

Smalle, J., Haegman, M., Kurepa, J., van Montagu, M. & van der Straeten, D. (1997) Ethylene can stimulate *Arabidopsis* hypocotyl elongation in the light. *Proc. Natl. Acad. Sci. USA*, **94**, 2756–2761.

Smertenko, A., Saleh, N., Igarashi, H., *et al.* (2000) A new class of microtubule-associated proteins in plants, *Nat. Cell Biol.*, **2**, 750–753.

Smith, L.G., Gerttula, S.M., Han, S. & Levy, J. (2001) TANGLED1: a microtubule binding protein required for the spatial control of cytokinesis in maize, *J. Cell Biol.*, **152**, 231–236.

Steinborn, K., Maulbetsch, C., Priester, B., *et al.* (2002) The *Arabidopsis PILZ* group genes encode tubulin-folding cofactor orthologs required for cell division but not cell growth, *Genes Dev.*, **16**, 959–971.

Steinmann, T., Geldner, N. & Grebe, M. (1999) Coordinated polar localization of auxin efflux carrier PIN1 by GNOM ARF GEF, *Science*, **286**, 316–318.

Stoppin-Mellet, V., Gaillard, J. & Vantard, M. (2002) Functional evidence for *in vitro* microtubule severing by the plant katanin homologue, *Biochem. J.*, **365**, 337–342.

Strompen, G., El Kasmi, F., Richter, S., *et al.* (2002) The *Arabidopsis HINKEL* gene encodes a kinesin-related protein involved in cytokinesis and is expressed in a cell cycle-dependent manner, *Curr. Biol.*, **12**, 153–158.

Sugimoto, K., Williamson, R.E. & Wasteneys, G.O. (2000) New techniques enable comparative analysis of microtubule orientation, wall texture, and growth rate in intact roots of arabidopsis, *Plant Physiol.*, **124**, 1493–1506.

Sugimoto, K., Williamson, R.E. & Wasteneys, G.O. (2001) Wall architecture in the cellulose-deficient *rsw1* mutant of *Arabidopsis thaliana*: microfibrils but not microtubules lose their transverse alignment before microfibrils become unrecognizable in the mitotic and elongation zones of roots, *Protoplasma*, **215**, 172–183.

Sugimoto, K., Himmelspach, R., Williamson, R.E. & Wasteneys, G.O. (2003) Mutation or drug-dependent microtubule disruption causes radial swelling without altering parallel cellulose microfibril deposition in arabidopsis root cells, *Plant Cell*, **15**, 1414–1429.

Sugimoto-Shirasu, K., Stacey, N.J., Corsar, J., Roberts, K. & McCann, M. (2002) DNA topoisomerase VI is essential for endoreduplication in *Arabidopsis*, *Curr. Biol.*, **12**, 1782–1786.

Swarup, R., Friml, J., Marchant, A., *et al.* (2001) Localization of the auxin permease AUX1 suggests two functionally distinct hormone transport pathways operate in the *Arabidopsis* root apex, *Genes Dev.*, **15**, 2648–2653.

Šamaj, J., Peters, M., Volkmann, D. & Baluška, F. (2000) Effects of myosin ATPase inhibitor 2,3-butaneione 2-monoxime on distributions of myosins, F-actin, microtubules, and cortical endoplasmic reticulum in maize root apices, *Plant Cell Physiol.*, **41**, 571–582.

Taguchi, T., Uraguchi, A. & Katsumi, M. (1999) Auxin- and acid-induced changes in the mechanical properties of the cell wall, *Plant Cell Physiol.*, **40**, 743–749.

Takeda, T., Furuta, Y., Awano, T., Mizuno, K., Mitsuishi, Y. & Hayashi, T. (2002) Suppression and acceleration of cell elongation by integration of xyloglucans in pea stem segments, *Proc. Natl. Acad. Sci. USA*, **99**, 9055–9060.

Takesue, K. & Shibaoka, H. (1999) Auxin-induced longitudinal-to-transverse reorientation of cortical microtubules in nonelongating epidermal cells of azuki bean epicotyls, *Protoplasma*, **206**, 27–30.

Tao, L.-Z., Cheung, A.Y. & Wu, H.-M. (2002) Plant Rac-like GTPases are activated by auxin and mediate auxin-responsive gene expression, *Plant Cell*, **14**, 1–16.

Thitamadee, S., Tuchihara, K. & Hashimoto, T. (2002) Microtubule basis for left-handed helical growth in *Arabidopsis*, *Nature*, **417**, 193–196.

Tournebize, R., Popov, A., Kinoshita, K., *et al.* (2000) Control of microtubule dynamics by the antagonistic activities of XMAP215 and XKCM1 in *Xenopus* egg extracts, *Nature Cell Biol.*, **2**, 13–19.

Traas, J., Bellini, C., Nacry, P., Kronenberger, J., Bouchez, D. & Caboche, M. (1995) Normal differentiation patterns in plants lacking microtubular preprophase bands, *Nature*, **375**, 676–677.

Twell, D., Park, S.K., Hawkins, T.J., *et al.* (2002) MOR1/GEM1 has an essential role in the plant-specific cytokinetic phragmoplast, *Nature Cell Biol.*, **4**, 711–714.

Valster, A.H., Pierson, E.S., Valenta, R., Hepler, P.K. & Emons, A.M.C. (1997) Probing the plant actin cytoskeleton during cytokinesis and interphase by profilin microinjection, *Plant Cell*, **9**, 1815–1824.

van der Straeten, D., Zhou, Z.Y., Prinsen, E., van Onckelen, H.A. & van Montagu, M.C. (2001) A comparative molecular-physiological study of submergence response in lowland and deepwater rice, *Plant Physiol.*, **125**, 955–968.

van Hengel, A.J. & Roberts, K. (2002) Fucosylated arabinogalactan-proteins are required for full root cell elongation in arabidopsis, *Plant J.*, **32**, 105–113.

Vaughn, K.C., Hoffman, J.C., Hahn, M.G. & Staehelin, L.A. (1996) The herbicide dichlobenil disrupts cell plate formation: immunogold characterization, *Protoplasma*, **194**, 117–132.

Vos, J.W., Safadi, F., Reddy, A.S.N. & Hepler, P.K. (2000) The kinesin-like calmodulin binding protein is differentially involved in cell division, *Plant Cell*, **12**, 979–990.

Wang, Q.-Y. & Nick, P. (1998) The auxin response of actin is altered in the rice mutant *Yin-Yang*, *Protoplasma*, **204**, 22–33.

Wasteneys, G.O. (2000) The cytoskeleton and growth polarity, *Curr. Op. Plant Biol.*, **3**, 503–511.

Wasteneys, G.O. (2002) Microtubule organization in the green kingdom: chaos or self-order?, *J. Cell Sci.*, **115**, 1345–1354.

Wasteneys, G.O. & Galway, M.E. (2003) Remodeling the cytoskeleton for growth and form: an overview with some new views, *Ann. Rev. Plant Biol.*, **54**, 691–722.

Wasteneys, G.O., Collings, D.A., Gunning, B.E.S., Hepler, P.K. & Menzel, D. (1996) Actin in living and fixed characean internodal cells: identification of a cortical array of fine actin strands and chloroplast actin rings, *Protoplasma*, **190**, 25–38.

Webb, M., Jouannic, S., Foreman, J., Linstead, P. & Dolan, L. (2002) Cell specification in the *Arabidopsis* root epidermis requires the activity of *ECTOPIC ROOT HAIR 3* – a katanin-p60 protein, *Development*, **129**, 123–131.

Wenzel, C.L., Williamson, R.E. & Wasteneys, G.O. (2000) Gibberellin-induced changes in growth anisotropy precede gibberellin-dependent changes in cortical microtubule orientation in developing epidermal cells of barley leaves. Kinematic and cytological studies on a gibberellin-responsive dwarf mutant, M489, *Plant Physiol.*, **124**, 813–822.

Whittington, A.T., Vugrek, O., Wei, K.J., *et al.* (2001) MOR1 is essential for organizing cortical microtubules in plants, *Nature*, **411**, 610–613.

Wiedemeier, A.M.D., Judy-March, J.E., Hocart, C.H., Wasteneys, G.O., Williamson, R.E. & Baskin, T.I. (2002) Mutant alleles of *Arabidopsis RADIALLY SWOLLEN 4* and *7* reduce growth anisotropy without altering the transverse orientation of cortical microtubules or cellulose microfibrils, *Development*, **129**, 4821–4830.

Yasuhara, H., Muraoka, M., Shogake, H., Mori, H. & Sonobe, S. (2002) TMBP200, a microtubule bundling polypeptide isolated from telophase tobacco BY-2 cells is a MOR1 homologue, *Plant Cell Physiol.*, **43**, 595–603.

Yin, Y., Cheong, H., Friedrichsen, D., *et al.* (2002) A crucial role for the putative *Arabidopsis* topoisomerase VI in plant growth and development, *Proc. Natl. Acad. Sci. USA*, **99**, 10191–10196.

Zhang, D., Wadsworth, P. & Hepler, P.K. (1993) Dynamics of microfilaments are similar, but distinct from microtubules during cytokinesis in living, dividing plant cells, *Cell Motil. Cytoskeleton*, **24**, 151–155.

Zuo, J., Niu, Q.-W., Nishizawa, N., Wu, Y., Kost, B. & Chua, N.-H. (2000) KORRIGAN, an arabidopsis endo-1,4-β-glucanase, localizes to the cell plate by polarized targeting and is essential for cytokinesis, *Plant Cell*, **12**, 1137–1152.

4 Re-staging plant mitosis

Magdalena Weingarner, Laszlo Bogre and John H. Doonan

4.1 Introduction

The cell division cycle is a fundamental feature of all living organisms, allowing cells to grow and proliferate. Cell division is complex, involving the coordination of diverse biochemical and biophysical processes. It is also inherently very dangerous, as mistakes can lead to defective or unviable daughter cells. To avoid such mistakes, the processes making up the cell division cycle proceed in a highly ordered fashion with numerous checks to ensure all processes are completed in the correct order and at high fidelity. In eukaryotes, the standard cell cycle (Fig. 4.1) consists of an alternation of genome replication, occurring in S-phase, and genome segregation in M-phase. This ensures that the genome remains the same size across generations. Normally, S- and M-phase are separated by two gap phases, G1 before S-phase and G2 before M-phase. The gap phases allow cell growth to occur and their duration can vary depending on the cell type. Fidelity and other checks operate during gap phases. For example, during G2, the completion of DNA replication is

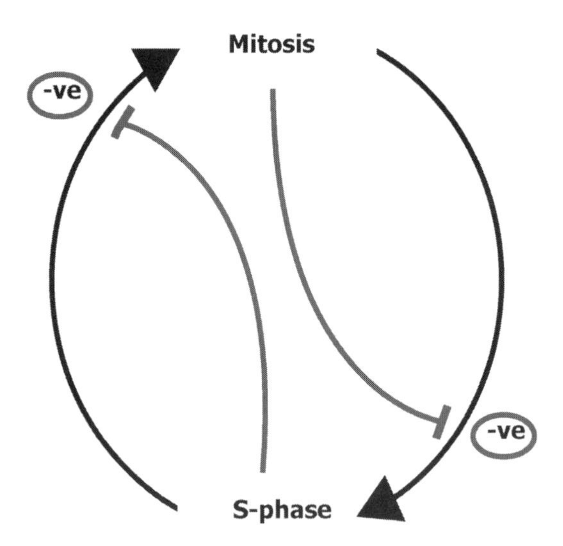

Fig. 4.1 The eukaryotic cell cycle consists of an alternation between genome duplication (S-phase) and genome separation (mitosis), which are separated by gap phases 1 and 2. Checkpoint mechanisms operate during the gap phases to ensure the correct alternation of S- and M-phases, the correction of errors and the coupling of cell cycle events to growth, development and the environment.

assessed before mitotic entry can occur. These checkpoints are important for ensuring that the cell and the cell's environment are optimal to undertake the next stage of the division cycle. Checkpoints provide an opportunity to assess growth and check for damage. If growth is insufficient or damage has occurred, checkpoints prevent nuclear and cell division and give time for corrective processes to occur. Developmental signals can also activate checkpoints, coupling cell growth and division with morphogenesis.

Profound changes in cellular organisation occur during the cell division cycle. The most dramatic of these is mitosis when the nucleus visibly alters due to chromosome condensation and microtubules reorganise to produce a spindle that separates the condensed chromatin. These complex changes in cellular structure are coordinated, also by checkpoint mechanisms. The mechanisms that coordinate cellular organisation during mitosis is the topic of this chapter. With particular emphasis on plant cells, we will discuss the mechanisms that drive eukaryotic cell cycle progression and coordinate nuclear and cytoplasmic re-organisation.

4.2 The cyclin dependent protein kinases

Cyclin dependent protein kinases (Cdks) are perhaps the best understood of the proteins that coordinate mitotic progression. Cdks can be considered as molecular switches that collate information on the status of the cell and then transmit this information, via protein phosphorylation, leading to profound changes in cellular function and architecture (Fig. 4.2). They are conserved multimeric complexes, minimally consisting of a cyclin dependent kinase (Cdk), which is periodically activated by its cyclin partner. The timed production and destruction of the cyclin partner mediates the proper temporal order of these events.

4.2.1 Cdk structure and diversity

Cdks belong to the serine/threonine kinases, and the prototypical cdc2 protein is characterised by a cyclin binding motif containing the highly conserved PSTAIRE sequence which binds cyclin, leading to a reorientation of key residues which allow the correct positioning of the ATP phosphates for the phosphotransfer reaction (Jeffrey *et al.*, 1995; Morgan, 1995). In yeast, both the G1/S and G2/M transition are induced by a single Cdk kinase, known as Cdc2p/CDC28p, which binds to S-phase and M-phase-specific cyclins. Animals have evolved several classes of Cdks, which act in combination with different cyclins at the major control steps of the cell cycle: S-phase is induced by Cdk2 bound to the S-phase-specific E- and A-type cyclins, and M-phase is triggered by Cdk1 associated with the mitotic A- and B-type cyclins.

Cdk-related proteins from plants can be grouped into six classes that have recently been renamed CdkA to CdkF (Joubes *et al.*, 2000). CdkA seems to be the orthologue of the yeast Cdc2p/CDC28p and animal Cdk1 and Cdk2, and contains a perfectly conserved PSTAIRE sequence within its cyclin interaction motif. Plant

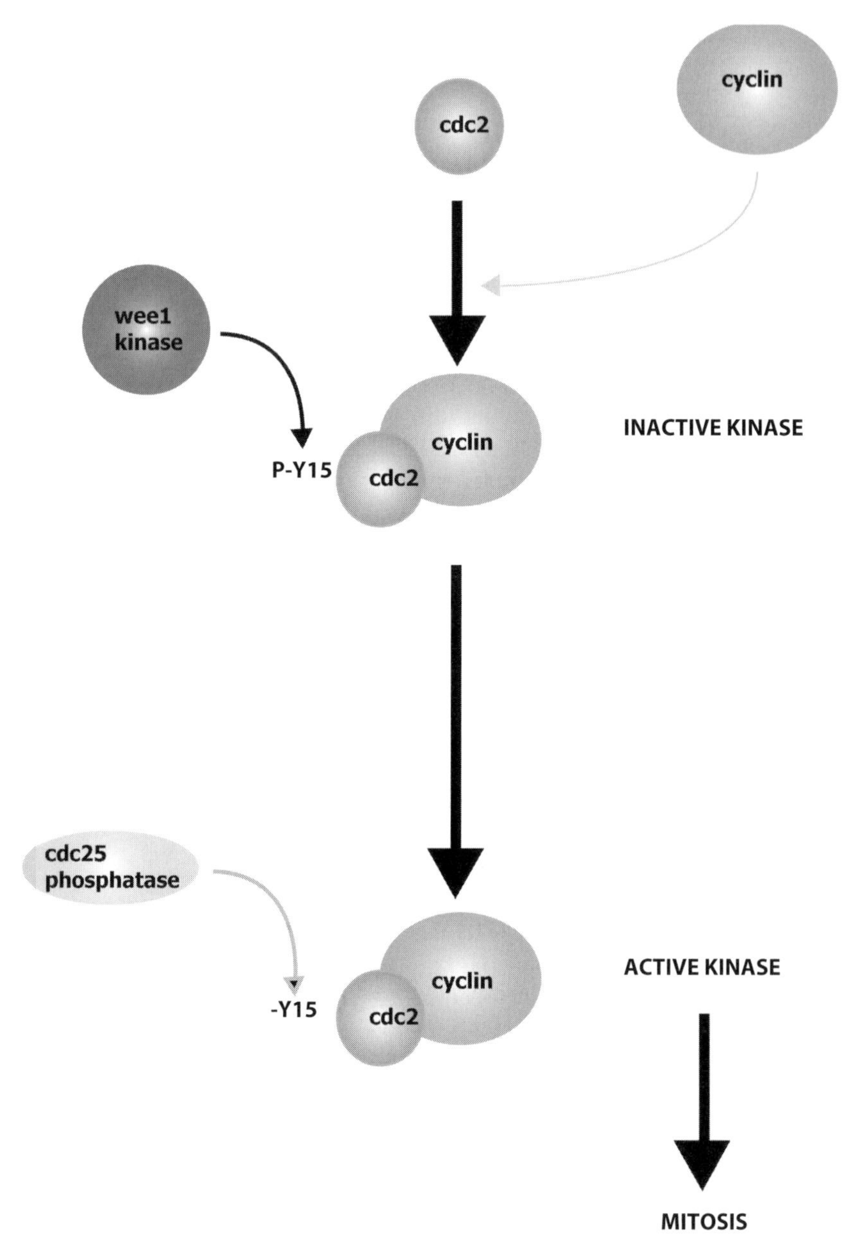

Fig. 4.2 Cyclin dependent protein kinases as integrators of cellular information. In yeast, Cdc2p receives signals in the form of interacting proteins, and phosphorylation/dephosphorylation modifies its activity and substrate specificity. When activated, the cdc2 kinase permits the cell to enter mitosis.

CdkA genes are able to complement temperature-sensitive mutants in yeast *cdc2/ CDC28* genes, demonstrating their functional conservation (Fobert *et al.*, 1996; Hirt *et al.*, 1991, 1993).

The CdkB proteins are characterised by a variant form of the PSTAIRE motif, suggesting that they may interact with a different set of cyclins. CdkB genes are only found in plants and phylogenetic comparisons of Cdk sequences suggests that the CdkB gene family may have arisen from gene duplication of a CdkA gene in the plant lineage (Doonan & Fobert, 1997). This class is divided into two subclasses, CdkB1 and CdkB2, that are differentially expressed (Fobert *et al.*, 1996), but both seem to function during G2 and mitosis (Magyar *et al.*, 1997). The CdkC class has a different variant motif, PITAIRE, that is also found in human CDK-related kinases. A member of this class has been studied in tomato (*Lycopsersicon esculentum*) and its expression was found to be not cell cycle regulated but is associated preferentially with dividing tissues. In a two-hybrid assay CdkC did not interact with mitotic or G1 cyclins, indicating that this kinase is not directly involved in cell division (Joubes *et al.*, 2000). Only very preliminary information about kinases belonging to the classes D and E exists and their functions still remain to be determined. CdkF is also known as cdk activating kinase, CAK.

There are at least 30 cell cycle-related cyclin genes in the *Arabidopsis* genome and, based on sequence analysis, most have been allocated to one of three groups: cyclins A, B or D (Vandepoele *et al.*, 2002). These groupings approximately parallel the mammalian cyclins, but experimental evidence for functional conservation between animal and plant cyclins is sparse (Renaudin *et al.*, 1996). Based on sequence comparisons, the plant cyclin A and cyclin B genes each form three subclasses, while the cyclin D genes may contain as many as seven.

4.2.2 Regulation of Cdk activity

Cdk activity can respond to a diversity of incoming signals and can provide a number of outputs (Fig.4.2). In this regard, it acts like a very sophisticated molecular switch. Regulation of Cdks occurs at a number of levels, including transcription, translation, protein stability, protein location, interactions with cyclins and other regulatory proteins, and phosphorylation status. Even in simple organisms such as yeast, cdc2 regulation and activity is still not fully understood and still provides much scope for further research. In multicellular organisms with multiple cyclins and Cdks, the additional complexity means that we have barely begun.

Early studies using antibodies indicated that differentiated tissues generally contain lower levels of CdkA transcript and protein than proliferating tissues (John *et al.*, 1990). While the protein level of CdkA does not change dramatically during the cell cycle, its kinase activity peaks at the G1/S and the G2/M transition, suggesting post-translational activation at both transition steps (Bögre *et al.*, 1997).

Cdks are multimeric, containing at least two subunits, a catalytic kinase and a regulatory cyclin. Since the kinase subunit usually has little activity alone, the amount of available cyclin can limit activity. Indeed most cyclin proteins are intermittently produced in response to either cell cycle cues or extracellular signals. The activity of Cdk–cyclin complexes is controlled at a further level by specific phosphorylation and dephosphorylation events, binding of cyclin–Cdk inhibitors and by their tightly controlled subcellular localisation. Phosphorylation on Y15 inhibits the mitotic activity of cdc2-related proteins and is mediated by the wee1 kinase (Morgan, 1995). This prevents premature activation of Cdc2p and couples mitosis to cell growth. The Cdc25 protein phosphatase removes this phosphate and activates the kinase, permitting mitotic entry. DNA damage or incomplete DNA replication prevents Cdc25p action. Thus, the balance of the wee1 kinase and the Cdc25 phosphatase activities control mitotic entry via cdc2 phosphorylation on a single residue. This mechanism is conserved in fungi and animals and seems likely to also be conserved in plants: wee1 has been identified and the Y15 residue is present in most CdkA and CdkB genes (Sun *et al.*, 1999). A plant version of Cdc25p has yet to be isolated.

4.3 Sequence of events during mitosis

Separating the replicated genome occurs during mitosis and involves a dramatic and potentially lethal rearrangement of the cell. Mitosis proceeds in a strictly choreographed series of phases (Fig. 4.3) that are defined by visible changes in the morphology of both nucleus and cytoplasm. Prophase is the first visible phase when chromosomes become condensed within the nucleus. During prometaphase, the nuclear envelope breaks down and chromosomes become attached to and positioned on the nascent mitotic spindle. Once the bipolar spindle is established and all chromosomes are aligned in the equatorial plane, the cell is considered to be in metaphase. Anaphase initiates when sister chromatids separate and begin to move towards the opposing spindle poles, as the spindle elongates. Finally, during telophase, the two daughter nuclei are formed. Cytokinesis is achieved by actin mediated cell cleavage in animals, but in plants, a microtubule-based phragmoplast assembles near the equator of the cell and participates in cytokinesis, while actin functions in positional signalling (Gimenez-Abian *et al.*, 1998; Pines & Rieder, 2001).

Plant cells are, in contrast to animal cells, surrounded by a rigid cell wall that immobilises them within tissues. It has been argued, therefore, that the orientation in which plant cells divide is critical for morphogenesis. While this may not be universally true, certain critical steps in plant development involve asymmetric divisions that give rise to daughter cells differing in both size and fate. Perturbing the cytoskeleton can eliminate such asymmetry and lead to profound developmental consequences. Particular examples include cell divisions in early embryogenesis. Two unique cytoskeletal arrays play central roles in cell plate orientation and cytokinesis in plants: the preprophase band (PPB), a transient ringlike structure of cortical microtubules, that is established during G2 and prophase, and marks the future division site (Staiger & Lloyd, 1991) and the phragmoplast, a microtubule structure, that arises in late anaphase and directs the formation of a new cell wall between the daughter nuclei during cytokinesis (Staehelin & Hepler, 1996). The location at

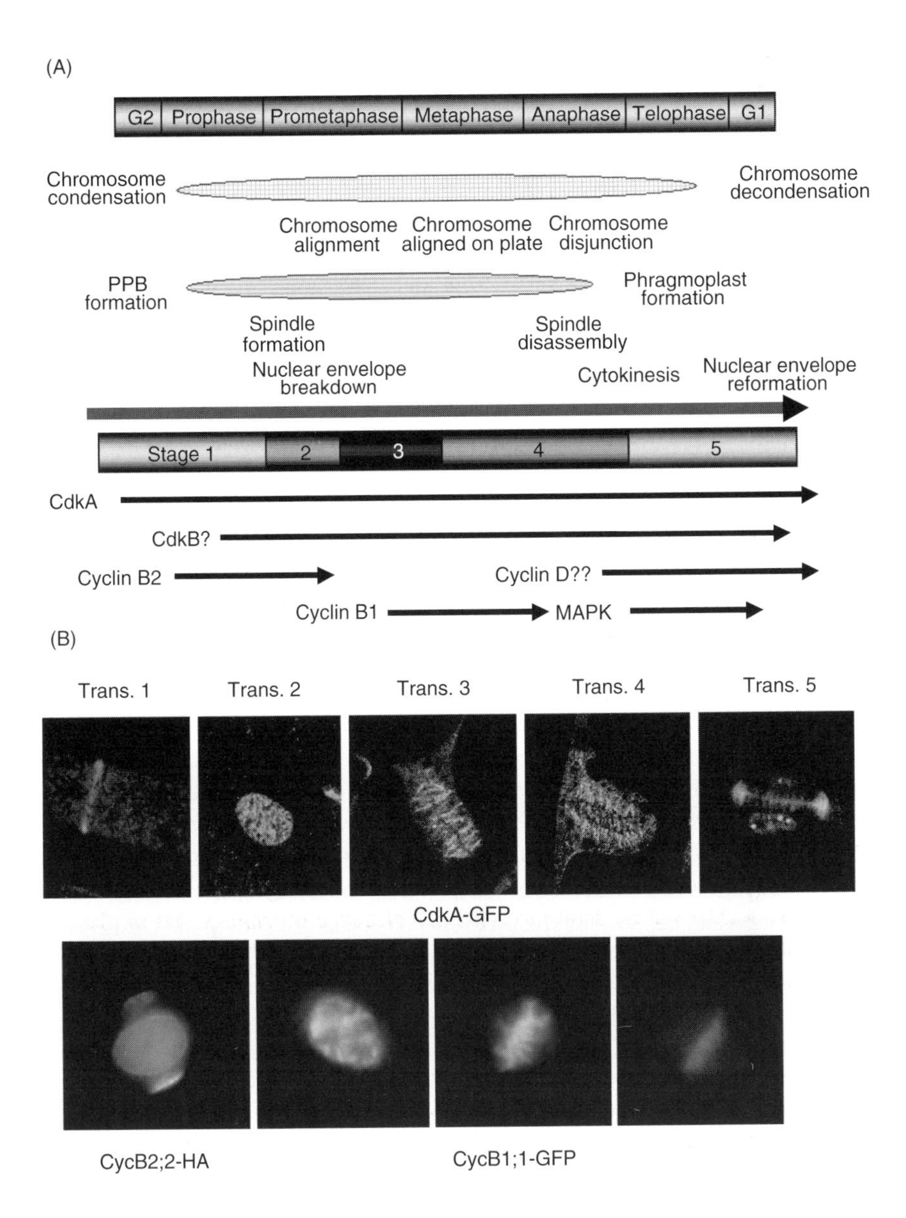

Fig. 4.3 Stages of mitosis. (A) Mitosis has been staged based on microscopical observation of chromatin and cytoskeleton structures. The realisation of the conserved regulation of mitotic processes, but alteration in the consecution of events led to the proposition that staging of mitosis according to molecular events might be more generally applicable across species. The plant regulators of mitosis are depicted as arrows. (B) Localisation of CDKA-GFP throughout mitosis, and the localisation of cyclin B2 and cyclin B1.

which the new cell wall will fuse with the parental wall is governed by an interaction between the phragmoplast and the cortical division site previously occupied by the PPB (Gunning, 1980).

Traditionally, mitotic progression is staged based on the morphology of the chromosomes and spindle. However, Pines and Rieder (2001) have used the state and activity of cell cycle regulators to define biochemical transitions during mitosis. Keeping in mind that the term *transition* is used for an irreversible process, while *stages* also include reversible steps within cell cycle progression, we subdivided plant mitosis into five stages. This method seems useful in comparing plants and animals, which share conserved regulators but show differences in their structural features.

The key stages are briefly summarised here and discussed in detail below:

4.3.1 Stage 1: preparation for mitosis

In animal cells, the molecular events that prepare the cell for mitosis are mainly controlled by the cyclin A-associated Cdk kinase (den Elzen & Pines, 2001), and other mitotic kinases like the polo-like kinases (Plks) and the aurora B kinase but cyclin B-associated protein kinases seem to be inactive at this stage (Nigg, 2001). In plants, some of these events are controlled by cyclin B2-associated Cdk (Weingartner *et al.*, 2003), and most probably also by cyclin A-associated Cdk functions while no other kinases are known to be involved.

4.3.2 Stage 2: commitment to mitosis

The final commitment to mitosis is mediated by the rapid activation of cyclin B–Cdk1. This process is highly regulated and involves several levels of control like phosphorylation and dephosphorylation events on conserved sites at the Cdk kinase, binding of inhibitors and the subcellular relocation of the Cdk–cyclin complex (Lew & Reed, 1993; Ohi & Gould, 1999). There are at least two (irreversible) transitions in early mitosis that are controlled by negative feedback regulation, known as checkpoint pathways. They sense whether requirements needed after such a transition are fulfilled or not (Elledge, 1996; Weinert & Hartwell, 1989). If not, they delay the transition to the next cycle phase by interfering with Cdk activation (Walworth, 2000; Weinert, 1997). The first is the transition from prophase to prometaphase, marked by nuclear envelope breakdown and existing in both plant and animal cells (Gimenez-Abian *et al.*, 2001; Sluder *et al.*, 1995). Second in midprophase, a checkpoint is located at the fringe between reversible to irreversible chromosome condensation (Garcia-Herdugo *et al.*, 1974; Rieder & Cole, 1998) again in plant and animal cells, while in yeast, the chromosome condensation cycle is also reversible (Lavoie *et al.*, 2002).

In plants, cyclin B1 seems to be responsible for the irreversible portion of the mitotic chromosome condensation (Criqui *et al.*, 2001). Thus, when nuclear envelope breakdown (Gimenez Martin *et al.*, 1971) or the metaphase to anaphase transition

(Sans *et al.*, 1989) is delayed in plant cells, the chromosomes overcondense in the additional time provided by the corresponding checkpoint. In yeast, cohesin-dependent linking of sister chromatids is an irreversible process underlying the mitotic chromosomal cycle (Lavoie *et al.*, 2002).

4.3.3 Stage 3: preventing premature genome separation

Before the onset of anaphase, a signal transduction pathway, known as the spindle assembly checkpoint, delays mitosis until all chromosomes are properly attached to the kinetochores and aligned on the metaphase plate. This checkpoint prevents activation of the anaphase-promoting complex (APC) and thus maintains cyclin B–Cdk activity (Kallio *et al.*, 1998; Rieder & Salmon, 1998). During preparation for mitosis and mitosis itself, most of the checkpoints also control the DNA decatenation produced by topoisomerase II. Thus, the removal of non-replicative catenations are needed for premitotic chromosome individualisation to occur (Gimenez-Abian *et al.*, 2000). Moreover, removal of replicative DNA catenations is required for the separation of sister chromatid cores taking place shortly after nuclear envelope breakdown (NEB) has taken place (early prometaphase), while removal of sister centromeres is needed for the segregation of sisters in anaphase (Gimenez-Abian *et al.*, 2002).

4.3.4 Stage 4: separating the genome

Once all kinetochores are attached to the spindle, APC becomes activated and mediates separation of sister chromatids (Nasmyth *et al.*, 2000) and degradation of cyclin B leading to inactivation of the Cdk kinase.

4.3.5 Stage 5: exit from mitosis

Return into interphase occurs after all cyclin B has been degraded and Cdk1 has been inactivated. This leads to the reversal of chromosome condensation allowing exit from mitosis and cytokinesis to occur (Holloway *et al.*, 1993; Surana *et al.*, 1993). Only the APC-mediated degradation of cyclin B leads to the reversal of chromosome condensation from the metaphase to anaphase fringe onwards (Hirano, 2000). Several other mitotic kinases like members of the families of plks (Carmena *et al.*, 1998; Nigg, 1998) or B-type aurora kinases (Schumacher *et al.*, 1998; Terada *et al.*, 1998) are also implicated in chromosome segregation and cytokinesis.

4.4 Preparing for mitosis

The structural changes that precede mitosis are often subtle and vary from cell to cell and between phylogenetic groups. The cell may have grown during the first gap (G1) phase and the nucleus generally increases in size due to DNA replication. The second gap (G2) phase, between the end of S-phase and the beginning of mitosis, provides time for a check of DNA integrity and for further cell growth, if required. In some cells, including many plant cells, decisions about the position and plane of cell division are made during G2. In yeasts, the presumptive cortical site of cell division acquires a unique molecular identity, allowing the attachment of cytoskeletal elements, which is usually actin based (Chant & Pringle, 1995). This is followed by nuclear migration to the division site and nuclear division. In many plant cells, the division site is marked by a cortical band of microtubules called the preprophase band. The position of the band predicts the future site of cytokinesis and, although the functional relationship is unclear, the two are strictly correlated (Gimenez-Abian *et al.*, 1998). A number of Cdk complexes are present during this phase of the cycle, but their identity depends on cell type.

4.4.1 Animal A-type cyclins

A-type cyclins have been found in plant, animal and human cells but are absent in yeast and fungi. In *Xenopus*, mice and humans, two versions exist, cyclin A_1 and cyclin A_2. While cyclin A_1 is only expressed in meiosis and early embryos, cyclin A_2 is found in all proliferating somatic cells and, together with its *Drosophila* cyclin A counterpart, has been studied more extensively (Lehner & O'Farrell, 1990). Animal cyclin A proteins form complexes with Cdk1 and Cdk2, which accumulate within the nucleus on assembly, and function in both S-phase and mitosis (Rosenblatt *et al.*, 1992; Roy *et al.*, 1991).

During mitosis cyclin A_2–Cdk activity has been found to peak at the time of nuclear envelope breakdown(Pines & Hunter, 1990). When exogenous cyclin A is microinjected into G2 cells, it rapidly induces chromosome condensation. Inhibitors of the cyclin A–Cdk-complex, like $p21^{cip1/waf1}$, prevent G2 cells from entering prophase, and induce early prophase cells to return back to interphase (Furuno *et al.*, 1999) suggesting a rate-limiting role for cyclin A–Cdk during this transition step. Live cell imaging in mammalian cells following a cyclin A–GFP fusion protein during mitosis revealed that cyclin A begins to be degraded just after nuclear envelope breakdown (den Elzen & Pines, 2001). Cyclin A_2 was found to be degraded before cyclin B but the underlying mechanism is not fully understood yet (Geley *et al.*, 2001; Whitfield *et al.*, 1990). The importance of the timely degradation of cyclin A during mitosis is illustrated by the consequences of expressing stable cyclin A in human and *Drosophila* cells.

Overexpression of non-degradable cyclin A in human cells was found to slow mitosis or block cells in anaphase with separated chromosomes, and cytokinesis does not occur (Geley *et al.*, 2001). A similar effect was also seen in *Drosophila* where overexpression of stable cyclin A led to metaphase arrest and this phenotype was enhanced by coexpression of Cdk1 (Parry & O'Farrell, 2001). These experiments show that the action of cyclin A is, in part, limited by the availability of a kinase partner and that the arrest is mediated by a CycA–Cdk complex. It was also suggested that cyclin A directly inhibits chromosome disjunction probably by acting synergistically with other proteins involved in this process (Parry & O'Farrell, 2001; Sigrist *et al.*, 1995). Among these are Pim, the *Drosophila* securin, which stabilises sister chromosome cohesion until it is degraded at the metaphase to anaphase transition (Leismann *et al.*, 2000). The Cdk inhibitor, Rux, has been shown to interact genetically and physically with cyclin A and inhibit its *in vitro* kinase activity (Foley *et al.*, 1999). Rux mutants exhibit a significantly longer metaphase than wild type and they are impaired in their ability to overcome the transient metaphase arrest induced by expression of stable cyclin A (Foley & Sprenger, 2001). In contrast to its overexpression, expression of stable cyclin A at endogenous levels had little effect on mitosis (Jacobs *et al.*, 2001).

A functional role for cyclin A in preventing premature anaphase may be part of a surveillance mechanism that preserves genomic integrity in the *Drosophila* embryo: a metaphase delay, induced by DNA damage, is accompanied by stabilisation of cyclin A. Furthermore, mutant cells lacking cyclin A are unable to delay in mitosis in response to DNA damage and enter anaphase with an increased number of lagging chromosomes. The delay in metaphase normally serves to reduce chromosome breakage in subsequent anaphase and telophase (Su & Jaklevic, 2001).

4.4.2 Plant A-type cyclins

Three distinct A-type cyclins have been described in various plant species, named cyclin A1, A2, and A3, which are further divided into several subgroups. Transcripts are present for much of the cell cycle. In synchronised tobacco BY2 suspension cultured cells, CycA3 transcript is expressed from the G1–S transition until the onset of mitosis. CycA1 and A2 transcripts are present between mid S-phase until mid-mitosis.

In contrast to A-type cyclins in other eukaryotes, the protein levels of plant cyclin A1 and cyclin A3 are found to be not significantly regulated during the cell cycle (Chaubet-Gigot, 2000). Immunolocalisation studies in maize root tips and tobacco cells indicate that cyclin A1 and cyclin A3 proteins are diffusely distributed in interphase cells and concentrate around the nucleus in G2-phase (Mews *et al.*, 1997). At nuclear envelope breakdown, the two cyclins relocate around the chromatin and persist around the mitotic spindle zone throughout mitosis but do not associate with chromosomes. The maize cyclin A1, but not the tobacco cyclin A1, was also detected along all microtubule structures throughout all stages of mitosis: the preprophase band in G2-phase, the spindle microtubules in metaphase and anaphase and the phragmoplast in telophase (Chaubet-Gigot, 2000; Mews *et al.*, 1997).

The prolonged stability of the plant A-type cyclins (until the end of mitosis) is intriguing since the destruction box, which is usually required for destruction of mitotic cyclins during mitosis, is present in all A-type cyclins. Moreover, this motif has been shown to be functional in directing proteolysis to the CAT reporter protein, which was fused to an N-terminal fragment of the tobacco cyclin A3 gene containing the destruction box motif (Genschik *et al.*, 1998). A CycA3-GFP fusion protein was found to become degraded at the onset of mitosis when stably expressed in transgenic tobacco By2 cells. The CycA3-GFP fusion localises to the nucleus and the nucleoli only in interphase cells and is not detected during any stages of mitosis (Criqui *et al.*, 2000). Similar results were obtained by following the localisation of an A2-type cyclin from *Medicago sativa* by immunofluorescence and immunogold labelling: the cyclin A2 protein is found in the nucleus during interphase and prophase while it is undetectable during later stages of mitosis (Roudier *et al.*, 2000).

Information about the Cdk partner of plant A-type cyclins is very preliminary: The tobacco cyclin A1 was shown to be able to interact with CdkA *in vitro*, after production of both proteins from a baculovirus expression system in insect cells (Nakagami *et al.*, 1999). Using a two-hybrid system in yeast the MedsaCycA2 has been shown to interact with Cdk A (Roudier *et al.*, 2000).

The PPB appears in late G2-phase and disassembles in prometaphase before formation of the spindle. There is some evidence that active Cdk kinase complexes are involved in the organisation of the PPB structure: Activation of both Cdk A and Cdk B kinases occurs in late G2-phase at a time when both kinases were found to be associated with the PPB. CdkA fused to green fluorscent protein (GFP) locates to a subset of microtubules within the PPB, probably at a late stage in its development (Weingartner *et al.*, 2001). Assembly and disassembly of the PPB is sensitive to kinase inhibitors (Shibaoka & Nagai, 1994) and microinjection of active Cdk-complexes into late G2 cells results in PPB disassembly (Hush *et al.*, 1996).

4.4.3 The DNA damage checkpoint

Checkpoint mechanisms ensure proper order and correct execution of cell cycle events, and checkpoints are thought to monitor passage through mitosis at several stages. Best understood is the DNA damage checkpoint that operates during G2 and ensures that entry into mitosis can only occur upon the successful completion of DNA replication or DNA repair events. Two groups of kinases that are widely conserved play a central role in response to DNA damage: the ATM/ATR group and the Chk1 and Chk2 group. Mutations in both groups of checkpoint kinases confers susceptibility to cancer and their homozygous deletion results in embryonic lethality in mice (Liu *et al.*, 2000; Takai *et al.*, 2000). The ATM/ATR group consists of proteins homologues to lipid kinases or the PI3-K family (phosphatidylinositol-3-kinase). Yeast counterparts are Rad3 in *S. pombe* (Furnari *et al.*, 1999) or Mec1 in budding yeast (Murray *et al.*, 1997) and the *Drosophila* homologue has been identified as Mei-41 (Sibon *et al.*, 1999). The Chk1 and Chk2 kinases look like classic serine/threonine kinases. Chk1 was first identified in yeast (Weinert & Hartwell, 1989) with homologues in mammals, *Xenopus* and flies (O'Connell *et al.*, 1997; Walworth & Bernards, 1996). Chk2 is the mammalian homologue to Cds1 in *S. pombe* and *Xenopus* and Rad53 in *S. cerevisiae* (Raleigh & O'Connell, 2000).

The molecular mechanism underlying this checkpoint pathway has been elucidated in fission yeast: In response to DNA damage the checkpoint kinase Chk1 becomes activated before entry into mitosis and inhibits Cdc25p activation by phosphorylating it at a site that promotes its association with a family of proteins known as 14-3-3 proteins thus preventing Cdc25p from dephosphorylating and activating cyclin B/cdc2 (Dalal *et al.*, 1999; Kumagai *et al.*, 1998). Only when Cdc25p is released from the 14-3-3 complex is it able to translocate into the nucleus and activate Cdc2–cyclin B. The basic mechanism is conserved but additional as yet unidentified levels of control are operating in animal and plant cells.

Drugs like caffeine or okadaic acid are known to interfere with checkpoint function in animal cells like the replication checkpoint in S-phase (Schlegel & Pardee, 1986) and to release the DNA damage checkpoint during G2-phase by inhibiting the ATM kinase (Blasina *et al.*, 1999; Bordello *et al.*, 1999). In plants the ATM kinase has been conserved (Garcia *et al.*, 2000) and caffeine is also able to cancel the G2 DNA damage checkpoint but it is unable to cancel replication checkpoints in S-phase (Amino & Nagata, 1996; Pelayo *et al.*, 2001). Recently it was shown that cells ectopically expressing cyclin B2 are able to overcome a hydroxyurea induced replication checkpoint block by treatment with caffeine, which induces also an increased CycB2-associated kinase activity (Weingartner *et al.*, 2003). These results suggest that at least one of the final targets of the replication checkpoint pathway in plants is a CycB2-associated Cdk.

4.5 Commitment to mitosis

4.5.1 Commitment to mitosis in animal cells

Entry into mitosis is irreversible and is triggered by the abrupt activation of Cdk1–cyclin B1 complex. The Cdk1–cyclin B complex is kept inactive by phosphorylation on two conserved phosphorylation sites, threonine 14 and tyrosine 15, by kinases known as Wee1p or Myt1p and related kinases (Mueller *et al.*, 1995; Russell & Nurse, 1987). Complete and sudden activation of the Cdk–cyclin complex as a mitotic kinase is achieved by dephosphorylation of these two inhibitory sites by the Cdc25 phosphatase, a tyrosine phosphatase. Cdc25p, itself, is activiated by a positive feedback loop between Cdc2p and Cdc25p. This process depends upon a two step mechanism: First the Cdc25 phosphatase activates the Cdc2 kinase which in turn phosphorylates Cdc25p leading to increased Cdc2 kinase activity which follows a linear kinetics. Second, the Plx1 kinase catalyses hyperphosphorylation of Cdc25 leading to full activation of the Cdc2–cyclin B complex. For this second step, the PP2A phosphatase has to be inhibited since PP2A antagonises the Plx1 mediated phosphorylation of Cdc25 (Dunphy & Newport, 1989; Morgan *et al.*, 1998; Nigg, 2001).

In fission yeast, dephosphorylation of Cdc2–cyclin B by Cdc25p is the rate-limiting step for the onset of mitosis (Millar *et al.*, 1991; Russell & Nurse, 1986). In mammalian cells, dephosphorylation of Cdk1 alone is not sufficient to trigger entry into mitosis and additional control mechanisms seem to be required. Nuclear import and export of Cdk–cyclin complexes and its regulators might be another mechanism contributing to the control of mitotic entry. Recent advances suggest that Cdk–cyclin activity might be locally regulated depending on its subcellular localisation and cues from the immediate environment (Pines, 1999).

4.5.2 Commitment to mitosis in plant cells

Phosphorylation mediated positive and negative regulatory pathways were also found to contribute to mitotic entry in plants. A Wee1 homologue that can inhibit Cdk activity has been identified in maize (Sun *et al.*, 1999). Although no direct homologues of the *Cdc25* gene have been found in the fully sequenced genome of *Arabidopsis*, all CdkA proteins contain a conserved tyrosine near the N-terminus. Moreover, *Nicotiana plumbagifolia* suspension cultured cells that have been arrested in G2 by cytokinin starvation, can be induced to resume mitosis by inducible expression of yeast *Cdc25* (John, 1996; Zhang *et al.*, 1996) indicating that Tyr15 phosphorylation of Cdks is an important control mechanism in plant mitosis. Taken together, these data suggest that cytokinin can regulate dephosphorylation of Cdks. Overexpression of *Cdc25* in fission yeast results in a decreased cell size in mitosis (Moreno *et al.*, 1990). Constitutive expression of the same fission yeast gene in tobacco plants results in precocious flowering, aberrant leaves and flowers and a smaller size of dividing cells in lateral roots compared to control plants (Bell *et al.*, 1993). These results reinforce the significance of a Cdc25-like phosphatase for mitotic entry in plants, which might be under control of the plant hormone cytokinin.

4.5.3 The role of animal B-type cyclins

Active cyclin B–Cdk1 initiates profound changes in the cellular architecture by phosphorylating a number of substrates like lamins and vimentin to reorganise the karyoskeletal and cytoskeletal intermediate filament networks and it promotes microfilament reorganisation by phosphorylating caldesmon (Yamashiro & Matsumura, 1991). It also changes the nucleating activity of centrosomes and hence the dynamics of microtubule organisation (Verde *et al.*, 1992). Five B-type cyclins have been isolated from *Xenopus laevis* (Minshull *et al.*, 1990) and at least two subtypes, B1 and B2, from mammalian cells (Jackman & Pines, 1997) and B2 and B3 from chicken cells (Gallant & Nigg, 1992). All B-type cyclins have three functional domains, comprising a cyclin box at the C-terminal half required for the binding to Cdk1, an N-terminal destruction box which is responsible for its degradation during mitosis and a cytoplasmic retention signal in between the two boxes which mediates the cytoplasmic localisation in interphase (Pines & Hunter, 1994).

The distinct functions of cyclin B1 and cyclin B2 have been most extensively studied in mammalian cells: both cyclins bind to Cdk1, are active only in mitosis and *in vitro* their associated kinase activity shows similar substrate specifities (Nigg, 1991). In knockout mice, only cyclin B1 was found to be essential, since mice lacking cyclin B1 die *in utero*, while mice lacking cyclin B2 are smaller than normal indicating that cyclin B2 confers a growth advantage (Brandeis *et al.*, 1998). These results indicate that a cyclin B2–Cdk complex operates as a target of a checkpoint mechanism that delays but does not block cell cycle progression. Studies in human cells have revealed that the two proteins differ in their subcellular localisation throughout cell cycle progression: Cyclin B1 is mainly cytoplasmic and shuttles between the nucleus and the cytoplasm during interphase. At the end of prophase it rapidly translocates to the nucleus and then binds to the mitotic apparatus (Clute & Pines, 1999; Hagting *et al.*, 1999). In contrast, cyclin B2 is continuously cytoplasmic, associated with the Golgi apparatus, does not translocate to the nucleus and remains associated with membrane vesicles until it is degraded in mitosis (Jackman *et al.*, 1995). By exchanging the region that is responsible for the localisation of the two mammalian B-type cyclins it has been demonstrated that it is not the intrinsic properties of the cyclin protein itself but rather the distinct subcellular distributions of cyclin B1 and cyclin B2 that seem to specify the biological activities of their associated kinase activity. Cdk1 activated by cyclin B1 is able to induce solubilisation of the nuclear lamina and, together with other factors, initiates chromosome condensation and microtubule reorganisation, whereas when activated by cyclin B2 it is only able to initiate disassembly of the Golgi apparatus (Draviam *et al.*, 2001). Expression of a stable version of cyclin B1 in *Drosophila* was shown to result in an arrest in anaphase with chromosomes chaotically arranged around the midpoint of the spindle. Real-time analysis revealed that the earliest defect following cyclin B expression is the slowing of chromosome movements during anaphase A, and at a later stage it blocks pole separation. These results suggest that the timely degradation of cyclin B is essential for normal chromosome movement and initiation of pole separation, implying a role for cyclin B in regulating motor proteins involved in the stability of kinetochore-spindle interactions and spindle movements (Parry & O'Farrell, 2001).

4.5.4 The role of plant B-type cyclins

Plant B-type cyclins form at least two groups, CycB1 and CycB2 with four and three subgroups respectively. These groups arise within the plant B-type cyclin class and are not related to the groups of animal B-type cyclins. Information about the distinct functions of plant B-type cyclins can only be guessed from the subcellular localisation of their gene products and the effect of their overexpression in plants and cultured cells. The expression pattern and localisation of three different B-type cyclins, ZmcycB1;1, ZmCycB1;2, ZmCycB2;1 during developmental changes in the maize root was studied by immunolocalisation experiments. ZmCycB1;2 is the only cyclin whose protein is only found in dividing tissues and its localisation closely resembles that of human cyclin B1 in that it relocates to the nucleus, associates with condensed chromosomes in prophase and becomes degraded in anaphase. ZmCycB1;2 and ZmCycB2;1 are found in the nucleus, or close to the nuclear material during all stages of mitosis (Mews *et al.*, 1997). In non-dividing cells ZmCycB1;2 declines to hardly detectable levels, while protein levels of ZmCycB1;1 and ZmCycB2;1 remain significantly high in the nucleus. This indicates that some plant cyclins are tightly linked to cell division while others can persist, but only in the nuclei, in non-dividing cells. The transition from division to differentiation might therefore be partly triggered by the decrease of cell cycle proteins and especially the decline of cyclins in the cytoplasm. After resuming cell division, which was followed by lateral-root formation and wound response, the reappearance of cyclin B2;1 with a high nuclear and a low cytoplasmic accumulation, is correlated with the first signs of dedifferentiation. Since later during this process ZmCycB2;1 is also found at places of high tubulin concentration it may function in the rearrangement of microtubules for cell division (Mews *et al.*, 1997). This study supports the possibility that mitotic cyclins are involved in the process of resumption of cell division during root morphogenesis. A similar conclusion came from a previous study in which the effect of ectopic cyclin B1 expression on root development was analysed in transgenic *Arabidopsis* plants expressing an Arath;CycB1;2 under the control of a CdkA promotor. While the overall root morphology was not changed, ectopic Arath;CycB1;2 expression enhanced root growth by increasing the rate of cell division without altering the organisation of the meristem. Thus it was concluded that cyclin expression might be a limiting factor in the regulatory network governing meristem activity, organised growth, and indeterminate development of the root system (Doerner *et al.*, 1996).

The *in vivo* sub-cellular localisation of two different plant B1-type cyclins, CycB1;1 and CycB1;3 was analysed by visualising their proteins as GFP-tagged fusion proteins in cultured tobacco BY2 cells. Both cyclins accumulate in the nucleus during late G2, associate with the condensing chromosomes during prophase and remain on the chromosomes until metaphase. Time-lapse images revealed that both cyclins become rapidly degraded at the onset of anaphase. A stable variant of cyclin B1;1-GFP, in which the destruction box has been mutated, maintains its association with the chromosomes during all stages of mitosis until telophase and during cytokinesis. Activation of the spindle assembly checkpoint by treatment with microtubule depolymerising drugs stabilises both B1 type cyclins and they remain associated with the chromosomes in metaphase arrested cells. These results are similar to the human cyclin B1, which was analysed as a GFP-fusion protein in HeLa cells and associates with the condensed chromosomes until prometaphase (Clute & Pines, 1999). But, in contrast to plant cyclin B1, animal cyclin B1 is also found along the microtubules of the mitotic spindle and the spindle poles. While a mutated non-destructible form of the human cyclin B1 is only associated with the spindle poles and unable to bind the condensed chromosomes (Clute & Pines, 1999), the mutant plant cyclin B1;1-GFP protein remains associated with the chromosomes along all the mitotic events (Criqui *et al.*, 2001). The identity of the Cdk-partner of plant cyclin B1 proteins has not been determined yet in plants (Meszaros *et al.*, 2000). Visualisation of the *in vivo* localisation of a CdkA-GFP fusion protein in cultured cells locates the CdkA-GFP to the nucleus in G2, along the condensed chromosomes until mid-prophase and along all microtubule arrays engaged in mitosis including the preprophase band, the mitotic spindle and the phragmoplast (Weingartner *et al.*, 2001). The colocalisation of CdkA-GFP and cyclin B1-GFP along the condensed chromosomes during prophase suggests that they transiently interact and trigger early events of mitosis. The mechanism translocating CdkA from the condensed chromosomes to the microtubules in metaphase remains elusive and so does the the identity of the Cdk partner of cyclin B1 during metaphase.

A plant cyclin B2 gene was found to play a critical role during the transition from G2- to M-phase. Ectopic expression of cyclin B2 in cultured tobacco cells during G2-phase accelerates entry into mitosis and allows cells to overcome checkpoint control pathways like the replication checkpoint in S-phase and the topoisomerase checkpoint during G2. The localisation of the cyclin B2 protein is confined to the nucleus and during mitosis it becomes efficiently degraded during mid-prophase. In this way it resembles much more animal A-type cyclins, since its degradation is not inhibited by activation of the spindle checkpoint. When ectopically expressed in transgenic plants, cyclin B2 was found to interfere with differentiation events and block root regeneration from leaf explants indicating control mechanisms operating at the G2 to M transition during plant developmental processes (Weingartner *et al.*, 2003). Evidence for specialised functions of closely related plant B-type cyclins came from the finding that ectopic expression of the *Arabidopsis cyclin B1;2* gene, but not of *cyclin B1;1* inhibits endoreduplication and induces ectopic cell divisions, resulting in multicellular trichomes (Schnittger *et al.*, 2002).

4.6 Condensation of chromatin

A universal feature of mitosis is the condensation of chromatin. Even in budding yeast, the structure of the genome changes as the cells progress through mitosis. Condensation involves the compaction of the chromosomes from an extended threadlike organisation to a highly organised and compact state, and is normally associated with the establishment of a spindle composed of microtubules.

4.6.1 Condensation of chromatin in animal cells

Several protein complexes are required for interconversion between extended interphase chromatin and highly compacted mitotic chromosomes. The condensins, originally identified in *Xenopus*, are involved in the topological changes in the DNA template (Hirano *et al.*, 1997). These complexes contain a SMC heterodimer comprising Smc2 (Cut14/XCAP-E) and Smc4 (Cut3/XCAP-C) and three non-SMC subunits, Cnd 1, 2, 3. In *Xenopus*, condensin activity is regulated by a Cdk–cyclin complex that phosphorylates the non-SCM subunits just after nuclear envelope breakdown. This mechanism explains the dependence of chromosome condensation on Cdk activation and decondensation on degradation of mitotic cyclins observed *in vivo* (Zachariae & Nasmyth, 1999).

Phosphorylation of histone H3 on its N-terminal tail correlates with mitosis and meiosis in a wide range of organisms but its functional relationship is not yet clear. Blocking H3 phosphorylation prevents initiation of chromatin condensation whereas inhibiting dephosphorylation of H3 induces abnormally condensed chromosomes and problems during exit from mitosis. One hypothesis is that mitotic histone H3 phosphorylation opens up chromatin fibres for nuclear factors that promote chromosome condensation (Cheung *et al.*, 2000). Using budding yeast and *Caenorhabditis* as models the IpP1/aurora kinase and its genetically interacting partner Glc7/PP1 phosphatase have been identified as the enzymes governing phosphorylation and dephosphorylation of histone H3 during mitosis (Hsu *et al.*, 2000).

Aurora, originally identified in *Drosophila*, is highly homologous to Ipl1 in budding yeast and both genes play an important function in cell division and chromosome segregation (Goepfert & Brinkley, 2000). Data on aurora-related kinases in other species suggest that this is an evolutionary conserved kinase, which is generally involved in progression through mitosis.

4.6.2 Condensation of chromatin in plant cells

In plants phosphorylation of histone H3 is also found to correlate with mitotic chromosome condensation. Immunolocalisation studies in maize using an antibody that recognises a Ser10 phosphoepitope on histone H3 revealed some differences in the pattern of histone H3 phosphorylation compared to animal cells. During mitosis, histone H3 phosphorylation occurs late in prophase and is maintained until telophase; but while in plants it is restricted to pericentromeric regions, in animal cells histone phosphorylation is found along the entire length of the chromosomes (Kaszas & Cande, 2000). During plant meiosis, histone H3 phosphorylation does not correlate with chromosome condensation *per se* but rather with the maintenance of sister chromatid cohesion suggesting a role for phosphorylated histone H3 during this process (Kaszas & Cande, 2000).

4.7 Spindle formation

4.7.1 Spindle formation in animal cells

During each cell cycle the microtubules have to organise into a bipolar spindle that segregates the replicated chromosomes into two daughter nuclei. In animal cells, the spindle microtubules are nucleated from the centrosomes which eventually form the spindle poles. The minus (less stable) ends of microtubules are embedded in amorphous material surrounding the centrosomes, while the plus ends grow towards the spindle equator to interact with microtubules emanating from the other pole and with the condensed chromosomes.

Spindle assembly and mitotic movements rely on three parameters. First the inherent dynamic properties of microtubule polymers that allow individual microtubules to grow and shrink. Mitotic microtubules are markedly more dynamic than interphase ones and this transition is triggered *in vitro* by several kinases including Cdk1–cyclin A and MAP kinase (Andersen, 1999). Second, a balance of microtubule stabilising and destabilising accessory proteins mediates local changes in microtubule stability and third, the action of microtubule-dependent motors of the dynein and kinesin families is mainly responsible for microtubule movements.

Most of these accessory and motor proteins are regulated by phosphorylation involving Cdk1–cyclin B and other kinases. A model has emerged from *in vitro* studies in *Xenopus* extracts suggesting that chromatin-induced spindle formation depends on gradients of differentially phosphorylated microtubule-associated proteins (MAPs) and these gradients seem to arise from immobilised kinases and diffusible phosphatases or vice versa (Andersen, 1998). This model highlights also the importance of the mutual localisation of kinases and phosphatases and their substrates in the control of mitotic reactions (Andersen, 1998). Microtubule-dependent motors are essential for spindle assembly and function throughout mitosis. Several distinct kinesins and kinesin-related proteins, also called KRPs, and cytoplasmic dynein have been isolated in diverse organisms. Although coordination between different motor activities would seem critical for the correct execution of chromosome segregation, little is known about how the individual motors are targeted to particular structures and how their local activities are controlled. Phosphorylation is certainly a key factor in the spatial and temporal control of motor activity, emphasising the predominant role of Cdk1–cyclin B complexes in spindle microtubule dynamics (Nigg, 2001).

The polo kinase was originally identified in *Drosophila* where its mutated form causes defects in centrosome behaviour leading to monopolar spindles (Sunkel & Glover, 1988). Polo kinase is a serine-threonine kinase that is highly conserved from yeast to human. All polo-like kinases (plks) share a closely related catalytic domain at the amino terminus and a characteristic sequence motif, the polo box. While in yeast and *Drosophila* a single plk gene has been identified, higher vertebrates have at least three different plk genes with plk1 being most similar to *Drosophila* polo (Lane & Nigg, 1996; Lee & Erikson, 1997). So far, polo and pololike kinases have not been identified in plants. The timing of activity of plks peaks during mitosis, and animal Plk1 was shown to bind to all mitotic microtubule arrays suggesting a role in the spatial organisation of mitosis (Arnaud *et al.*, 1998; Golsteyn *et al.*, 1995). Functional studies in a variety of organisms have pointed to key roles of the polo kinase family at several steps of mitosis: A *Xenopus* Plk copurifies with and can activate cdc25, and thus may play a role in the positive feedback loop that operates during cdc2 activation (Kumagai & Dunphy, 1996; Qian *et al.*, 1998). As inferred from microinjection studies in human cells (Lane & Nigg, 1996) and *Xenopus* embryos (Qian *et al.*, 1998) centrosome maturation and separation requries the action of pololike kinases. A role for the enzyme in activating the APC has been suggested in yeast and vertebrate cells (Kotani *et al.*, 1999; Shirayama *et al.*, 1998).

4.7.2 *Spindle formation in plant cells*

The plant spindle has a similar overall organisation to the animal spindle but it lacks centrosomes and spindle poles are usually less focused. Consistent with these observations, the most active MTOC (microtubule organising centre) in early mitosis is spread around the nuclear surface (Lambert, 1993). Kinetochores may also serve as MTOCs in that microtubules appear to accumulate around kinetochores early in spindle formation (Chan & Cande, 1998) and at least two plant kinetochore proteins have homology to animal centrosome components. One of these is γ-tubulin (Binarova *et al.*, 1998, 2000; Drykova *et al.*, 2003), an important component of centrosomes in animals that functions primarily in microtubule nucleation and the organisation of microtubule dynamics.

Microtubule motors and MAPs (see Chapter 1) are likely to be involved in organising the plant spindle and aiding the separation of chromatids. Several plant MAPs have been isolated with little homology to animal MAPS. For example, a group of at least four different proteins known as MAP65 were found to bind to a subset of interphase and mitotic microtubule arrays and to bundle microtubules *in vitro* (Jiang & Sonobe, 1993; Chan *et al.*, 1999; Smertenko *et al.*, 2000). These proteins share a remote ancestor with a yeast protein involved in cytokinesis, but have diversified in plants and perhaps acquired new functions (Pellman *et al.*, 1995).

Most of the kinesins or kinesin-related proteins found in plants have been shown to be involved in phragmoplast formation rather than spindle movements. But the complete sequence of the *Arabidopsis* genome has revealed a list of putative kinesin proteins whose functions remain to be elucidated (Lloyd & Hussey, 2001).

CdkA activity is required for normal entry into mitosis and the reorganisation of mitotic microtubule arrays. Treatment of root tip cells with the cdk-specific inhibitor, roscovitine, disrupts mitotic spindle formation leading to the formation of abnormal or monopolar spindles (Binarova *et al.*, 1998). A direct interaction between CdkA and microtubules was found by immunolocalisation experiments in isolated root tip cells of *Mediacago sativa* and dynamic colocalisation of CdkA with several microtubule arrays during mitosis (Stals *et al.*, 1997; Weingartner *et al.*, 2001).

4.8 The spindle assembly checkpoint pathway

A multimeric complex containing the products of the BUB genes delays sister-chromatid separation until all chromosomes are properly aligned on the spindle and has been identified by genetic studies in yeast (Gorbsky, 1997). This checkpoint monitors the attachment of microtubules to kinetochores and/or the generation of tension that results from bipolar attachment of sister chromatids, and is therefore also referred to as the kinetochore attachment checkpoint. In yeast, there are at least six gene products involved in this checkpoint and include the two kinases Mps1p and Bub1p and its partner Bub3p, and the three proteins Mad1p, 2p, 3p. Homologues of several of these proteins have been shown to associate preferentially to unattached kinetochores in animal cells, suggesting a role for phosphorylation in the generation of an anaphase-inhibitory signal at kinetochores (Nigg, 2001). The spindle assembly checkpoint is not merely activated in response to spindle damage but contributes to the timing of anaphase onset during every cell division (Geley *et al.*, 2001). According to a current model, structural changes induced by microtubule attachment are translated into a biochemical signal through phosphorylation. In vertebrate cells, this mechano-chemical coupling involves a molecular interaction between the kinesin-related protein CENP-E, which acts in tethering kinetochores to the plus ends of microtubules, and the kinase BubR1. How this regulates the kinetochore association of Mad proteins is unknown (Abrieu *et al.*, 2000). Unattached kinetochores are thought to function as sites of continuous assembly and release of Mad2–Cdc20 complexes that prevent the activation of APC–Cdc20. Upon attachment of the last kinetochore, the production of Mad2–Cdc20 complexes ceases and thus allows Cdc20 to dissociate from Mad2 and bind to and activate the anaphase promoting complex (APC) (Hardwick, 1998; Kallio *et al.*, 1998; Rieder & Salmon, 1998).

4.8.1 Regulation of APC

The APC is a large ubiquitin ligase complex that triggers the ubiquitin-dependent proteolysis of key cell cycle regulators. Since the degradation of different substrates occurs at different times, an exquisite regulation of the APC is required. The APC is sequentially activated by the association with two distinct WD40 repeat proteins known as Cdc20 and Cdh1. The Cdc20- and Cdh1-bound forms of the APC demonstrate different substrate specifities and are active during different stages of mitosis: Cdc20–APC recognises destruction box containing proteins and Cdh1–APC recognises proteins containing a destruction box or a KEN box motif (Fang *et al.*, 1999; Pfleger & Kirschner, 2000). Cdc20–APC is active at the metaphase/anaphase transition and responsible for anaphase onset by initiating the degradation of securin. Cdh1–APC is turned on later in mitosis but remains active throughout the subsequent G1-phase and targets mitotic cyclins and several other mitotic regulators for destruction (Morgan, 1999), perhaps providing a mechanism to prevent accidental mitotic entry during G1. Mitotic Cdks and probably other mitotic kinases regulate the two forms of APC in an opposite fashion thus establishing the temporal order of APC activity: phosphorylation of APC core subunits is required for activation of Cdc20–APC whereas phosphorylation of Cdh1 prevents activation of Cdh1–APC (Kramer *et al.*, 2000). The current model is that activation of Cdks in early mitosis leads first to Cdc20 activation, which then brings on cyclin destruction and Cdk inactivation, leading finally to stable Cdh1 activation in late mitosis and G1 (Morgan, 1999). Many of the proteins involved in the spindle checkpoint are highly conserved and several have been found in plants (Starr *et al.*, 1997). Among them is the maize homologue of MAD2, which is abundant at kinetochores during mitosis. In early prometaphase MAD2 is barely detectable at kinetochores after the microtubules have attached (Yu *et al.*, 1999). The existence of a spindle checkpoint mechanism in plants is further indicated by pharmacological studies. Treatment of synchronised plant cell cultures with microtubule destabilising drugs leads to a transient metaphase-like arrest with highly condensed chromosomes scattered throughout the cell (Planchais *et al.*, 2000).

4.9 Separating the genome

4.9.1 Onset of APC-mediated proteolysis in animal cells

Once activated, the APC–Cdc20 complex mediates the simultaneous separation of all sister chromatids, which in yeast depends on the degradation of an inhibitor, called securin (Nasmyth *et al.*, 2000; Sutani *et al.*, 1999). Securin prevents a protease, termed separase, from abolishing sister-chromatid cohesion by eliminating a component of a multiprotein complex known as cohesin (Uhlmann *et al.*, 2000).

Although timed destruction of cohesion has been broadly conserved during evolution, different mechanisms exist for destroying cohesion at chromosome arms and centromeres. In animal cells, the bulk of cohesin is removed from chromosome arms in prophase, perhaps to permit extensive condensation. This first wave of cohesin removal does not depend on the APC but instead requires phosphorylation of cohesin (Losada *et al.*, 2000; Sumara *et al.*, 2000). One kinase able to phosphorylate cohesin is Cdk1, but other kinases are probably also involved. The small amount of cohesin remaining at centromeres is removed at the metaphase to anaphase transition and this second step is dependent on the APC, presumably following the same securin-separase pathway described for sister-chromatid separation in yeast (Waizenegger *et al.*, 2000b).

Degradation of both cyclin A and cyclin B is known to be mediated by APC–Cdc20 (Morgan *et al.*, 1998) but the precise mechanism regulating cyclin A degradation is still obscure. Cyclin A degradation is initiated shortly after nuclear envelope breakdown, at a time when securin or cyclin B degradation is still blocked by the spindle checkpoint that keeps Cdc20–APC inactive. While degradation of cyclin B1 is dependent only on the highly conserved destruction box motif, degradation of human or *Drosophila* cyclin A requires an extended sequence including a KEN box motif. One possibility is that Cdc20 has different binding sites for A- and B-type cyclins and in complex with MAD2 is only inhibited to bind to cyclin B but is still able to bind and target cyclin A for degradation (den Elzen & Pines, 2001; Geley *et al.*, 2001).

4.9.2 Onset of APC-mediated proteolysis in plant cells

Mitotic progression in plants also depends on APC mediated proteolysis. The N-terminal domains of both A and B cyclin confer cell-cycle-stage specific instability on reporter proteins (Genschik *et al.*, 1998). Mutations in the destruction box motif abolished cell-cycle specific proteolysis. Further evidence that protein degradation is involved in the metaphase/anaphase transition in plants was obtained by pharmacological studies with the proteasome inhibitor MG132. MG132 is a peptide aldehyde that functions as a substrate analogue and was shown in animals to effectively halt cells in metaphase by inhibiting APC-dependent proteolysis of cohesion proteins responsible for sister-chromatid separation (Sherwood *et al.*, 1993). Treatment of synchronised BY2 cell cultures with MG132 was shown to block cells in metaphase with elevated levels of Cdk–kinase activity and leaving cyclin–CAT fusion proteins undegraded (Genschik *et al.*, 1998). Whether the metaphase arrest observed in plant cells is induced by a similar mechanism as in yeast and in animal cells, involving stabilisation of chromatid cohesion protein, is not known yet.

Plant cyclin B1 becomes degraded at the onset of anaphase and is stabilised during activation of the spindle checkpoint pathway by treatment with microtubule-disrupting drugs, similar to animal B-type cyclins (Criqui *et al.*, 2001). In contrast, plant cyclin B2 seems to use a degradation mechanism similar to that of animal cyclin A since it becomes degraded during mid-prophase, is not stabilised by activation of the spindle checkpoint, and the arrangement of its N-terminal destruction box contains multiple destruction box elements (Weingartner *et al.*, 2003).

4.10 Exit from mitosis and cytokinesis

As mitosis ends, the separated chromosomes decondense, the nuclear envelope reassembles around the chromatin and nuclear sub-compartments such as the nucleolus are established. Mitotic exit also involves the disassembly of the mitotic spindle and establishment of replication competent origins on the DNA.

In most cells, mitosis is followed immediately by cell division or cytokinesis. The mechanism of cytokinesis is substantially different in animal and plant cells. In animal cells, a ring of actin and myosin anchored to the cell membrane pulls the existing plasma membrane inward to the centre of the plane of cell division (Field *et al.*, 1999). At the end of cytokinesis, the midbody is formed containing the remains of the two sets of polar microtubules along which vesicles may traffic to complete the plasma membrane between the daughter cells (Nacry *et al.*, 2000). During plant cytokinesis, a phragmoplast forms within the late anaphase spindle and progresses from the centre outwards to the periphery of the cell. The origins of the phragmoplast are obscure but it arises in the midzone between the two daughter nuclei. A predominant role for the phragmoplast is targeting membrane vesicles to the centre of the division plane where they fuse with one another to form a transient membrane compartment, the cell plate. The cell plate expands laterally by continuous vesicle fusion to its margin until it fuses with the parental cell wall at the predicted division site.

4.10.1 Regulators of late mitotic events in animal cells

Most of these events are dependent on complete inactivation of the cyclin B–Cdk complex, which is mediated by APC–Cdh1-dependent degradation of cyclin B. Other substrates of APC–Cdh1 are found among kinesin-related motor proteins required for spindle function like CENP-E (Brown *et al.*, 1996) in mammalian cells or Ase1 in yeast (Juang *et al.*, 1997) whose destruction presumably contributes to spindle disassembly. Studies in yeast and recent findings in animals indicate that after inactivation of mitotic Cdks, several other mitotic kinases are needed for the execution of the cytokinetic pathway.

A role for Plks in cytokinesis has originally been found in *S. pombe*, where septation is impaired in the absence of Plo1p, while overexpression of the kinase triggers additional rounds of septation at any point of the cell cycle (Ohkura *et al.*, 1995). Polo mutants in *Drosophila* fail in cytokinesis (Carmena *et al.*, 1998) and data from mammalian cells support the idea that Plks are upstream regulators of cytokinesis (Nigg, 1998). Relevant substrates in Plk mutants have not been identified yet, but the observed cytokinesis defects might result from mislocalisation or impaired function of KRPs or cytoskeleton-associated proteins whose association with the polo kinase is essential for cytokinesis (Adams *et al.*, 1998; Bahler *et al.*, 1998; Lee *et al.*, 1995).

A role for B-type aurora kinases in cytokinesis is supported by the finding that overexpression of a catalytically inactive aurora B disrupts cleavage furrow formation in mammalian cells (Terada *et al.*, 1998) and RNAi suppression of its homologue, AIR, in *Caenorhabditis elegans* leads to missegregation of chromosomes and cytokinesis defects (Schumacher *et al.*, 1998). Members of the human aurora kinase family have also been identified as kinases, which are overexpressed in colon carcinomas, and some of them have been found to be required during late mitosis for the completion of cytokinesis (Bischoff & Plowman, 1999).

Sak is a polo-like kinase found in *Drosophila* and mammals, which is localised to the nucleolus in late G2, to the centrosomes in G2/M and the cleavage furrow during cytokinesis. In Sak deficient mice, cells arrested in anaphase with high levels of cyclin B and phosphorylated histone H3 indicating that Sak is required for APC-dependent destruction of cyclin B1 and exit from mitosis. Since Sak is localised to the premitotic nucleolus where activators of Cdh1 may be sequestered, this enzyme might be involved in the assembly and activation of Cdh1–APC complexes (Hudson *et al.*, 2001).

4.10.2 Late mitotic events in plant cells

Screening of mutants has revealed a list of genes implicated in different steps of cytokinesis and cell plate maturation in plants. Among these are a number of proteins that function during different events of vesicle trafficking like vesicle formation, transport, and fusion revealing a highly controlled vesicle trafficking machinery implicated in plant cytokinesis (Nacry *et al.*, 2000). The phragmoplast consists of short bundles of antiparallel microtubules (MTs) that are believed to mediate the delivery of Golgi-derived vesicles to the plane of division during the process of cell plate formation. Because the plus ends of phragmoplast MTs overlap at the equator, a plus end-directed motor such as kinesin is believed to mediate vesicle transport, although no candidate has been identified from among the large family of kinesins described (Nacry *et al.*, 2000). Two proteins related to the animal large GTPase dynamin, which is involved in endocytosis of synaptic vesicles, associate with the cell plate in plant cells: phragmoplastin (Gu & Verma, 1996) and its *Arabidopsis* homologue ADL1 (Lauber *et al.*, 1997). Both proteins may be involved in the formation of fusion tubes during the initial stage of cell-plate formation. The products of two genes, KNOLLE and KEULE, have been found to concertedly mediate membrane fusion events during cytokinesis in the *Arabidopsis* embryo. KNOLLE encodes a cytokinesis-specific syntaxin expressed in vesicle-like structures during mitosis and at the phragmoplast (Lauber *et al.*, 1997; Lukowitz *et al.*, 1996). KEULE encodes a member of the Sec1 superfamily of proteins that are capable of inducing conformational changes in syntaxins and priming them for interaction with target proteins on vesicle membranes. KEULE has been shown to bind KNOLLE *in vitro* and the synthetic lethality of *knolle keule* double mutants indicates that the two proteins interact *in vivo* (Waizenegger *et al.*, 2000a; Assaad *et al.*, 2001). The precise function of the KEULE/KNOLLE complex is not known yet but it might be involved in integrating cell cycle signals and transduce them into the cytokinetic vesicle fusion machinery (Assaad *et al.*, 2001).

No homologues of polo-like kinases or aurora kinases, which are all involved in microtubule organisation during mitosis and cytokinesis in animals and yeast, have been described in plants. But evidence for a MAP kinase cascade that is required for plant cytokinesis comes from a recent study showing that kinase negative mutations in NPK1, a MAP kinase kinase kinase, disrupt cytokinesis in tobacco cells. NPK1 is upregulated during late M-phase and the protein is present at the equatorial zone of the phragmoplast where it may be required for phragmoplast expansion towards the cell cortex (Nishihama *et al.*, 2001). NPK1 was also found to interact with a tobacco MAP kinase kinase NtMEK1, which is known to interact and activate the tobacco MAP kinase Ntf6 (Calderini *et al.*, 2001). Ntf6 is regulated in a cell-cycle specific manner, activated during anaphase and telophase and localises to the phragmoplast (Calderini *et al.*, 1998). These data establish a MAP kinase module composed of NPK1, NtMEK1, and Ntf6 that might function in plant cytokinesis whose final downstream target remains to be found. Another MAP kinase, MMK3, which is the alfalfa (*Medicago sativa*) orthologue of the tobacco Ntf6 has also been detected at the phragmoplast in alfalfa suspension cultured cells (Bögre *et al.*, 1999).

Furthermore, two mitotic plant Cdks, CdkA and CdkB (Meszaros *et al.*, 2000; Stals *et al.*, 1997) and some plant cyclins, like ZmCycA1;1, ZmCycB1;1 and ZmCycB2;1 are present and associated with the phragmoplast and the midline of cell division through anaphase, telophase and cytokinesis (Mews *et al.*, 1997). They might be involved in the regulation of MAPs like MAP65 or molecular motors like TPRK125 (Asada *et al.*, 1997; Barroso *et al.*, 2000), which are also located at the midplane of cell division and contain potential Cdc2 phosphorylation sites. Several other proteins and among them kinases have been localised to the growing cell plate in plants whose precise functions remain to be determined but they indicate a highly regulatory network functioning during this final step of mitosis which ultimately severs a cell into two.

4.11 Concluding remarks and perspectives

There are at least three positively regulated processes that run in parallel during G2 and mitosis: (i) the chromosomal condensation cycle; (ii) mitotic spindle assembly and function, responsible for the segregation of both sets of sister chromatids in anaphase; and (iii) division of the mitotic cell cytoplasm to produce two cells at the end of mitosis. There are at least three (irreversible) transitions in early mitosis that are controlled by checkpoints: (i) in midprophase, from reversible to irreversible chromosome condensation, (ii) the prophase to prometaphase transition, marked by nuclear envelope breakdown, and (iii) the metaphase to anaphase transition, kept by the spindle checkpoint, that licenses the segregation of sister chromatids in anaphase. The relative timing as well as the mechanisms of these processes are very different in various organisms as well as in different cell types. However, it is beginning to be realised, that most key regulators are conserved among yeasts, plants and animals. In plants, how cell cycle regulators function is far from being fully understood, but from the available information, mitotic progression can be re-staged based on the activity of these regulators. This will aid comparisons across species and also ascertain what the plant-specific targets of these regulators are, and how they coordinate plant-specific mitotic events such as the formation and dissassembly of the preprophase band, nucleation of microtubules and formation of the spindle without defined centrosomes, and cytokinesis with the aid of phragmoplast.

Acknowledgements

We are very grateful to Consuelo de la Torre for comments and suggestions on this manuscript.

References

Abrieu, A., Kahana, J.A., Wood, K.W. & Cleveland, D.W. (2000) CENP-E as an essential component of the mitotic checkpoint in vitro, *Cell*, **102**, 817–826.

Adams, R.R., Tavares, A.A., Salzberg, A., Bellen, H.J. & Glover, D.M. (1998) Pavarotti encodes a kinesin-like protein required to organize the central spindle and contractile ring for cytokinesis, *Genes Dev.*, **12**, 1483–1494.

Amino, S. & Nagata, T. (1996) Caffeine-induced uncoupling of mitosis from DNA replication in tobacco BY2-cells, *J. Plant Res.*, **109**, 219–222.

Andersen, S.S. (1998) Xenopus interphase and mitotic microtubule-associated proteins differentially suppress microtubule dynamics in vitro, *Cell Motil. Cytoskeleton*, **41**, 202–213.

Andersen, S.S. (1999) Balanced regulation of microtubule dynamics during the cell cycle: a contemporary view, *Bioessays*, **21**, 53–60.

Arnaud, L., Pines, J. & Nigg, E.A. (1998) GFP tagging reveals human Polo-like kinase 1 at the kinetochore/centromere region of mitotic chromosomes, *Chromosoma*, **107**, 424–429.

Asada, T., Kuriyama, R. & Shibaoka, H. (1997) TKRP125, a kinesin-related protein involved in the centrosome-independent organization of the cytokinetic apparatus in tobacco BY-2 cells, *J. Cell Sci.*, **110**(Pt 2), 179–189.

Assaad, F.F., Huet, Y., Mayer, U. & Jurgens, G. (2001) The cytokinesis gene KEULE encodes a Sec1 protein that binds the syntaxin KNOLLE, *J. Cell Biol.*, **152**, 531–543.

Bahler, J., Steever, A.B., Wheatley, S., *et al.* (1998) Role of polo kinase and Mid1p in determining the site of cell division in fission yeast, *J. Cell Biol.*, **143**, 1603–1616.

Barroso, C., Chan, J., Allan, V., Doonan, J., Hussey, P. & Lloyd, C. (2000) Two kinesin-related proteins associated with the cold-stable cytoskeleton of carrot cells: characterization of a novel kinesin, DcKRP120-2, *Plant J.* **24**, 859–868.

Bell, M.H., Halford, N.G., Ormrod, J.C. & Francis, D. (1993) Tobacco plants transformed with cdc25, a mitotic inducer gene from fission yeast, *Plant Mol. Biol.* **23**, 445–451.

Binarova, P., Cenklova, V., Hause, B., *et al.* (2000) Nuclear gamma-tubulin during acentriolar plant mitosis, *Plant Cell*, **12**, 433–442.

Binarova, P., Dolezel, J., Draber, P., Heberle-Bors, E., Strnad, M. & Bogre, L. (1998) Treatment of *Vicia faba* root tip cells with specific inhibitors to cyclin-dependent kinases leads to abnormal spindle formation, *Plant J.*, **16**, 697–707.

Bischoff, J.R. & Plowman, G.D. (1999) The Aurora/Ipl1p kinase family: regulators of chromosome segregation and cytokinesis, *Trends Cell Biol.*, **9**, 454–459.

Blasina, A., Price, B.D., Turenne, G.A. & McGowan, C.H. (1999) Caffeine inhibits the checkpoint kinase ATM, *Curr. Biol.* **9**, 1135–1138.

Bögre, L., Calderini, O., Binarova, P., *et al.* (1999) A MAP kinase is activated late in plant mitosis and becomes localized to the plane of cell division, *Plant Cell*, **11**, 101–114.

Bögre, L., Zwerger, K., Meskiene, I., *et al.* (1997) The cdc2Ms kinase is differently regulated in the cytoplasm and in the nucleus, *Plant Physiol.*, **113**, 841–852.

Brandeis, M., Rosewell, I., Carrington, M., *et al.* (1998) Cyclin B2-null mice develop normally and are fertile whereas cyclin B1-null mice die in utero, *Proc. Natl. Acad. Sci. USA*, **95**, 4344–4349.

Brown, K.D., Wood, K.W. & Cleveland, D.W. (1996) The kinesin-like protein CENP-E is kinetochore-associated throughout poleward chromosome segregation during anaphase-A, *J. Cell Sci.*, **109** (Pt 5), 961–969.

Calderini, O., Bogre, L., Vicente, O., Binarova, P., Heberle-Bors, E. & Wilson, C. (1998) A cell cycle regulated MAP kinase with a possible role in cytokinesis in tobacco cells, *J. Cell Sci.*, **111**(Pt 20), 3091–3100.

Calderini, O., Glab, N., Bergounioux, C., Heberle-Bors, E. & Wilson, C. (2001) A novel tobacco mitogen-activated protein (MAP) kinase kinase, NtMEK1, activates the cell cycle-regulated p43Ntf6 MAP kinase, *J. Biol. Chem.*, **276**, 18139–18145.

Carmena, M., Riparbelli, M.G., Minestrini, G., *et al.* (1998) Drosophila polo kinase is required for cytokinesis, *J. Cell Biol.*, **143**, 659–671.

Chan, A. & Cande, W.Z. (1998) Maize meiotic spindles assemble around chromatin and do not require paired chromosomes, *J. Cell Sci.*, **111**(Pt 23), 3507–3515.

Chan, J., Jensen, C.G., Jensen, L.C., Bush, M. & Lloyd, C.W. (1999) The 65-kDa carrot microtubule-associated protein forms regularly arranged filamentous cross-bridges between microtubules, *Proc. Natl. Acad. Sci. USA*, **96**, 14931–14936.

Chant, J. & Pringle, J.R. (1995) Patterns of bud-site selection in the yeast *Saccharomyces cerevisiae*, *J. Cell Biol.*, **129**, 751–765.

Chaubet-Gigot, N. (2000) Plant A-type cyclins, *Plant Mol. Biol.*, **43**, 659–675.

Cheung, P., Tanner, K.G., Cheung, W.L., Sassone-Corsi, P., Denu, J.M. & Allis, C.D. (2000) Synergistic coupling of histone H3 phosphorylation and acetylation in response to epidermal growth factor stimulation, *Mol. Cell*, **5**, 905–915.

Clute, P. & Pines, J. (1999) Temporal and spatial control of cyclin B1 destruction in metaphase, *Nat. Cell Biol.*, **1**, 82–87.

Criqui, M.C., Parmentier, Y., Derevier, A., Shen, W.H., Dong, A. & Genschik, P. (2000) Cell cycle-dependent proteolysis and ectopic overexpression of cyclin B1 in tobacco BY2 cells, *Plant J.*, **24**, 763–773.

Criqui, M.C., Weingartner, M., Capron, A., *et al.* (2001) Sub-cellular localisation of GFP-tagged tobacco mitotic cyclins during the cell cycle and after spindle checkpoint activation, *Plant J.*, **28**, 569–581.

Dalal, S.N., Schweitzer, C.M., Gan, J. & DeCaprio, J.A. (1999) Cytoplasmic localization of human cdc25C during interphase requires an intact 14–3–3 binding site, *Mol. Cell Biol.*, **19**, 4465–4479.

den Elzen, N. & Pines, J. (2001) Cyclin a is destroyed in prometaphase and can delay chromosome alignment and anaphase, *J. Cell Biol.*, **153**, 121–136.

Doerner, P., Jorgensen, J.E., You, R., Steppuhn, J. & Lamb, C. (1996) Control of root growth and development by cyclin expression, *Nature*, **380**, 520–523.

Doonan, J. & Fobert, P. (1997) Conserved and novel regulators of the plant cell cycle, *Curr. Opin. Cell Biol.*, **9**, 824–830.

Draviam, V.M., Orrechia, S., Lowe, M., Pardi, R. & Pines, J. (2001) The localization of human cyclins b1 and b2 determines cdk1 substrate specificity and neither enzyme requires mek to disassemble the golgi apparatus, *J. Cell Biol.*, **152**, 945–958.

Drykova, D., Cenklova, V.V., Sulimenko, V., Volc, J., Draber, P. & Binarova, P. (2003) Plant gamma-tubulin interacts with alphabeta-tubulin dimers and forms membrane-associated complexes, *Plant Cell*, **15**, 465–480.

Dunphy, W.G. & Newport, J.W. (1989) Fission yeast p13 blocks mitotic activation and tyrosine dephosphorylation of the Xenopus cdc2 protein kinase, *Cell*, **58**, 181–191.

Elledge, S.J. (1996) Cell cycle checkpoints: preventing an identity crisis, *Science*, **274**, 1664–1672.

Fang, G., Yu, H. & Kirschner, M.W. (1999) Control of mitotic transitions by the anaphase-promoting complex, *Philos. Trans. R Soc. Lond. B Biol. Sci.*, **354**, 1583–1590.

Field, C., Li, R. & Oegema, K. (1999) Cytokinesis in eukaryotes: a mechanistic comparison, *Curr. Opin. Cell Biol.*, **11**, 68–80.

Fobert, P.R., Gaudin, V., Lunness, P., Coen, E.S. & Doonan, J.H. (1996) Distinct classes of cdc2-related genes are differentially expressed during the cell division cycle in plants, *Plant Cell*, **8**, 1465–1476.

Foley, E., O'Farrell, P.H. & Sprenger, F. (1999) Rux is a cyclin-dependent kinase inhibitor (CKI) specific for mitotic cyclin-Cdk complexes, *Curr. Biol.*, **9**, 1392–1402.

Foley, E. & Sprenger, F. (2001) The cyclin-dependent kinase inhibitor Roughex is involved in mitotic exit in Drosophila, *Curr. Biol.*, **11**, 151–160.

Furnari, B., Blasina, A., Boddy, M.N., McGowan, C.H. & Russell, P. (1999) Cdc25 inhibited in vivo and in vitro by checkpoint kinases Cds1 and Chk1, *Mol. Biol., Cell*, **10**, 833–845.

Furuno, N., den Elzen, N. & Pines, J. (1999) Human cyclin A is required for mitosis until mid prophase, *J. Cell Biol.*, **147**, 295–306.

Gallant, P. & Nigg, E.A. (1992) Cyclin B2 undergoes cell cycle-dependent nuclear translocation and, when expressed as a non-destructible mutant, causes mitotic arrest in HeLa cells, *J. Cell Biol.*, **117**, 213–224.

Garcia, V., Salanoubat, M., Choisne, N. & Tissier, A. (2000) An ATM homologue from *Arabidopsis* thaliana: complete genomic organisation and expression analysis. *Nucleic Acids Res.*, **28**, 1692–1699.

Garcia-Herdugo, G., Fernandez-Gomez, M.E., Hidalgo, J. & Lopez-Saez, J.F. (1974) Effects of protein synthesis inhibition during plant mitosis, *Exp. Cell Res.*, **89**, 336–342.

Geley, S., Kramer, E., Gieffers, C., Gannon, J., Peters, J.M. & Hunt, T. (2001) Anaphase-promoting complex/cyclosome-dependent proteolysis of human cyclin A starts at the beginning of mitosis and is not subject to the spindle assembly checkpoint, *J. Cell Biol.*, **153**, 137–148.

Genschik, P., Criqui, M.C., Parmentier, Y., Derevier, A. & Fleck, J. (1998) Cell cycle-dependent proteolysis in plants. Identification of the destruction box pathway and metaphase arrest produced by the proteasome inhibitor mg132, *Plant Cell*, **10**, 2063–2076.

Gimenez-Abian, M.I., Utrilla, L., Canovas, J.L., Gimenez-Martin, G., Navarette, M.H. & De la Torre, C. (1998) The positional control of mitosis and cytokinesis in higher plant cells, *Planta*, **204**, 37–43.

Gimenez-Abian, J.F., Clarke, D.J., Devlin, J., *et al.* (2000) Premitotic chromosome individualization in mammalian cells depends on topoisomerase II activity, *Chromosoma*, **109**, 235–244.

Gimenez-Abian, J.F., Clarke, D.J., Gimenez-Abian, M.I., de la Torre, C. & Gimenez-Martin, G. (2001) Synchronous nuclear-envelope breakdown and anaphase onset in plant multinucleate cells, *Protoplasma*, **218**, 192–202.

Gimenez-Abian, J.F., Clarke, D.J., Gimenez-Martin, G., *et al.* (2002) DNA catenations that link sister chromatids until the onset of anaphase are maintained by a checkpoint mechanism, *Eur. J. Cell Biol.*, **81**, 9–16.

Gimenez-Martin, G., Gonzalez Fernandez, A., De la Torre, C. & Fernandez Gomez, M.E. (1971) Partial initiation of endomitosis by 3′-deoxyadenosine, *Chromosoma*, **33**, 361–371.

Goepfert, T.M. & Brinkley, B.R. (2000) The centrosome-associated Aurora/Ipl-like kinase family, *Curr. Top Dev. Biol.*, **49**, 331–342.

Golsteyn, R.M., Mundt, K.E., Fry, A.M. & Nigg, E.A. (1995) Cell cycle regulation of the activity and subcellular localization of Plk1, a human protein kinase implicated in mitotic spindle function, *J. Cell Biol.*, **129**, 1617–1628.

Gorbsky, G.J. (1997) Cell cycle checkpoints: arresting progress in mitosis, *Bioessays*, **19**, 193–197.

Gu, X. & Verma, D.P. (1996) Phragmoplastin, a dynamin-like protein associated with cell plate formation in plants, *Embo. J.*, **15**, 695–704.

Gunning, B.S. (1980) Spatial and temporal regulation of nucleating sites for arrays of cortical microtubules in root tip cells of the water fern Azolla pinnata, *Eur. J. Cell Biol.*, **23**, 53–65.

Hagting, A., Jackman, M., Simpson, K. & Pines, J. (1999) Translocation of cyclin B1 to the nucleus at prophase requires a phosphorylation-dependent nuclear import signal, *Curr. Biol.*, **9**, 680–689.

Hardwick, K.G. (1998) The spindle checkpoint, *Trends Genet.*, **14**, 1–4.

Hirano, T. (2000) Chromosome cohesion, condensation, and separation, *Annu. Rev. Biochem.*, **69**, 115–144.

Hirano, T., Kobayashi, R. & Hirano, M. (1997) Condensins, chromosome condensation protein complexes containing XCAP-C, XCAP-E and a Xenopus homolog of the Drosophila Barren protein, *Cell*, **89**, 511–521.

Hirt, H., Pay, A., Bogre, L., Meskiene, I. & Heberle-Bors, E. (1993) cdc2MsB, a cognate cdc2 gene from alfalfa, complements the G1/S but not the G2/M transition of budding yeast cdc28 mutants, *Plant J.*, **4**, 61–69.

Hirt, H., Pay, A., Gyorgyey, J., *et al.* (1991) Complementation of a yeast cell cycle mutant by an alfalfa cDNA encoding a protein kinase homologous to p34cdc2, *Proc. Natl. Acad. Sci. USA*, **88**, 1636–1640.

Holloway, S.L., Glotzer, M., King, R.W. & Murray, A.W. (1993) Anaphase is initiated by proteolysis rather than by the inactivation of maturation-promoting factor, *Cell*, **73**, 1393–1402.

Hsu, J.Y., Sun, Z.W., Li, X., *et al.* (2000) Mitotic phosphorylation of histone H3 is governed by Ipl1/aurora kinase and Glc7/PP1 phosphatase in budding yeast and nematodes, *Cell*, **102**, 279–291.

Hudson, J.W., Kozarova, A., Cheung, P., *et al.* (2001) Late mitotic failure in mice lacking Sak, a polo-like kinase, *Curr. Biol.*, **11**, 441–446.

Hush, J., Wu, L., John, P.C., Hepler, L.H. & Hepler, P.K. (1996) Plant mitosis promoting factor disassembles the microtubule preprophase band and accelerates prophase progression in *Tradescantia*, *Cell Biol. Int.*, **20**, 275–287.

Jackman, M., Firth, M. & Pines, J. (1995) Human cyclins B1 and B2 are localized to strikingly different structures: B1 to microtubules, B2 primarily to the Golgi apparatus, *Embo. J.*, **14**, 1646–1654.

Jackman, M.R. & Pines, J.N. (1997) Cyclins and the G2/M transition, *Cancer Surv.*, **29**, 47–73.

Jacobs, H.W., Keidel, E. & Lehner, C.F. (2001) A complex degradation signal in Cyclin A required for G1 arrest, and a C-terminal region for mitosis, *Embo. J.*, **20**, 2376–2386.

Jeffrey, P.D., Russo, A.A., Polyak, K., *et al.* (1995) Mechanism of CDK activation revealed by the structure of a cyclinA-CDK2 complex, *Nature*, **376**, 313–320.

Jiang, C.-J. & Sonobe, S. (1993) Identification and preliminary characterization of a 65 kDa higher-plant microtubule-associated protein, *J. Cell Sci.*, **105**(Pt 4), 891–901.

John, P.C. (1996) The plant cell cycle: conserved and unique features in mitotic control, *Prog. Cell Cycle Res.*, **2**, 59–72.

John, P.C., Sek, F.J., Carmichael, J.P. & McCurdy, D.W. (1990) p34cdc2 homologue level, cell division, phytohormone responsiveness and cell differentiation in wheat leaves, *J. Cell Sci.*, **97** (Pt 4), 627–630.

Joubes, J., Chevalier, C., Dudits, D., *et al.* (2000) CDK-related protein kinases in plants, *Plant Mol. Biol.*, **43**, 607–620.

Juang, Y.L., Huang, J., Peters, J.M., McLaughlin, M.E., Tai, C.Y. & Pellman, D. (1997) APC-mediated proteolysis of Ase1 and the morphogenesis of the mitotic spindle, *Science*, **275**, 1311–1314.

Kallio, M., Weinstein, J., Daum, J.R., Burke, D.J. & Gorbsky, G.J. (1998) Mammalian p55CDC mediates association of the spindle checkpoint protein Mad2 with the cyclosome/anaphase-promoting complex, and is involved in regulating anaphase onset and late mitotic events, *J. Cell Biol.*, **141**, 1393–1406.

Kaszas, E. & Cande, W.Z. (2000) Phosphorylation of histone H3 is correlated with changes in the maintenance of sister chromatid cohesion during meiosis in maize, rather than the condensation of the chromatin, *J. Cell Sci.*, **113**(Pt 18), 3217–3226.

Kotani, S., Tanaka, H., Yasuda, H. & Todokoro, K. (1999) Regulation of APC activity by phosphorylation and regulatory factors, *J. Cell Biol.*, **146**, 791–800.

Kramer, E.R., Scheuringer, N., Podtelejnikov, A.V., Mann, M. & Peters, J.M. (2000) Mitotic regulation of the APC activator proteins CDC20 and CDH1, *Mol. Biol. Cell*, **11**, 1555–1569.

Kumagai, A. & Dunphy, W.G. (1996) Purification and molecular cloning of Plx1, a Cdc25-regulatory kinase from Xenopus egg extracts, *Science*, **273**, 1377–1380.

Kumagai, A., Yakowec, P.S. & Dunphy, W.G. (1998) 14–3–3 proteins act as negative regulators of the mitotic inducer Cdc25 in Xenopus egg extracts, *Mol. Biol. Cell*, **9**, 345–354.

Lambert, A.M. (1993) Microtubule-organizing centers in higher plants, *Curr. Opin. Cell Biol.*, **5**, 116–122.

Lane, H.A. & Nigg, E.A. (1996) Antibody microinjection reveals an essential role for human polo-like kinase 1 (Plk1) in the functional maturation of mitotic centrosomes, *J. Cell Biol.*, **135**, 1701–1713.

Lauber, M.H., Waizenegger, I., Steinmann, T., *et al.* (1997) The *Arabidopsis* KNOLLE protein is a cytokinesis-specific syntaxin, *J. Cell Biol.*, **139**, 1485–1493.

Lavoie, B.D., Hogan, E. & Koshland, D. (2002) In vivo dissection of the chromosome condensation machinery: reversibility of condensation distinguishes contributions of condensin and cohesin, *J. Cell Biol.*, **156**, 805–815.

Lee, K.S. & Erikson, R.L. (1997) Plk is a functional homolog of Saccharomyces cerevisiae Cdc5, and elevated Plk activity induces multiple septation structures, *Mol. Cell Biol.*, **17**, 3408–3417.

Lee, K.S., Yuan, Y.L., Kuriyama, R. & Erikson, R.L. (1995) Plk is an M-phase-specific protein kinase and interacts with a kinesin-like protein, CHO1/MKLP-1, *Mol. Cell Biol.*, **15**, 7143–7151.

Lehner, C.F. & O'Farrell, P.H. (1990) The roles of Drosophila cyclins A and B in mitotic control, *Cell*, **61**, 535–547.

Leismann, O., Herzig, A., Heidmann, S. & Lehner, C.F. (2000) Degradation of Drosophila PIM regulates sister chromatid separation during mitosis, *Genes Dev.*, **14**, 2192–2205.

Lew, D.J. & Reed, S.I. (1993) Morphogenesis in the yeast cell cycle: regulation by Cdc28 and cyclins, *J. Cell Biol.*, **120**, 1305–1320.

Liu, Q., Guntuku, S., Cui, X.S., *et al.* (2000) Chk1 is an essential kinase that is regulated by Atr and required for the G(2)/M DNA damage checkpoint, *Genes Dev.*, **14**, 1448–1459.

Lloyd, C. & Hussey, P. (2001) Microtubule-associated proteins in plants – why we need a MAP, *Nat. Rev. Mol. Cell Biol.*, **2**, 40–47.

Losada, A., Yokochi, T., Kobayashi, R. & Hirano, T. (2000) Identification and characterization of SA/Scc3p subunits in the Xenopus and human cohesin complexes, *J. Cell Biol.*, **150**, 405–416.

Lukowitz, W., Mayer, U. & Jurgens, G. (1996) Cytokinesis in the *Arabidopsis* embryo involves the syntaxin-related KNOLLE gene product, *Cell*, **84**, 61–71.

Magyar, Z., Meszaros, T., Miskolczi, P., *et al.* (1997) Cell cycle phase specificity of putative cyclin-dependent kinase variants in synchronized alfalfa cells, *Plant Cell*, **9**, 223–235.

Meszaros, T., Miskolczi, P., Ayaydin, F., *et al.* (2000) Multiple cyclin-dependent kinase complexes and phosphatases control G2/M progression in alfalfa cells, *Plant Mol. Biol.*, **43**, 595–605.

Mews, M., Sek, F.J. & Moore, R. (1997) Mitotic cyclin distribution during maize cell division: implications for the sequence diversity and function of cyclins in plants, *Protoplasma*, **200**, 128–145.

Millar, J.B., McGowan, C.H., Lenaers, G., Jones, R. & Russell, P. (1991) p80cdc25 mitotic inducer is the tyrosine phosphatase that activates p34cdc2 kinase in fission yeast, *Embo. J.*, **10**, 4301–4309.

Minshull, J., Golsteyn, R., Hill, C.S. & Hunt, T. (1990) The A- and B-type cyclin associated cdc2 kinases in Xenopus turn on and off at different times in the cell cycle, *Embo. J.*, **9**, 2865–2875.

Moreno, S., Nurse, P. & Russell, P. (1990) Regulation of mitosis by cyclic accumulation of p80cdc25 mitotic inducer in fission yeast, *Nature*, **344**, 549–552.

Morgan, D.O. (1995) Principles of CDK regulation, *Nature*, **374**, 131–134.

Morgan, D.O. (1999) Regulation of the APC and the exit from mitosis, *Nat. Cell Biol.*, **1**, E47–E53.

Morgan, D.O., Fisher, R.P., Espinoza, F.H., *et al.* (1998) Control of eukaryotic cell cycle progression by phosphorylation of cyclin-dependent kinases, *Cancer J. Sci. Am.*, **4**(Suppl. 1), S77–S83.

Mueller, P.R., Coleman, T.R., Kumagai, A. & Dunphy, W.G. (1995) Myt1: a membrane-associated inhibitory kinase that phosphorylates Cdc2 on both threonine-14 and tyrosine-15, *Science*, **270**, 86–90.

Murray, J.M., Lindsay, H.D., Munday, C.A. & Carr, A.M. (1997) Role of Schizosaccharomyces pombe RecQ homolog, recombination, and checkpoint genes in UV damage tolerance, *Mol. Cell Biol.*, **17**, 6868–6875.

Nacry, P., Mayer, U. & Jurgens, G. (2000) Genetic dissection of cytokinesis, *Plant Mol. Biol.*, **43**, 719–733.

Nakagami, H., Sekine, M., Murakami, H. & Shinmyo, A. (1999) Tobacco retinoblastoma-related protein phosphorylated by a distinct cyclin-dependent kinase complex with Cdc2/cyclin D in vitro, *Plant J.*, **18**, 243–252.

Nasmyth, K., Peters, J.M. & Uhlmann, F. (2000) Splitting the chromosome: cutting the ties that bind sister chromatids, *Science*, **288**, 1379–1385.

Nigg, E.A. (1991) The substrates of the cdc2 kinase, *Semin. Cell Biol.*, **2**, 261–270.

Nigg, E.A. (1998) Polo-like kinases: positive regulators of cell division from start to finish, *Curr. Opin. Cell Biol.*, **10**, 776–783.

Nigg, E.A. (2001) Mitotic kinases as regulators of cell division and its checkpoints, *Nat. Rev. Mol. Cell Biol.*, **2**, 21–32.

Nishihama, R., Ishikawa, M., Araki, S., Soyano, T., Asada, T. & Machida, Y. (2001) The NPK1 mitogen-activated protein kinase kinase kinase is a regulator of cell-plate formation in plant cytokinesis, *Genes Dev.*, **15**, 352–363.

O'Connell, M.J., Raleigh, J.M., Verkade, H.M. & Nurse, P. (1997) Chk1 is a wee1 kinase in the G2 DNA damage checkpoint inhibiting cdc2 by Y15 phosphorylation, *Embo. J.*, **16**, 545–554.

Ohi, R. & Gould, K.L. (1999) Regulating the onset of mitosis, *Curr. Opin. Cell Biol.*, **11**, 267–273.

Ohkura, H., Hagan, I.M. & Glover, D.M. (1995) The conserved Schizosaccharomyces pombe kinase plo1, required to form a bipolar spindle, the actin ring, and septum, can drive septum formation in G1 and G2 cells, *Genes Dev.*, **9**, 1059–1073.

Parry, D.H. & O'Farrell, P.H. (2001) The schedule of destruction of three mitotic cyclins can dictate the timing of events during exit from mitosis, *Curr. Biol.*, **11**, 671–683.

Pelayo, H.R., Lastres, P. & De la Torre, C. (2001) Replication and G2 checkpoints: their response to caffeine, *Planta*, **212**, 444–453.

Pellman, D., Bagget, M., Tu, Y.H., Fink, G.R. & Tu, H. (1995) Two microtubule-associated proteins required for anaphase spindle movement in Saccharomyces cerevisiae, *J. Cell. Biol.*, **130**, 1373–1385.

Pfleger, C.M. & Kirschner, M.W. (2000) The KEN box: an APC recognition signal distinct from the D box targeted by Cdh1, *Genes Dev.*, **14**, 655–665.

Pines, J. (1999) Four-dimensional control of the cell cycle, *Nat. Cell Biol.*, **1**, E73–E79.

Pines, J. & Hunter, T. (1990) Human cyclin A is adenovirus E1A-associated protein p60 and behaves differently from cyclin B, *Nature*, **346**, 760–763.

Pines, J. & Hunter, T. (1994) The differential localization of human cyclins A and B is due to a cytoplasmic retention signal in cyclin B, *Embo. J.*, **13**, 3772–3781.

Pines, J. & Rieder, C.L. (2001) Re-staging mitosis: a contemporary view of mitotic progression, *Nat. Cell Biol.*, **3**, E3–E6.

Planchais, S., Glab, N., Inze, D. & Bergounioux, C. (2000) Chemical inhibitors: a tool for plant cell cycle studies, *FEBS Lett.*, **476**, 78–83.

Qian, Y.W., Erikson, E., Li, C. & Maller, J.L. (1998) Activated polo-like kinase Plx1 is required at multiple points during mitosis in Xenopus laevis, *Mol. Cell Biol.*, **18**, 4262–4271.

Raleigh, J.M. & O'Connell, M.J. (2000) The G(2) DNA damage checkpoint targets both Wee1 and Cdc25, *J. Cell Sci.*, **113**, 1727–1736.

Renaudin, J.P., Doonan, J.H., Freeman, D., *et al.* (1996) Plant cyclins: a unified nomenclature for plant A-, B- and D-type cyclins based on sequence organization, *Plant Mol. Biol.*, **32**, 1003–1018.

Rieder, C.L. & Cole, R.W. (1998) Entry into mitosis in vertebrate somatic cells is guarded by a chromosome damage checkpoint that reverses the cell cycle when triggered during early but not late prophase, *J. Cell Biol.*, **142**, 1013–1022.

Rieder, C.L. & Salmon, E.D. (1998) The vertebrate cell kinetochore and its roles during mitosis, *Trends Cell Biol.*, **8**, 310–318.

Rosenblatt, J., Gu, Y. & Morgan, D.O. (1992) Human cyclin-dependent kinase 2 is activated during the S and G2 phases of the cell cycle and associates with cyclin A, *Proc. Natl. Acad. Sci. USA*, **89**, 2824–2828.

Roudier, F., Fedorova, E., Gyorgyey, J., *et al.* (2000) Cell cycle function of a Medicago sativa A2-type cyclin interacting with a PSTAIRE-type cyclin-dependent kinase and a retinoblastoma protein, *Plant J.*, **23**, 73–83.

Roy, L.M., Swenson, K.I., Walker, D.H., *et al.* (1991) Activation of p34cdc2 kinase by cyclin A, *J. Cell Biol.*, **113**, 507–514.

Russell, P. & Nurse, P. (1986) cdc25+ functions as an inducer in the mitotic control of fission yeast, *Cell*, **45**, 145–153.

Russell, P. & Nurse, P. (1987) Negative regulation of mitosis by wee1+, a gene encoding a protein kinase homolog, *Cell*, **49**, 559–567.

Sans, J., Utrilla, L. & De la Torre, C. (1989) The maintenance of colchicine-arrested metaphases in plants requires protein synthesis, *Cell Tissue Kinet.*, **22**, 319–331.

Schlegel, R. & Pardee, A.B. (1986) Caffeine-induced uncoupling of mitosis from the completion of DNA replication in mammalian cells. *Science*, **232**, 1264–1266.

Schnittger, A., Schobinger, U., Stierhof, Y.D. & Hulskamp, M. (2002) Ectopic B-type cyclin expression induces mitotic cycles in endoreduplicating *Arabidopsis* trichomes, *Curr. Biol.*, **12**, 415–420.

Schumacher, J.M., Golden, A. & Donovan, P.J. (1998) AIR-2: An Aurora/Ipl1-related protein kinase associated with chromosomes and midbody microtubules is required for polar body extrusion and cytokinesis in Caenorhabditis elegans embryos, *J. Cell Biol.*, **143**, 1635–1646.

Sherwood, S.W., Kung, A.L., Roitelman, J., Simoni, R.D. & Schimke, R.T. (1993) In vivo inhibition of cyclin B degradation and induction of cell-cycle arrest in mammalian cells by the neutral cysteine protease inhibitor N-acetylleucylleucylnorleucinal, *Proc. Natl. Acad. Sci. USA*, **90**, 3353–3357.

Shibaoka, H. & Nagai, R. (1994) The plant cytoskeleton, *Curr. Opin. Cell Biol.*, **6**, 10–15.

Shirayama, M., Zachariae, W., Ciosk, R. & Nasmyth, K. (1998) The Polo-like kinase Cdc5p and the WD-repeat protein Cdc20p/fizzy are regulators and substrates of the anaphase promoting complex in Saccharomyces cerevisiae, *Embo. J.*, **17**, 1336–1349.

Sibon, O.C., Laurencon, A., Hawley, R. & Theurkauf, W.E. (1999) The Drosophila ATM homologue Mei-41 has an essential checkpoint function at the midblastula transition, *Curr. Biol.*, **9**, 302–312.

Sigrist, S., Jacobs, H., Stratmann, R. & Lehner, C.F. (1995) Exit from mitosis is regulated by Drosophila fizzy and the sequential destruction of cyclins A, B and B3, *Embo. J.*, **14**, 4827–4838.

Sluder, G., Thompson, E.A., Rieder, C.L. & Miller, F.J. (1995) Nuclear envelope breakdown is under nuclear not cytoplasmic control in sea urchin zygotes, *J. Cell Biol.*, **129**, 1447–1458.

Smertenko, A., Saleh, N., Igarashi, H., *et al.* (2000) A new class of microtubule-associated proteins in plants, *Nat. Cell Biol.*, **2**, 750–753.

Staehelin, L.A. & Hepler, P.K. (1996) Cytokinesis in higher plants, *Cell*, **84**, 821–824.

Staiger, C.J. & Lloyd, C.W. (1991) The plant cytoskeleton, *Curr. Opin. Cell Biol.*, **3**, 33–42.

Stals, H., Bauwens, S., Traas, J., Van Montagu, M., Engler, G. & Inze, D. (1997) Plant CDC2 is not only targeted to the pre-prophase band, but also co-localizes with the spindle, phragmoplast, and chromosomes, *FEBS Lett.*, **418**, 229–234.

Starr, D.A., Williams, B.C., Li, Z., Etemad-Moghadam, B., Dawe, R.K. & Goldberg, M.L. (1997) Conservation of the centromere/kinetochore protein ZW10, *J. Cell Biol.*, **138**, 1289–1301.

Su, T.T. & Jaklevic, B. (2001) DNA damage leads to a Cyclin A-dependent delay in metaphase-anaphase transition in the Drosophila gastrula, *Curr. Biol.*, **11**, 8–17.

Sumara, I., Vorlaufer, E., Gieffers, C., Peters, B.H. & Peters, J.M. (2000) Characterization of vertebrate cohesin complexes and their regulation in prophase, *J. Cell Biol.*, **151**, 749–762.

Sun, Y., Dilkes, B.P., Zhang, C., *et al.* (1999) Characterization of maize (Zea mays L.) Wee1 and its activity in developing endosperm, *Proc. Natl. Acad. Sci. USA*, **96**, 4180–4185.

Sunkel, C.E. & Glover, D.M. (1988) polo, a mitotic mutant of Drosophila displaying abnormal spindle poles, *J. Cell Sci.*, **89**(Pt 1), 25–38.

Surana, U., Amon, A., Dowzer, C., McGrew, J., Byers, B. & Nasmyth, K. (1993) Destruction of the CDC28/CLB mitotic kinase is not required for the metaphase to anaphase transition in budding yeast, *Embo. J.*, **12**, 1969–1978.

Sutani, T., Yuasa, T., Tomonaga, T., Dohmae, N., Takio, K. & Yanagida, M. (1999) Fission yeast condensin complex: essential roles of non-SMC subunits for condensation and Cdc2 phosphorylation of Cut3/SMC4, *Genes Dev.*, **13**, 2271–2283.

Takai, H., Tominaga, K., Motoyama, N., *et al.* (2000) Aberrant cell cycle checkpoint function and early embryonic death in Chk1(-/-) mice, *Genes Dev.*, **14**, 1439–1447.

Terada, Y., Tatsuka, M., Suzuki, F., Yasuda, Y., Fujita, S. & Otsu, M. (1998) AIM-1: a mammalian midbody-associated protein required for cytokinesis, *Embo. J.*, **17**, 667–676.

Uhlmann, F., Wernic, D., Poupart, M.A., Koonin, E.V. & Nasmyth, K. (2000) Cleavage of cohesin by the CD clan protease separin triggers anaphase in yeast, *Cell*, **103**, 375–386.

Vandepoele, K., Raes, J., De Veylder, L., Rouze, P., Rombauts, S. & Inze, D. (2002) Genome-wide analysis of core cell cycle genes in *Arabidopsis*, *Plant Cell*, **14**, 903–916.

Verde, F., Dogterom, M., Stelzer, E., Karsenti, E. & Leibler, S. (1992) Control of microtubule dynamics and length by cyclin A- and cyclin B-dependent kinases in Xenopus egg extracts (1992), *J. Cell Biol.*, **118**, 1097–1108.

Waizenegger, I., Lukowitz, W., Assaad, F., Schwarz, H., Jurgens, G. & Mayer, U. (2000a) The *Arabidopsis* KNOLLE and KEULE genes interact to promote vesicle fusion during cytokinesis, *Curr. Biol.*, **10**, 1371–1374.

Waizenegger, I.C., Hauf, S., Meinke, A. & Peters, J.M. (2000b) Two distinct pathways remove mammalian cohesin from chromosome arms in prophase and from centromeres in anaphase, *Cell*, **103**, 399–410.

Walworth, N.C. (2000) Cell-cycle checkpoint kinases: checking in on the cell cycle, *Curr. Opin. Cell Biol.*, **12**, 697–704.

Walworth, N.C. & Bernards, R. (1996) Rad-dependent response of the chk1-encoded protein kinase at the DNA damage checkpoint, *Science*, **271**, 353–356.

Weinert, T. (1997) A DNA damage checkpoint meets the cell cycle engine, *Science*, **277**, 1450–1451.

Weinert, T. & Hartwell, L. (1989) Control of G2 delay by the rad9 gene of Saccharomyces cerevisiae, *J. Cell Sci.*, **12** (Suppl.), 145–148.

Weingartner, M., Binarova, P., Drykova, D., *et al.* (2001) Dynamic recruitment of Cdc2 to specific microtubule structures during mitosis, *Plant Cell*, **13**, 1929–1943.

Weingartner, M., Pelayo, H.R., Binarova, P., *et al.* (2003) A plant cyclin B2 is degraded early in mitosis and its ectopic expression shortens G2-phase and alleviates the DNA-damage checkpoint, *J. Cell Sci.*, **116**, 487–498.

Whitfield, W.G., Gonzalez, C., Maldonado-Codina, G. & Glover, D.M. (1990) The A- and B-type cyclins of Drosophila are accumulated and destroyed in temporally distinct events that define separable phases of the G2-M transition, *Embo. J.*, **9**, 2563–2572.

Yamashiro, S. & Matsumura, F. (1991) Mitosis-specific phosphorylation of caldesmon: possible molecular mechanism of cell rounding during mitosis, *Bioessays*, **13**, 563–568.

Yu, H.G., Muszynski, M.G. & Kelly Dawe, R. (1999) The maize homologue of the cell cycle checkpoint protein MAD2 reveals kinetochore substructure and contrasting mitotic and meiotic localization patterns, *J. Cell Biol.*, **145**, 425–435.

Zachariae, W. & Nasmyth, K. (1999) Whose end is destruction: cell division and the anaphase-promoting complex, *Genes Dev.*, **13**, 2039–2058.

Zhang, K., Letham, D.S. & John, P.C. (1996) Cytokinin controls the cell cycle at mitosis by stimulating the tyrosine dephosphorylation and activation of p34cdc2-like H1 histone kinase, *Planta*, **200**, 2–12.

5 Organelle movements: transport and positioning

Franz Grolig

5.1 Introduction

While short-range movements of supramolecular aggregates in cells can be driven effectively by diffusion, the translocation of such aggregates is clearly diffusion-limited in large and polarized cells. In such cells, more or less extensive long-range transport systems in the form of dynamic cytoskeletal *tracks* (microfilaments (MFs) and microtubules (MTs)) permit effective translocation of cargo hooked on track-associated molecular motor proteins (myosins, kinesins, dyneins) fuelled by ATP. Such translocation is supposed to have evolved primarily as a means of intracellular convection, to compensate for the limited efficiency of mere diffusion to support metabolism in cells larger than 10 μm (Hochachka, 1999).

In the case of so-called cytoplasmic streaming, continual movement of a large pool of intracellular particles that migrate during a shorter or longer period of time in a similar direction and with seemingly uniform velocity give rise to the impression of streaming or bulk flow. Cytoplasmic streaming in plant cells is achieved by the actomyosin motility system, with a few exceptions of MT-based translocations (e.g. Kuroda & Manabe, 1983; Menzel & Schliwa, 1986; Menzel, 1987). The intensity and covered distances of intracellular movements are most prominent within the thin peripheral layer of cytoplasm in fully expanded plant cells during interphase, declining markedly during mitosis. In most interphase plant cells, three contiguous MF-arrays, which support particle movement throughout the cell, can be discerned: (i) short cortical, probably non-bundled MFs; (ii) long, subcortical unipolar MF-bundles which support long-range translocations and extend also within the transvacuolar strand; and (iii) a perinuclear network. The fastest (subcortical) translocation velocities in plant cells range up to 10^{-4} m s^{-1} (*Chara* & *Nitella*, Fig. 5.1), with this upper extreme being the highest velocity observed in eukaryotic cells.

Besides being a means of convection, intracellular motility often accomplishes (re)distribution and positioning of cellular constituents (organelles) as governed by internal (cell-cycle-related) or external stimuli. Positioning can be considered as the result of an ongoing redistribution process leading to a stochastic non-random position as, for example, in the case of the light-governed relocation of chloroplasts in cells exposed to *variant* light conditions. Premitotic relocation of the nucleus, a process which determines the site of the division wall, is one of the most intriguing examples of precise positioning in plant cells (where both, cargo and its destination, are highly specified), involving the interaction of MT- and MF-based forces.

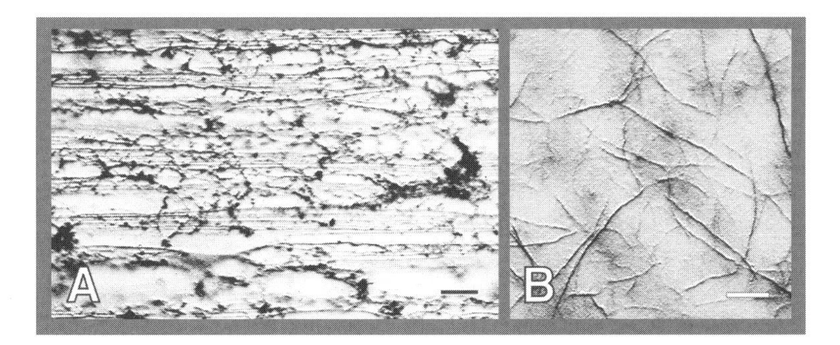

Fig. 5.1 The F-actin cytoskeleton underlying organelle transport by rotational cytoplasmic streaming (cyclosis) in the internodal cell of *Chara corallina* (A) and by agitational cytoplasmic streaming in *Spirogyra crassa* (B). Both images are derived from fluorescence micrographs in which contrast has been enhanced by inversion and shadowcasting; they are oriented with the horizontal axis parallel to the cell long axis. In *Chara* (A), the thick, parallel MF-bundles at the ectoplasmic interface to the moving endoplasm have been marked by indirect immunofluorescence of the MF-associated myosin in a perfused, ATP-depleted cell and are viewed from "inside" in the cut open and spread cell (Grolig *et al.*, 1988). Numerous small organelles and entangled membrane tubules (probably ER), labeled by the myosin antibody, are attached to the MF-bundles, with the latter reaching into the endoplasmic space. In *Spirogyra* (B) short cortical MFs and thicker long subcortical MF-bundles have been labeled by rhodamine phalloidin (Grolig, 1990). Bars: 10 μm.

The following cytoskeletal inhibitors, used for depolymerization of the cytoskeletal polymers, have been used to implicate MFs (cytochalasin B or D [CB or CD], Latrunculin A or B [Lat A or Lat B]) or MTs (colchicine, nocodazole, amiprophos-methyl, oryzalin, cremart) in plant cell organelle movements (Williamson, 1993).

Progress in the field of organelle movements in plant cells has been reviewed extensively in general (Kamiya, 1960, 1981, 1986; Kuroda, 1990; Wick, 1991; Shimmen, 1992; Williamson, 1992, 1993; Shimmen & Yokota, 1994) and under various aspects (Haupt, 1982; Haupt & Scheuerlein, 1990; Wagner & Grolig, 1992; Nagai, 1993; Wada *et al.*, 1993; Asada & Collings, 1997; Takagi, 1997; Takagi, 2000; Grolig & Pierson, 2000; Shimmen *et al.*, 2000; Wada & Kagawa, 2001, 2002). This chapter focusses on progress made over approximately the past decade, with particular emphasis on light-governed plastid relocation and nuclear positioning/cytokinesis.

5.2 Transport and positioning of particular organelles

Tagging of GFP (green fluorescent protein) to organelle-specific (marker) proteins or simple delivery of GFP to the respective compartment by a GFP-construct containing the respective signal sequence has allowed the corroboration and refinement of previous phase or interference contrast microscopy studies because the GFP technology allows the unequivocal identification of the respective organelles. Moreover, targeted and random GFP-tagging appear to be promising approaches in unravelling previously unknown structural constituents and their dynamic changes within the cell (Cutler & Ehrhardt, 2000).

5.2.1 Peroxisome

An organelle close to the size resolution limit of the light microscope is the peroxisome. Over the last year (2002) a number of publications have described the motility of GFP-labeled peroxisomes in plant cells (Collings *et al.*, 2002, *Allium porrum* L.; Jedd & Chua, 2002, *A. thaliana*; Mano *et al.*, 2002, *A. thaliana*; Mathur *et al.*, 2002, *Allium cepa, A. thaliana*). All reports demonstrate that peroxisome motility is inhibited by MF-depolymerization (by CD or by Lat B) and by the myosin inhibitor 2,3 butanedione monoxime (BDM), and that peroxisome translocations follow the MF-paths of the other organelles taking part in cytoplasmic streaming. Based on incomplete inhibition of peroxisome motility by BDM, Mathur *et al.* (2002) have suggested that a motility process based on actin polymerization similar to that observed in pathogenic microorganisms like *Listeria* and *Shigella* (Cossart, 2000), and possibly for the movement of mitochondria in yeast (Boldogh *et al.*, 2001), might propel peroxisomes through the plant cytoplasm.

5.2.2 Endoplasmic reticulum

The membrane system of the endoplasmic reticulum (ER) in plant cells (Staehelin, 1997) is spread throughout the cell cortex in the form of an irregular network of membrane tubules with occasional smaller lamellar membrane sheets (cisternae) parallel to the plasma membrane. ER membrane tubules are extended along MF-tracks in the subcortical and transvacuolar cytoplasm. All extended (i.e. tense) tubules appear to be associated with F-actin, as indicated by other organelles (small vesicles, Golgi stacks and mitochondria) moving along extended ER tubules (Quader *et al.*, 1987, 1989; Allen & Brown, 1988; Lichtscheidl & Weiss, 1988; Grolig, 1990; review by Lichtscheidl & Baluska, 2000).

The elements of the cortical ER network change in position, shape and size, but with immobile, fixed sites, which persist for more than 30 min in onion bulb scale epidermal cells. They are arranged in helicoidal rows of knot- or ring-like structures that enlarge upon application of oryzalin (Knebel *et al.*, 1990). The dynamic structure of the network appears to derive mainly from (actomyosin-based) pulling forces, counteracting membrane surface tension and sites of fixation. Polygons mainly form when newly extended tubules join to other parts of the network; they disappear upon contraction and fusion of tubules. Application of CD causes loss of all actomyosin-extended ER structures, which fuse into cisternae and accumulate into patches. Probably due to the fixed sites, the architecture of the cortical ER network remains unaffected. In animal cells, where ER-dynamics is usually MT-based (Waterman-Storer & Salmon, 1998), no peripheral attachment of the ER has been reported.

Targeting of GFP by specific leader sequences to the ER permitted prolonged observation of fluorescence without apparent disturbance of the structure of the organelle (Boevink *et al.*, 1998; Hawes *et al.*, 1999; Nebenführ *et al.*, 1999). The arrangement of the cortical ER shows intriguing changes as cells mature in the hypocotyl and root epidermis of *Arabidopsis*: expanding cells show extensive perforated sheets of cortical ER which transform into the loose reticulum at the basipetal end of the elongation zone (Ridge *et al.*, 1999).

5.2.3 Golgi

The Golgi apparatus (Staehelin & Moore, 1995) in higher plant cells exists as a large number of independent Golgi stacks (dictyosomes) distributed throughout the cytoplasm. Golgi stacks move rapidly and consistently along the polygonal cortical ER of leaf epidermal cells (Boevink *et al.*, 1998). F-actin precisely matches the arrangement of the ER-network; CD stops the movement of stacks which gather on small islands of cortical lamellar ER. Each stack can travel as fast as 2–4 $\mu m\ s^{-1}$ on the same subcortical actin network along which the dynamic, tubular ER is translocated (Boevink *et al.*, 1998; Nebenführ *et al.*, 1999). As both organelles rely on the same actin tracks for their dynamics, both remain in close association. Each Golgi stack travels independently, with periods of vectorial movement alternating with periods of oscillatory (Brownian) motion.

The question of how the membrane traffics between the ER and the Golgi apparatus in higher plants in such a dynamic system has led to the formulation of two models. Both models imply the existence of a vesicular intermediate (though such vesicles have not been identified as yet). The *vacuum cleaner* model (Boevink *et al.*, 1998) postulates that Golgi stack movement is essential to collect transport vesicles still attached to the ER strand which extends along the same actin track. Alternatively, the interruptions in vectorial movement (resulting in oscillatory motion) may serve as periods of vesicle acquisition from particular, specialized sites of the ER (*recruitment* model, Nebenführ *et al.*, 1999; Nebenführ & Staehelin, 2001), while the vectorial movement supports the spreading of Golgi-derived secretory vesicles around the cell. Transgene expression of *AtRab1b(N121I)*, predicted to be a dominant inhibitory mutant of the *Arabidopsis* Rab GTPase AtRab1b, resulted in the reduction or cessation of vectorial Golgi movement (Batoko *et al.*, 2000), which was reversed by coexpression of the wild type protein. AtRab1b is needed for vesicle traffic from the ER to the Golgi, but it also affects Golgi movement – the latter observation possibly supporting the *recruitment* model. Neither the retrograde redistribution of membrane proteins between the Golgi apparatus and the ER (upon treatment with Brefeldin A), nor the reformation of Golgi stacks (after removal of the drug) require either MFs or MTs (Saint-Jore *et al.*, 2002).

With GFP-tagged Golgi stacks it has become possible to follow the redistribution of the stacks during cytokinesis (Nebenführ *et al.*, 2000). Redistribution occurs long before phragmoplast formation. Prior to cell division, when cytoplasmic streaming and also the movement of Golgi stacks stops, about one-third of the peripheral

Golgi stacks redistributes to the perinuclear cytoplasm, the phragmosome, thereby reversing the ratio of perinuclear to peripheral Golgi from 2:3 to 3:2. About, 20% of all Golgi stacks aggregate in the immediate vicinity of the mitotic spindle and a similar number become concentrated in an equatorial belt under the plasma membrane, which had previously been marked by the preprophase band (PPB). In spite of increased Golgi stack frequency at the PPB site, inhibition of Golgi secretion by Brefeldin A during PPB formation revealed that Golgi secretion is not actively involved in the determination of the cell division site (Dixit & Cyr, 2002). During telophase and cytokinesis, many Golgi stacks redistribute around the phragmoplast so that at the end of cytokinesis the daughter cells have very similar Golgi stack densities. The sites of Golgi stack localization are preferential for this organelle and largely exclude mitochondria and plastids. Organelle segregation begins in metaphase and ends with the completion of cytokinesis. While the mitotic spindle appears to play a major role in organizing the organelle redistribution, persistence of the distribution pattern then appears independent of MTs or MFs. The distribution of Golgi stacks during mitosis and cytokinesis is consistent with the hypothesis that Golgi stacks are repositioned to ensure equal partitioning between the daughter cells as well as rapid cell plate assembly (Nebenführ *et al.*, 2000).

The dynamics of Golgi vesicles involved in wound wall secretion as visualized by video-enhanced microscopy could be related to reorganization of the actin cytoskeleton in characean internodal cells (Foissner *et al.*, 1996). Upon wounding-induced local inhibition of unidirectional endoplasmic (subcortical) cytoplasmic streaming, previously fast moving vesicles perform Brownian motion while following passive endoplasmic flow. After a few minutes, the vesicles then show intermittent short-range, directional movements, with their trajectories correlating with a fine-meshed MF-network that forms at the wound site. Vesicle fusion as indicated by effusion of their contents was observed only when vesicles had access to the plasma membrane during the period of short-range movements.

5.2.4 Vacuoles

Plant cell vacuoles are multifunctional organelles that are central to cellular strategies of plant development, serving physical and metabolic functions that are essential to plant life. Vacuoles are widely diverse in form, size, content and dynamics; a single cell may contain more than one type of vacuole (Marty, 1999). Their functions in cell morphogenesis and cell division are not clear. Complex architectural remodeling of the vacuole membrane system has been observed by use of transgenic *A. thaliana* expressing a vacuolar syntaxin-related protein fused with GFP (Uemura *et al.*, 2002): Cylindrical and sheet-like structures are observed moving in the vacuolar lumen and sometimes penetrate it, just like transvacuolar strands. CD abolishes these movements. In tobacco (and other species) a set of ripple-shaped protrusions of the vacuole into the surrounding cytoplasm was identified (Verbelen & Tao, 1998); the motility of these structures also depends on F-actin. By using transgenic tobacco BY-2 cells expressing both the vacuolar GFP-AtVam3p and a GFP-tubulin fusion protein, the dynamic relation of vacuoles and MTs could be followed in highly synchronized cells (Kutsuna & Hasezawa, 2002): at late G2- phase, tubular vacuolar structures (developing probably in the phagmosome) surround the mitotic apparatus. These tubules invade the phragmoplast from anaphase to telophase, and some of them expand rapidly between cell plate and daughter nuclei, developing into large vacuoles at interphase. Vacuoles in plant cells can be eliminated by centrifugation of protoplasts through a density gradient, yielding so-called miniprotoplasts (Sonobe, 1996). Miniprotoplasts, prepared from tobacco BY-2 cells whose cell-cycle had been synchronized at late anaphase, continue to divide to form two daughter cells.

5.2.5 *Mitochondria*

Small, rodshaped plant mitochondria, vitally stained with the oxocarbocyanine dye DiOC6(3) have been found to move on the same MF-tracks as the subcortical ER-strands in onion (*Allium cepa* L.) epidermal cells (Quader *et al.*, 1987, 1989; Allen & Brown, 1988; Liebe & Quader, 1994; Olyslaegers & Verbelen, 1998) and in the green alga *Spirogyra crassa* (Grolig, 1990). Stably transformed plant lines with GFP-targeted mitochondria have been established for *Arabidopsis* (Logan & Leaver, 2000) and tobacco (*Nicotiana*, Köhler *et al.*, 1997).

Van Gestel *et al.* (2002) observed in elongated tobacco cells that mitochondria in the cortical cytoplasm are less mobile than in the subcortical and transvacuolar MF-tracks. Prolonged application of oryzalin (several hours) increased slightly, but significantly, both mitochondrial mobility and the number of transvacuolar strands. After MF-disruption by Lat B, the (eventually immobile) mitochondria accumulated in the cell cortex and aligned in often slightly helical, transverse arrays. This puzzling arrangement was lost in cells treated with oryzalin. Uncoupling of the mitochondria with 2,4-dinitrophenol (DNP; in order to deplete cytoplasmic ATP), strongly amplified this intriguing phenomenon, but not in oryzalin-pretreated cells. The authors argue that the cortical arrangement of mitochondria depends on the cortical microtubules in elongating cells. Unfortunately, the effect of DNP (ATP-depletion) on the fate of the persistance and dynamics of the cortical MTs has not been tested.

5.2.6 *Chloroplasts*

Chloroplasts as the genuine light harvesting organelles of plant cells either are positioned (more or less *randomly* and *stationary*) within the cell or they can undergo directional or non-directional active movements. Both types of movement may be under light control, in particular in plant cells which have to perform highly efficient photosynthesis under both low fluence rate (LFR) and high fluence rate (HFR) light conditions. Blue (B) and/or red light (RL) photoreceptors sense the quality, fluence rate and direction of the incident light. Directional movement in single-chloroplast cells (some algae) and in multi-chloroplast cells (seed plants, ferns and mosses) leads to specific orientation/relocation of the randomly oriented/distributed chloroplasts in darkness. In multi-chloroplast cells, the chloroplasts move under LFR conditions to the periclinal cell walls and to the bottom of the cell; in HFR they locate near anticlinal walls. Although light-controlled chloroplast movement is a fundamental phenomenon in plant life, the details behind the particular phenomenon appear to vary considerably between taxonomic units (review Wada & Kagawa, 2002).

5.2.6.1 Algae

In the cylindrical cells of the green algae *Mougeotia* and *Mesotaenium* (Charophyceae, the group of the Chlorophyta that is closest to the ancestor of the higher plants) an axial ribbon-shaped chloroplast (*megaplast* with multiple genomes) is sandwiched between two large vacuoles. In darkness, the chloroplast takes a slightly twisted, random position. Only few particles move in the peripheral cytoplasm of *Mougeotia*, and over much shorter distances than in cells of the closely related genus *Spirogyra* (Grolig, 1990). Highly refractive globules with a phenolic, Ca^{2+}-sequestering matrix (Grolig & Wagner, 1987, 1989) are mainly found at the chloroplast edge. In LFR light, the flat chloroplast turns its face to the incident light, in HFR light, its edge. The LFR response is inducible (i.e. it can proceed after a pulse irradiation in darkness) by red (R) or weak white light, and is cancelled by subsequent far red (FR) light, indicative of the (photochromic, i.e. light-revertible) action of the pigment phytochrome. Chloroplast photo-orientation is restricted to the irradiated part of the cell (Haupt, 1972). Upon irradiation or after changing the light direction, the chloroplast starts to rotate virtually without a lag period (Haupt & Übel, 1975) with the chloroplast edges, which are in contact with the cortical cytoplasm, moving randomly in either direction. Segments of the chloroplast are free to turn in either direction, which may result in a twisted shape of the chloroplast. After occasional primary bending of the edges toward each other, one edge stops and turns back all the way, resulting finally in synchronized movement of the opposing edges in the respective cell segment (Haupt & Übel, 1975). The entire chloroplast appears soft; while its edges often become lobed during movement (Wagner & Klein, 1981), the reoriented chloroplast eventually straightens out. The overall motile activity of particles in the cortex of *Mougeotia* increases upon irradiation of dark-adapted cells, but returns to low activity after chloroplast photo-orientation (Schönbohm, 1987).

Sophisticated microbeam irradiation with polarized red light to study the location and the transition moment of active phytochrome molecules led to a model of phytochrome distribution which appears to explain all effects of experimental irradiation conditions on plastid photoorientation: phytochrome localizes to the plasma membrane, with the transition moment of the red absorbing form (P_r) parallel and that of the far-red absorbing form (P_{fr}) normal to the plasma membrane (Haupt, 1972). Because of the dichroic orientation of phytochrome, a tetrapolar gradient of Pfr in the cell periphery reflects the direction of incident (polarized and non-polarized) light. While in the LFR response the chloroplast moves to the sites of high P_r (low P_{fr}), the response is inverted in HFR B, applied together with R: the chloroplast edges are moved within the tetrapolar P_{fr}-gradient to the sites of low Pr (high Pfr). The direction of light in the case of HFR B again is sensed by phytochrome, however, with HFR-B inverting the polarity of the gradient (Schönbohm, 1987). The (non-pigment–pigment) interactions of the photoreceptors in the case of the HFR response probably involve a chemical mediator interacting with phytochrome.

Treatment with CB, CD and NEM indicate that the actomyosin system mediates the chloroplast rotational movement. In *Mougeotia*, the P_{fr}-gradient as such stores the information on light direction: the chloroplast starts to reorient in darkness according to the direction of the initial light signal as soon as CB is washed out after photostimulation. A short FR pulse applied in the presence of CB to a previously photostimulated cell abolishes the ability of the chloroplasts to reorientate after removal of CB (Wagner *et al.*, 1972). The memory is less persistent in *Mesotaenium* and in the protonema cells of the fern *Adiantum* (see below). Fringes of seemingly non-bundled F-actin at the chloroplast edges could be visualized with rhodamine phalloidin in white light after optimized fixation. In contrast, *Mougeotia* protoplasts exhibited bundles of F-actin without structural relation to the chloroplast edge (Grolig *et al.*, 1990). Dynamic changes of the F-actin at the chloroplast edge could be followed in the course of the HFR response. F-actin appears at the chloroplast edge at the start of chloroplast rotation, increases during movement and vanishes after completion of photoorientation (Mineyuki *et al.*, 1995). Depolymerization of the cortical microtubules accelerates the LFR-R induced movement twofold (Serlin & Ferrel, 1989) up to the speed of rotation in HFR-B. HFR-B diminishes the abundance of cortical MTs, as indicated by quantitation of indirect immunofluorecence. Taxol counteracts the HFR-B elicited decrease of MTs, which may be Ca^{2+}-mediated (Russ *et al.*, 1991; Al Rawass *et al.*, 1997).

The angular velocity of chloroplast rotation was measured by near infrared laser diffractometry (Ytow *et al.*, 1992). A constant velocity of 1 mrad s^{-1} was reached within less than 30 s; 90° rotation was complete within, 20 min. Constant velocity indicates that the net torque acting on the chloroplast is close to zero. The driving force acting on the chloroplast was estimated to be 1–10 pN, which is about the force exerted by a single myosin motor molecule.

A stochastic model (of tensional integrity), intending to satisfy the results and mechanical constraints outlined above, has been developed to explain how the as such unidirectional translational movement of the actomyosin system can be arranged and controlled to result in a rotational, light-governed movement of the chloroplast (reviewed in detail in Wada *et al.*, 1993).

5.2.6.2 Mosses

Light-controlled chloroplast relocation (accumulation and depletion) is mediated by dichroic phytochrome as well as dichroic BL photoreceptor(s) in protonemata of *Physcomitrella patens* (Kadota *et al.*, 2000). Most interesting – and in contrast to all other systems presented here – this moss uses both MTs and MFs for chloroplast translocation. In darkness, rapid movement of plastids in the longitudinal direction (cell axis) uses MTs as the tracks, while MFs serve as the tracks for slow random movement. As tested by cytoskeletal inhibitors, phytochrome-mediated plastid relocation uses only the MT-tracks, while BL-induced movements are suppressed completely only by the combined depolymerization of MTs and MFs. Differences in CB-treated cells indicate different types of MT-based systems or different use of the same system in RL and BL (Sato *et al.*, 2001a).

5.2.6.3 Ferns

Blue and red light cause chloroplast relocation in the cells of fern sporophytic protonemata (filamentous) and gametophytic prothallium (two-dimensional, single cell layer). While in *Adiantum capillus-veneris* both, B and R, are effective, in *Pteris vittata* only B induces relocation (Kadota *et al.*, 1989). In *Adiantum* protonema cells relocation can be induced by microbeam irradiation or by whole cell irradiation with polarized R or B (Yatsuhashi *et al.*, 1985). LFR R or B induces chloroplast accumulation in the microbeam irradiated area; HFR B or R causes avoidance of the irradiated area. In darkness, the chloroplasts in the peripheral cytoplasm of the highly vacuolated cylindrical cells move independently from each other in random directions; their (long range) translocation rate is much slower than that of the subcortical cytoplasmic streaming. Movement of the chloroplasts is inhibited by CB and NEM (Kadota & Wada, 1992a). In darkness, the actin cytoskeleton comprises thick subcortical MF bundles running predominantly about parallel to the cell axis, and abundant fine MF(-bundle?)s throughout the cortex (Kadota & Wada, 1989, 1992b).

As concluded from experiments with both B and R polarized light, also in *Adiantum* protonemata the chloroplasts appear to orient in a tetrapolar gradient of dichroic phytochrome and a dichroic cryptochrome. However, in contrast to *Mougeotia/Mesotaenium*, the chloroplasts are translocated toward high Pfr. Kinetic studies suggest that Pfr does not only determine the direction of movement but also how long the chloroplasts reside at their new location in subsequent darkness.

Chloroplast translocation within a strong azimuthal gradient of light absorption was traced by video-tracking. Randomly oriented chloroplasts show (more pronounced in RL than in BL) about 15 min after the onset of light, a prominent bias of their tracks perpendicular to the long axis of the cell (Kadota & Wada, 1992a), with no change of net velocity. At the relocation site, the gathering chloroplasts show random movements with reduced mobility. Cytoplasmic streaming along the thick subcortical MF bundles does not change during chloroplast relocation. Concomitant to reduction of the fine cortical MF(-bundle?)s, 4 h after chloroplast relocation in polarized RL or BL, only within the area of chloroplast accumulation a ring of F-actin forms on the plasmalemma-facing side of the plastid along the edge of each chloroplast. The rings are not always closed, but were never found without chloroplast (Kadota & Wada, 1989). They appear with delay after chloroplast relocation and disappear before dispersal of the chloroplasts in subsequent darkness. MF-reorganization is similar in both LFR-R and -B and HFR-B (Kadota & Wada, 1992b) and suggests an anchoring effect on the chloroplasts.

RL-aphototropic (rap) mutants of *A. capillus-veneris*, which lack RL polarotropism, also lack RL-induced chloroplast relocation, suggesting that both processes share the same dichroic phytochrome and have a signal transduction pathway different from those responses which rely on non-dichroic phytochrome (Kadota & Wada, 1999).

In prothallial cells of *Adiantum capillus-veneris*, a short pulse of B microbeam irradiation induces chloroplast accumulation, with transfer of a signal (in the dark) within the particular cell (Kagawa & Wada, 1996, 1999). Effects of polarized B on chloroplast relocation patterns indicate a dichroic arrangement of the photoreceptors at the plasma membrane (Yatsuhasi *et al.*, 1985). Sequential irraditions with R and B at subthreshold fluence rates add up to elicit chloroplast relocation, suggesting that R and B share parts of the transduction pathway (Kagawa & Wada, 1996). Phytochrome 3, a chimeric gene with an N-terminal chromophore-binding domain of phytochrome and a C-terminal full-length sequence of phototropin (the B photoreceptor for phototropism) could serve well as a photoreceptor for both the BL and RL responses (Nozue *et al.*, 1998; Christie *et al.*, 1999).

The movement of chloroplasts to avoid HFR light can be best analyzed by partial irradiation of a cell with a microbeam. Central HFR irradiation of *A. capillus-veneris* cells, adapted to LFR white light with the chloroplasts located at the periclinal cell wall (cell surface), makes the plastids move out of the irradiated area. In dark-adapted cells, however, with the plastids staying at the anticlinal cell walls (cell periphery), the chloroplasts move *toward* the irradiated area, but they do not enter it (Kagawa & Wada, 1999). Apparently, HFR-B elicits signals for both accumulation and depletion, but the first signal spreads throughout the cell while the latter is restricted to the irradiated area (Kagawa & Wada, 1999; Tlalka *et al.*, 1999). The depletion signal appears to be retained, because depletion occurs after short HFR-B pulses in darkness in *Lemna trisulca* (Zurzycki *et al.*, 1983) and in *A. capillus-veneris* (Kagawa & Wada, 1999). LFR-R does not influence HFR-B elicited depletion, indicating that the HFR-B generated depletion signals are different from the LFR-R generated accumulation signals (Kagawa & Wada, 1999). So far nothing is known about any spatially different organization of the actin cytoskeleton mediating the different responses in prothallial cells.

By gently pressing a short segment of protonemal cells of *A. capillus-veneris* for 20 s, Sato *et al.* (1999) could elicit chloroplast relocation. The avoidance movement of chloroplasts, beginning within 30 min, led to maximal chloroplast depletion at the stimulation site within 2 h. Chloroplast motility was inhibited by CB and BDM, but not by MT inhibitors. In contrast to light-induced chloroplast relocation responses, the mechano-response appears to depend on the influx of external Ca^{2+} (Sato *et al.*, 2001b), because it can be inhibited by the Ca^{2+}-channel blocker La^{3+} and by Gd^{3+}, an inhibitor of stretch-activated channels .

5.2.6.4 *Seed plants*

In leaf mesophyll cells of the aquatic angiosperms *Elodea* and *Vallisneria* irradiation or application of various chemicals induces rotational streaming of the cytoplasm (cyclosis; photodinesis and chemodinesis, respectively), similar to the persistant type of cyclosis, e.g. in characean internodal cells (Williamson, 1993). In *Vallisneria*, subcortical MF bundles are the structural basis for unipolar, belt-like cytoplasmic streaming (Yamaguchi & Nagai, 1981) along two of the longer side walls and the end walls of the elongated cells. Treatment of cells with external EGTA induces cytoplasmic streaming, and this streaming can be subsequently inhibited by FR, but only if Ca^{2+} are present in the external medium. The amount of cytoplasmic Ca-pyroantimonate precipitate increased after FR-irradiation and decreased after extracellular application of EGTA. In R-irradiated cells, considerable precipitate was found within the chloroplasts (Takagi & Nagai, 1985). In the presence of La^{3+}, the inhibition of cytoplasmic streaming by FR decreases to almost that of the dark control. Light-dependent fluxes of Ca^{2+} across the plasmalemma of *Vallisneria* protoplasts were monitored with the Ca^{2+}-sensitive dye murexide. While R stimulated Ca^{2+}-efflux, FR caused a rapid influx. No Ca^{2+}-influx was observed in the presence of the Ca^{2+}-channel blocker nifedipine. R-induced cytoplasmic streaming has an apparent latency of about 2.5×10^2 s independent of R intensity. However, after photosynthesis has taken place for more than 6 min, streaming starts within approximately 70 s after R irradiation. The cellular energy charge needed for cytoplasmic streaming is apparently maintained even in the dark, as indicated by EGTA-induced streaming in darkness. Thus, phytochrome and photosynthesis in cooperation appear to regulate cytoplasmic streaming via modulation of the cytoplasmic Ca^{2+} concentration. Based on photometric measurements, the content of phytochrome in *Vallisneria* was estimated to be about one tenth of that from light grown pea seedlings (Takagi *et al.*, 1990).

The association of the plasma membrane with the cell wall at the end of the cell is indispensable for maintaining the organization of the MF-bundles. The underlying linkage between plasma membrane and cell wall is trypsin-sensitive (Masuda *et al.*, 1991). While the MFs along the side walls are readily depolymerized by CD, at the end walls only partially disrupted MFs remain even after prolonged drug treatment. After drug removal, in most cells cytoplasmic streaming along the reassembled MF-bundles has the same polarity as before drug treatment. However, when the MFs at the end walls have been completely disrupted by trypsin treatment prior to application of CD, the direction of reinitiated cytoplasmic streaming is reversed in 50% of the cells, suggesting that the polarity of reassembled side wall MFs is determined by the polarity of the remnant MFs at the end walls (Ryu *et al.*, 1995). Application of synthetic hexapeptides, which include either a RGD or RYD motif (known as recognition sites in molecules that are required for adhesion to the substratum at focal contact sites), within 24 h caused (compared to control peptides) a strikingly altered MF arrangement and cytoplasmic streaming. The RYD-peptide was localized by immunofluorescence to the cell wall; the same polyclonal antibody detected polypeptides of 54 and 27 kDa on western blots of a total extract from *Vallisneria* leaves (Ryu *et al.*, 1997).

Liebe and Menzel (1995) observed the distribution and movement of the ER and mitochondria (stained by $DiOC_6$ (3)) and plastids in *Vallisneria* mesophyll cells both in darkness and after light-induced cyclosis. In darkness, the immobile plastids appeared to be enclosed in pockets of subcortical lamellar ER-membranes adjacent to the cortical ER-network, while ER-tubules, mitochondria and proplastids moved (at about 5 $\mu m\, s^{-1}$) on restricted subcortical tracks. After 30 min irradiation, at first the ER-tubules became rearranged into small strands translocated along linear tracks. The chloroplasts, still trapped in the cell periphery, began short-distance translocations. With transformation of the lamellar into tubular ER, directionality of chloroplast movement toward the anticlinal cell walls increased. In full cyclosis, the chloroplasts streaming with a velocity of up to 15 $\mu m\, s^{-1}$ were still surrounded by a fine network of tubular ER. Plastids and smaller organelles/tubular ER were moving subcortically as a complex with the same directionality, however, the latter moving faster, with up to 20 $\mu m\, s^{-1}$. Corresponding to the changing pattern of movements, different arrays of MFs were found: in darkness, a fine network of MFs is present at the periclinal wall, together with thick MF-bundles at the anticlinal walls. Transiently, before full cyclosis is reached, short MFs form at the plastid surface, a process possibly accompanying (Ca^{2+}-dependent?) release of the previously immobile plastids from the ER sheets. Indirect immunofluorecence of myosin revealed that in darkness myosin localizes preferentially to membranes of the ER and hardly to the surface of plastids. Prominent labeling of the plastid surface, however, is observed when cylosis has fully developed, suggesting that (a specific?) myosin binds to this cargo under the particular cell physiological conditions of cyclosis.

In epidermal cells of *Vallisneria gigantea*, the chloroplasts in the peripheral cytoplasm are distributed randomly in darkness. They migrate to the periclinal cell wall in LFR light (R is most effective) and to the anticlinal walls in B-HFR light (Izutani *et al.*, 1990). Even in darkness, the plastids move around, but randomly. The migration rate (of each individual chloroplast) increases rapidly within minutes of saturating R irradiation (with FR completely reversing the effect of R). Migration of the chloroplasts to the periclinal walls then, however, declines more rapidly than that to the periclinal wall, resulting in progressive accumulation of plastids at the periclinal wall. The reduction of motility at the periclinal wall (and thus chloroplast accumulation) does not occur, if photosynthesis is inhibited, indicating cooperative action of phytochrome and photosynthesis in light-governed chloroplast relocation.

As in mesophyll cells, plastid motility in *Vallisneria* epidermal cells relies on the actomyosin-system (Dong *et al.*, 1996, 1998) and is inhibited by CB and BDM. Investigation of dynamic changes of the actin cytoskeleton revealed that a suspicious reorganization of the actin cytoskeleton at the periclinal cell wall coincides with the decline of chloroplast net translocation leading to plastid redistribution (Dong *et al.*, 1998). The arrangement of MFs changes from a randomly oriented network of subcortical MF bundles in darkness to a *honeycomb*-like array, in which the chloroplasts appear embedded (immobilized) in RL. Such plastids at the periclinal wall showed increased (CB-sensitive) resistance to centrifugational displacement. Thus, MFs not only appear to support the transport to, but also the local anchorage of, the chloroplast at their light controlled destiny.

Chloroplast accumulation in leaves of *Arabidopsis thaliana* shows essentially the same features as that observed in ferns (Wada & Kagawa, 2001). Avoidance and accumulation responses were detected photometrically in *Arabidopsis* mesophyll cells (Trojan & Gabrys, 1996) and were confirmed by microscopical observation: chloroplasts move out of the microbeam spot in continuous HFR-B within a few minutes (and stay outside); they gather in the beam spot in continuous LFR-B within 10 min (Kagawa & Wada, 2000).

Screening mutants of *A. thaliana* by partial irradiation of leaves yielded several mutants defective in BL-induced chloroplast depletion of the periclinal cell wall. *PHOT2*, a gene homologous to *PHOT1*, the phototropin gene, was identified by map-based cloning (Kagawa *et al.*, 2001) and reverse genetics (Jarillo *et al.*, 2001) as the mutagenized gene. The two LOV domains of *PHOT2* expressed in *E. coli* bound FMN (flavin mononucleotide) and showed an absorption spectrum (Sakai *et al.*, 2001) similar to the action spectrum for the chloroplast depletion response (Zurzycki, 1980). Interestingly, in *phot2* mutants, chloroplasts accumulate in HFR-B, indicating that HFR-B generates the signal for both, chloroplast depletion and accumulation. The depletion signal apparently overcomes the accumulation signal in wild type. No chloroplast relocation occurred in double mutants (*phot1* and *phot2*) (Sakai *et al.*, 2001). As *phot1* mutants show both relocation responses, PHOT1 and PHOT2 seem to be the photoreceptors for chloroplast accumulation, and PHOT2 is the photoreceptor for depletion.

The structural relationship of MFs, MTs and plastids has been studied in cryo-fixed and freeze-substituted *Arabidopsis* leaf cells (Kandasamy & Meagher, 1999). Immunolabeling with two plant-actin specific antibodies revealed longitudinal arrays of thick MF-bundles and randomly oriented thin MFs that extend from the bundles. Chloroplasts either align along the MF-bundles or associate with the fine MFs that can form baskets around them. Such MF-baskets may anchor the chloroplasts and allow positional control with respect to light direction. Lat B disrupted MFs and their association with chloroplasts. MTs exhibited no apparent association with chloroplasts, and MT-depolymerization did not affect the distribution of chloroplasts.

The importance of light-governed plastid relocation for optimized photosynthetic yield and plant growth was nicely demonstrated by use of transgenic tobacco plants hampered in plastid division (Jeong *et al.*, 2002), with leaf mesophyll cells containing only one to three enlarged chloroplasts. In these plants no light-governed plastid relocation occurred, leading to either decreased light absorbance (under LFR conditions) or photodamage (under HFR conditions), which both resulted in reduced plant growth.

So-called stromules (stroma-filled tubules), enclosed by the inner and outer plastid envelope membranes, interconnect plastids and extend as highly dynamic structures, which continuously and rapidly change shape, from the surface of plastids in many cell types and plant species. Actomyosin-based movement and extension of stromules provides a means to enormously increase the plastid surface area in cells containing a relatively small plastid volume (Gray *et al.*, 2001).

5.2.7 *Nucleus*

5.2.7.1 *Premitotic nuclear positioning*

In vacuolated higher plant cells preparing for division, the nucleus migrates into the center of the cell, suspended by transvacuolar strands. Movement of the nucleus to its premitotic position, which in turn determines the site of cytokinesis, appears to involve MTs in that MTs, radiating from the nuclear surface, first mobilize the nucleus and then stabilize it in the plane of division (Bakhuizen *et al.*, 1985). Accordingly, movement of the nucleus to the cell center is inhibited by MT-depolymerization and not by CB (Venverloo & Libbenga, 1987). Premitotic nuclear migration in cultured BY-2 cells (Katsuta *et al.*, 1990) was also not impaired by CD but it could be inhibited completely *only* by simultaneous application of both propyzamide and CD. The transvacuolar strands, which form during translocation of the premitotic nucleus to the cell center, comprise both MT-bundles and MF(bundle?)s and initially radiate throughout the cell. The arrangement of the transvacuolar nucleus-suspending strands, which are under tension (Goodbody *et al.*, 1991), can be modeled by springs held in two-dimensional hexagonal frames, and by soap bubbles in three-dimensional hexagonal frames, suggesting that their arrangement follows minimal path criteria of the respective cell geometry (Flanders *et al.*, 1990), thus giving clues on how the cytoskeleton might align the division plane of plant cells (Lloyd, 1991a). A few hours later, the membranes of the transvacuolar strands anastomose into a diaphragm, building the phragmosome, which then persists throughout mitosis (Sinnott & Bloch, 1940). A few transvacuolar strands extend from the nucleus perpendicular to the plane of the diaphragm (Flanders *et al.*, 1990; Katsuta *et al.*, 1990), giving the phragmosome the overall appearance of a wheel on a hub (Lloyd, 1991b). Concomitant (or slightly delayed) to formation of the phragmosome, the preprophase band (PPB), a circumferential band of both MTs and MFs, arises at the site, where the phragmosome connects to the cell cortex, and, though disappearing (like the MTs of the phragmosome) at late prophase, marks the site where the young cross wall fuses with the parental wall (review Mineyuki, 1999; Murata, 2000; Hasezawa & Kumagai, 2002). This MT-band, initially fairly broad, narrows concomitant to progressive transformation of the three-dimensional arrangement of the transvacuolar strands into the phragmosome. If formation of the PPB is suppressed (Katsuta *et al.*, 1990), premitotic migration of the nucleus nevertheless takes place. The MTs of the phragmosome and PPB are suggested to serve as scaffolds for the construction of a MF-array (Katsuta *et al.*, 1990) which after prophase keeps the mitotic apparatus in position (Traas *et al.*, 1987; Venverloo & Libbenga, 1987; Lloyd & Traas, 1988) and appears to direct the margin of the expanding phragmoplast to the cell cortex (Kakimoto & Shibaoka, 1988).

The premitotic nucleus of *Spirogyra crassa*, a particularly large and translucent filamentous charophycean green alga, is tethered precisely in the center of the cell by an extensive perinuclear scaffold that resembles the higher plant phragmosome (Grolig, 1992). Cytoplasmic strands of enhanced flexural rigidity and fasciate appearance radiate from the rim of the lenticular nucleus through the vacuole, frequently split once or twice and attach to chloroplast bands arranged helically in the peripheral cytoplasm. Nuclear centering in the cylindrical *Spirogyra* cell implies positioning in both the *longitudinal* direction (attained by postmitotic nuclear migration) and in the *transverse* direction. Indirect anchorage of the nucleus via the peripheral chloroplast bands allows direct monitoring of the balance of forces acting within the scaffold: cytoskeletal perturbation results either in nuclear or in plastid dislocation. MF- and/or MT-depolymerization affect the transverse nuclear position slowly, but in distinct ways: upon application of oryzalin, the scaffold stalks shorten and the chloroplasts are drawn toward the nucleus only in the presence of intact, antiparallel MFs, which convey tension within the stalks. The tension appears to be balanced by MT-bundles within the stalks, which seem to act within the scaffold like the struts of a tent. Randomization of the transversal nuclear position after disruption of MFs demonstrates that (i) MFs are indispensible to attain the central position of the nucleus and that (ii) non-MF-associated forces exist which can dislocate the nucleus out of the cell center. Loss of (MF-mediated) tension conveyed along the scaffold stalks may lead to uncontrolled, and with respect to central nuclear positioning, unbalanced tubulin polymerization. The differential effects of cytoskeletal inhibitors suggest that (at least) transverse nuclear centering depends on tensional integrity of the perinuclear scaffold, with microfilaments conveying tension along stabilized microtubules and the actin cytoskeleton integrating the translocation forces generated within the scaffold (Grolig, 1998).

Protonemal cells of the fern *Adiantum capillus-veneris* usually do not develop a phragmosome prior to cell division. However, when the nucleus and the endoplasm are displaced from the presumptive division site (as indicated by a PPB) by centrifugation, a phragmosome-like structure develops at the site of the PPB (Murata & Wada, 1997). The unusual structure, containing MTs, MFs, oil droplets and mitochondria, is not built if the cell has been centrifuged prior to PPB formation, or if formation of the PPB has been inhibited by MT-depolymerization. It vanishes after cytokinesis (which proceeds at the site of the displaced nucleus), but persists if the cell cycle is arrested at M-phase by MT-deploymerization.

5.2.7.2 Nuclear migrations elicited by external stimuli

The asymmetrical divisions in the four epidermal subsidiary mother cells of *Tradescantia virginiana* are preceded by MF-based nuclear migrations from random locations to positions adjacent to the central guard mother cells, with polarization occurring more than 20 h prior to mitosis (Kennard & Cleary, 1997). The nuclei are able to reposition after experimental displacement by centrifugation. Most interestingly, the nuclei can move in a similar way toward a needle pressed against the epidermal cell surface. Such localized pressure caused an almost immediate accumulation of cytoplasm adjacent to the needle tip. The nuclear response occurs more quickly when the pressure is increased or when a larger area of the cell wall is deformed. The nucleus retains its position adjacent to the needle tip, as long as force is exerted (30 min), and it moves away upon removal of the needle.

MFs and MTs extensively reorganize (and cytoplasmic constituents relocate) in resisting mesophyll cells of flax (*Linum usitatissimum* L.) during attempted infection by the flax rust fungus (Kobayashi *et al.*, 1994). Treatment of non-host tissues with MF- and/or MT-depolymerizing or MT-stabilizing inhibitors made such plants susceptible even for non-pathogen penetration (Kobayashi *et al.*, 1997a). The extent of MF-depolymerization correlates significantly with increasing efficiency of fungal penetration; depolymerized MTs have an additive effect. Simultaneous depolymerization of MTs and MFs suppresses the polarization (with respect to the site of fungal attack) of defense-related responses, as for example, massive cytoplasmic aggregation (Kobayashi *et al.*, 1997b). Cell death and nuclear movements were inhibited by cytochalasin E (and not by oryzalin and taxol), suggesting that MFs are required for the hypersensitive response of cowpea (*Vigna unguiculata*) during the infection by the cowpea rust fungus *Uromyces vignae* (Skalamera & Heath, 1998).

5.2.7.3 Light-governed nuclear migration

In prothallial cells of the fern *Adiantum capillus-veneris*, nuclear relocation can be induced by R or B light irradiation (Kagawa & Wada, 1993). Depending on the status of light adaptation of the cells and on polarization of the light, nuclear migration shows different speeds (Kagawa & Wada, 1995), probably reflecting gradients in the distribution of dichroic photoreceptors. Although the photoreceptors for the relocation of chloroplasts and nuclei are the same, the organelles behave differently: the fast chloroplast relocation (close to the plasma membrane) can be induced by brief irradiation, while the slow nuclear migration (beneath the chloroplast layer) needs long-term irradiation. So far, nothing is known about the cytoskeletal basis for this nuclear movement.

5.2.8 Phragmoplast/cytokinesis

The shape and relative position of cells in plant tissues are constrained by their cell walls. The cellular architecture of plant tissues is therefore defined by the pattern of cell division during development. While variant strategies for premitotic establishment of the plane of cell division exist in other plant taxa (Pickett-Heaps *et al.*, 1999), in most cells of flowering plants the plane of division (as the final result of the spatial regulation of cytokinesis) is predicted by the PPB, which girdles the cell during G2 and early prophase at the prospective fusion site of the new cell plate with the parental cell wall. Cytokinesis in higher plant cells is achieved by a unique cytoskeletal structure arising after mitosis in between the daughter nuclei, the so-called phragmoplast. The barrel-shaped phragmoplast is built from two opposing arrays of parallel MTs and MFs, which both interdigitate at their plus ends at the equator of the phragmoplast (Kakimoto & Shibaoka, 1988) and incorporate tubulin subunits (Asada *et al.*, 1991). Golgi vesicles with polysaccharides, proteins and membrane material, which appear to be guided to/concentrated at the interface of the opposing arrays, give rise to a new cross wall initiated by centrifugal outgrowth of a cell plate. To complete cytokinesis, the phragmoplast expands laterally (concomitant with disappearance of MTs and MFs from the center of the phragmoplast) until the cell plate reaches and fuses with the parental cell wall (reviews by Wick, 1991; Staehelin & Hepler, 1996; Smith, 1999, 2002; Otegui & Staehelin, 2000; Brown & Lemon, 2001; Verma, 2001). The details and the molecules involved in particular aspects of this sophisticated and highly orchestrated process are addressed in other chapters of this book (see Chapters 1 and 3). Here, an evolutionary perspective is given with respect to functional clues possibly provided by phylogenetic precursors.

The mechanisms underlying the sequential changes (preprophase band, phragmosome, phragmoplast) of the higher plant MT- and actin cytoskeleton accompanying cytokinesis are still largely obscure. Comparative studies of cytokinesis in related organisms provide a chance to discover the evolutionary sequence leading from a primitive precursor structure to the derived structures, and therefore may help to identify functionally significant elements in the advanced cytokinetic apparatus. Cytokinesis in the charophycean green alga *Spirogyra* (Zygnematales) is characterized by centripetal growth of a septum, which impinges on a persistent, centrifugally expanding telophase spindle, leading to a phragmoplast-like structure of potential phylogenetic significance (Fowke & Pickett-Heaps, 1969). That a phragmoplast occurs in some, but not all, charophycean green algae (Pickett-Heaps, 1975) suggests that evolution of the phragmoplast took place in the course of establishment of this advanced lineage.

Cytokinetic progression in *Spirogyra* can be divided into three functional stages with respect to the contribution of MFs and MTs (McIntosh *et al.*, 1995; Sawitzky & Grolig, 1995): (i) In early prophase, a cross wall initial forms independently of MFs and MTs at the presumptive site of wall growth; (ii) Numerous organelles accumulate at the cross wall initial (Fig. 5.2A) concomitant with reorganization of the extensive peripheral interphase MF-array (Fig. 5.1B) into a distinct circumferential MF-array (Fig. 5.2B). This array guides the ingrowing septum until it contacts the expanding interzonal MT array; (iii) MFs at the growing edge of the septum coalign with and extend along the interzonal MTs toward the daughter nuclei (Fig. 5.2C, D). Actin-based transportation of small organelles during the last stage occurs, in part, along a scaffold previously deployed in space by MTs. Displacement of the nuclei-associated interzonal MT-array by centrifugation and depolymerization of the phragmoplast-like structure demonstrate that the success of cytokinesis at the final stage depends on the interaction of both MF- and MT-cytoskeletons.

While the higher plant phragmoplast seems to derive from pre-existing interzone MTs by lateral coalescence (Zhang *et al.*, 1993), in *S. crassa* a similar MT-array perpendicular to the plane of cell division is built by persisting interzonal MTs which contact the ingrowing septum. In cells of the advanced charophycean green alga *Coleochaete* (Brown *et al.*, 1994) and in cells of higher plants the

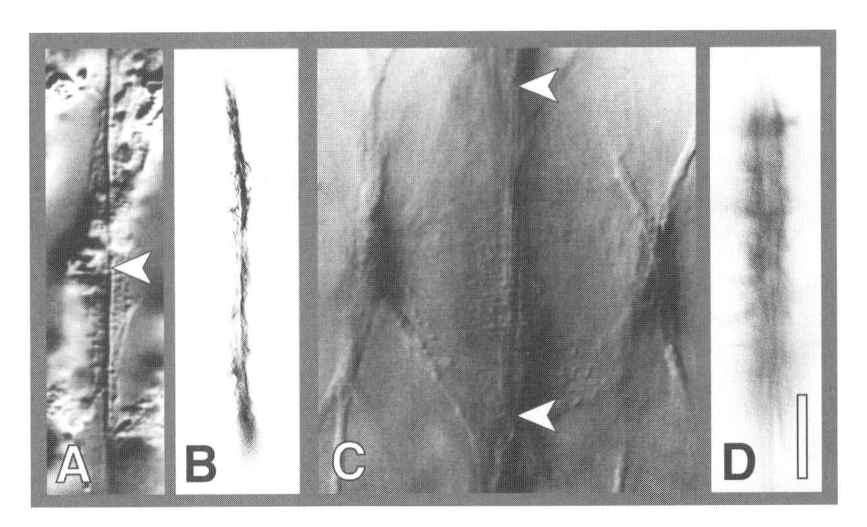

Fig. 5.2 Actin-associated organelle relocation/positioning during cytokinesis in *Spirogyra crassa* (Sawitzky & Grolig, 1995); organelles visualized in live cells by video-enhanced differential interference contrast (A, C), and MFs in fixed cells by rhodamine phalloidin (B, D). Fluorescence images were processed like in Fig. 5.1. All micrographs are oriented with the horizontal axis parallel to the cell long axis. (A) At late anaphase, organelles have accumulated for start of centripetal ingrowth at the circumferential line (arrowhead) previously marked by a cross wall initial. (B) The corresponding arrangement of MFs (cf. the interphase array in Fig. 5.1B); short MFs run at different angles to the division plane. (C) The persistent, expanding telophase spindle has contacted the ingrowing septum (edge in optical section marked by arrowheads); the previously septum-associated organelles arrange along the (MT-related) striations of the contacted telophase spindle. On both sides of the spindle the daughter nuclei (slightly out of focus) are suspended by their perinuclear scaffolds. (D) The corresponding arrangement of the MFs at the contact site of the ingrowing septum shows that these have aligned parallel to the striations of the telophase spindle. After contact of the centripetally advancing MF-array with the centrifugally expanding MT-array, completion of cell wall growth depends on both cytoskeletal components. Bar: 20 µm.

connection of the interzonal MTs to the daughter nuclei disappears, whereas that connection persists in *Spirogyra* (as in *Mougeotia* and *Zygnema*).

No preprophase band was found in *S. crassa*, but cytoplasmic strands with fairly weak MF-related fluorescence radiate from the mitotic figure close to the area of the cross wall initial. In addition, remnants of the interphase perinuclear scaffold persist at the spindle poles and continue to link the mitotic apparatus to the peripheral chloroplast bands. As mitosis proceeds unimpeded in the presence of CD, the MFs in the cytoplasmic strands appear to be less important than the residual MTs for keeping the mitotic apparatus in position. However, after contact with the expanding telophase spindle, the MFs of the ingrowing septum apparently contribute to proper orientation of this interzonal MT array.

In *Spirogyra*, inverse to the situation in the higher plant phragmoplast, the extensive peripheral MF-system of interphase disappears during formation of the cytokinetic array, while the cortical MTs diminish only gradually. Toward the end of cytokinesis, a striking local depletion of cortical MTs close to the ingrown cross wall is observed in *S. crassa*, but has not been described for other zygnematacean species (Sawitzky & Grolig, 1995). In higher plants, MFs instead of MTs disappear beneath the new cross wall (Cleary *et al.*, 1992).

5.3 Concluding remarks

The basic model of actomyosin-based movements of plant cell organelles has been impressively corroborated by a number of exciting (and very aestethic) studies employing the study of live cells with GFP-labeled organelles. The intracellular distribution of GFP-labeled organelles can be followed in live cells with high temporal resolution (limited by the performance of the confocal laser scan or the charge coupled device (CCD) detector system). GFP-based cytoskeletal and organellar markers certainly will further promote the study of organelle movements. Multi-color GFP variants allow simultaneous imaging of two and more organelles (Ellenberg *et al.*, 1999). Mutational analysis combined with GFP live cell approaches shall provide further important contributions to unravel mechanisms and functions of known (and yet to be discovered) subcellular structures/patterns.

Investigations on chloroplast movement have identified and partially characterized the photoreceptor molecules which control light-governed chloroplast relocation. The mostly actin-based mechanisms for rearrangement of the chloroplasts in the cell periphery appear – with respect to the details – diverse in the various taxa. Light-governed rotational movement of the single *Mougeotia* chloroplast and relocation of the multiple chloroplasts in *Adiantum* protonemal and prothallial cells (and probably also in most higher plant cells) result from a change from random movement in darkness to a biased movement (with respect to the quality and direction of the incident light) preceded/accompanied by specific rearrangement of the cortical actin cytoskeleton. Alternatively, in *Vallisneria* mesophyll cells, chloroplasts anchored in darkness at the periclinal walls under the control of phytochrome and cytosolic Ca^{2+}-concentration move to and hook on existing subcortical MF-bundles at the anticlinal walls for cyclosis. Chloroplast movements thus clearly demonstrate the distinct translocation functionality of the cortical and the subcortical actin cytoskeleton: In contrast to the long-range translocation along subcortical MF-bundles, the short cortical MFs allow sequential random short-range movements, which, however, appear to be biased by light-governed abundance of these MFs, finally resulting in accumulation of the chloroplasts at the sites of highest cortical MF density.

Most interesting, a first case of differential use (and thus differential regulation) of MF- and/or MT-based motility in a plant interphase cell has been described for chloroplast relocation in the moss *Physcomitrella patens* (Sato *et al.*, 2001a), suggesting that even in the case of transport and positioning of a single organelle intricate details of cargo attachment (Kamal & Goldstein, 2002) await their elucidation in plant cells.

Intracellular displacements of individual particles vary in direction and velocity of movement over time and in space, reflecting the respective forces involved in particle–cytoskeleton interaction. Careful description of structural organization and dynamics often elucidate important functional principles that underlie this organization. However, intracellular motility as a stochastic process is difficult to describe in detail and quantitatively. As human vision and brain cannot realize and memorize well trajectories (in particular of slow translocations), improved methods (with respect to spatial and temporal resolution) are needed for proper localization and tracking of intracellular particles in order to gain statistical parameters which permit quantitative comparison of differing motility scenarios underlying organelle (re)distributions. Finally, simulations with varied parameters may help to clarify the question how a particular arrangement of the cytoskeleton achieves by ongoing transportation a steady-state distribution of particles for (over time) even allocation of cargo units to multiple (but nevertheless non-random) sites of destination.

Just as cytoskeletal tension can be probed by optical displacement assays (Grabski *et al.*, 1994; Schindler, 1995; Simmons *et al.*, 1996; Grabski & Schindler, 1998), the quantitative analysis of organelle movements can be a tool to diagnose the endogenous status of a cell (as e.g. perturbed by pharmacological treatment) and to examine the impact of physiological factors. Quantitative motion analyses have been applied to trajectories reflecting the actomyosin based cytoplasmic dynamics (*streaming*) in tobacco pollen tubes (de Win *et al.*, 1997, 1998, 1999). Calculations of the regularity quotient (Q_r; a measure for the directionality of movement) and the progressiveness ratio (a measure for the straightness of a trajectory) permit statistical evaluation of how the movement becomes less organized and less vectorial toward the pollen tube tip. Arithmetical dissection in order to discern movement types with respect to persistance revealed that pollen tube organelles move primarily as individual elements, asynchronously and with heterogeneous patterns but with overall vectorial similarities in the respective regions of the tube.

Currently four basic types of mechanism have to be envisaged how force can be transmitted for organelle translocation: (i) motor proteins move cargo over a stationary cytoskeletal track; (ii) non-force-generating molecules link cargo to MFs or MTs which in turn are moved by molecular motors over a stationary cytoskeletal track; (iii) antiparallel, stationary cytoskeletal tracks are sliding along each other by linking motor proteins and either are set under tension (in particular MFs) or caused by compression to bend/buckle (possible in case of enhanced flexural rigidity of the cytoskeletal polymer like in case of MTs); (iv) independent of motor proteins, polymerization of MF(-bundle)s or MTs on the organelle surface can cause translocation. Prerequisite for elucidation of the (more or less complex) framework of forces, which yields the resultant force vector behind a particular organelle translocation, is definition (a) of the anchored and of the moving cytoskeletal components, and (b) the polarity of the cytoskeletal polymer(s) and the direction of movement of the associated motor proteins. Mechanistical models, integrating one or more of the above mentioned basic types of mechanism, provide a communicative basis to verify, refine or discard hypothetical implications – so that the basic question "How does it work?" can be progressively specified. However, the intriguing structural details of the supramolecular organization behind the more complex types of movement probably can be resolved only beyond the resolution limit of the fluorescence microscope and will need advanced ultrastructural investigation of properly preserved specimen.

The complexity of nuclear translocation and positioning processes, involving MF- and/or MT-based forces, and the puzzling sequence of orchestrated cytoskeletal rearrangements accompanying higher plant cytokinesis has brought into focus the architectural concept of cytoskeletal tensional integrity (tensegrity; Ingber, 1993) for structural features and dynamic remodeling of the plant cytoskeleton (Grolig, 1998; Pickett-Heaps *et al.*, 1999). Comparison of higher plant cytokinesis with putative phylogenetic precursor processes should profit from comparative localization of various cytoskeletal components (which have been identified in higher plant cells) in the presumed precursor structures.

Apart from optical tweezers (Felgner *et al.*, 1997; Grabski & Schindler, 1998), cytoskeletal forces in cell-walled plant cells unfortunately can hardly be probed locally and quantitatively. Exciting findings that chloroplasts of protonemal cells of *A. capillus-veneris* (Sato *et al.*, 2001) and nuclei of *Tradescantia* epidermal cells (Kennard & Cleary, 1997) relocate reversibly upon localized application of external force (strain and/or stress) have established readily accessible experimental systems to study in detail the mechanism(s) of cytoskeletal force-sensing and force-governed remodeling of the plant cytoskeleton, which presumably are also involved in cellular responses to fungal attack and tissue wounding (Kobayashi *et al.*, 1997b).

In view of the specific difficulties encountered with plant biochemistry, mutants and the wealth of (latent) information provided by sequenced plant genomes today provide the superior tools to trace down the many molecular components which contribute to the multiple functions and interrelations of the most dynamic and intervening organelle: the cytoskeleton.

Note added in proof: The anticipated, but so far not strictly proven mitigation of photodamage by means of chloroplast relocation (avoidance response) has been convincingly corroborated by the use of different classes of mutants defective in this response (Kasahara *et al.*, *Nature* 420 (2002) 829–832). Using mutational (and complementation) analysis, a function of phy3 (the chimaera of phytochrome and phototrophin) has been identified for RL-induced plastid relocation in *Adiantum capillus-veneris* (Kawai *et al.*, *Nature* 421 (2003) 287–290). Chloroplast relocation upon mechanical stimulation has been found also in bryophytes. In contrast to MF-based chloroplast *depletion* in *Adiantum* protonemata, in *Physcomitrella patens* protonemata MT-based accumulation of chloroplasts is observed. In both cases, mechanoperception involves influx of Ca^{2+} (Sato *et al.*, *Planta* 216 (2003) 772–777). Using sophisticated image analysis, Takagi *et al.* (*Plant Cell* 15 (2003) 331–345) have shown that already within 2.5 seconds phytochrome(II)-dependent induction of cytoplasmic motility in *Vallisneria gigantea* can be observed in scattered cytoplasmic patches.

Acknowledgements

I apologize to all authors who made significant contributions to the field but whose work is not cited due to space limitations. The author's work is supported by the Deutsche Forschungsgemeinschaft.

References

Allen, N.S. & Brown, D.T. (1988) Dynamics of the endoplasmic reticulum in living onion epidermal cells in relation to microtubules, microfilaments and intracellular particle movement, *Cell Motility Cytoskeleton*, **10**, 153–163.

Al Rawass, B., Grolig, F. & Wagner, G. (1997) High irradiance blue light affects cortical microtubules in the green alga *Mougeotia scalaris*, *Plant Cell Physiology*, **38**, 882–886.

Asada, T. & Collings, D. (1997) Molecular motors in higher plants, *Trends Plant Science*, **2**, 29–37.

Asada, T., Sonobe, S. & Shibaoka, H. (1991) Microtubule translocation in the cytokinetic apparatus of cultured tobacco cells, *Nature*, **350**, 239–241.

Bakhuizen, R., Van Spronsen, P.C., Sluiman-den Hertog, F.A.J., Venverloo, C.J. & Goosen-de Roo, L. (1985) Nuclear envelope radiating microtubules in plant cells during interphase mitosis transition, *Protoplasma*, **128**, 43–51.

Batoko, H., Zheng, H.-Q., Hawes, C. & Moore, I. (2000) A Rab1 GTPase is required for transport between the endoplasmic reticulum and Golgi apparatus and for normal Golgi movement in plants, *Plant Cell*, **12**, 2201–2218.

Boevink, P., Oparka, K., Cruz, S.S., Martin, B., Betteridge, A. & Hawes, C. (1998) Stacks on tracks – the plant Golgi apparatus traffics on an actin/ER network, *Plant Journal*, **15**, 441–447.

Boldogh, I.R., Yang, H.-C., Nowakowski, W.D., *et al.* (2001) Arp2/3 complex and actin dynamics are required for actin-based mitochondrial motility in yeast, *Proceedings National Academy Sciences USA*, 1362–1367.

Brown, R.C. & Lemon, B.E. (2001) The cytoskeleton and spatial control of cytokinesis in the plant life cycle, *Protoplasma*, **215**, 35–49.

Brown, R.C., Lemon, B.E. & Graham, L.E. (1994) Morphogenetic plastid migration and microtubule arrays in mitosis and cytokinesis in the green alga *Coleochaete orbicularis*, *American Journal Botany*, **81**, 127–133.

Christie, J.N., Reymond, P., Powell, G.K., *et al.* (1999) *Arabidopsis* NPH1: Flavoprotein with the properties of a photoreceptor for phototropism, *Science*, **282**, 1698–1701.

Collings, D.A., Harper, J.D.I., Marc, J., Overall, R.L. & Mullen, R.T. (2002) Life in the fast lane: actin-based motility of plant peroxisomes, *Canadian Journal Botany*, **80**, 430–441.

Cossart, P. (2000) Actin-based motility of pathogens: the Arp2/3 complex is a central player, *Cellular Microbiology*, **2**, 195–205.

Cutler, S. & Ehrhardt, D. (2000) Dead cells don't dance: insights from live-cell imaging in plants, *Current Opinion Plant Biology*, **3**, 532–537.

de Win, A.H.N., Pierson, E.S. & Derksen, J. (1999) Rational analyses of organelle trajectories in tobacco pollen tubes reveal characteristics of the actomyosin cytoskeleton, *Biophysical Journal*, **76**, 1648–1658.

de Win, A.H.N., Pierson, E.S., Timmer, C., Lichtscheidl, I.K. & Derksen, J. (1998) An interactive computer-assisted position acquisition procedure designed for the analysis of organelle movement in pollen tubes, *Cytometry*, **32**, 263–267.

de Win, A.H.N., Worring, M., Derksen, J. & Pierson, E.S. (1997) Classification of organelle trajectories using region-based curve analysis, *Cytometry*, **29**, 136–146.

Dixit, R. & Cyr, R. (2002) Golgi secretion is not required for marking the preprophase band site in cultured tobacco cells, *Plant Journal*, **29**, 99–108.

Dong, X.J., Nagai, R. & Takagi, S. (1998) Microfilaments anchor chloroplasts along the outer periclinal wall in *Vallisneria* epidermal cells through cooperation of Pfr and photosynthesis, *Plant Cell Physiology*, **39**, 1299–1306.

Dong, X.J., Ryu, J.H., Takagi, S. & Nagai, R. (1996) Dynamic changes in the organization of microfilaments associated with the photocontrolled motility of chloroplasts in epidermal cells of *Vallisneria*, *Protoplasma*, **195**, 18–24.

Ellenberg, J., Lippincott-Schwartz, J. & Presley, J.F. (1999) Dual color imaging with GFP variants, *Trends Cell Biology*, **9**, 52–56.

Felgner, H., Grolig, F., Müller, O. & Schliwa, M. (1997) In vivo manipulation of internal cell organelles, in *Laser tweezers in cell biology* (ed. M.P. Sheetz), *Methods Cell Biology*, **55**, 195–203.

Flanders, D.J., Rawlins, D.J., Shaw, P.J. & Lloyd, C.W. (1990) Nucleus-associated microtubules help determine the division plane of plant epidermal cells: avoidance of four-way junctions and the role of cell geometry, *Journal Cell Biology*, **110**, 1111–1122.

Foissner, I., Lichtscheidl, I.K. & Wasteneys, G.O. (1996) Actin-based vesicle dynamics and exocytosis during wound wall formation in characean internodal cells, *Cell Motility Cytoskeleton*, **35**, 35–48.

Fowke, L.C. & Pickett-Heaps, J.D. (1969) Cell division in *Spirogyra*. II. Cytokinesis, *Journal Phycology*, **5**, 273–281.

Goodbody, K.C., Venverloo, C.J. & Lloyd, C.W. (1991) Laser microsurgery demonstrates that cytoplasmic strands anchoring the nucleus across the vacuole of premitotic plant cells are under tension: implications for division plane alignment, *Development*, **113**, 931–939.

Grabski, S. & Schindler, M. (1998) Auxins and cytokinins as antipodal modulators of elasticity within the actin network of plant cells, *Plant Physiology*, **110**, 965–970.

Grabski, S., Xie, X.G., Holland, J.F. & Schindler, M. (1994) Lipids trigger changes in the elasticity of the cytoskeleton in plant cells: a cell optical displacement assay for live cell measurements, *Journal Cell Biology*, **126**(3), 713–726.

Gray, J.C., Sullivan, J.A., Hibberd, J.M. & Hansen, M.R. (2001) Stromules: Mobile protrusions and interconnections between plastids, *Plant Biology*, **3**, 223–233.

Grolig, F. (1990) Actin-based organelle movements in interphase *Spirogyra*, *Protoplasma*, 155, 29–42.

Grolig, F. (1992) The cytoskeleton of the Zygnemataceae, in *The cytoskeleton of the algae* (ed. D. Menzel), CRC Press, Boca Raton, pp. 165–193.

Grolig, F. (1998) Nuclear centering in *Spirogyra*: force integration by microfilaments along microtubules, *Planta*, **204**, 54–63.

Grolig, F. & Pierson, E.P. (2000) From flow to track, in *Actin: A dynamic framework for multiple plant cell functions* (eds C.J. Staiger, F. Baluska, D. Volkmann & P.W. Barlow), Kluwer Academic Publishers, Dordrecht Boston London, pp. 165–190.

Grolig, F. & Wagner, G. (1987) Vital staining permits isolation of calcium vesicles from the green alga *Mougeotia*, *Planta*, **171**, 433–437.

Grolig, F. & Wagner, G. (1989) Characterization of the isolated calcium-binding vesicles from the green alga *Mougeotia scalaris*, and their relevance to chloroplast movement, *Planta*, **177**, 169–177.

Grolig, F., Weigang-Köhler, K. & Wagner, G. (1990) Different extent of F-actin bundling in walled cells and protoplasts of *Mougeotia scalaris*, *Protoplasma*, **157**, 225–230.

Grolig, F., Williamson, R.E., Parke, J., Miller, C. & Anderton, B.H. (1988) Myosin and Ca^{2+}-sensitive streaming in the alga *Chara*: two polypeptides reacting with a monoclonal anti-myosin and their localization in the streaming endoplasm, *European Journal Cell Biology*, **47**, 22–31.

Hasezawa, S. & Kumagai, F. (2002) Dynamic changes and the role of the cytoskeleton during the cell cycle in higher plant cells, *International Review Cytology*, **214**, 161–191.

Haupt, W. (1972) Localisation of phytochrome within the cell, in *Phytochrome* (eds K. Mitrakos & W. Shropshire), Academic Press, London, pp. 553–569.

Haupt, W. (1982) Light-mediated movement of chloroplasts, *Annual Review Plant Physiology*, **33**, 205–233.

Haupt, W. & Scheuerlein, R. (1990) Chloroplast movement, *Plant Cell Environment*, **13**, 595–614.

Haupt, W. & Übel, H. (1975) Zum Mechanismus der Phytochromwirkung bei der Chloroplastenbewegung von *Mougeotia*, *Zeitschrift Pflanzenphysiologie*, **75**, 165–171.

Hawes, C., Brandizzi, F. & Andreeva, A.V. (1999) Endomembranes and vesicle trafficking, *Current Opinion Plant Biology*, **2**, 454–461.

Hochachka, P.W. (1999) The metabolic implications of intracellular circulation, *Proceedings National Academy Sciences USA*, **96**, 12233–12239.

Ingber, D.E. (1993) Cellular tensegrity: defining new rules of biological design that govern the cytoskeleton, *Journal Cell Science*, **104**, 613–627.

Izutani, Y., Takagi, S. & Nagai, R. (1990) Orientation movements of chloroplasts in *Vallisneria* epidermal cells: Different effects of light at low- and high-fluence rate, *Photochemistry Photobiology*, **51**, 105–111.

Jarillo, J.A., Gabrys, H., Capel, J., Alonso, J.M., Ecker, J.R. & Cashmore, A.R. (2001) Phototropin-related NPL1 controls chloroplast relocation induced by blue light, *Nature*, **410**, 952–954.

Jedd, G. & Chua, N.H. (2002) Visualization of peroxisomes in living plant cells reveals acto-myosin-dependent cytoplasmic streaming and peroxisome budding, *Plant Cell Physiology*, **43**, 384–392.

Jeong, W.J., Park, Y.I., Suh, K., Raven, J.A., Yoo, O.J. & Liu, J.R. (2002) A large population of small chloroplasts in tobacco leaf cells allows more effective chloroplast movement than a few enlarged chloroplasts, *Plant Physiology*, **129**, 112–121.

Kadota, A. & Wada, M. (1989) Photoinduction of circular F-actin on chloroplast in a fern protonemal cell, *Protoplasma*, **151**, 171–174.

Kadota, A. & Wada, M. (1992a) Photoorientation of chloroplasts in protonemal cells of the fern *Adiantum* as analyzed by use of a video-tracking system, *Botanical Magazin Tokyo*, **105**, 265–279.

Kadota, A. & Wada, M. (1992b) Photoinduction of formation of circular structures by microfilament on chloroplasts during intracellular orientation in protonemal cells of the fern *Adiantum capillus-veneris*, *Protoplasma*, **167**, 97–107.

Kadota, A. & Wada, M. (1999) Red light-aphototropic (rap) mutants lack red light-induced chloroplast relocation movement in the fern *Adiantum capillus-veneris*, *Plant Cell Physiology*, **40**, 238–247.

Kadota, A., Kohyama, I. & Wada, M. (1989) Polarotropism and photomovement of chloroplasts in the fern *Pteris* and *Adiantum* protonemata: Evidence for the possible lack of dichroic phytochrome in *Pteris*, *Plant Cell Physiology*, **30**, 523–531.

Kadota, A., Sato, Y. & Wada, M. (2000) Intracellular chloroplast relocation in the moss *Physcomitrella patens* is mediated by phytochrome as well as by a blue-light receptor, *Planta*, **210**, 932–937.

Kagawa, T. & Wada, M. (2000) Blue light-induced chloroplast relocation in *Arabidopsis thaliana* as analyzed by microbeam irradiation, *Plant Cell Physiology*, **41**, 84–93.

Kagawa, T. & Wada, M. (1993) Light-dependent nuclear positioning in prothallial cells of *Adiantum capillus-veneris*, *Protoplasma*, **177**, 82–85.

Kagawa, T. & Wada, M. (1995) Polarized light induces nuclear migration in prothallial cells of *Adiantum capillus-veneris* L., *Planta*, **196**, 775–780.

Kagawa, T. & Wada, M. (1996) Phytochrome- and blue light-absorbing pigment-mediated directional movement of chloroplasts in dark-adapted prothallial cells of fern *Adiantumas* analyzed by microbeam irradiation, *Planta*, **198**, 488–493.

Kagawa, T. & Wada, M. (1999) Chloroplast avoidance response induced by blue light of high fluence rate in prothallial cells of the fern *Adiantumas* analyzed by microbeam irradiation, *Plant Physio*logy, **119**, 917–923.

Kagawa, T., Sakai, T., Suetsugu, N., *et al.* (2001) *Arabidopsis* NPL1: A phototropin homolog controlling the chloroplast high-light avoidance response, *Science*, **291**, 2138–2141.

Kakimoto, T. & Shibaoka, H. (1988) Cytoskeletal ultrastructure of phragmoplast-nuclei complexes isolated from cultured tobacco cells, *Protoplasma*, **2**, 95–103.

Kamal, A. & Goldstein, L.S.B. (2002) Principles of cargo attachment to cytoplasmic motor proteins, *Current Opinion Cell Biology*, **14**, 63–68.

Kamiya, N. (1960) Physics and chemistry of protoplasmic streaming, *Annual Review Plant Physiology*, **11**, 323–340.

Kamiya, N. (1981) Physical and chemical basis of cytoplasmic streaming, *Annual Review Plant Physiology*, **32**, 205–236.

Kamiya, N. (1986) Cytoplasmic streaming in giant algal cells: a historical survey of experimental approaches, *Botanical Magazine Tokyo*, **99**, 441–467

Kandasamy, M.K. & Meagher, R.B. (1999) Actin-organelle interaction: Association with chloroplast in *Arabidopsis* leaf mesophyll cells, *Cell Motility Cytoskeleton*, **44**, 110–118.

Katsuta, J., Hashiguchi, Y. & Shibaoka, H. (1990) The role of the cytoskeleton in positioning of the nucleus in premitotic tobacco BY-2-cells, *Journal Cell Science*, **95**, 413–422.

Kennard, J.L. & Cleary, A.L. (1997) Pre-mitotic nuclear migration in subsidiary mother cells of *Tradescantia* occurs in G1 of the cell cycle and requires F-actin, *Cell Motility Cytoskeleton*, **36**, 55–67.

Knebel, W., Quader, H. & Schnepf, E. (1990) Mobile and immobile endoplasmic reticulum in onion bulb epidermis cells: Short- and long-term observations with a confocal laser scanning microscope, *European Journal Cell Biology*, **52**, 328–340.

Kobayashi, I., Kobayashi, Y. & Hardham, A.R. (1994) Dynamic reorganization of microtubules and microfilaments in flax cells during the resistance response to flax rust infection, *Planta*, **195**, 237–247.

Kobayashi, Y., Kobayashi, I., Funaki, Y., Fujimoto, S., Takemoto, T. & Kunoh, H. (1997a) Dynamic reorganization of microfilaments and microtubules is necessary for the expression of non-host resistance in barley coleoptile cells, *Plant Journal*, **11**, 525–537.

Kobayashi, Y., Yamada, M., Kobayashi, I. & Kunoh, H. (1997b) Actin microfilaments are required for the expression of nonhost resistance in higher plants, *Plant Cell Physiology*, **38**, 725–733.

Köhler, R.H., Zipfel, W.R., Webb, W.W. & Hanson, M.R. (1997) The green fluorescent protein as a marker to visualize plant mitochondria in vivo, *Plant Journal*, **11**, 613–621.

Kuroda, K. (1990) Cytoplasmic streaming in plant cells, *International Review Cytology*, **121**, 267–307.

Kuroda, K. & Manabe, E. (1983) Microtubule-associated cytoplasmic streaming in *Caulerpa*, *Proceedings Japanese Academy*, **59**, 131–134.

Kutsuna, N. & Hasezawa, S. (2002) Dynamic organisation of vacuolar and microtubule structures during cell cycle progression in synchronized tobacco BY-2 cells, *Plant Cell Physiology*, **43**, 965–973.

Lichtscheidl, I.K. & Baluska, F. (2000) Motility of endoplasmic reticulum in plant cells, in *Actin, a dynamic framework for multiple plant cell functions* (eds C.J. Staiger, F. Baluska, D. Volkmann & P.W. Barlow), Kluwer Academic Publishers, Dordrecht Boston London, pp. 191–201.

Lichtscheidl, I.K. & Weiss, D.G. (1988) Visualization of submicroscopic structures in the cytoplasm of *Allium cepa* inner epidermal cells by video-enhanced contrast light microscopy, *European Journal Cell Biology*, **46**, 376–382.

Liebe, S. & Menzel, D. (1995) Actomyosin-based motility of endoplasmic reticulum and chloroplasts in *Vallisneria* mesophyll cells, *Biology Cell*, **85**, 207–222.

Liebe, S. & Quader, H. (1994) Myosin in onion (*Allium cepa*) bulb scale epidermal cells: Involvement in dynamics of organelles and endoplasmic reticulum, *Physiologia Plantarum*, **90**, 114–124.

Lloyd, C.W. (1991a) Cytoskeletal elements of the phragmosome establish the division plane in vacuolated higher plant cells, in *The cytoskeletal basis of plant growth and form* (ed. C.W. Lloyd), Academic Press, London, pp. 245–257.

Lloyd, C.W. (1991b) How does the cytoskeleton read the laws of geometry in aligning the division plane of plant cells?, *Development*, **1**, 55–65.

Lloyd, C.W. & Traas, J.A. (1988) The role of F-actin determining the division plane of carrot suspension cells. Drug studies, *Development*, **102**, 211–221.

Logan, D.C. & Leaver, C.J. (2000) Mitochondria-targeted GFP highlights the heterogeneity of mitochondrial shape, size and movement within living plant cells, *Journal Experimental Botany*, **51**, 865–871.

Mano, S., Nakamori, C., Hayashi, M., Kato, A., Kondo, M. & Nishimura, M. (2002) Distribution and characterization of peroxisomes in *Arabidopsis* by visualization with GFP: dynamic morphology and actin-dependent movement, *Plant Cell Physiology*, **43**, 331–341.

Marty, F. (1999) Plant vacuoles, *Plant Cell*, **11**, 587–599.

Masuda, Y., Takagi, S. & Nagai, R. (1991) Protease-sensitive anchoring of microfilament bundles provides tracks for cytoplasmic streaming in *Vallisneria*, *Protoplasma*, **162**, 151–159.

Mathur, J., Mathur, N. & Hulskamp, M. (2002) Simultaneous visualization of peroxisomes and cytoskeletal elements reveals actin and not microtubule-based peroxisome motility in plants, *Plant Physiology*, **128**, 1031–1045.

McIntosh, K., Pickett-Heaps, J.D. & Gunning, B.E.S. (1995) Cytokinesis in *Spirogyra*: integration of cleavage and cell-plate formation, *International Journal Plant Sciences*, **156**, 1–8.

Menzel, D. (1987) The cytoskeleton of the giant coenocytic green alga *Caulerpa* visualized by immunocytochemistry, *Protoplasma*, **139**, 71–76.

Menzel, D. & Schliwa, M. (1986) Motility in the siphonous green alga *Bryopsis*. I. Spatial organisation of the cytoskeleton and organelle movements, *European Journal Cell Biology*, **40**, 275–285.

Mineyuki, Y. (1999) The preprophase band of microtubules: its function as a cytokinetic apparatus in higher plants, *International Review Cytology*, **187**, 1–49.

Mineyuki, Y., Kataoka, H., Masuda, Y. & Nagai, R. (1995) Dynamic changes in the actin cytoskeleton during the high-fluence rate response of the *Mougeotia* chloroplast, *Protoplasma*, **185**, 222–229.

Murata, T. (2000) Preprophase band in fern protonemata: Mechanism of development and comparison with flowering plants, *Journal Plant Research*, **113**, 111–118.

Murata, T. & Wada, M. (1997) Formation of a phragmosome-like structure in centrifuged protonemal cells of *Adiantum capillus-veneris* L., *Planta*, **201**, 273–280.

Nagai, R. (1993) Regulation of intracellular movements in plant cells by environmental stimuli, *International Review Cytology*, **145**, 251–309.

Nebenführ, A. & Staehelin, L.A. (2001) Mobile factories: Golgi dynamics in plant cells, *Trends in Plant Science*, **6**, 160–167.

Nebenführ, A., Frohlick, J.A. & Staehelin, L.A. (2000) Redistribution of Golgi stacks and other organelles during mitosis and cytokinesis in plant cells, *Plant Physiology*, **124**, 135–151.

Nebenführ, A., Gallagher, L.A., Dunahay, T.G., *et al.* (1999) Stop-and-go movements of plant Golgi stacks are mediated by the acto-myosin system, *Plant Physiology*, **121**, 1127–1141.

Nozue, K., Kanegae, T., Imaizumi, T., *et al.* (1998) A phytochrome from the fern *Adiantum* with features of the putative photoreceptor NPHl, *Proceedings National Academy Sciences USA*, **95**, 1582–15830.

Olyslaegers, G. & Verbelen, J.P. (1998) Improved staining of F-Actin and co-localization of mitochondria in plant cells, *Journal Microscopy*, **192**, 73–77.

Otegui, M. & Staehelin, L.A. (2000) Cytokinesis in flowering plants: more than one way to divide a cell, *Current Opinion Plant Biology*, **3**, 493–502.

Pickett-Heaps, J.D., Gunning, B.E.S., Brown, R.C., Lemmon, B.E. & Cleary, A.L. (1999) The cytoplast concept in dividing plant cells: Cytoplasmic domains and the evolution of spatially organized cell division, *American Journal Botany*, **86**, 153–172.

Pickett-Heaps, J.D. (1975) *Green Algae*, Sinauer Associates Inc., Sunderland, Mass.

Quader, H., Hofmann, A. & Schnepf, E. (1987) Shape and movement of the endoplasmic reticulum in onion bulb epidermis cells: possible involvement of actin, *European Journal Cell Biology*, **44**, 17–26.

Quader, H., Hofmann, A. & Schnepf, E. (1989) Reorganization of the endoplasmic reticulum in epidermal cells of onion bulb scales after cold stress: Involvement of cytoskeletal elements, *Planta*, **177**, 273–280.

Ridge, R.W., Uozomi, Y., Plazinski, J., Hurley, U.A. & Williamson, R.E. (1999) Developmental transitions and dynamics of the cortical ER of *Arabidopsis* cells seen with green fluorescent protein, *Plant Cell Physiology*, **40**, 1253–1261.

Russ, U., Grolig, F. & Wagner, G. (1991) Changes of cytoplasmic free calcium in the green alga *Mougeotia scalaris*: monitored with indo-1, and effect on the velocity of the chloroplast movement, *Planta*, **184**, 105–112.

Ryu, J.H., Mizuno, K., Takagi, S. & Nagai, R. (1997) Extracellular components implicated in the stationary organization of the actin cytoskeleton in mesophyll cells of *Vallisneria*, *Plant Cell Physiology*, **38**, 420–432.

Ryu, J.H., Takagi, S. & Nagai, R. (1995) Stationary organization of the actin cytoskeleton in *Vallisneria*: the role of stable microfilaments at the end walls, *Journal Cell Science*, **108**, 1531–1539.

Saint-Jore, C.M., Evins, J., Batoko, H., Brandizzi, F., Moore, I. & Hawes, C. (2002) Redistribution of membrane proteins between the Golgi apparatus and endoplasmic reticulum in plants is reversible and not dependent on cytoskeletal networks, *Plant Journal*, **29**, 661–678.

Sakai, T., Kagawa, T., Kasahara, M., *et al.* (2001) *Arabidopsis* nph1 and npl1: blue light receptors that mediate both phototropism and chloroplast relocation, *Proceedings National Academy Sciences USA*, **98**, 6969–6974.

Sato, Y., Kadota, A. & Wada, M. (1999) Mechanically induced avoidance response of chloroplasts in fern protonemal cells, *Plant Physiology*, **121**, 37–44.

Sato, Y., Wada, M. & Kadota, A. (2001a) Choice of tracks, microtubules and/or actin filaments for chloroplast photo-movement is differentially controlled by phytochrome and a blue light receptor, *Journal Cell Science*, **114**, 269–279.

Sato, Y., Wada, M. & Kadota, A. (2001b) External Ca^{2+} is essential for chloroplast movement induced by mechanical stimulation but not by light stimulation, *Plant Physiology*, **127**, 497–504.

Sawitzky, H. & Grolig, F. (1995) Phragmoplast of the green alga *Spirogyra* is functionally distinct from the higher plant phragmoplast, *Journal Cell Biology*, **130**, 1359–1371.

Schindler, M. (1995) The optical displacement assay (CODA): measurements of cytoskeletal tension in living plant cells with a laser optical trap, *Methods Cell Biology*, **49**, 69–82.

Schönbohm, E. (1987) Movement of *Mougeotia* chloroplasts under continuous weak and strong light, *Acta Physiologia Plantarum*, **9**, 109–135.

Serlin, B.S. & Ferrell, S. (1989) The involvement of microtubules in chloroplast rotation in the alga *Mougeotia*, *Plant Science*, **60**, 1–8.

Shimmen, T. (1992) The characean cytoskeleton: dissecting the streaming mechanism, in *The Cytoskeleton of the Algae* (ed. D. Menzel), CRC Press, Boca Raton, pp. 297–314.

Shimmen, T. & Yokota, E. (1994) Physiological and biochemical aspects of cytoplasmic streaming. *International Review Cytology*, **155**, 97–139.

Shimmen, T., Ridge, R.W., Lambiris, I., Plazinski, J., Yokota, E. & Williamson, R.E. (2000) Plant myosins, *Protoplasma*, **214**, 1–10.

Simmons, R.M., Finer, J.T., Chu, S. & Spudich, J.A. (1996) Quantitative measurements of force and displacement using an optical trap, *Biophysical Journal*, **70**, 1813–1822.

Sinnott, E.W. & Bloch, R. (1940) Cytoplasmic behaviour during division of vacuolate plant cells, *Proceedings National Academy Sciences USA*, **26**, 223–227.

Skalamera, D. & Heath, M.C. (1998) Changes in the cytoskeleton accompanying infection-induced nuclear movements and the hypersensitive response in plant cells invaded by rust fungi, *Plant Journal*, **16**, 191–200.

Smith, L.G. (1999) Divide and conquer: cytokinesis in plant cells, *Current Opinion Plant Biology*, **2**, 447–453.

Smith, L.G. (2002) Plant cytokinesis: Motoring to the finish, *Current Biology*, **12**, R206–R208.

Sonobe, S. (1996) Studies on the plant cytoskeleton using miniprotoplasts of tobacco BY-2 cells, *Journal Plant Research*, **109**, 437–448.

Staehelin, L.A. (1997) The plant ER: a dynamic organelle composed of a large number of discrete functional domains, *Plant Journal*, **11**, 1151–1165.

Staehelin, L.A. & Moore, I. (1995) The plant Golgi-Apparatus – structure, functional organization and trafficking mechanisms, *Annual Review of Plant Physiology Plant Molecular Biology*, **46**, 261–288.

Staehelin, L.A. & Hepler, P.K. (1996) Cytokinesis in higher plants, *Cell*, **84**, 821–824.

Takagi, S. (1997) Photoregulation of cytoplasmic streaming – cell biological dissection of signal transduction pathway, *Journal Plant Research*, **110**, 299–303.

Takagi, S. (2000) Roles for actin filaments in chloroplast motility and anchoring, in *Actin: a dynamic framework for multiple plant cell functions* (eds C.J. Staiger, F. Baluska, D. Volkmann & P.W. Barlow), Kluwer Academic Publishers, Dordrecht Boston London, pp. 203–212.

Takagi, S. & Nagai, R. (1985) Light-controlled cytoplasmic streaming in *Vallisneria* mesophyll cells, *Plant Cell Physiology*, **26**, 941–951.

Takagi, S., Yamamoto, K.T., Furuya, M. & Nagai, R. (1990) Cooperative regulation of cytoplasmic streaming and Ca^{2+}-fluxes by Pfr and photosynthesis in *Vallisneria* mesophyll cells, *Plant Physiology*, **94**, 1702–1708.

Tlalka, M., Runquist, M. & Fricker, M. (1999) Light perception and the role of the xanthophyll cycle in blue-light-dependent chloroplast movements in *Lemna trisulca* L., *Plant Journal*, **20**, 447–459.

Traas, J.A., Doonan, J.H., Rawlins, D.J., Shaw, P.J., Watts, J. & Lloyd, C.W. (1987) An actin network is present in the cytoplasm throughout the cell cycle of carrot cells and associates with the dividing nucleus, *Journal Cell Biology*, **105**, 387–395.

Trojan, A. & Gabrys, H. (1996) Chloroplast distribution in *Arabidopsis thaliana* (L.) depends on light conditions during growth, *Plant Physiology*, **111**, 419–425.

Uemura, T., Yoshimura, S.H., Takeyasu, K. & Sato, M.H. (2002) Vacuolar membrane dynamics revealed by GFP-AtVam3 fusion protein, *Genes to Cells*, **7**, 743–753.

Van Gestel, K., Köhler, R.H. & Verbelen, J.P. (2002) Plant mitochondria move on F-actin, but their positioning in the cortical cytoplasm depends on both F-actin and microtubules, *Journal Experimental Botany*, **53**, 659–667.

Venverloo, C.J. & Libbenga, K.R. (1987) Regulations of the plane of cell division in vacuolated cells. I. The function of nuclear positioning and phragmosome formation, *Journal Plant Physiology*, **131**, 267–284.

Verbelen, J.P. & Tao, W. (1998) Mobile arrays of vacuole ripples are common in plant cells, *Plant Cell Reports*, **17**, 917–920.

Verma, D.P. (2001) Cytokinesis and building of the cell plate in plants, *Annual Review Plant Physiology Plant Molecular Biology*, **52**, 751–784.

Wada, M. & Kagawa, T. (2001) Light-controlled chloroplast movement, in *Photomovement* (eds D.-P. Häder & M. Lebert), Comprehensive Series in Photosciences, Elsevier, Amsterdam, pp. 897–924.

Wada, M. & Kagawa, T. (2002) Blue light-induced chloroplast relocation, *Plant Cell Physiology*, **43**, 367–371.

Wada, M., Grolig, F. & Haupt, W. (1993) Light-oriented chloroplast positioning. Contribution to progress in photobiology, *Journal Photochemistry Photobiology B: Biology*, **17**, 3–25.

Wagner, G. & Grolig, F. (1992) Algal chloroplast movements, in *Algal Cell Motility* (ed. M. Melkonian), Chapmann and Hall, New York, pp. 39–72.

Wagner, G. & Klein, K. (1981) Mechanism of chloroplast movement in *Mougeotia*, *Protoplasma*, **109**, 169–185.

Wagner, G., Haupt, W. & Laux, A. (1972) Reversible inhibition of chloroplast movement by cytochalasin B in the green alga *Mougeotia*, *Science*, **176**, 808–809.

Waterman-Storer, C. & Salmon, E.D. (1998) Endoplasmic reticulum membrane-tubules are distributed by microtubules in living cells using three distinct mechanisms, *Current Biology*, **8**, 798–806.

Wick, S.M. (1991) Spatial aspects of cytokinesis in plant cells, *Current Opinion Cell Biology*, **3**, 253–260.

Williamson, R.E. (1992) Cytoplasmic streaming in characean algae: Mechanism, regulation by Ca^{2+} and organization, in *Algal Cell Motility* (ed. M. Melkonian), Chapmann and Hall, New York, pp. 73–98.

Williamson, R.E. (1993) Organelle movements, *Annual Review Plant Physiology Plant Molecular Biology*, **44**, 181–202.

Yamaguchi, Y. & Nagai, R. (1981) Motile apparatus in *Vallisneria* leaf cells. I. Organization of microfilaments, *Journal Cell Science*, **48**, 193–205.

Yatsuhashi, H., Kadota, A. & Wada, M. (1985) Blue- and red-light action in photoorientation of chloroplasts in *Adiantum* protonemata, *Planta*, **165**, 43–50.

Ytow, N., Yamada, T. & Ishizaka S. (1992) Mechanics of the chloroplast rotation in *Mougeotia*: Measurement of angular velocity by laser diffractometry, *Cell Motility and the Cytoskeleton*, **23**, 102–110.

Zhang, D., Wadsworth, P. & Hepler, P.K. (1993) Dynamics of microfilaments are similar, but distinct from microtubules during cytokinesis in living, dividing plant cells, *Cell Motility Cytoskeleton*, **24**, 151–155.

Zurzycki, J. (1980). Blue light-induced intracellular movement, in *Blue light syndrome* (ed. H. Senger), Springer-Verlag, Berlin, pp. 50–68.

Zurzycki, J., Walczak, T., Gabrys, H. & Kajfosz, J. (1983) Chloroplast translocation in *Lemna trisulca* L. induced by continuous irradiation and by light pulses, *Planta*, **157**, 502–510.

6 The cell wall: a sensory panel for signal transduction

Keiko Sugimoto-Shirasu, Nicholas C. Carpita and Maureen C. McCann

6.1 Introduction

Plant form derives from two processes: first, establishing the planes of division in cells, thus determining the positions in which new cell walls are created, and second, targeting new cell wall material to particular areas of the cell surface. The cell wall yields at particular sites to accommodate directional growth, generating the diversity of plant cell shapes. Thus, the cell wall, like the cytoskeleton, is both a scaffolding structure, providing shape for the protoplast, and a dynamic matrix, becoming modified as the cell divides, grows, and becomes specialized for function. Whilst the cell wall cannot be disassembled and reassembled as rapidly as microtubules (Mayer & Jurgens, 2002) or actin filaments (Vantard & Blanchoin, 2002), fairly rapid responses of less than one hour occur locally in response to pathogens (Grant & Mansfield, 1999) or to growth-promoting proteins (Cosgrove, 2000). The osmotic pressure exerted by the protoplast is necessary to drive cell expansion, but because growth can begin and end with almost imperceptible changes in turgor pressure, researchers have focused on wall loosening as the primary determinant of cell expansion. The cell wall architecture must be extensible, that is, mechanisms must exist that allow discrete biochemical loosening of the cell wall matrix, permitting microfibril separation and insertion of newly synthesized polymers.

Apart from the dynamics associated with growth, walls contain the elements of response mechanisms to biotic and abiotic stimuli, as well as developmental and positional signals, and, thus, they can be viewed as an intricate sensory panel for signal transduction to and from the cytoskeleton. In this chapter, we review the architecture of the cell wall and what mechanical properties are conferred on this structure by the component molecules. Then, we discuss the role(s) of the cytoskeleton in altering the physical parameters of the wall that are responsible for its material properties. We consider three mechanisms by which the cytoskeleton can influence these properties: first, by targeting secretion of wall molecules to specific sites that may alter wall extensibility and strength through compositional changes, second, by exerting mechanical forces on wall polymers or by perceiving such forces through direct molecular connections, and third, by being the downstream target of extracellular signals transduced by the cell wall. Whilst the molecular mechanisms of crosstalk between the cell wall and the cytoskeleton remain obscure in this under-researched area, the requirement for interactions between these internal and external scaffolding structures to achieve sustainable cell growth is clear.

6.2 Plant cell wall composition and architecture

Regardless of chemical composition, the primary wall is always defined as the structure that participates in irreversible expansion of the cell. The middle lamella forms the interface between the primary walls of neighboring cells. Finally, at differentiation, some cells elaborate within the primary wall a secondary cell wall, building complex structures uniquely suited to the cell's function.

The plant cell wall is a highly organized composite of many different polysaccharides, proteins, and aromatic substances. Some structural molecules act as fibers, others as a cross-linked matrix, analogous to the glass fibers and plastic matrix in fiberglass (McCann & Roberts, 1991; Carpita & Gibeaut, 1993). The primary cell wall is made up of two, sometimes three, structurally independent but interacting networks. The scaffolding framework of cellulose and cross-linking glycans lies embedded in a second network of matrix pectic polysaccharides. A third network consists of the structural proteins or a phenylpropanoid network.

As in animal cells, the plant Golgi apparatus is a factory for the synthesis, processing, and targeting of glycoproteins (Nebenfuhr & Staehelin, 2001). The Golgi has also been shown by autoradiography to be the site of synthesis of non-cellulosic polysaccharides. Thus, with the exception of cellulose, the polysaccharides, structural proteins and a broad spectrum of enzymes are coordinately secreted in Golgi-derived vesicles and targeted to the cell wall. Many polymers are modified by esterification, acetylation or arabinosylation for solubility during transport. Later, extracellular enzymes de-esterify, de-acetylate or de-arabinosylate to free sites along the polymers for cross-linking into the cell wall. These sites are determined by the long-range binding order of polysaccharides, permitting their assembly into very precise cell-wall architectures. A range of cross-linking possibilities exists, including hydrogen-bonding, ionic bonding with Ca^{2+} ions, covalent ester linkages, ether linkages, and van der Waals' interactions.

6.2.1 Cellulose

The only polymers known to be made at the outer plasma membrane surface of plants are cellulose and callose. Cellulose, a polymer of glucose, accounts for 15–30% of the dry mass of all primary cell walls, and an even larger percentage of secondary walls. Cellulose synthesis is catalyzed by multimeric enzyme complexes, imaged as rosette structures in freeze-fracture replicas of maize plasma membrane, located at the termini of growing cellulose microfibrils (Mueller & Brown, 1980). The appearance of rosettes in the plasma membrane coincides with active cellulose synthesis (Schneider & Herth, 1986), and rosette particles label with antibodies raised against the presumed catalytic subunit of cellulose synthase (Kimura *et al.*,

1999). UDP-Glc is the primary substrate for cellulose synthase. Isoforms of sucrose synthase, an enzyme that produces UDP-Glc directly from sucrose, are also associated with the plasma membrane (Salnikov *et al.*, 2001) where, in close association with cellulose synthase, they may contribute substrate directly into the catalytic site of the enzyme (Amor *et al.*, 1995). Cellulose synthase generates one of the most abundant biopolymers on earth, yet the plant enzyme has proven curiously difficult to purify in active form. Cellulose synthase activity from plants disappears even under conditions of gentle plasma membrane isolation, and no rosettes have been found in isolated membranes. The complex may require cytoskeletal or cell wall components to stabilize it in the membrane.

Cellulose forms microfibrils, which are paracrystalline assemblies of several dozen (1→4)-β-D-glucan chains hydrogen-bonded to one another along their length. Microfibrils of angiosperms have been measured to be between 5 and 12 nm wide in the electron microscope (McCann *et al.*, 1990). Each (1→4)-β-D-glucan chain is several thousand glucosyl units (~2–3 μm long), but individual chains begin and end at different places within the microfibril to allow a microfibril to reach lengths of hundreds of micrometers and contain thousands of individual glucan chains. By electron diffraction, the (1→4)-β-D-glucan chains of cellulose are arranged parallel to one another, that is, all of the reducing ends of the chains point in the same direction (Koyama *et al.*, 1997).

Callose is a glucan polymer with (1→3)-β-rather than (1→4)-β-glycosidic linkages, forming helical duplexes and triplexes rather than crystalline microfibrils. Although callose synthesis in isolated membranes may have resulted from damage to the cellulose synthase, callose is a natural component of the initial cell plate, the pollen tube wall, and in transitional stages of wall development of certain cell types (Stone & Clarke, 1991), and callose synthase activities are encoded by different genes from those that encode cellulose synthases (Verma & Hong, 2001). Glucan synthesis in ordered arrays on the plasma membrane requires cortical microtubules attached to the membrane (Hirai *et al.*, 1998). When the microtubules are disrupted by treatment with propyzamide, glucans are deposited in masses.

6.2.2 Cross-linking glycans

Cross-linking glycans are a class of polysaccharides that can hydrogen-bond to cellulose microfibrils: they may coat microfibrils but are also long enough to span between microfibrils, and link them together to form a network. The two major cross-linking glycans of all primary cell walls of flowering plants are xyloglucans (XyGs) and glucuronoarabinoxylans (GAXs) (Carpita & Gibeaut, 1993). There are two distinct wall types that differ in composition and are associated with distinct plant taxa (Carpita & Gibeaut, 1993). The XyGs cross-link the Type I walls of all dicots and about one-half of the monocots, but in the Type II cell walls of the commelinoid line of monocots, which includes bromeliads, palms, gingers, cypresses, and grasses, the major cross-linking glycan is GAX (Carpita, 1996).

In the Type I wall, the XyGs consist of linear chains of (1→4)-β-D-glucan with numerous α-D-Xyl units linked at regular sites to the O-6 position of the Glc units. Some of the xylosyl units are substituted further with α-L-Ara or β-D-Gal, depending on species, and sometimes the Gal is substituted further with α-L-Fuc. The XyGs are constructed in block-like unit structures containing six to eleven sugars, the proportions of which vary among tissues and species (Sims *et al.*, 1996).

XyGs occur in two distinct locations in the wall. They bind tightly to the exposed faces of glucan chains in the cellulose microfibrils, and they lock the microfibrils into the proper spatial arrangement by spanning the distance between adjacent microfibrils or by linking to other XyGs. With an average length of ~200 nm, XyGs are long enough to span the distance between two microfibrils and bind to each of them.

All angiosperms also contain at least small amounts of GAX, but their structure may vary considerably with respect to the degree of substitution and position of attachment of α-L-Ara residues. In Type I walls, the α-L-Ara units are more commonly found at the O-2 position. However, in the Type II walls of commelinoid monocots, where GAXs are the major cross-linking polymers, the Ara units are invariably on the O-3 position. In all GAXs, the α-D-GlcA units are attached to the O-2 position.

Unbranched GAXs can hydrogen-bond to cellulose or to each other. The attachment of the α-L-Ara and α-D-GlcA side groups to the xylan backbone of GAXs prevents the formation of hydrogen bonds, and therefore blocks cross-linking between two branched GAX chains, or GAX to cellulose. In contrast, the α-D-Xyl units attached at the O-6 of XyG, away from the binding plane, stabilize the linear structure and permit binding to one side of the glucan backbone. The number of side groups of Ara and GlcA along the GAX chains varies markedly, from GAXs whose Xyl units are nearly all branched to those with only 10% or less of the Xyl units bearing side groups. Side groups not only prevent hydrogen-bonding but also render the GAX water-soluble. In dividing and elongating cells, highly branched GAXs are abundant, whereas after elongation and differentiation, more and more unbranched GAX accumulates. Cleavage of the Ara and other side groups from contiguously branched Xyl units could yield runs of unbranched xylan capable of binding to other unbranched xylans or to cellulose microfibrils (Carpita, 1996).

In the order Poales, which contains the cereals and grasses, a third major cross-linking glycan, called mixed-linkage (1→3),(1→4)-β-D-glucans, distinguishes these species from the other commelinoid species (Smith & Harris, 1999). Unbranched polymers consist of 90% cellotriose and cellotetraose units in a ratio of about 2.5:1 and connected by (1→3)-β-D-linkages (Carpita, 1996). The cellotriosyl and cellotetraosyl units together make up helices about 50 residues long that are spaced by oligomers of four or more contiguous (1→4)-β-Glc units. Absent from meristems and dividing cells, (1→3),(1→4)-β-D-glucans accumulate to almost 30% of the non-cellulosic cell wall material during the peak of cell elongation, and then are largely hydrolyzed by the cells during differentiation (Carpita *et al.*, 2001a). The appearance of (1→3),(1→4)-β-D-glucans during cell expansion and the acceleration of their hydrolysis by growth regulators all implicate direct involvement of the polymer in growth.

Other, much less abundant non-cellulosic polysaccharides, such as glucomannans and galactoglucomannans, and galactomannans potentially interlock the microfibrils in some primary walls. These mannans are found in virtually all angiosperms examined.

6.2.3 Pectins

Pectins comprise a mixture of heterogeneous, branched and highly hydrated polysaccharides rich in D-galacturonic acid. Two classes of pectic polymer have been distinguished: galacturonans, including homogalacturonan (HG), xylogalacturonan, apiogalacturonan and rhamnogalacturonan (RG II), and rhamnogalacturonan (RG I) (Carpita & Gibeaut, 1993; Mohnen, 1999; Ridley *et al.* 2001; Willats *et al.*, 2001). HG is a polymer of (1→4)-linked α-D-GalA residues, modified by either methyl-esterification or substitution with acetyl groups and xylose or apiose. The methyl-esterification of HG determines to a large extent the gelation characteristics of pectin: both the amount and the distribution of methyl groups on the HG backbone are important. RG II is actually a modified HG bearing complex branched side-chains of 11 sugars including apiose, aceric acid (3-C′-carboxy-5-deoxy-L-xylose), 2-O-methyl fucose, 2-O-methyl xylose, Kdo (3-deoxy-D-*manno*-2-octulosonic acid), and Dha (3-deoxy-D-*lyxo*-2-heptulosaric acid) (O'Neill *et al.*, 1996; Vidal *et al.*, 2000).

Another constituent polysaccharide of pectin, RG I, is composed of a repeating disaccharide unit $(\rightarrow 2)$-α-L-Rha*p*-$(1\rightarrow 4)$-α-D-Gal*p*A-$(1\rightarrow)_n$, where n can be larger than 100 (McNeil *et al.*, 1980; Albersheim *et al.*, 1996). The galacturonosyl residues can carry acetyl groups on *O*-2 and *O*-3. The Rha residues can be substituted at *O*-4 with neutral sugars. The proportion of branched Rha residues generally varies from ~20 to ~80% depending on the source of the polysaccharide, although essentially unbranched RG I molecules have also been reported in seed mucilages (Penfield *et al.*, 2001). The side chains can be single-unit [β-D-Gal*p*-(1→4)], but also polymeric such as arabinogalactan I (AG-I) and arabinan (50 glycosyl residues or more). AG-I is composed of a (1→4)-linked β-D-Gal*p* backbone with α-L-Ara*f* residues attached to the *O*-3 of the galactosyl residues (Mohnen, 1999; Ridley *et al.*, 2001). The arabinans consist of a (1→5)-linked α-L-Ara*f* backbone, which can be substituted with α-L-Ara*f*-(1→2)-, α-L-Ara*f*-(1→3)-, and/or α-L-Ara*f*-(1→3)-α-L-Ara*f*-(1→3)-side chains (Schols & Voragen, 1996). Neutral polymers (arabinans or galactans) are pinned at one end to the pectic backbone, but extend into, and are highly mobile in, the wall pores. At some stages of cell development, hydrolases are released that trim these neutral polymers, potentially increasing pore size. The precise structures of these side-chains and their distribution along the backbone of RG I are not known.

Pectins form a cross-linked three-dimensional hydrated network that may regulate cell wall porosity, hydration, and pH, in turn regulating the access or activities of cell wall modifying enzymes to either the pectin network or to the cellulose – cross-linking glycan network. Three of the potential cross-links, in addition to glycosidic linkages between different pectin molecules are detailed below:

Ca^{2+}-gels. HGs can adopt different conformations in solution of which the right-handed two-(2_1) and three-fold (3_1) helices seem to be most favorable in terms of minimal energy (Braccini *et al.*, 1999). HG is thought to be secreted as highly methyl esterified polymers, and the enzyme pectin methylesterase (PME) located in the cell wall cleaves some of the methyl groups to initiate binding of the carboxylate ions to Ca^{2+}. The helical chains of HGs can condense by cross-linking with Ca^{2+} to form junction zones, linking two anti-parallel chains. Maximally strong junctions occur between two chains of at least seven unesterified GalA units each (Daas *et al.*, 2001). If sufficient Ca^{2+} is present, some methyl esters can be tolerated in the junction and the HGs can bind in both parallel and antiparallel orientation. The spacing of the junctions is postulated to create a cell-specific pore-size. Rha units of RG I and their side chains interrupt the Ca^{2+} junctions and contribute to the pore definition. The extent of methyl esterification may remain high in the walls of some cells, and a type of gel may form with highly esterified parallel chains of HGs. In meristems and elongating cells, where Ca^{2+} concentrations are kept quite low, significant de-esterification of HGA can occur without Ca^{2+} binding. While this may not contribute to pore-size dynamics, it alters charge density and local pH.

Borate-diol esters. Two molecules of RG II can complex with boron, forming a borate-diol ester (Kobayashi *et al.*, 1996; Ishii *et al.*, 1999). Only the apiofuranosyl residues of the 2-O-methyl-D-xylose-containing side chains in each of the subunits of the dimer participate in the cross-linking (Ishii *et al.*, 1999). Because RG II is covalently linked to HG or RG I (Ishii & Matsunaga, 2001), borate-diol esters can cross-link two pectin chains. The widespread occurrence of RG II in the plant kingdom, and its structural conservation indicate a distinct role in wall integrity for this constituent of pectin. RG II polymers may also be scattered along the pectic backbone to space out sites of borate di-diester cross-linking.

Uronyl esters. Some HGs and RGs are cross-linked by ester linkages to pectins or other polymers held more tightly in the wall matrix and can only be released from the wall by de-esterifying agents. Kim and Carpita (1992) suggested that HG could be cross-linked to other components by uronyl esters, and Brown and Fry (1993) provided some evidence for their chemical synthesis. Up to approximately 2% of the GalA residues could be cross-linked in this way.

In general, Type II walls are pectin-poor, but an additional contribution to the charge density of the wall is provided by the α-L-GlcA units on GAX. These walls have very little structural protein compared with dicots and other monocots, but have extensive interconnecting networks of phenylpropanoids that can accumulate, particularly as the cells stop expanding.

6.2.4 *Structural proteins*

Although the structural framework of the cell wall is largely carbohydrate, structural proteins may also form networks in the wall. There are three major classes of structural proteins named for their uniquely enriched amino acid: the hydroxyproline-rich proteins (HRGPs), the proline-rich proteins (PRPs), and the glycine-rich proteins (GRPs). A fourth class, arabinogalactan proteins (AGPs) are more aptly named proteoglycans, as they can consist of more than 95% carbohydrate (Du *et al.*, 1996). AGPs constitute a broad class of molecules that are located in Golgi-derived vesicles, the plasma membrane, and the cell wall. Of the few core proteins that have been characterized, they are enriched in Pro(Hyp), Ala, and Ser/Thr. They possess no distinguishing common motifs, but contain sequence runs with similarity to some PRPs, extensins, and the Solanaceous lectins (Gaspar *et al.*, 2001). All four of these large multigene families are developmentally regulated, with relative amounts varying among tissues and species (Cassab, 1998).

Many structural cell wall proteins are specifically associated with secondary thickenings. In bean, GRPs are synthesized in the xylem parenchyma cells and exported into the walls of protoxylem vessels (Ryser & Keller, 1992). Arabinogalactan proteins (AGPs) (Schindler *et al.*, 1995), an extensin-like protein (Bao *et al.*, 1992) and a tyrosine and lysine-rich protein (Domingo *et al.*, 1994) have been found in maize, loblolly pine and tomato xylem, respectively. Some PRPs concentrate in the secondary walls of protoxylem elements of bean.

Extensin, encoded by a multigene family, is one of the best-studied HRGPs of plants. Extensin consists of repeating Ser-$(Hyp)_4$ and Tyr-Lys-Tyr sequences that are important for secondary and tertiary structure: the repeating Hyp units predict a polyproline II rod-like molecule. HRGPs are generally found at low levels in the primary walls of all tissues, although they are particularly abundant in phloem. Glycosylation patterns of HRGPs are dependent upon the primary sequence of the protein (Kieliszewski & Schpak, 2001). Synthesis, deposition and cross-linking of extensins is thought to help increase the mechanical strength of the cell wall. Both HRGPs and PRPs are also considered to be involved in the responses of plants to environmental factors, such as wounding and infection (Sheng *et al.*, 1993), and specific extensins are expressed in elongating cells (Dubreucq *et al.*, 2000).

6.2.5 Aromatic substances

The Type II primary walls of the commelinoid orders of monocots, and the Chenopodiaceae (such as sugar beet and spinach) contain significant amounts of aromatic substances in their unlignified cell walls – a feature that makes them fluorescent under UV light. A large fraction of plant aromatics consists of hydroxycinnamic acids, such as ferulic and *p*-coumaric acids. In grasses, these hydroxycinnamates are attached as carboxyl esters to the O-5 position of a few of the Ara units of GAX. A small proportion of the ferulic acid units of neighboring GAXs may cross-link by phenyl-phenyl or phenyl-ether linkages to interconnect the GAX into a large network. In the Chenopodiaceae, ferulic acids are attached to Gal or Ara units on side-chains subtending some RG I molecules.

The most obvious distinguishing feature of secondary walls is the incorporation of lignins, complex networks of aromatic compounds called phenylpropanoids. The synthesis of lignin is initiated solely when secondary wall deposition commences. The phenylpropanoids, hydroxycinnamoyl alcohols or monolignols, *p*-coumaryl, coniferyl and sinapyl alcohol comprise most of lignin networks (Hatfield *et al.*, 1999). The monolignols are linked by way of ester, ether, or carbon–carbon bonds. The diversity of monolignols and their possible intra-linkages impart remarkable complexity of structure.

6.3 Cell growth and wall extensibility

Despite the marked differences in the composition of Type I and Type II walls, the biophysics of growth of grasses and other flowering plants are similar (Carpita & Gibeaut, 1993). Although the chemical complexity of the wall is daunting, the similarity of the physiological responses to acid, auxin and light of different quality by all flowering plants indicates that a few common mechanisms of wall expansion exist, regardless of the kinds of molecules that cross-link the microfibrils. Features of cell wall architecture that influence mechanical properties, and, therefore, the ability of cells to expand, include the length and stiffness of the component fibers, their entanglement and physical interactions, the extent and nature of the cross-links, the hydrogen bonds and ionic interactions between polymers, three-dimensional architecture and the water potential. In addition to these cellular or sub-cellular levels of organization it is also clear that tissue architecture (cell shape and orientation) and overall anatomy (distribution of vascular bundles and fibers) also play a major role in determining mechanical properties of plant organs (Jarvis & McCann, 2000). Primary cell walls are adapted to withstand tensile stresses while secondary walls also need to withstand compression. To understand how the tensile stress is resisted, we need to know how walls deform under load and how rigid are the specific constituents of each type of cell wall.

6.3.1 The biophysics of growth underpins cell wall dynamics

Three classes of cell wall polymers constitute nearly independent determinants of strength in elongating cells: (1) the microfibrils arranged in the transverse axis; (2) the cross-linking glycans in the longitudinal axis; (3) putative networks involving structural proteins or phenylpropanoid compounds, or elements of the pectin network. When plant growth regulators, such as auxin and gibberellin, change the dimension of growth, they do so through changes in the orientation of cortical microtubules and cellulose microfibrils. When they change the rate of growth, their mechanisms include dissociation or breakage of the cross-linking glycans from microfibrils. A mathematical formulation that describes the growth rate of plant cells allows us to define the cell wall properties that must be modified to permit growth.

The driving force for water uptake in elongating cells can be quantified by Equation 6.1:

$$dl/dt = Lp(\Delta\Psi_w) \tag{6.1}$$

where dl/dt is the change in length per unit time, Lp is hydraulic conductivity, i.e. the rate at which water can flow across the membrane, and $\Delta\Psi_w$ is the water potential difference between the cell and the external medium. The difference in water potential is the driving force for water movement and comprises two components, $\Psi_\pi + \Psi_p$, which are the osmotic potential and pressure potential (turgor), respectively. The equation is revised to include any type of growth by:

$$dV/dt = A \cdot Lp(\Delta\Psi_w) \tag{6.2}$$

where growth is defined as a change in volume (V) per unit time, and dependent on the surface area (A) of the plasma membrane available for water uptake. Thus, the rate of growth is proportional to membrane surface area, the conductivity of the membrane, and the water potential difference driving water uptake. In non-growing cells, $\Delta\Psi_w$ and thus dV/dt, equals zero, because the rigid cell wall prevents water uptake and the turgor pressure rises to a value equal to that of the cell's osmotic potential. By contrast, in growing cells, the $\Delta\Psi_w$ does not reach zero because the cross-links have been loosened. As a result, cell volume increases irreversibly. This wall-localized event, called stress relaxation, serves as the fundamental difference between growing and non-growing cells.

When turgor is reduced in growing cells by an increase in the external osmotic potential, growth ceases before turgor reaches zero. This value is called the yield threshold, the pressure potential above which expansion can occur. The increment of growth rate change above the yield threshold is dependent not only on turgor but also on a factor called wall extensibility, which is the slope (m) of a general equation:

$$rate = m(\Psi p - Y) \tag{6.3}$$

where Y is the yield threshold.

6.3.2 *The biochemical determinants of yield threshold and extensibility*

The acid-growth hypothesis proposes that auxin activates a plasma membrane proton-pump that acidifies the cell wall (Taiz, 1984). The low pH in turn activates apoplast-localized growth-specific hydrolases, which cleave the load-bearing bonds that tether cellulose microfibrils to other polysaccharides. Cleavage of these bonds results in loosening of the wall, and the water potential difference causes uptake of water. Relaxation of the wall, i.e. separation of the microfibrils, passively leads to an increase in cell size.

The basic tenets of the acid-growth hypothesis have stood the test of time, but three problems persist. First, no enzymes have been found that hydrolyze wall cross-linking glycans exclusively at pH lower than 5.0. Second, no reasonable explanation exists for how growth is kept in check once the hydrolases are activated. Third, no hydrolases extracted from the wall and added back to the isolated tissue sections, regardless of external pH, cause extension *in vitro*.

Two candidate wall-loosening enzymes are currently being studied. One of these is xyloglucan endotransglycosylase (XET), which carries out a transglycosylation of XyG where one chain of XyG is cleaved and reattached to the non-reducing terminus of another XyG chain (Fry *et al.*, 1992). Given such a mechanism, microfibrils could undergo a transient slippage but overall tensile strength of the interlocking XyG matrix would not diminish. XETs may also function in the realignment of XyG chains in different strata during growth, and in assembly of the wall as newly synthesized XyGs are incorporated. In some cases, the correlation between growth and XET activity is not clear, and some XETs can function hydrolytically.

Other proteins, called expansins, catalyze wall extension *in vitro* without any detectable hydrolytic or transglycolytic events. These proteins probably catalyze breakage of hydrogen bonds between cellulose and the load-bearing cross-linking glycans (McQueen-Mason & Rochange, 1999; Cosgrove, 2000). Such an activity could disrupt the tethering of cellulose by XyGs in Type I walls and by GAXs in Type II walls and by GAXs and (1→3),(1→4)-β-D-glucans in grass walls.

On the basis of sequence similarities, expansins are classified into two multigene families, the α- and β-expansins. A notable difference between the two groups is the extensive glycosylation of β-expansins which appears to be absent in the α-expansins. Analysis of the complete *Arabidopsis* genome suggests that there might be more than 24 members of the α-expansin family in *Arabidopsis* (http://www.bio.psu.edu/expansins; Cosgrove, 2000). XET and expansin may not be the only wall-loosening agents, and work continues to determine the roles played by hydrolases.

Whilst XETs, expansins and hydrolases all affect extensibility, a new class of cell wall proteins have been identified that directly affect Y, the yield threshold. Yieldins are homologous to acidic class III endochitinases and concanavalin B, and promote wall extension under acidic conditions (Okamoto-Nakazato *et al.*, 2000).

6.4 Functional architecture revealed by mutation and transgenic approaches

The *Arabidopsis mur* mutants, mapping to 11 different loci, were identified in a screen for cell wall mutants because one or several specific sugars were over- or under-represented compared with the sugar composition in wild-type plants (Reiter *et al.*, 1997). However many cell wall mutants have also been selected on the basis of a growth or developmental phenotype. As the cell wall constrains the final size and shape that plant cells achieve, it is perhaps not surprising that such screens have identified lesions in cell wall-related genes. Additionally, transgenic approaches are being used to take advantage of genes of known function in cell walls to reveal their roles in cell and plant growth and differentiation. Over 2500 gene products are probably involved in cell wall biosynthesis, assembly and turnover (Carpita *et al.*, 2001b) and so the current range of materials with genetically defined variation represents the tip of the iceberg for cell wall biology.

6.4.1 Cellulose

Arabidopsis contains a family of ten similar *CESA* sequences, thought to encode the catalytic subunit of the cellulose synthase complex. Genetic proof of one *CESA* homolog has been obtained by complementing the *Arabidopsis rsw1* mutant, which cannot make cellulose at the restricted temperature (Arioli *et al.*, 1998). Transformation of the temperature-sensitive mutant with the wild-type *CESA* gene restores the normal phenotype, providing the first evidence that a plant *CESA* gene functions in the formation of cellulose microfibrils. However, direct proof of the function of the gene product is still lacking, because cellulose synthase cannot presently be synthesized *in vitro*. The *rsw1* mutant was selected by a root radial swelling phenotype at restrictive temperatures (Arioli *et al.*, 1998). *Procuste*, mutated in a second *CESA* gene, was selected in a screen for hypocotyls that are dwarf when grown in the dark (Fagard *et al.*, 2000). Cellulose synthesis and deposition is essential for normal cell elongation as shown by the cellulose-deficient and dwarf phenotypes of *rsw1* (Arioli *et al.*, 1998) and *prc* (Fagard *et al.*, 2000). Two other mutants with growth defects are also cellulose-deficient, including a novel protein called *cobra* with a glycophosphatidylinositol (GPI) lipid anchor signal sequence (Schindelman *et al.*, 2001) and *korrigan*, an endo-(1→4)-β-D-glucanase (Nicol *et al.*, 1998).

In a more direct screen for cellulose mutants, Somerville and colleagues screened for resistance to isoxaben, a cellulose synthesis inhibitor, and identified a second allele of *PRC* and a mutation in a third *CESA* gene (Scheible *et al.*, 2001). Three secondary wall cellulose synthase mutants, *irx1*, *irx3* and *irx5*, were selected by a collapsed xylem phenotype (Turner & Somerville, 1997; Taylor *et al.*, 2003) in which the cellulose-deficient secondary thickenings of the xylem result in reduced stem strength.

6.4.2 The cellulose – cross-linking glycan network

Two of the *mur* mutants are defective in the fucosyl transferase (*mur2*; Vanzin *et al.*, 2002) and galactosyl transferase (*mur3*; Madson *et al.*, 2003) that add these subtending units to the xyloglucan backbone. Both mutants have the appearance of wild-type plants. However, mutation in XET, that uses xyloglucan as a substrate, indicates that the polymer is involved in growth mechanisms. TOUCH4 is required for the normal dwarfing response of *Arabidopsis* to a touch stimulus by gentle wind or hand (Xu *et al.*, 1995). This phenotype is difficult to interpret in terms of a role for XET in wall loosening during growth but it does indicate that its substrate may be load-bearing in the wall.

Expansins may act at the interface between cellulose and cross-linking glycans to reduce hydrogen-bonding between them. The response of meristematic cells to expansin is particularly interesting. Exogenous application of beads loaded with purified expansins induces bulging of the meristem at the sites of application, some bulges later producing leaf-like structures (Fleming *et al.*, 1997). Pien *et al.* (2000) have shown that localized induction of expansion within the meristem results in the full developmental program that leads to leaf formation. Local transient induction of expansion expression on the flank of developing primordia leads to the induction of ectopic lamina tissue and modulation of leaf shape (Pien *et al.*, 2000). Thus, morphogenesis may be manipulated by alteration of the biophysical properties of the cell wall.

6.4.3 Pectins

Pectins may modulate the elasticity of the wall and regulate accessibility of enzymes with roles in modifying the cellulose/cross-linking glycan network. In work with onion epidermis, Wilson *et al.* (2000) obtained evidence for the independence of the cellulose and pectin networks and showed that pectin is mechanically important, affecting the viscoelastic properties of the cell wall through its modification of cellulose hydration. These observations are consistent with earlier results with suspension-cultured cells of tomato adapted to growth on the cellulose synthesis inhibitor 2,6-dichlorobenzonitrile (DCB), which make a pectin-rich wall that virtually lacks a cellulose-xyloglucan network (Shedletzky *et al.*, 1990). When the formation of one structural network within the wall is inhibited, then the other is still able to provide a cell wall with at least some of its usual functions (Wells *et al.*, 1994).

A major goal in pectin biology is to determine the functions of the individual pectic domains. There is evidence that RG II affects wall porosity and thickness by forming borate di-ester-linked dimers (Ishii *et al.*, 2001) and is important for growth (O'Neill *et al.*, 2001). *mur1*, a fucose-deficient mutant, has a tensile strength of the stem less than 50% that of wild-type and exhibits a slightly dwarfed growth habit (Reiter *et al.*, 1993; Bonin *et al.*, 1997). The growth phenotype can be rescued by supplying excess boron to the plants and compensating for the lowered binding efficiency of RG II, a fucose-containing polymer, for boron to form di, di-ester cross-links (O'Neill *et al.*, 2001). A pectin glucuronyltransferase that transfers glucuronic acid to RG II (Iwai *et al.*, 2002) is the first gene involved in decorating a pectic backbone. The activity of this gene is required for intercellular attachment.

HG chains can be interlocked by Ca^{2+} bridges to contribute to the mechanical strength of the pectin network within the wall (Jarvis, 1984; Jarvis & McCann, 2000). Changes in the esterification of HG have been correlated in many different systems with growth. For example, the degree of methyl esterification is highest in maize cells (Kim & Carpita, 1992) and tobacco cells (McCann *et al.*, 1994) that are actively elongating. Transgenic approaches have given a variety of phenotypes. Pilling *et al.* (2000) expressed a petunia pectin esterase in potato and observed enhanced stem elongation. Inhibition of gene expression of pectin esterase in tomato plants had no effect on growth or ripening characteristics (Hall *et al.*, 1993) but, in another study, did affect tissue integrity during fruit senescence (Tieman & Handa, 1994). Inhibition of gene expression of polygalacturonase in tomato decreased the solubility of pectins from the wall (Carrington *et al.*, 1993). *Quasimodo 1*, which has weak cell–cell adhesion, encodes a putative membrane-bound glycosyltransferase that has reduced galacturonic acid content, and is therefore a candidate for a pectin biosynthetic gene (Bouton *et al.*, 2002).

RG I and its side-chains has been implicated in the mechanical properties of walls (McCartney *et al.*, 2000), the softening of tissues during fruit ripening (Jones *et al.*, 1997), wall porosity (Orfila & Knox, 2000; Oxenbøll-Sørensen *et al.*, 2000) and cell elongation and proliferation (Willats *et al.*, 1998; Ermel *et al.*, 2000; Bush *et al.*, 2001). It is likely that the functions of RG I will vary with the wall or tissue type and involve complex interactions with other wall components.

6.4.4 Structural proteins

The phenylglucoside Yariv reagent binds specifically to AGPs and has been used to perturb AGP function. The use of this inhibitor has shown that AGPs are associated with cell expansion in roots, cell growth in tissue culture, and pollen-tube growth (Willats & Knox, 1996; Yang, 1998; Roy *et al.*, 1998). The *root epidermal bulger* (*reb1*) mutant in *Arabidopsis* is allelic to *ROOT HAIR DEFICIENT 1*, recently shown to encode an isoform of UDP-D-glucose-4-epimerase, and lacks galactosylated AGP and also galactosylated xyloglucan (Seifert *et al.*, 2002). Microtubules in the swollen trichoblasts of *reb1* are disordered or absent, further suggesting a role for AGPs in orienting cortical microtubules (Andeme-Onzighi *et al.*, 2002). AGPs of the fucose-deficient mutant *mur1* lack fucose: the *mur1* roots are about half as long as wild-type and this can be phenocopied by the addition of eel lectin, which specifically binds to a fucosylated AGP population in the cell wall of differentiating root cells (Van Hengel & Roberts, 2002).

In total, 13 classical AGPs, 10 AG-peptides, three basic AGPs that include a short lysine-rich region, and 21 fasciclin-like AGPs have been identified in the *Arabidopsis* genome (Schultz *et al.*, 2002). Fasciclin-like domains are typically found in animal cell–cell adhesion proteins. The *SALT OVERLY SENSITIVE* (*SOS*) 5 gene is required for normal cell expansion – when the mutant is under salt stress, root tips swell and root growth is arrested (Shi *et al.*, 2003). The predicted SOS5 protein contains an N-terminal signal sequence for plasma membrane localization, two AGP-like domains, two fasciclin-like domains and a C-terminal GPI anchor sequence.

The ROOT-SHOOT-HYPOCOTYL-DEFECTIVE (RSH) protein is an HRGP and is essential for embryogenesis (Hall & Cannon, 2002). The RSH mutation results in aberrant positioning of the cell plate during cytokinesis, even to the first asymmetrical division of the zygote. Such a protein may be a candidate for an anchoring protein within the wall for the forming cell plate.

6.5 The role of the cytoskeleton

Chapter 3 describes the dynamic changes in the cytoskeleton that may drive or accompany cell expansion. Above, we have laid out the biophysical basis of cell expansion and the requirement for cell wall loosening. The question then remains of how these two processes might be coupled to achieve cell growth. The concept of a cell wall-plasma membrane-cytoskeleton continuum in which physical connections are made such that microtubules may exert forces on microfibrils is attractive. However, many cells may be plasmolyzed without harm, with points of adhesion between membrane and wall occurring only in an inducible fashion (Wyatt & Carpita, 1993). Recent progress suggests that the coupling of these two scaffolding frameworks might involve both physical interaction and signal transduction. The wall is a complex sensory panel of receptors that can transmit information to the cytoskeleton for signaling and regulation.

6.5.1 Targeting of cell wall components

There are at least three processes in which the cytoskeletal machinery targets wall polymers to specific sites – the establishment of wall domains of distinct compositions around a cell, the formation of the cell plate, and during tip growth.

The molecular composition and arrangements of the wall polymers differ among species, among tissues of a single species, among individual cells, and even among regions of the wall around a single protoplast. At the single-cell level, modifications occur that distinguish between transverse and longitudinal walls. For example, in sieve tube members a preferential digestion of end walls occurs. Some substances, such as waxes, are only secreted to a cell's outer epidermal face. Within a single wall there are zones of different architecture; the middle lamella, plasmodesmata, thickenings, channels, pit-fields and the cell corners, and there are also domains within the thickness of a wall in which the degree of pectin esterification and the abundance of RG I side-chains differ. This indicates the existence of precise targeting mechanisms for the delivery of polymers to sites at the plasma membrane by trafficking of Golgi-derived vesicles. Golgi vesicles travel along actin filaments to the plasma membrane in plants (Hawes & Satiat-Jeunemaitre, 2001). How plant cells deliver those vesicles to the specific site of the plasma membrane, e.g. how ten-fold the amount of wall material is delivered to the outer, rather than the inner, epidermal wall, is not known. However, the discovery that intact Golgi stacks can travel along actin filaments using myosin motors (Boevink *et al.*, 1998) raises the possibility that the Golgi stacks themselves may be viewed as delivery vehicles for their products (Nebenfuhr & Staehelin, 2001).

Cell walls originate in the developing cell plate. The phragmoplast is commonly thought to initiate in the centre of the cell and grow outward to the parental cell membrane (Verma, 2001). As plant nuclei complete division during telophase of the mitotic cell cycle, the phragmosome, a flattened membranous vesicle containing cell wall components, forms across the cell within a cytoskeletal array called the phragmoplast. The non-cellulosic cell wall polysaccharides synthesized in the Golgi apparatus and packaged in vesicles fuse with the growing cell plate (Samuels *et al.*, 1995). The plate grows outward until the edges of the membranous vesicle fuse with the plasma membrane, creating two cells. Finally, the new cell wall fuses with the existing primary wall. However, three-dimensional, live-cell imaging reveals that the mitotic spindle and phragmoplast are laterally displaced in *Arabidopsis* shoot cells, and that the growing cell plate anchors on one side of the cell at an early stage of cytokinesis. Growth of the phragmoplast across the cell creates a new partition in its wake (Cutler & Ehrhardt, 2002). Such polarized cytokinesis is not solely a function of cell size, being observed even in some small root cells. Thus, short-range interactions between the phragmoplast and the plasma membrane may play important roles in guiding the cell plate throughout its development (Cutler & Ehrhardt, 2002). Cleary (2001) showed that plasmolysed leaf epidermal cells had patches of plasma membrane – cell wall adhesion at sites involved in cytoskeletal arrays during asymmetric cell divisions – the cortical actin patch and the division site preprophase band.

The *Arabidopsis TONNEAU2* gene encodes a protein phosphatase regulatory subunit that may be involved in phosphorylation cascades that control the dynamic organization of the cortical cytoskeleton in plant cells. Lack of a preprophase band (PPB) in the *ton2* mutant leads to mis-positioning of new cell walls after mitosis, affecting cell shape and arrangement and overall plant morphology (Camilleri *et al.*, 2002). Similarly, in maize, *TANGLED1* encodes a microtubule-associated protein (MAPs) (Smith *et al.*, 2001) that has an altered cell division pattern associated with defective PPB positioning and phragmoplast guidance (Cleary & Smith, 1998). These mutations affect cytoskeletal elements specifically but also potentially the interaction of the cytoskeleton with specific sites on the walls of the parent cells.

In the vast majority of plant cells, growth and the deposition of new wall material occur uniformly along the entire expanding wall. However, tip growth, the growth and deposition of new wall material strictly at the tip of a cell, occurs in some plant cells, notably root hairs and pollen tubes. These cells grow at prodigious rates for plant cells (in excess of 200 nm/s for pollen tubes) (Hepler *et al.*, 2001). The site of expansion in tip-growing cells is focused to the dome-shaped apex, which is filled with secretory vesicles by myosin-mediated delivery on the tracks of actin microfilaments. Intracellular gradients of calcium and protons are spatially localized at the growing apex. In root hairs, actin has been visualized *in vivo* by using a green fluorescent-protein-talin reporter and shown to form a dense mesh in the apex of the growing tip (Baluska *et al.*, 2000). Disruption by drug treatments of actin but not microtubules blocks tip growth. Expansins are located specifically to the region of the trichoblast (hair-forming) cell wall that bulges out to initiate the root hair (Baluska *et al.*, 2000). Installation of the actin-based tip growth machinery may take place only after expansin-associated bulge formation and requires assembly of profilin-supported dynamic F-actin meshworks (Baluska *et al.*, 2000). The localized wall loosening in the bulge by expansin may activate plasma membrane stretch-activated Ca^{2+} channels, transducing the biomechanical signal into a signal that can be interpreted by the cytoskeleton.

Depolymerizing or stabilizing the microtubule cytoskeleton of apically growing root hairs leads to a loss of directionality of growth and the formation of multiple independent growth points in a single root hair. Microtubules occur as axial arrays parallel to the growth axis and are absent from the apical dome (Geitmann & Emons, 2000). Disturbing microtubule functions in *Arabidopsis* root hairs by drug treatments (Bibikova *et al.*, 1999) or a *mor1* mutation (Whittington *et al.*, 2001) results in wavy, branched cell growth, indicating the loss of growth polarity. *ZWICHEL* (*ZWI*) encodes a kinesin-like calmodulin-binding protein (KCBP) which can act as a microtubule-bundling protein. The *zwi* mutant has reduced number of trichome branches (Oppenheimer *et al.*, 1997). Antisense α-tubulin gene expression also causes root hair branching in *Arabidopsis* (Bao *et al.*, 2001). These results suggest that microtubules are involved in restricting vesicle delivery to a single site of growth but how they do this remains unclear. *ECTOPIC ROOT HAIR 3* encodes a katanin-p60 protein and a mutation in the gene alters the specification of cell fates within the root promoting root hair growth in atrichoblast cells (Webb *et al.*, 2002). Katanin proteins are known to sever microtubules and also have a role in organization of the cell wall (Burk & Ye, 2002). Microtubules may be required for the localization of positional cues in the wall that have previously been shown to operate in the development of the root epidermis (Webb *et al.*, 2002).

6.5.2 Mechanical connections

Integrins are a family of cell surface adhesion receptors involved in proliferation, differentiation, migration and adhesion (Hynes, 1987; Schwartz *et al.*, 1995). With an extracellular domain on the cell surface and a cytoplasmic region, integrins link the extracellular matrix directly to the cytoskeletal proteins and actin filaments on the cytoplasmic side of the membrane. Animal integrins recognize the sequence Arg-Gly-Asp (RGD) which is a distinct motif conserved in several proteins that comprise the extracellular matrix in animal cells (Hynes, 1992). Don Ingber and colleagues tugged on the outside of an animal cell using optical tweezers to hold RGD-coated beads that were bound to integrin molecules on the cell surface and observed the movements of microtubules inside the cell in response to tensile and compressive forces (Wang *et al.*, 2001). They proposed a microstructural model of the living animal cell as a simple mechanical continuum or tensegrity network that maintains a stabilizing prestress through incorporation of discrete structural elements that bear compression. The microtubules bear compression forces and cells behave as discrete structures composed of an interconnected network of actin microfilaments and microtubules to bear tensile forces.

This inherent connectivity of extracellular matrix and cytoskeleton has yet to be demonstrated in plant cells. RGD-binding proteins have been found in soybean root-cell culture (Schindler *et al.*, 1989), tobacco suspension cells adapted to grow in high NaCl (Zhu *et al.*, 1993) and the plasma membrane of *Arabidopsis* (Canut *et al.*, 1998), and there are numerous reports of cross-reactivity of plant proteins with antibodies to various animal matrix proteins (Wyatt & Carpita, 1993). A cDNA clone isolated from a cDNA expression library using an antibody to a conserved region of a vertebrate β1 integrin is localized partly in the plasma membrane and has a transmembrane domain and a small region with sequence similarity to integrins (Nagpal & Quatrano, 1999). Two bean cell wall proteins, that accumulate during water deficit, bind to a plasma membrane protein. This binding is competed out by a peptide that contains the RGD motif, as well as with fibronectin, which also includes this sequence (Garcia-Gomez *et al.*, 2000). Animal proteins that form direct links between the cytoskeleton and the extracellular matrix, including integrins, talin, spectrin, α-actinin, vitronectin or vinculin, have not been found in the *Arabidopsis* genome – (The *Arabidopsis* Genome Initiative, 2000; Brownlee, 2002). However, the possibility remains that analogous functions are provided by different molecules.

Plant cells possess arrays of cortical microtubules underlying the plasma membrane. During primary wall formation, orientation of the cortical microtubule array often pre-stages the orientation of cellulose microfibril deposition, aligning the cellulose synthase complexes, either by direct protein-mediated connection or by defining channels in the membrane in which the synthase can move (Wymer & Lloyd, 1996). In cells that grow by expansion, layers of microfibrils lie unaligned in the wall matrix, but in cells that grow by elongation, microfibrils align in transverse or helical orientations to the axis of elongation. Several *Arabidopsis* mutants with cell elongation defects provided the genetic evidence that the transverse alignment of cortical microtubules and cellulose microfibrils is essential for establishing growth polarity (see Chapter 3). However, recent studies on *mor1* mutants and drug-treated *Arabidopsis* roots suggest that the alignment of cortical microtubules and cellulose microfibrils is not always coupled during cell elongation, raising a question over the role of cortical microtubules (Sugimoto *et al.*, 2003). Cellulose microfibrils appear to have the ability to largely self-align in the absence of cortical microtubules and/or pre-existing cellulose microfibrils (Wasteneys, personal comm., and Chapter 3). Therefore cortical microtubules may only be required for directing cellulose into a new orientation, for example, in response to various environmental stimuli.

When tobacco protoplast or culture cells were treated with isoxaben, an inhibitor of cellulose synthesis, the cells failed to elongate and also the cortical microtubules failed to become organized. The effects of isoxaben were reversible, and after its removal microtubules reorganized and cells elongated (Fisher & Cyr, 1998). Depolymerization of microtubules *in vivo* and polymerization of tubulin *in vitro* are unaffected by isoxaben. There may be a bidirectional flow of information allowing cellulose microfibrils to provide spatial cues for cortical microtubule organization.

For many types of plant cell, the process of differentiation is accompanied by the assembly, on the plasma-membrane side of the primary wall, of a distinct secondary wall. The deposition of the secondary wall begins as cells stop growing. Secondary walls often exhibit elaborate specializations. The cotton fiber, for example, consists of nearly 98% cellulose at maturity. In some cells, like sclereids and vascular fibers, and the stone cells of pear, the secondary wall becomes uniformly thick, composed largely of cellulose microfibrils that almost fill the entire lumen of the cell. The secondary wall may, however, contain additional non-cellulosic polysaccharides, proteins and aromatic substances, and become lignified. In tracheary elements (TEs), secondary walls can display annular or helical coils or reticulate and pitted patterns (Esau, 1977). The production of a thick secondary wall of carbohydrate, structural protein and lignin is essential for TE function, as the wall must be reinforced to resist compressive forces from the other cells that arise as a consequence of the extreme negative pressures that may develop in actively transpiring plants.

Cortical microtubules play a key role in secondary wall formation. Bands of cellulose microfibrils form one of the major elements of the TE wall thickenings. In the *Zinnia* system, secondary wall deposition is a hierarchical process in which the deposition of cellulose reflects the patterning of cortical microtubules. Disassembly of microtubules by depolymerizing agents results in random deposition of secondary wall material over the entire surface of the inner wall. Plasma membrane-associated sucrose synthase, thought to channel the immediate substrate for cellulose synthase (UDP-Glc), has been immuno-localized to sites of secondary thickenings (Salnikov *et al.*, 2001), indicating that the cellulose synthase complexes may be restricted to regions of the plasma membrane by cortical microtubules. Deposition of lignin in ordered patterns depends on the prior deposition of cellulose, and inhibition of cellulose synthesis by DCB disrupts lignin patterning and also causes the loss of xylans from the cellulose-depleted thickenings (Taylor *et al.*, 1992).

Fibers are a good system to analyze cellulose deposition because microfibrils are deposited transversely in a highly oriented fashion. The *fragile fiber 2* mutant (*fra2*) in the *Arabidopsis* katanin p60 homolog demonstrated that cortical microtubules contribute to the wall thickening of interfascicular fiber cells (Burk *et al.*, 2001; Burk & Ye, 2002). In the *fra2* mutant, deposition of cellulose microfibrils does not occur in a transverse orientation and the cortical microtubules are disorganized (Burk & Ye, 2002). Similarly, the FRA1 mutation, in a kinesin-like motor protein, causes a dramatic reduction in fiber mechanical strength without altering wall composition but with an alteration in the oriented deposition of cellulose microfibrils in the fiber cell walls (Zhong *et al.*, 2002). However, the orientation of cortical microtubules was not affected. Because the aberrant cellulose microfibril deposition was not caused by an alteration in the organization of cortical microtubules, the FRA1 kinesin-like protein may play a role in a step between microtubules and microfibrils (Zhong *et al.*, 2002). The *fragile fiber* phenotype, despite normal cell wall composition, reduces stem strength to under half that of wild-type.

In small deformation rheology measurements of isolated cell walls, the cellulose network dominates the mechanical response but the manner in which the cellulose is deposited is conditioned by the presence of other molecules leading to gross mechanical differences (Chanliaud & Gidley, 2002).

6.5.3 Sensing through the plasma membrane

Not all specialized functions of cell walls are structural. Some contain molecules that affect patterns of development and mark a cell's position within the plant. Walls contain signaling molecules that participate in cell–cell and wall-nucleus communication. Fragments of wall polysaccharides may elicit the secretion of defense molecules, and the wall may become impregnated with protein and lignin to armor it against invading fungal and bacterial pathogens. In other instances, the walls participate in early recognition of symbiotic nitrogen-fixing bacteria. Cell wall surface molecules allow plant cells to recognize their own kind in pollen-style interactions. These receptors provide the opportunity for crosstalk between the extracellular environment, perhaps modulated by local matrix composition, and the cytoskeleton, through indirect signal transduction cascades. We will discuss three classes of receptor molecule – leucine-rich repeat (LRR) receptor kinases, wall-associated kinases (WAKs) and AGPs.

LRR kinases and other kinases populate the plasma membrane and allow signal transduction across this topological barrier. LRR kinases are transmembrane proteins with extracellular LRRs and cytoplasmic protein kinase domains, and comprise a gene family of 174 members in *Arabidopsis* (Jones, 2001). In all living organisms, LRR domains are specialized for interactions with protein ligands (Kobe & Kajava, 2001). In plants, the majority of known resistance R gene products are LRR proteins (Jones, 2001). Interactions between the extracellular domains of plasma membrane receptor kinases and their ligands are potentially regulated by the immediate matrix environment.

A chimeric protein, LRX1, has LRR and extensin sequences and is specifically localized in root hair walls (Baumberger *et al.*, 2001). Mutant *lrx1* plants have few fully elongated root hairs and, occasionally, the root hairs have swollen bases. LRX1 becomes cross-linked into the root hair wall as hair elongation progresses, suggesting that this protein has a structural role.

The distinct signaling and recognition components of the *XA21* gene, a rice gene for bacterial blight resistance, were visualized in a domain swap with BRI 1 (BRASSINOSTEROID INSENSITIVE, 1) another transmembrane LRR kinase, which is required for brassinosteroid perception (He *et al.*, 2000). He *et al.* (2000) made chimeric genes in which the BRI1 LRRs were fused to the protein kinase domain of XA21, and assayed the fusions in rice suspension cells. One such fusion, which retained all of the extracellular amino acids of BRI1 plus the transmembrane domain and 65 amino acids of the intracellular BRI1 protein, activated defense responses upon treatment with brassinosteroid. One implication of this result is that BRI1 and XA21 recognize their distinct ligands and initiate signaling through similar mechanisms, and that the protein kinase domain specifies which specific signal transduction pathway ensues.

Signaling via the cell wall has been shown to underpin fate decisions during the patterning of the shoot apical meristem. The *CLAVATA* (*CLV*) genes encode an LRR receptor kinase (*CLV1*), a receptor-interacting protein (CLV2) and a small secreted peptide (*CLV3*) that is the ligand for CLV2 (Clarke, 2001). CLV3 produced by stem cells interacts with CLV1 causing repression of WUSCHEL, a homeodomain transcription factor, to maintain meristem identity. Cells outside the range of CLV3 are able to differentiate into organs at the edge of the meristem.

The cell wall-associated kinases (WAKs) lie in a tight cluster of five highly similar genes (WAK1–5) within a 30 kb region (He *et al.*, 1999). WAKs have a highly conserved serine/threonine kinase cytoplasmic domain, a plasma membrane-spanning region, and a more variable extracellular domain containing EGF (epidermal growth factor) repeats, likely to be involved in ligand recognition (Kohorn, 2001). The extracellular regions also contain limited amino acid identities to the tenascin superfamily, collagen, or the neurexins (He *et al.*, 1999). A large amount of WAK is covalently linked to pectin, and most WAK that is bound to pectin is also phosphorylated (Anderson *et al.*, 2001). In addition, WAK1 has been shown to interact with the glycine-rich protein, AtGRP3 (Park *et al.*, 2001), to form a complex in conjunction with the kinase-associated protein phosphatase (KAPP) that has been shown to interact with a number of plant receptor-like kinases. WAK1 will bind GRP3 and pectin but WAK2 will not bind GRP3. Inducible antisense expression of *WAK4* showed that cell elongation was impaired and lateral root development blocked, in proportion to the concentration of the applied inducer, dexamethasone (Lally *et al.*, 2000). Similarly, Wagner and Kohorn (2001), showed that leaves expressing an antisense *WAK* gene showed a loss of cell expansion. Coupled to these effects is a dramatic reduction in the levels of expansin transcript, indicating a possible feedback mechanism from the cell wall environment to expression of cell wall genes involved in wall loosening. WAKs are expressed in different tissues and serve a variety of functions related to the coordination of wall properties and signaling (He *et al.*, 1999; Lally *et al.*, 2000; Wagner & Kohorn, 2001).

Many AGPs are GPI-anchored in common with a large proportion of plant plasma membrane proteins (Borner *et al.*, 2002). The presence of a GPI anchor is frequently associated with polar protein sorting in animal cells in which such proteins are organized into microdomains at the cell surface (Friedrichson & Kurzchalia, 1998; Varma & Mayor, 1998). Secreted N-acetylglucosamine and glucosamine-containing AGPs have been shown to promote somatic embryogenesis from carrot protoplast cultures (Van Hengel *et al.*, 2001). Treatment of carrot AGPs with endochitinase enhanced their embryo-promoting activity, suggesting the involvement of chitinase in the regulation of AGP activity in the control of cell fate. A stylar transmitting tract AGP from *Nicotiana alata* has been found to promote pollen growth *in vitro* (Wu *et al.*, 2000). Motose *et al.* (2001) published a sensitive bioassay using the Zinnia mesophyll cell system, in which cells are induced to trans-differentiate to TEs *in vitro*, and found that AGPs precipitated by Yariv are capable of promoting TE formation. The AGP-binding Yariv reagent alters the immunolabeling patterns of an antibody specific for pectic galactan in elongating root cells of *Arabidopsis*: the implication is that modulation of pectic galactan may be an event downstream of AGP function during cell expansion (McCartney *et al.*, 2003). Thus, AGPs are implicated in a diverse range of developmental roles including embryogenesis, growth and differentiation.

6.6 Concluding remarks

We can view the cell wall as a sensory panel of receptors with specificity for a diverse range of ligands and the cytoskeleton as the internal scaffolding for signaling and response. The most likely targets for the developmental and environmental signals that induce changes in cell shape reside in components of the cytoskeleton, including the organization of the cortical microtubules and actin filaments, and their transport and targeting capacities. Downstream integrated changes in the extensibility and resistance of cell walls, the coordinated synthesis of new wall material and its localized reorganization, are necessary for plant cells to become specialized for their functions within the plant.

Acknowledgements

The authors thank all colleagues who provided preprints of publications and are grateful to the BBSRC and The Royal Society for financial support.

References

Albersheim, P., Darvill, A.G., O'Neill, M.A., Schols, H.A. & Voragen, A.G.J. (1996) An hypothesis: the same six polysaccharides are components of the primary cell walls of all higher plants, in *Progress in Biotechnology 14: Pectins and Pectinases* (eds J. Visser & A.G.J. Voragen), Elsevier, Dordrecht, The Netherlands, pp. 47–53.

Amor, Y., Haigler, C.H., Johnson, S., Wainscott, M. & Delmer, D.P. (1995) A membrane-associated form of sucrose synthase and its potential role synthesis of cellulose and callose in plants, *Proc. Natl. Acad. Sci. USA*, **92**, 9353–9357.

Andeme-Onzighi, C., Siveguru, M., Judy-March, J., Baskin, T.I. & Driouich, A. (2002) The *reb1-1* mutation of *Arabidopsis* alters the morphology of trichoblasts, the expression of arabinogalactan-proteins and the organization of cortical microtubules, *Planta*, **215**, 949–958.

Anderson, C.M., Wagner, T.A., Perret, M., He, Z.H., He, D.Z. & Kohorn, B.D. (2001) WAKs: cell wall-associated kinases linking the cytoplasm to the extracellular matrix, *Plant Mol. Biol.*, **47**, 197–206.

Arabidopsis Genome Initiative (2000) Analysis of the genome sequence of the flowering plant *Arabidopsis thaliana*, *Nature*, **408**, 796–815.

Arioli, T., Peng, L.C., Betzner, A.S., *et al.* (1998) Molecular analysis of cellulose biosynthesis in *Arabidopsis*, *Science*, **279**, 717–720.

Baluska, F., Salaj, J., Mathur, J., *et al.* (2000) Root hair formation: F-actin-dependent tip growth is initiated by local assembly of profiling-supported F-actin meshworks accumulated within expansin-enriched bulges, *Dev. Biol.*, **227**, 618–632.

Bao, W.L., O'Malley, D.M. & Sederoff, R.R. (1992) Wood contains a cell wall structural protein, *Proc. Natl. Acad. Sci. USA*, **89**, 6604–6608.

Bao, Y.Q., Kost, B. & Chua, N.H. (2001) Reduced expression of α-tubulin genes in *Arabidopsis thaliana* specifically affects root growth and morphology, root hair development and root gravitropism, *Plant J.*, **28**, 145–157.

Baumberger, N., Ringli, C. & Keller, B. (2001) The chimeric leucine-rich repeat/extensin cell wall protein LRX1 is required for root hair morphogenesis in *Arabidopsis thaliana, Genes Dev.*, **15**, 1128–1139.

Bibikova, T.N., Blancaflor, E.B. & Gilroy, S. (1999) Microtubules regulate tip growth and orientation in root hairs of *Arabidopsis thaliana*, *Plant J.*, **17**, 657–665.

Boevink, P., Oparka, K., Cruz, S.S., Martin, B., Betteridge, A. & Hawes, C. (1998) Stacks on tracks: the plant Golgi apparatus traffics on an actin/ER network, *Plant J.*, **15**, 441–447.

Bonin, C.P., Potter, I., Vanzin, G.F. & Reiter, W.-D. (1997) The *MUR1* gene of *Arabidopsis thaliana* encodes an isoform of GDP-D-mannose-4,6-dehydratase, catalyzing the first step in the de novo synthesis of GDP-L-fucose, *Proc. Natl. Acad. Sci. USA*, **94**, 2085–2090.

Borner, G.H.H., Sherrier, D.J., Stevens, T.J., Arkin, I.T. & Dupree, P. (2002) Prediction of glycosyl-phosphatidylinositol-anchored proteins in *Arabidopsis*. A genomic analysis, *Plant Physiol.*, **129**, 486–499.

Bouton, S., Leboeuf, E., Mouille, G. *et al.* (2002) *Quasimodo1* encodes a putative membrane-bound glycosyltransferase required for normal pectin synthesis and cell adhesion in *Arabidopsis*, *Plant Cell*, **14**, 2577–2590.

Braccini, I., Grasso, R.P. & Pérez, S. (1999) Conformational and configurational features of acidic polysaccharides and their interactions with calcium ions: a molecular modeling investigation, *Carbohydr. Res.*, **317**, 119–130.

Brown, J.A. & Fry, S.C. (1993) Novel O-D-galacturonoyl esters in the pectic polysaccharides of suspension-cultured plant cells, *Plant Physiol.*, **103**, 993–999.

Brownlee, C. (2002) Role of the extracellular matrix in cell–cell signaling: paracrine paradigms, *Curr. Opin. Plant Biol.*, **5**, 396–401.

Burk, D.H., Lin, B., Zhong, R., Morrison, W.H. & Ye, Z.H. (2001) A katanin-like protein regulates normal cell wall biosynthesis and cell elongation, *Plant Cell*, **13**, 807–827.

Burk, D.H. & Ye, Z.H. (2002) Alteration of oriented deposition of cellulose microfibrils by mutation of a katanin-like microtubule-severing protein, *Plant Cell*, **14**, 2145–2160.

Bush, M.S., Marry, M., Huxham, I.M., Jarvis, M.C. & McCann, M.C. (2001) Developmental regulation of pectic epitopes during potato tuberisation, *Planta*, **213**, 869–880.

Camilleri, C., Azimzadeh, J., Pastuglia, M., Bellini, C., Grandjean, O. & Bouchez, D. (2002) The *Arabidopsis TONNEAU2* gene encodes a putative novel protein phosphatase 2A regulatory subunit essential for the control of the cortical cytoskeleton, *Plant Cell*, **14**, 833–845.

Canut, H., Carrasco, A., Galaud, J.P., *et al.* (1998) High affinity RGD-binding sites at the plasma membrane of *Arabidopsis thaliana* links the cell wall, *Plant J.*, **16**, 63–71.

Carpita, N.C. (1996) Structure and biogenesis of the cell walls of grasses, *Annu. Rev. Plant Physiol. Plant Mol. Biol.*, **47**, 445–476.

Carpita, N.C. & Gibeaut, D.M. (1993) Structural models of primary cell walls in flowering plants: consistency of molecular structure with the physical properties of the walls during growth, *Plant J.*, **3**, 1–30.

Carpita, N.C., Defernez, M., Findlay, K., *et al.* (2001a) Cell wall architecture of the elongating maize coleoptile, *Plant Physiol.*, **127**, 551–565.

Carpita, N., Tierney, M. & Campbell, M. (2001b) Molecular biology of the plant cell wall: searching for the genes that define structure, architecture and dynamics, *Plant Mol. Biol.*, **47**, 1–5.

Carrington, C.M.S., Greve, L.C. & Labavitch, J.M. (1993) Cell-wall metabolism in ripening fruit. 6. Effect of the antisense polygalacturonase gene on cell-wall changes accompanying ripening in transgenic tomatoes, *Plant Physiol.*, **103**, 429–434.

Cassab, G.I. (1998) Plant cell wall proteins. *Annu. Rev. Plant Physiol. Plant Mol. Biol.*, **49**, 281–309.

Chanliaud, E., Burrows, K.M., Jeronimidis, G. & Gidley, M.J. (2002) Mechanical properties of primary plant cell wall analogues, *Planta*, **215**, 989–996.

Clarke, S.E. (2001) Cell signaling at the shoot meristem. *Nat. Rev. Mol. Cell Biol.*, **2**, 276–284.

Cleary, A.L. (2001) Plasma membrane-cell wall connections: roles in mitosis and cytokinesis revealed by plasmolysis of *Tradescantia virginiana* leaf epidermal cells, *Protoplasma*, **215**, 21–34.

Cleary, A.L. & Smith, L.G. (1998) The *Tangled1* gene is required for spatial control of cytoskeletal arrays associated with cell division during maize leaf development, *Plant Cell*, **10**, 1875–1888.

Cosgrove, D.J. (2000) Loosening of plant cell walls by expansins, *Nature*, **407**, 321–326.

Cutler, S.R. & Ehrhardt, D.W. (2002) Polarized cytokinesis in vacuolated cells of *Arabidopsis*, *Proc. Natl. Acad. Sci. USA*, **99**, 2812–2817.

Daas, P.J.H., Boxma, B., Hopman, A.M.C.P., Voragen, A.G.J. & Schols, H.A. (2001) Nonesterified galacturonic acid sequence homology of pectins, *Biopolymers*, **58**, 1–8.

Domingo, C., Gomez, M.D., Canas, L., Hernandez-Yago, J., Conejero, V. & Vera, P. (1994) A novel extracellular matrix protein from tomato associated with lignified secondary cell walls, *Plant Cell*, **6**, 1035–1047.

Du, H., Clarke, A.E. & Bacic, A. (1996) Arabinogalactan-proteins: a class of extracellular matrix proteoglycans involved in plant growth and development, *Trends Cell Biol.*, **6**, 411–414.

Dubreucq, B., Berger, N., Vincent, E., *et al.* (2000) The *Arabidopsis* AtEPR1 extensin-like gene is specifically expressed in endosperm during seed germination, *Plant J.*, **23**, 643–652.

Ermel, F.F., Follet-Gueye, M.L., Cibert, C., *et al.* (2000) Differential localization of arabinan and galactan side chains of rhamnogalacturonan 1 in cambial derivatives, *Planta*, **210**, 732–740.

Esau, K. (1977) *Anatomy of Seed Plants*, John Wiley & Sons, New York, pp. 75–95.

Fagard, M., Desnos, T., Desprez, T., *et al.* (2000) *Procuste1* encodes a cellulose synthase required for normal cell elongation specifically in roots and dark-grown hypocotyls of *Arabidopsis*, *Plant Cell*, **12**, 2409–2423.

Fisher, D.D. & Cyr, R.J. (1998) Extending the microtubule/microfibril paradigm – Cellulose synthesis is required for normal cortical microtubule alignment in elongating cells, *Plant Physiol.*, **116**, 1043–1051.

Fleming, A.J., McQueenMason, S., Mandel, T. & Kuhlemeier, C. (1997) Induction of leaf primordia by the cell wall protein expansion, *Science*, **276**, 1415–1418.

Friedrichson, T. & Kurzchalia, T.V. (1998) Microdomains of GPI-anchored proteins in living cells revealed by cross-linking, *Nature*, **394**, 802–805.

Fry, S.C., Smith, R.C., Renwick, K.F., Martin, D.J., Hodge, S.K. & Matthews, K.J. (1992) Xyloglucan endotransglycosylase, a new wall-loosening enzyme activity from plants, *Biochem. J.*, **282**, 821–828.

Garcia-Gomez, B.I., Campos, F., Hernandez, M. & Covarrubias, A.A. (2000) Two bean cell wall proteins more abundant during water deficit are high in proline and interact with a plasma membrane protein, *Plant J.*, **22**, 277–288.

Gaspar, Y., Johnson, K.L., McKenna, J.A., Bacic, A. & Schultz, C.J. (2001) The complex structures of arabinogalactan-proteins and the journey towards understanding function, *Plant Mol. Biol.*, **47**, 161–176.

Geitmann, A. & Emons, A.M.C. (2000) The cytoskeleton in plant and fungal cell growth, *J. Microscopy*, **198**, 218–245.

Grant, M. & Mansfield, J. (1999) Early events in host-pathogen interactions, *Curr. Opin. Plant Biol.*, **2**, 312–319.

Hall, Q. & Cannon, M.C. (2002) The cell wall hydroxyproline-rich glycoprotein RSH is essential for normal embryo development in *Arabidopsis*, *Plant Cell*, **14**, 1161–1172.

Hall, L.N., Tucker, G.A., Smith, C.J.S., *et al.* (1993) Antisense inhibition of pectin esterase gene expression in transgenic tomatoes, *Plant J.*, **3**, 121–129.

Hatfield, R.D., Ralph, J. & Grabber, J.H. (1999) Molecular basis for improving forage digestibilities, *Crop Science*, **39**, 27–37.

Hawes, C.R. & Satiat-Jeunemaitre, B. (2001) Trekking along the cytoskeleton, *Plant Physiol.*, **125**, 119–122.

He, Z., Wang, Z.Y., Li, J., *et al.* (2000) Perception of brassinosteroids by the extracellular domain of the receptor-kinase BRI1, *Science*, **288**, 2360–2363.

He, Z.H., Cheeseman, I., He, D.Z. & Kohorn, B.D. (1999) A cluster of five cell wall-associated receptor kinase genes, *Wak1–5*, are expressed in specific organs of *Arabidopsis*, *Plant Mol. Biol.*, **39**, 1189–1196.

Hepler, P.K., Vidali, L. & Cheung, A. Y. (2001) Polarized cell growth in higher plants, *Annu. Rev. Cell Devel. Biol.*, **17**, 159–187.

Hirai, N., Sonobe, S. & Hayashi, T. (1998) In situ synthesis of β-glucan microfibrils on tobacco plasma membrane sheets, *Proc. Natl. Acad. Sci. USA*, **95**, 15102–15106.

Hynes, R.O. (1987) Integrins: a family of cell surface receptors, *Cell*, **48**, 549–554.

Hynes, R.O. (1992) Integrins: versatility, modulation and signaling in cell adhesion, *Cell*, **69**, 11–25.

Ishii, T., Matsunaga, T., Pellerin, P., O'Neill, M.A., Darvill, A. & Albersheim, P. (1999) The plant cell wall polysaccharide rhamnogalacturonan II self-assembles into a covalently cross-linked dimer, *J. Biol. Chem.*, **274**, 13098–13104.

Ishii, T. & Matsunaga, T. (2001) Pectic polysaccharide rhamnogalacturonan II is covalently linked to homogalacturonan, *Phytochemistry*, **57**, 969–974.

Ishii, T., Matsunaga, T. & Hayashi, N. (2001) Formation of rhamnogalacturonan II-borate dimer in pectin determines cell wall thickness of pumpkin tissue, *Plant Physiol.*, **126**, 1698–1705.

Iwai, H., Masaoka, N., Ishii, T. & Satoh, S. (2002) A pectin glucuronyltransferase gene is essential for intercellular attachment in the plant meristem, *Proc. Natl. Acad. Sci. USA*, **99**, 16319–16324.

Jarvis, M.C. (1984) Structure and properties of pectin gels in plant cell walls, *Plant Cell Environ.*, **7**, 153–164.

Jarvis, M.C. & McCann, M.C. (2000) Macromolecular biophysics of the plant cell wall: concepts and methodology, *Plant Physiol. Biochem.*, **38**, 1–13.

Jones, J.D. (2001) Putting knowledge of plant disease resistance genes to work, *Curr. Opin. Plant Biol.*, **4**, 281–287.

Jones, L., Seymour, G.B. & Knox, J.P. (1997) Localization of pectic galactan in tomato cell walls using a monoclonal antibody specific to (1,4)-β-D-galactan, *Plant Physiol.*, **113**, 1405–1412.

Kieliszewski, M.J. & Shpak, E. (2001) Synthetic genes for the elucidation of glycosylation codes for arabinogalactan-proteins and other hydroxyproline-rich glycoproteins, *Cell. Mol. Life Sci.*, **58**, 1386–1398.

Kim, J.-B. & Carpita, N.C. (1992) Esterification of maize cell wall pectins related to cell expansion, *Plant Physiol.*, **98**, 646–653.

Kimura, S., Laosinchai, W., Itoh, T., Cui, X., Linder, C.R. & Brown, Jr, R.M. (1999) Immunogold labeling of rosette terminal cellulose-synthesizing complexes in the vascular plant *Vigna angularis*, *Plant Cell*, **11**, 2075–2086.

Kobayashi, M., Matoh, T. & Azuma, J. (1996) Two chains of rhamnogalacturonan II are cross-linked by borate-diol ester bonds in higher plant cell walls, *Plant Physiol.*, **110**, 1017–1020.

Kobe, B. & Kajava, A.V. (2001) The leucine-rich repeat as a protein recognition motif, *Curr. Opin. Struct. Biol.*, **11**, 725–732.

Kohorn, B. (2001) WAKs: cell wall associated kinases, *Curr. Opin. Cell Biol.*, **13**, 529–533.

Koyama, M., Helbert, W., Imai, T., Sugiyama, J. & Henrissat, B. (1997) Parallel-up structure evidences the molecular directionality during biosynthesis of bacterial cellulose, *Proc. Nat. Acad. Sci. USA*, **94**, 9091–9095.

Lally, D., Ingmire, P., Tong, H.Y. & He, Z.H. (2000) Antisense expression of a cell wall-associated protein kinase, WAK4, inhibits cell elongation and alters morphology, *Plant Cell*, **13**, 1317–1331.

Madson, M., Dunand, C., Li, X., *et al.* (2003) The MUR3 gene of *Arabidopsis thaliana* encodes a galactosyltransferase that is evolutionarily related to animal exostosins, *Plant Cell*, in press.

Mayer, U. & Jurgens, G. (2002) Microtubule cytoskeleton: a track record, *Curr. Opin. Plant Biol.*, **5**, 494–501.

McCann, M.C., Wells, B. & Roberts, K. (1990) Direct visualisation of cross-links in the primary plant cell wall, *J. Cell Sci.*, **96**, 323–334.

McCann, M.C. & Roberts, K. (1991) Architecture of the primary cell wall, in *The Cytoskeletal Basis of Plant Growth and Form* (ed. C.W. Lloyd), Academic Press, New York, pp. 109–129.

McCann, M.C., Shi, J., Roberts, K., & Carpita, N.C. (1994) Changes in pectin structure and localisation during the growth of unadapted and NaCl-adapted tobacco cells, *Plant J.*, **5**, 773–785.

McCartney, L., Ormerod, A.P., Gidley, M.J. & Knox, P. (2000) Temporal and spatial regulation of pectic (1→4)-β-D-galactan in cell walls of developing pea cotyledons: implications for mechanical properties, *Plant J.*, **22**, 105–113.

McCartney, L., Steele-King, C.G., Jordan, E. & Knox, J.P. (2003) Cell wall pectic (1→4)-β-D-galactan mark the acceleration of cell elongation in the *Arabidopsis* seedling root meristem, *Plant J.*, **33**, 447–454.

McNeil, M., Darvill, A.G. & Albersheim, P. (1980) Structure of plant cell walls. X. Rhamnogalacturonan I, a structurally complex pectic polysaccharide in the walls of suspension-cultured sycamore cells, *Plant Physiol.*, **66**, 1128–1134.

McQueen-Mason, S.J. & Rochange, F. (1999) Expansins in plant growth and development: an update on an emerging topic, *Plant Biol.*, **1**, 19–25.

Mohnen, D. (1999) Biosynthesis of pectins and galactomannans, in *Comprehensive Natural Products Chemistry*, Vol 3 (eds D. Barton, K. Nakanishi & O. Meth-Cohn), Elsevier, Dordrecht, The Netherlands, pp. 497–527.

Motose, H., Sugiyama, M. & Fukuda, H. (2001) An arabinogalactan protein(s) is a key component of a fraction that mediates local intercellular communication involved in tracheary element differentiation of Zinnia mesophyll cells, *Plant Cell Physiol.*, **42**, 129–137.

Mueller, S.C. & Brown, R.M. (1980) Evidence for an intramolecular component associated with a cellulose microfibril synthesizing complex in higher plants, *J. Cell Biol.*, **84**, 315–326.

Nagpal, P. & Quatrano, R.S. (1999) Isolation and characterization of a cDNA clone form *Arabidopsis thaliana* with partial sequence similarity to integrins, *Gene*, **230**, 33–40.

Nebenfuhr, A. & Staehelin, L.A. (2001) Mobile factories: Golgi dynamics in plant cells, *Trends Plant Sci.*, **6**, 160–167.

Nicol, F., His, I., Jauneau, A., Vernhettes, S., Canut, H. & Höfte, H. (1998) A plasma membrane-bound putative endo-1,4-β-D-glucanase is required for normal wall assembly and cell elongation in *Arabidopsis*, *EMBO J.*, **17**, 5563–5576.

Okamoto-Nakazato, A., Takahashi, K., Kido, N., Owaribe, K. & Katou, K. (2000) Molecular cloning of yieldins regulating the yield threshold of cowpea cell walls: cDNA cloning and characterization of recombinant yieldin, *Plant Cell Environ.*, **23**, 155–164.

O'Neill, M.A., Warrenfeltz, D., Kates, K., *et al.* (1996) Rhamnogalacturonan-II, a pectic polysaccharide in the walls of growing plant cell, forms a dimer that is covalently-linked by a borate ester, *J. Biol. Chem.*, **271**, 22923–22930.

O'Neill, M.A., Eberhard, S., Albersheim, P. & Darvill, A.G. (2001) Requirement of borate cross-linking of cell wall rhamnogalacturonan II for *Arabidopsis* growth, *Science*, **294**, 846–849.

Oppenheimer, D.G., Pollock, M.A., Vacik, J., *et al.* (1997) Essential role of a kinesin-like protein in *Arabidopsis* trichome morphogenesis, *Proc. Natl. Acad. Sci. USA*, **94**, 6261–6266.

Orfila, C. & Knox, J.P. (2000) Spatial regulation of pectic polysaccharides in relation to pit fields in cell walls of tomato fruit pericarp, *Plant Physiol.*, **122**, 775–781.

Oxenbøll-Sørensen, S., Pauly, M., Bush, *et al.* (2000) Pectin engineering: modification of potato pectin by in vivo expression of an endo-1,4-β-D-galactanase, *Proc. Natl. Acad. Sci. USA*, **97**, 7639–7644.

Park, A.R., Cho, S.K., Yun, U.J., *et al.* (2001) Interaction of the *Arabidopsis* receptor protein kinase Wak1 with a glycine-rich protein, AtGRP3, *J. Biol. Chem.*, **276**, 26688–26693.

Penfield, S., Meissner, R.C., Shoue, D.A., Carpita, N.C. & Bevan, M.W. (2001) MYB61 is required for mucilage deposition and extrusion in the *Arabidopsis* seed coat, *Plant Cell*, **13**, 2777–2791.

Pien, S., Wyrzykowska, J., McQueen-Mason, S., Smart, C. & Fleming, A. (2000) Local expression of expansin induces the entire process of leaf development and modifies leaf shape, *Proc. Natl. Acad. Sci. USA*, **98**, 11812–11817.

Pilling, J., Willmitzer, L. & Fisahn, J. (2000) Expression of a *Petunia inflata* pectin methyl esterase in *Solanum tuberosum* L. enhances stem elongation and modifies cation distribution, *Planta*, **210**, 391–399.

Reiter, W.-D., Chapple, C.C.S. & Somerville, C.R. (1993) Altered growth and cell walls in a fucose-deficient mutant of *Arabidopsis*, *Science*, **261**, 1032–1035.

Reiter, W.-D., Chapple, C. & Somerville, C.R. (1997) Mutants of *Arabidopsis thaliana* with altered cell wall polysaccharide composition, *Plant J.*, **12**, 335–345.

Ridley, B.L., O'Neill, M.A. & Mohnen, D. (2001) Pectins: structure, biosynthesis, and oligogalacturonide-related signaling, *Phytochemistry*, **57**, 929–967.

Roy, S., Jauh, G.Y., Hepler, P.K. & Lord, E.M. (1998) Effects of Yariv phenylglucoside on cell wall assembly in the lily pollen tube, *Planta*, **204**, 450–458.

Ryser, U. & Keller, B. (1992) Ultrastructural localization of a bean glycine-rich protein in unlignified primary walls of protoxylem cells, *Plant Cell*, **4**, 773–783.

Salnikov, V.V., Grimson, M.J., Delmer, D.P. & Haigler, C.H. (2001) Sucrose synthase localizes to cellulose synthesis sites in tracheary elements, *Phytochemistry*, **57**, 823–833.

Samuels, A.L., Giddings, T.H. & Staehelin, L.A. (1995) Cytokinesis in tobacco BY-2 and root tip cells – A new model of cell plate formation in higher plants, *J. Cell Biol.*, **130**, 1345–1357.

Scheible, W.R., Eshed, R., Richmond, T., Delmer, D. & Somerville, C. (2001) Modifications of cellulose synthase confer resistance to isoxaben and thiazolidinone herbicides in *Arabidopsis Ixr1* mutants, *Proc. Natl. Acad. Sci. USA*, **98**, 10079–10084.

Schindelman, G., Morikami, A., Jung, J., *et al.* (2001) *COBRA* encodes a putative GPI-anchored protein, which is polarly localized and necessary for oriented cell expansion in *Arabidopsis*, *Gen. Dev.*, **15**, 1115–1127.

Schindler, M., Meiners, S. & Cheresh, D.A. (1989) RGD-dependent linkage between plant cell wall and plasma membrane: consequences for growth, *J. Cell Biol.*, **108**, 1955–1965.

Schindler, T., Bergfeld, R. & Schopfer, P. (1995) Arabinogalactan proteins in maize coleoptiles – Developmental relationship to cell death during xylem differentiation but not to extension growth, *Plant J.*, **7**, 25–36.

Schneider, B. & Herth, W. (1986) Distribution of plasma membrane rosettes and kinetics of cellulose formation in xylem development of higher plants, *Protoplasma*, **131**, 142–152.

Schols, H.A. & Voragen, A.G.J. (1996) Complex pectins: structure elucidation using enzymes, in *Progress in Biotechnology 14: Pectins and Pectinases* (eds J. Visser & A.G.J. Voragen), Elsevier, Dordrecht, The Netherlands, pp. 3–19.

Schultz, C.J., Rumsewicz, M.P., Johnson, K.L., Jones, B.J., Gaspar, Y.M. & Bacic, A. (2002) Using genomic resources to guide research directions. The arabinogalactan protein gene family as a test case, *Plant Physiol.*, **129**, 1448–1463.

Schwartz, M.A., Schaller, M.D. & Ginsberg, M.H. (1995) Integrins: emerging paradigms of signal transduction, *Annu. Rev. Cell Dev. Biol.*, **11**, 549–599.

Seifert, G.J., Barber, C., Wells, B., Dolan, L. & Roberts, K. (2002) Galactose biosynthesis in *Arabidopsis*: genetic evidence for substrate channeling from UDP-D-galactose into cell wall polymers, *Curr. Biol.*, **12**, 1840–1845.

Shedletzky, E., Shmuel, M., Delmer, D.P. & Lamport, D.T.A. (1990) Adaptation and growth of tomato cells on the herbicide 2,6-dichlorobenzonitrile leads to production of unique cell walls virtually lacking a cellulose xyloglucan network, *Plant Physiol.*, **94**, 980–987.

Sheng, J.S., Jeong, J.M. & Mehdy, M.C. (1993) Developmental regulation and phytochrome-mediated induction of messenger RNAs encoding a proline-rich protein, glycine-rich proteins, and hydroxyproline-rich proteins in *Phaseolus vulgaris* L., *Proc. Natl. Acad. Sci. USA*, **90**, 828–832.

Shi, H., Kim, Y.-S., Guo, Y., Stevenson, B. & Zhu, J.-K. (2003) The *Arabidopsis* SOS5 locus encodes a putative cell surface adhesion protein and is required for normal cell expansion, *Plant Cell*, **15**, 19–32.

Sims, I.M., Munro, S.L.A., Currie, G., Craik, D. & Bacic, A. (1996) Structural characterisation of xyloglucan secreted by suspension-cultured cells of *Nicotiana plumbaginifolia*, *Carbohydr. Res.*, **293**, 147–172.

Smith, B.G. & Harris, P.J. (1999) The polysaccharide composition of Poales cell walls: Poaceae cell walls are not unique, *Biochem. Syst. Ecol.*, **27**, 33–53.

Smith, L.G., Gerttula, S.M., Han, S. & Levy, J. (2001) TANGLED1: a microtubule binding protein required for the spatial control of cytokinesis in maize, *J. Cell Biol.*, **152**, 231–236.

Stone, B.A. & Clarke, A.E. (1991) *Chemistry and Biology of (1→3)-β-Glucans*. La Trobe University Press, Bundoora, Victoria, Australia.

Sugimoto, K., Hinnelspack, R., Williamson, R.E. & Wasteneys, G.O. (2003) Mutation or drug-dependent microtubule disruption causes radial swelling without albering parallel cellulose microfibril deposition in *Arabidopsis* root cells, *Plant Cell*, **15**, 1414–1429.

Taiz, L. (1984) Plant cell expansion: regulation of cell-wall mechanical properties, *Annu. Rev. Plant Physiol.*, **35**, 585–657.

Taylor, J.G., Owen, T.P., Koonce, L.T. & Haigler, C.H. (1992) Dispersed lignin in tracheary elements treated with cellulose synthesis inhibitors provides evidence that molecules of the secondary cell wall mediate wall patterning, *Plant J.*, **2**, 959–970.

Taylor, N.G., Howells, R.M., Huttly, A.K., Vickers, K. & Turner, S.R. (2003) Interactions among three distinct CesA proteins essential for cellulose synthesis, *Proc. Natl. Acad. Sci. USA*, **100**, 1450–1455.

Tieman, D.M. & Handa, A.K. (1994) Reduction in pectin methylesterase activity modifies tissue integrity and cation levels in ripening tomato (*Lycopersicon esculentum* Mill.) fruits, *Plant Physiol.*, **106**, 429–436.

Turner, S.R. & Somerville, C.R. (1997) Collapsed xylem phenotype of *Arabidopsis* identifies mutants deficient in cellulose deposition in the secondary cell wall, *Plant Cell*, **9**, 689–701.

Van Hengel, A.J., Tadess, Z., Immerzeel, P., Schols, H., van Kammen, A. & de Vries, S.C. (2001) N-acetylglucosamine and glucosamine-containing arabinogalactan proteins control somatic embryogenesis, *Plant Physiol.*, **125**, 1880–1890.

Van Hengel, A.J. & Roberts, K. (2002) Fucosylated arabinogalactan-proteins are required for full root cell elongation in *Arabidopsis*, *Plant J.*, **32**, 105–113.

Vantard, M. & Blanchoin, L. (2002) Actin polymerization processes in plant cells, *Curr. Opin. Plant Biol.*, **5**, 502–506.

Vanzin, G.F., Madson, M., Carpita, N.C., Raikhel, N.V., Keegstra, K. & Reiter, W.-D. (2002) The MUR2 mutant of *Arabidopsis thaliana* lacks fucosylated xyloglucan because of a lesion in fucosyltransferase AtFUT1, *Proc. Natl. Acad. Sci. USA*, **99**, 3340–3345.

Varma, R. & Mayor, S. (1998) GPI-anchored proteins are organized in submicron domains at the cell surface, *Nature*, **394**, 798–801.

Verma, D.P.S. (2001) Cytokinesis and building of the cell plate in plants, *Annu. Rev. Plant Physiol. Plant Mol. Biol.*, **52**, 751–784.

Verma, D.P.S. & Hong, Z.L. (2001) Plant callose synthase complexes, *Plant Mol. Biol.*, **47**, 693–701.

Vidal, S., Doco, T., Williams, P., *et al.* (2000) Structural characterization of the pectic polysaccharide rhamnogalacturonan II: evidence for the backbone location of the aceric acid-containing oligoglycosyl side chain, *Carbohydr. Res.*, **326**, 227–294.

Wagner, T.A. & Kohorn, B.D. (2001) Wall-associated kinases are expressed throughout plant development and are required for cell expansion, *Plant Cell*, **13**, 303–318.

Wang, N., Naruse, K., Stamenovic, D., *et al.* (2001) Mechanical behavior in living cells consistent with the tensegrity model, *Proc. Natl. Acad. Sci. USA*, **98**, 7765–7770.

Webb, M., Jouannic, S., Foreman, J., Linstead, P. & Dolan, L. (2002) Cell specification in the *Arabidopsis* root epidermis requires the activity of ECTOPIC ROOT HAIR 3 – a katanin-p60 protein, *Development*, **129**, 123–131.

Wells, B., McCann, M.C., Shedletzky, E., Delmer, D. & Roberts, K. (1994) Structural features of cell walls from tomato cells adapted to grow on the herbicide 2,6-dichlorobenzonitrile, *J. Microscopy*, **173**, 155–164.

Whittington, A.T., Vugrek, O., Wei, K.J., *et al.* (2001) MOR1 is essential for guiding cortical microtubules in plants, *Nature*, **411**, 610–613.

Willats, W.G.T. & Knox, J.P. (1996) A role for arabinogalactan proteins in plant cell expansion: evidence from studies on the interaction of β-glucosyl Yariv reagent with seedlings of *Arabidopsis thaliana*, *Plant J.*, **9**, 919–925.

Willats, W.G.T., Marcus, S.E. & Knox, J.P. (1998) Generation of a monoclonal antibody specific to (1→5)-α-L-arabinan, *Carbohydr. Res.*, **308**, 149–152.

Willats, W.G.T., Orfila, C., Limberg, G., *et al.* (2001) Modulation of the degree and pattern of methyl-esterification of pectic homogalacturonan in plant cell walls: implications for pectin methyl esterase action, matrix properties and cell adhesion, *J. Biol. Chem.*, **276**, 19404–19413.

Wilson, R.H., Smith, A.C., Kacurakova, M., Saunders, P.K., Wellner, N. & Waldron, K.W. (2000) The mechanical properties and molecular dynamics of plant cell wall polysaccharides studies by Fourier transform infrared spectroscopy, *Plant Physiol.*, **124**, 397–406.

Wu, H.-M., Wong, E., Ogdahl, J. & Cheung, A.Y. (2000) A pollen tube growth-promoting arabino-galactoprotein from *Nicotiana alata* is similar to the tobacco TTS protein, *Plant J.*, **22**, 165–176.

Wyatt, S.E. & Carpita, N.C. (1993) The plant cytoskeleton-cell wall continuum, *Trends Cell Biol.*, **3**, 413–417.

Wymer, C. & Lloyd, C. (1996) Dynamic microtubules: Implications for cell wall patterns, *Trends Plant Sci.*, **1**, 222–228.

Xu, W., Purugganan, M.M., Polisensky, D.H., Antosiewicz, D.M., Fry, S.C. & Braam, J. (1995) *Arabidopsis* TCH4, regulated by hormones and the environment, encodes a xyloglucan endotransglycosylase, *Plant Cell*, **7**, 1555–1567.

Yang, Z. (1998) Signaling tip growth in plants, *Curr. Opin. Plant Biol.*, **1**, 525–530

Zhong, R.Q., Burk, D.H., Morrison, W.H. & Ye, Z.H. (2002) A kinesin-like protein is essential for oriented deposition of cellulose microfibrils and cell wall strength, *Plant Cell*, **14**, 3101–3117.

Zhu, J.-K., Shi, J., Singh, U., *et al.* (1993) Enrichment of vitronectin- and fibronectin-like proteins in NaCl-adapted plant cells and evidence for their involvement in plasma membrane-cell wall adhesion, *Plant J.*, **3**, 637–646.

Part 3

The cytoskeleton and plant cell morphogenesis

7 Development of root hairs

Claire Grierson and Tijs Ketelaar

7.1 Introduction

Extensive evidence across kingdoms shows that the cytoskeleton coordinates, mediates and controls cell growth. In this chapter we shall discuss how the microtubule and actin cytoskeletons contribute to the development of root hairs on higher plants, and we shall describe work to identify mechanisms that regulate and mediate the effects of the cytoskeleton on this process.

Root hairs are long tubes produced by special root epidermal cells. Root hairs have a very long (≥1 mm) and narrow (typically 10–20 μm) shape. They comprise 75% of the surface area of the roots of many crop plants, where they are the major point of contact between roots and the environment, and take part in most of the important functions that roots perform. These include taking up nutrients (Bates & Lynch, 2000a,b) and water, as well as interactions with symbiotic microorganisms and pathogens. Root hairs are important for nitrogen fixation in legumes because nitrogen-fixing *Rhizobium* bacteria enter host plants by manipulating root hair growth (Lhuissier *et al.*, 2001).

Root hairs are an excellent experimental system. Developing root hair cells are transparent, and development takes place on the surface of the plant, so the entire growth process can be observed at high resolution under a microscope. Root hair growth is tightly localized to one region of the cell, and is very reproducible, predictable, and easy to manipulate with molecular, cellular, and genetic techniques. Unlike pollen tube mutants, plants with genetically manipulated root hair growth are usually fertile, so genetic research is straightforward.

The patterning and cell fate of root hair cells is controlled by developmental and environmental factors. This is best understood in *Arabidopsis,* as summarized in Fig.7.1. In *Arabidopsis*, hairs are only produced by epidermal cells overlying the anticlinal wall between two adjacent cortical cells; this is called the H position. The other cells in the epidermis only contact the periclinal wall of a single cortical cell; this is the N position. Cells in the N position do not usually form hairs.

Once an epidermal cell is committed to forming a hair, a growth site is selected on the surface of the cell. This process is called initiation (Fig. 7.2 (A) and (B)). During initiation, a swelling forms at a precise location toward the apical end of the cell (the end nearest the root tip). Before the swelling forms, an acidophilic xyloglucan endotransglycosylase (XET) is active at the growth site (see Chapter 6). This remains active throughout root hair formation, and might be partly responsible for loosening the cell wall (Vissenberg *et al.*, 2001). The wall begins to bulge

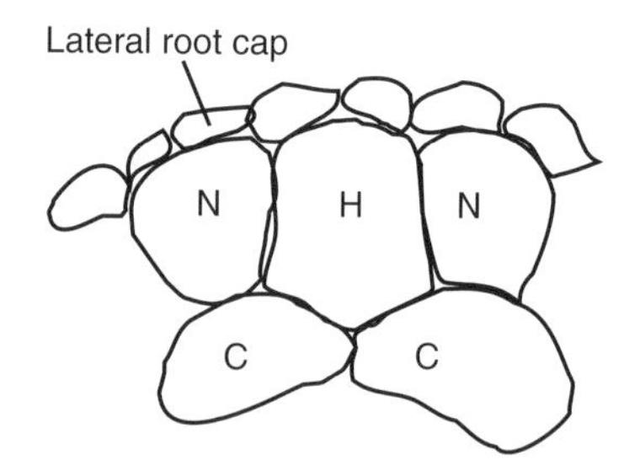

Fig. 7.1 Root hair patterning in *Arabidopsis*. Part of a cross section of an *Arabidopsis* root meristem showing the positions of young hair and non-hair cells relative to underlying cortical (C) cells. In this part of the root, a layer of lateral root cap cells covers the epidermis. Cells in the H position (H) overlie the anticlinal wall between two adjacent cortical cells and will usually form hairs. Cells in the N position (N) overlie the periclinal wall of a single cortical cell and will not usually form hairs.

out. At the same time the pH of this part of the wall falls. This pH change might activate expansin proteins that also contribute to wall loosening (Bibikova *et al.*, 1998; Baluska *et al.*, 2000). The mechanism responsible for the pH change is uncertain; it could be local changes in wall structure and ion exchange capacity, or else local activation of a proton ATPase or other proton transport activity (Bibikova *et al.*, 1998). As the bulge grows, the endoplasmic reticulum within it condenses (Ridge *et al.*, 1999). Under optimal conditions, it takes about 30 min for the swelling to form (Bibikova *et al.*, 1998).

Once a swelling has formed, the rest of the root hair develops by tip growth. Tip growth is extremely tightly focused growth that occurs by targeted secretion. Many of the mechanisms of tip growth are conserved across kingdoms (for other examples see Chapter 8). During root hair tip growth, new cell wall material is synthesized and precisely secreted in a dome-shaped region at the tip (Derksen & Emons, 1990; Shaw *et al.*, 2000). The cytoplasm inside the tube is highly organized (Fig. 7.2 (C)). The dome is packed with vesicles full of cell wall matrix and cell wall synthesizing enzymes. These vesicles are produced by smooth and rough endoplasmic reticulum and processed in Golgi bodies, which are very abundant in the part of the tube just behind the growing tip. The vesicles supply the essentials for cell wall synthesis, and the vesicle membrane delivers cellulose-synthesizing complexes into the plasma membrane. Root hairs typically grow at $1\,\mu m\,min^{-1}$ (Wymer *et al.*, 1997, 1–$2\,\mu m\,min^{-1}$; Bibikova *et al.*, 1999, $1.22\,\mu m\,min^{-1}$; Ketelaar *et al.*, 2002, $0.93\,\mu m\,min^{-1}$), rapidly producing and incorporating vesicles. Cytoplasmic contents are moved rapidly around the hair by cytoplasmic streaming. The nucleus enters the shaft of the growing root hair and moves along behind the tip (Fig. 7.2D). At a typical growth rate ($1\,\mu m\,min^{-1}$), *Arabidopsis* root hairs increase their volume by approximately $50\,fL\,min^{-1}$ (Lew, 2000). To remain turgid whilst increasing in volume, the total amount of osmotic ions such as K^+ and Cl^- gradually increases (Lew, 1991, 1998).

Root hair tip growth requires calcium (Schiefelbein *et al.*, 1992). When *Arabidopsis* root hairs are 5–10 μm long, the Ca^{2+} concentration at the tip of the swelling rises

A

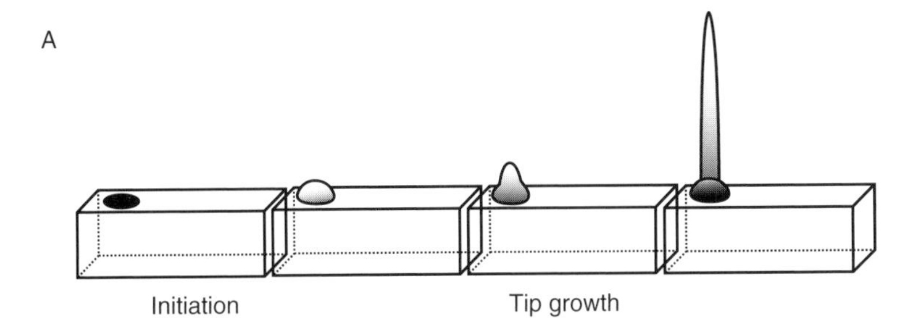

B

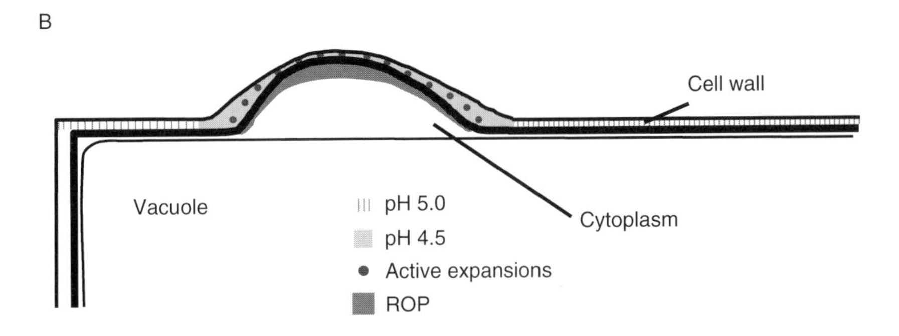

C

K⁺
Vacuole
Ca²⁺
Golgi body
Vesicle
Calcium channel
ROP
Cytoplasm
Cl⁻
Cell wall
Endoplasmic reticulum

D

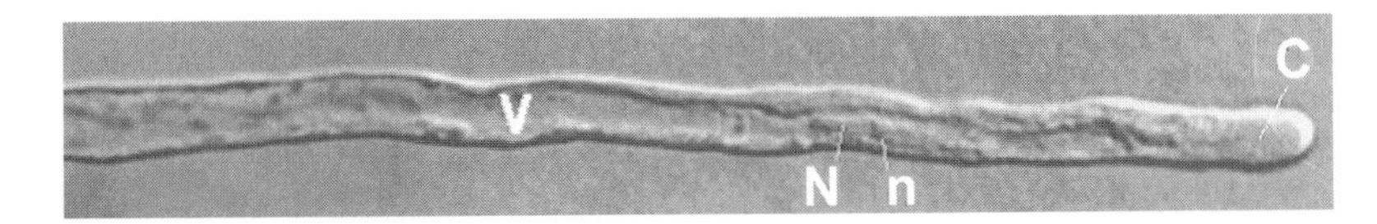

Fig. 7.2 Stages of root hair development. (A) Drawing of part of a row of root hair cells, showing hairs at different stages of development; (B) Diagram summarizing root hair initiation; (C) Diagram summarizing root hair tip growth; (D) Differential interference contrast image of a tip-growing root hair of *Arabidopsis thaliana*. C – cytoplasm, N – nucleus, n – nucleolus, V – vacuole.

from about 200 nM to at least 1 μM, and remains very high throughout tip growth (Wymer *et al.*, 1997; Bibikova *et al.*, 1999). Calcium is imported across the membrane at the tip of the hair by channels that are activated by the negative potential across the membrane at the tip of the hair (−160 to −200 mV, Lew, 1996; Very & Davies, 2000). This is sufficiently negative to activate the channels. This potential is probably generated by a plasma membrane H^+ ATPase. Root hairs grow well in moist air and are abundant in air pockets in soil (Grierson, unpublished observations; Ryan *et al.*, 2001). Under these conditions, calcium for tip growth must be (1) released from the newly deposited wall; or (2) transported through the apoplast, and deposited near the tip of the hair; or (3) released from intracellular stores (Ryan *et al.*, 2001).

Tip growth ends in a precisely coordinated manner, producing a smooth dome at the tip with the same maximum diameter as the hair shaft. The calcium gradient dissipates, the vacuole enlarges into the dome, and the nucleus retreats from the tip of the hair.

7.2 Roles of the cytoskeleton in root hair morphogenesis

Visualizing the cytoskeleton in root hairs is technically demanding. Advances in fixation techniques, and the use of cytoskeleton binding proteins fused to jellyfish fluorescent proteins, in conjunction with very high quality light microscopy, are producing new data about the arrangements and dynamics of microtubules and actin in root hair cells. Work using chemical inhibitors has suggested roles for the cytoskeleton at many stages of root hair development. These results are being complemented and enhanced by work using mutants and transgenic plants with altered cytoskeletal composition or dynamics.

7.2.1 Microtubules

7.2.1.1 Microtubules affect root hair cell fate

Microtubules mediate cell division (see Chapters 1, 3 and 4) and cell wall formation (see Chapters 3 and 6) in plant tissues. Abnormalities in these processes can affect root hair development. Transgenic lines with reduced α-tubulin expression affect the patterning of root hairs, and produce neighboring hair-bearing cells (Bao *et al.*, 2001). It is not clear whether these are truly ectopic hairs (hair-bearing cells in the N position) or extra epidermal cells in the H position produced by erroneous cell division patterns. However, a katanin subunit involved in microtubule organization, the small p60 subunit encoded by the *ECTOPIC ROOT HAIR 3/AtKSS/BOTERO1/FRA2* gene, has a strong effect on cell fate that is not caused by altering cell division patterns. Webb *et al.* (2002) found patches on *erh3* mutant roots where the arrangement of cells was normal, but cell fate was altered; cells in the N position sometimes made hairs, and cells in the H position were sometimes hairless and sometimes expressed non-hair-specific marker genes. The details of the mechanism of katanin action remain to be resolved. There are two alternative hypotheses. The first is that root hair cell fate is fixed by molecules embedded in cell walls, and that the correct localization of these molecules depends on katanin activity, probably via its affect on microtubule organization. The second is that microtubule structures themselves somehow specify cell fate. There are precedents for this in animal cells, where microtubule structures, such as asters, and the polarized localization of microtubules from cell center to cell periphery affect the asymmetrical distribution of proteins, and hence cell fate (Webb *et al.*, 2002).

7.2.1.2 Microtubules and root hair initiation

There is inconclusive evidence that microtubules might affect the location, number, or size, of root hair initiation sites on root hair cells. Transgenic lines with reduced α-tubulin expression sometimes have cells with multiple root hairs emerging from a single swelling, but it is not clear whether it is the swellings themselves or tip growth from the swellings that is defective (Bao *et al.*, 2001). Evidence from the *mor1* mutant and from drug treatments suggests that microtubules do not contribute significantly during root hair initiation. The *mor1* mutant has disturbed microtubule organization but initiates root hairs normally (Whittington *et al.*, 2001). Treatments with drugs that destabilize (oryzalin) or polymerize (paclitaxel) microtubules do not affect root hair initiation, although they do affect later stages of hair development.

7.2.1.3 Microtubules control direction of root hair tip growth and prevent hairs from branching

Microtubules play an important role during root hair tip growth. Evidence from studies with drugs that stabilize or depolymerize microtubules, mutants with altered microtubule organization, and transgenic lines with reduced α-tubulin gene expression, all support the conclusion that microtubules control the direction of root hair tip growth and prevent branching.

Figure 7.3 shows the arrangement of microtubules in tip-growing hairs of *Arabidopsis thaliana*. Large numbers of microtubules are seen around the periphery of the cytoplasm, running parallel to the long axis, or in a gentle helix along the hair shaft.

Figure 7.4 shows the phenotypes of *Arabidopsis* root hairs produced when microtubule organization is disrupted by drug treatments, in mutants, and in transgenic lines. Importantly, disrupting microtubules does not significantly alter the rate of growth of *Arabidopsis* root hairs (measured as an increase in surface area), and does not change root hair diameter, suggesting that microtubules are not required for vesicle transport or vesicle fusion per se, although there is ample evidence that they do affect where these processes take place (Bibikova *et al.*, 1999; Ketelaar *et al.*, 2003).

Elegant work by Bibikova *et al.* (1999) showed that microtubules control the direction of root hair growth and ensure that there is only one site for tip growth. When the microtubule cytoskeleton is depolymerized or stabilized using the drugs oryzalin and paclitaxel (taxol) respectively, root hair growth becomes wavy, and

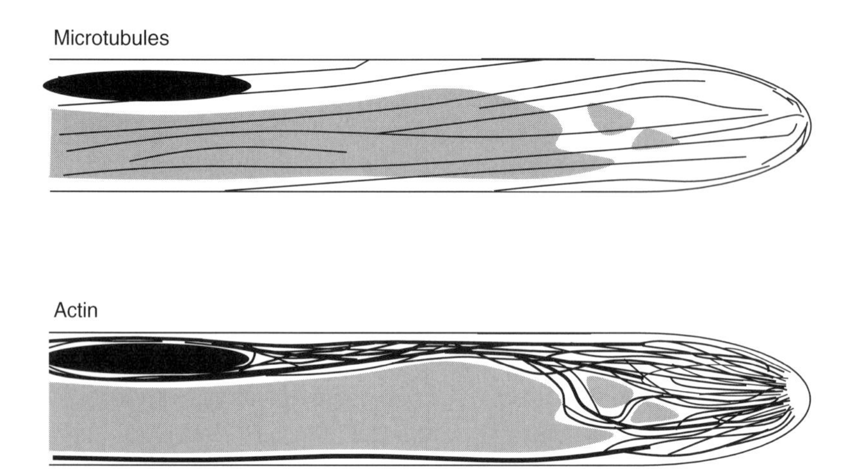

Fig. 7.3 Diagram illustrating typical microtubule and actin organization in tip-growing root hairs of *Arabidopsis thaliana*.

hairs branch to form two or more tips (Fig. 7.4 (H), (I), and (J)). Treatment with oryzalin causes a weak wavy phenotype of *Arabidopsis* root hairs, whereas treatment with low concentrations of paclitaxel leads to a severe wavy phenotype (Bibikova *et al.*, 1999), indicating that microtubule dynamics play an important role in the proper localization of the growth site. The *mor1* mutant, which has disturbed microtubule organization, has slightly wavy hairs, similar to root hairs grown in the presence of oryzalin (Fig. 7.4 (H), Whittington *et al.*, 2001). *Arabidopsis* plants with reduced α-tubulin expression often have multiple hairs emerging at the same initiation site, and hairs that branch during tip growth (Fig. 7.4 (K), Bao *et al.*, 2001). Branching during tip growth is also seen in hairs treated with paclitaxel and oryzalin (Bibikova *et al.*, 1999). Taken together these results show that normal tubulin levels and dynamics are required to maintain tip growth of a single tip in a straight line.

Root hair tip growth can be reoriented by touch stimulus or by increasing the concentration of calcium using localized release of a caged ionophore. The region at the root hair tip where growth can be reoriented by these stimuli is larger when microtubules are depolymerized or stabilized (Bibikova *et al.*, 1999). This suggests that microtubules usually focus on the area at the tip where growth can take place. There is no evidence that microtubules limit the size of the growth area; if they did, then microtubule disruption would lead to changes in root hair diameter, and this has not been demonstrated. Evidence from Tominaga *et al.* (1997) and Ketelaar *et al.* (2003) suggests that microtubules focus growth by influencing the direction of polymerization of actin filaments (see below).

Root hair cell walls are reinforced with cellulose, which is synthesized by cellulose synthase complexes in the plasma membrane (see Chapter 6). Heath (1974)

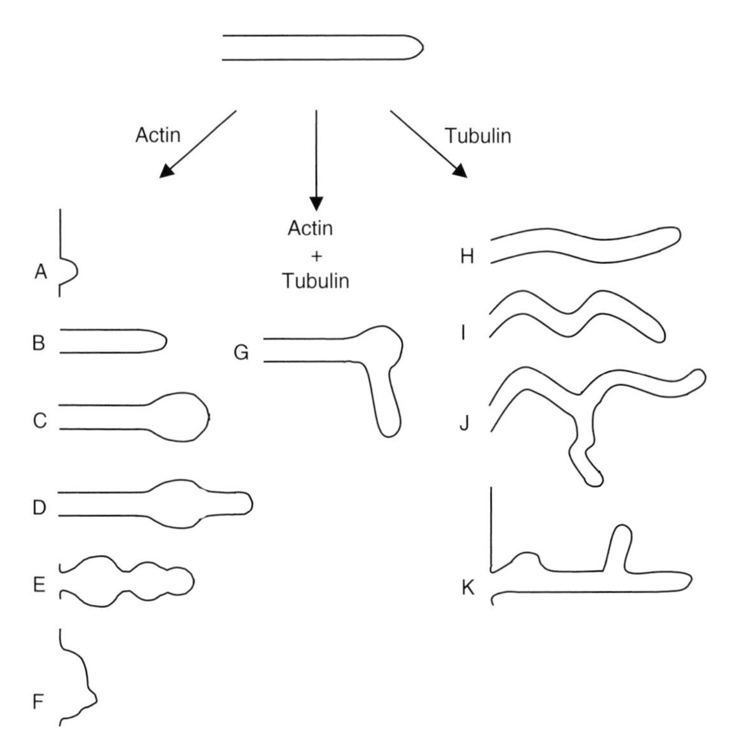

Fig. 7.4 Diagram summarizing the phenotypes of wild type root hairs grown in the presence of actin- or microtubule-perturbing drugs, and of root hairs on mutants or transgenic plants with altered actin or microtubules; (A) Root hairs do not develop past the bulge stage in plants grown in high concentrations of actin-depolymerizing drugs; (B) Growth is inhibited in developing root hairs treated with moderate or high concentrations of actin-depolymerizing drugs; (C) Low concentrations of actin-depolymerizing drugs, applied to growing root hairs cause an increase in diameter; (D) After washing out low concentrations of cytochalasin D, the diameter of a growing root hair recovers to normal values; (E) and (F) Moderate and severe phenotypes of plants with mutations in the *ACTIN2* gene (*der1* and *act2-1* mutants); (G) When microtubules are depolymerized, perturbing the actin cytosketelon with a pulse of cytochalasin D leads to an increase in diameter, which recovers to growth with a normal diameter in a random direction, whereas growth continues in the same direction in root hairs with an intact microtubule cytoskeleton after a similar cytochalasin D treatment (see (D)); (H) In plants with depolymerized microtubules, produced either by pharmacological agents (oryzalin) or by a mutation (e.g. *mor1* and *erh3*), root hair growth is slightly wavy; (I) and (J) When microtubules are stabilized (paclitaxel treatment), root hairs are wavy and occasionally branch; (K) In transgenic plants with a reduced amount of α-tubulin, occasionally branches and multiple root hairs from one initiation site occur.

hypothesized that cortical microtubules determine the orientation of the cellulose microfibrils in plant cell walls, and thus the direction of cell expansion. This might be true in the expanding part of the root hair cell (in the apical dome), although it is difficult to tell because both microtubules and cellulose microfibrils are randomly arranged in this part of the hair (Emons, 1989; Ketelaar & Emons, 2000). In the hair shaft, which is no longer expanding, and is being reinforced with cellulose, cortical microtubules are not always in the same orientation as the cellulose microfibrils (Ketelaar & Emons, 2000).

7.2.1.4 Microtubules help to move the nucleus during tip growth in some species, but not in others

In *Arabidopsis*, endoplasmic microtubules have rarely been found (Ketelaar *et al.*, 2002). However, endoplasmic microtubules have been seen in the tips of legume root hairs. It is not clear whether these differences between species reflect differences in cell biology, or differences in the visibility of some types of microtubules between the two species. Lloyd *et al.* (1987) reported that endoplasmic microtubules connect the nucleus to the tip of growing *Vicia hirsute* root hairs. Depolymerization of the microtubules with colchicine leads to basipetal migration of the nucleus. However, the growth rate of the root hairs was not checked after application of colchicine and therefore the basipetal migration could have been caused by growth termination. Sieberer *et al.* (2002) detected endoplasmic microtubules in the subapex of *Medicago trunculata* root hairs. Depolymerization with oryzalin decreased the growth speed of the hairs and the nucleus moved backwards, away from the tip. In contrast, Ketelaar *et al.* (2002) showed that treatment with oryzalin had no effect on the position of the nucleus or the growth rate of *Arabidopsis* root hairs.

7.2.2 Actin filaments

7.2.2.1 Actin limits the size of the initiation site

Actin has no proven role at the very beginning of root hair formation. The actin depolymerizing drugs latrunculin B and cytochalasin D, have no effect on the position of the initiation site (Miller *et al.*, 1999; Ovecka *et al.*, 2000; Molendijk *et al.*, 2001). Nonetheless, actin accumulates in developing swellings soon after hairs start to form (Baluska *et al.*, 2000). Evidence from promoter-reporter gene studies suggests that at least three actin genes, *ACT2*, *ACT7*, and *ACT8*, are expressed in root hairs (An *et al.*, 1996; McDowell *et al.*, 1996). Three EMS mutants (*der1-1*, *der1-2*, and *der1-3*) and one T-DNA insertion mutant (*act2-1*) in *ACT2* have root hair phenotypes. The swellings on *der1-2*, *der1-3*, and *act2-1* mutants are often very large (Fig. 7.4 (F), Ringli *et al.*, 2002; Gilliland *et al.*, 2002). These results implicate actin in mechanisms that control swelling size.

Gilliland *et al.* (2002) demonstrated that the root hair defects in *act2-1* mutants are due to a reduction in the total amount of actin monomer in the cell, and do not reflect functional differences between actin monomers with different amino acid sequences. The *act2-1* root hair phenotype could be completely restored to wild type by overexpressing another vegetative actin, ACT7, even though the ACT7 amino acid sequence differs from that of ACT2 by 6.8%. The reproductive actin, ACT1, which is the most divergent actin compared to ACT2, and has many non-conservative differences in amino acid sequence, can also replace ACT2 in root hairs (Gilliland *et al.*, 2002). This evidence suggests that root hair development does not require any specific properties of particular actin isovariants.

7.2.2.2 Actin mediates tip growth by targeting vesicle delivery

Figure 7.3 shows a picture of the actin organization in a tip-growing root hair. The role of actin in root hair tip growth is quite well understood.

In the dense cytoplasm of the growing tip, actin bundles branch, becoming thinner and thinner. The tip contains a high density of fine bundles of actin filaments that are predominantly oriented along the long axis of the hair (net-axial FB-actin) (Miller *et al.*, 1999). Other bundles of actin filaments loop back through cytoplasmic strands running through the vacuole. This is where villin-mediated bundling takes place (Tominaga *et al.*, 2000). Most FB-actin points to the tip, which contains few detectable actin filaments (Miller *et al.*, 1999). In this area, Emons (1987) found a cytoplasmic streaming speed of 0 $\mu m\, s^{-1}$ in *Equisetum* root hairs, suggesting that indeed no actin filaments were present.

Subapical FB-actin has been hypothesized to fulfill three functions as follows: first, to deliver vesicles containing new plasma membrane and cell wall material to the tip; second, as a buffer to retain the vesicles in the tip; third, as a sieve to inhibit penetration of other organelles into the tip (De Ruijter & Emons, 1999). Miller *et al.* (1999) demonstrated that subapical FB-actin is more sensitive to cytochalasin D than more basal actin filament bundles. Continued application of cytochalasin D leads to disappearance of the FB-actin, and growth stops, but cytoplasmic streaming is normal. This strongly suggests that FB-actin is required for root hair growth, i.e. for vesicle delivery. Ketelaar *et al.* (2003) took this work a stage further by applying pulses of low concentrations of cytochalasin D or latrunculin A to tip-growing *Arabidopsis* root hairs. This increased the diameter of the hairs without reducing the growth rate (measured as an increase in root hair surface area per minute). These results indicate that the amount and location of fine F-actin determines where tip growth takes place. One prediction that arises from these observations is that mutations or transgenes that depolymerize actin should broaden root hair diameter. Mutants in the *ACTIN2* gene produce root hairs whose diameters vary widely (Fig. 7.4 (E)) suggesting that the rates and locations of actin polymerization and depolymerization are defective in these mutants and vary as growth proceeds.

Ketelaar *et al.* (2003) also showed that root hairs recovering from treatment with cytochalasin D produced new growth in the original direction, but this depended on microtubules. If microtubules were depolymerized with a low concentration of oryzalin, which on its own had no effect on root hair elongation, then root hairs recovering from cytochalasin D treatment grew in random directions. This concurs with the earlier work of Bibikova *et al.* (1999) which showed that microtubules control the direction of root hair tip growth.

Mutants in the *ACTIN2* gene (*der* and *act2-1* mutants) often produce multiple tip-growing structures from each swelling (Gilliland *et al.*, 2002; Ringli *et al.*, 2002). This suggests that mechanisms that normally prevent the formation of multiple hairs from a single swelling depend on precise control of actin levels or dynamics. Similar phenotypes are found in root hairs with disrupted microtubules (e.g. Fig. 7.4 (K)).

7.2.2.3 F-actin is essential for the Arabidopsis *nucleus to move during and after tip growth*

As root hairs grow, the nucleus enters the hair and migrates at a fixed distance behind the growing tip (about 75 microns in *Arabidopsis*). Whilst microtubules are involved in nuclear movement in legume root hairs, *Arabidopsis* root hairs use actin to control nuclear position (Ketelaar *et al.*, 2002). The connection between the nucleus and the tip strongly correlates with tip growth. If tip growth in *Arabidopsis* is stopped by carefully depolymerizing F-actin in the hair tip with cytochalasin D (without affecting thicker actin bundles), tip growth stops and the nucleus moves away from the tip. Trapping the nucleus in a laser beam for 10–15 min so that it cannot move disrupts the link between the nucleus and the tip and growth ceases. Microinjection of anti-villin antibody unbundles the actin bundles in the hair, and the nucleus moves toward the tip, suggesting that actin bundles usually help to keep the nucleus away from the tip. Taken together, these results support a model where actin dynamics regulate nuclear position in *Arabidopsis*, with fine F-actin connecting the nucleus to the tip and actin bundles preventing the nucleus from approaching the tip (Ketelaar *et al.*, 2002).

7.2.2.4 Actin mediates cytoplasmic streaming in roots hairs

In plant cells, although some transport may involve microtubules (Romagnoli *et al.*, 2003), actin filaments appear to be the backbone of cytoplasmic strands and the basis of most cytoplasmic streaming (Valster *et al.*, 1997; Esseling *et al.*, 2000). In growing root hairs, cytoplasmic streaming takes place in a reversed fountain-like pattern (Sieberer & Emons, 2000). Thick bundles of longitudinally orientated actin filaments run through the cortical cytoplasm and the transvacuolar strands, and are thought to mediate streaming; if all of the F-actin in a root hair is depolymerized using drug treatment, cytoplasmic streaming stops. By carefully controlling the treatment, it is possible to stop growth by depolymerizing F-actin without affecting the thick actin bundles, or cytoplasmic streaming (Miller *et al.*, 1999; Ketelaar *et al.*, 2002). An *Arabidopsis* mutant with negligible cytoplasmic streaming still makes long root hairs (Parker & Grierson, unpublished), supporting the idea that cytoplasmic streaming is not essential for root hair growth.

7.2.2.5 Actin at the end of tip growth

Root hair tip growth stops when hairs reach a predictable length. When hairs stop growing, the cytoarchitecture at the tip changes. Thick bundles of actin filaments loop through the tip (Miller *et al.*, 1999). The cytoplasmic streaming pattern reflects the actin organization, shifting the route of cytoplasmic streaming to rotational streaming, which takes place through the tip of the hair (Sieberer & Emons, 2000).

As tip growth is ceasing, legume root hairs become susceptible to changes induced by nodulation factors, signaling molecules of *Rhizobium* bacteria (Heidstra *et al.*, 1994; Cárdenas *et al.*, 1998). In response to these molecules, tip growth can re-initiate in these hairs (De Ruijter *et al.*, 1998). The actin cytoskeleton seems to be the target of these signaling molecules, since in all root hairs of *Vicia sativa*, Nod factor application leads to an increase in the density of subapical FB-actin (De Ruijter *et al.*, 1999). In growth terminating root hairs, the density of FB-actin decreases. After Nod factor application, the formation of a new outgrowth only takes place when the density of FB-actin increases to the level seen in growing root hairs. This suggests that tip growth cannot re-establish unless sufficient fine bundles of actin are present.

At the end of tip growth, the nucleus retreats back toward the base of the hair, eventually adopting a random position in the hair cell. This movement depends on F-actin bundles because it does not take place when all the F-actin has been depolymerized. The basipetal movement does not happen if transcription or protein synthesis has been inhibited, suggesting that it depends on gene products that are produced during tip growth (Ketelaar *et al.*, 2002).

7.3 Mechanisms that regulate the cytoskeleton during root hair development

A number of regulators of root hair development have been identified. However, we are only just beginning to connect most of this knowledge to our understanding of the cytoskeleton.

7.3.1 Mechanisms regulating root hair patterning

The evidence that katanin influences root hair cell fate (see Section 2.1 a) indicates that microtubules are involved in the control of root hair patterning. Root hair development is controlled by a hierarchy of gene activity. The first step is specification of the epidermis, as root hairs are normally only produced by epidermal cells. No positive regulators of epidermal cell fate have been identified, but *SCHIZORHIZA* (*SCZ*) prevents cells in the subepidermal layer (ground tissue) from developing into epidermis; *scz* mutants have root hairs on ground tissue cells as well as epidermal cells (Mylona *et al.*, 2002). The molecular identity of *SCZ* has not been reported. Once the epidermis has been specified, an unidentified positional signal controls which epidermal cells form hairs. This signal seems to act by influencing the level of WEREWOLF (WER) MYB transcription factor so that cells in the H position have less WER than cells in the N position (Lee & Schiefelbein, 2002). WER induces another MYB protein, CAPRICE (CPC), and a homeodomain-leucine zipper transcription factor called GLABRA2 (GL2). GL2 is believed to be a repressor of root hair genes, and cells with high levels of GL2 do not make root hairs (DiCristina *et al.*, 1996; Masucci & Schiefelbein, 1996). CPC is exported from the N cells, where it is synthesized, to neighboring epidermal cells (Wada *et al.*, 2002), where it represses the expression of GL2, WER, and itself. This system of lateral inhibition and feedback fixes epidermal cell fates; any slight differences in the level of WER in neighboring cells are rapidly reinforced, fixing the relative levels of GL2 and CPC as well (Lee & Schiefelbein, 2002). ERH3 somehow affects this mechanism, probably in several distinct ways. ERH3 does not just affect the cell fate of epidermal cells; the expression of a lateral root cap marker (some cells that should express the marker do not) and a cortical/endodermal marker (which is sometimes misexpressed in cells in the epidermal layer of *erh3* mutants) are also affected. Genetic analysis suggests that ERH3 acts in the same pathway as CPC, but independently of WER. Mutation of both *ERH3* and *CPC* (in an *erh3 cpc* double mutant) has a synergistic effect on the number of non-hairs in the H position, and has a *cpc* phenotype for ectopic hairs in the N position (Webb *et al.*, 2002). These results could be partly explained if ERH3 facilitates the movement of CPC from N cells to H cells, and it is worth testing the hypothesis that the movement of CPC requires microtubules. *RHD6* is a positive regulator of root hair differentiation whose molecular identity has not been reported; *rhd6* is epistatic to *erh3*, so these two genes probably act in the same pathway (Webb *et al.*, 2002).

The patterning of root hairs can be overridden by environmental signals, fine tuning root hair production to suit the environment immediately surrounding the root. Contact with solid growth medium (Okada & Shimura, 1994; Cho & Cosgrove, 2002), or local Fe or P concentrations (Schmidt & Schikora, 2001), and mutations or treatments that affect auxin or ethylene signaling (Masucci & Schiefelbein, 1996; Cao *et al.*, 1999) can all increase or decrease root hair production. Subtle mechanisms can be involved. For example, the auxin response *AUX/IAA* genes, *AXR3* and *SHY2*, have opposing effects on root hair development. A gain-of-function, dominant allele of *axr3* prevents root hair formation, whereas a gain-of-function, dominant *shy2* allele promotes root hair initiation. Loss-of-function *axr3* alleles don't have a root hair phenotype, so either *AXR3* does not play a role in wild type root hair development, or this function can be fulfilled by another *AUX/IAA* gene. Loss-of-function *shy2* mutants have fewer root hairs, suggesting that SHY2 plays a role in the timing of root hair initiation on wild type roots. The amino acid sequences of AXR3 and SHY2 are similar, and there is evidence that the relative amounts of these two proteins have a strong effect on root hair production (Knox *et al.*, in press). In *rhd6* mutants, root hair production is greatly reduced, but responds strongly to auxin, ethylene, and environmental signals. Cho and Cosgrove (2002) showed that auxin treatment or contact with solid growth medium has no effect on *rhd6* if ethylene perception is blocked by the competitive inhibitor 1-methylcyclopropene (1-MCP). It is theoretically possible that other environmental factors that affect root hair production act via ethylene perception, and this should be tested. At the moment, we do not know how environmental signals are linked to changes in the root hair cytoskeleton.

7.3.2 Mechanisms that regulate initiation

Besides their role in governing the presence or absence of root hairs, auxin and ethylene signaling sets up the polarity of root hair cells and controls where on each cell root hair initiation takes place. As a general rule, increased auxin or ethylene signaling, caused by chemical treatments or mutant genes, moves the initiation site toward the apical end of the cell (the end nearest the root tip), and decreased signaling moves the initiation site toward the center of the cell (away from the root tip, Masucci & Schiefelbein, 1994, 1996). Treatment with Brefeldin A (BFA) disrupts secretion by interfering with the function of Sec7-type GTP-exchange factors (GEFs, Nebenfuhr *et al.*, 2002). At low concentrations of BFA, the localization of the auxin influx carrier AUX1 to the plasma membrane is disrupted, reducing the polarity of the root hair cell, and moving the initiation site toward the center of the cell (Grebe *et al.*, 2002). PIN auxin efflux carriers are expected to play an important role in root hair cell polarity, but no link between a PIN and root hair initiation has been reported so far.

Cho and Cosgrove (2002) showed that the effect of auxin on the presence or absence of root hairs can be mediated via ethylene perception (see above). It will be important to show whether this is also the case for the effect of auxin on the positions of root hairs on hair cells.

Before root hairs begin to grow, ROP2, and possibly other related ROPs, appear at the growth site (Fig. 7.5, Molendijk *et al.*, 2001; Jones *et al.*, 2002). ROPs are unique to plants, but are related to Rho small GTPases that control many processes including morphogenesis in animal and yeast cells (Eaton *et al.*, 1996; Chant, 1999; see Chapters 8 and 10). Applying high concentrations of BFA prevents ROP localization, suggesting that either ROP itself or a molecule that localizes ROP is placed at the future site of hair formation by targeted secretion (Molendijk *et al.*, 2001). ROPs continuously cycle between GTP- and GDP-bound forms. The amount of ROP, and its ability to cycle, both affect root hair initiation. Plants overexpressing ROP2, or a mutant CArop2 that is permanently in the GTP-bound form produce extra initiation sites (Fig. 7.5 (A)). Overexpressed CArop2 has a weaker effect than overexpressed wild type ROP2, suggesting that cycling between the GTP- and GDP-bound forms is important for efficient initiation. Plants expressing a mutant DNrop2 that is permanently in the GDP-bound form have fewer root hairs (Jones *et al.*, 2002). In other systems, such as budding yeast, ROP-like proteins (Ras, cdc42, and Rho1) are involved in signaling cascades that select, establish, and maintain polar growth sites. In this case, the localization of cortical actin patches is part of the mechanism. Ringli *et al.* (2002) reported that the position of root hairs on hair cells was occasionally altered in *der* (*ACTIN2*) mutants of *Arabidopsis*, and Braun *et al.* (1999) reported actin accumulation very early in bulge formation in maize, cress, and *Arabidopsis* root hairs. These results implicate actin in root hair initiation. However, treatment with cytochalasin D had no effect on ROP localization, or root hair initiation (Baluska *et al.*, 2000;

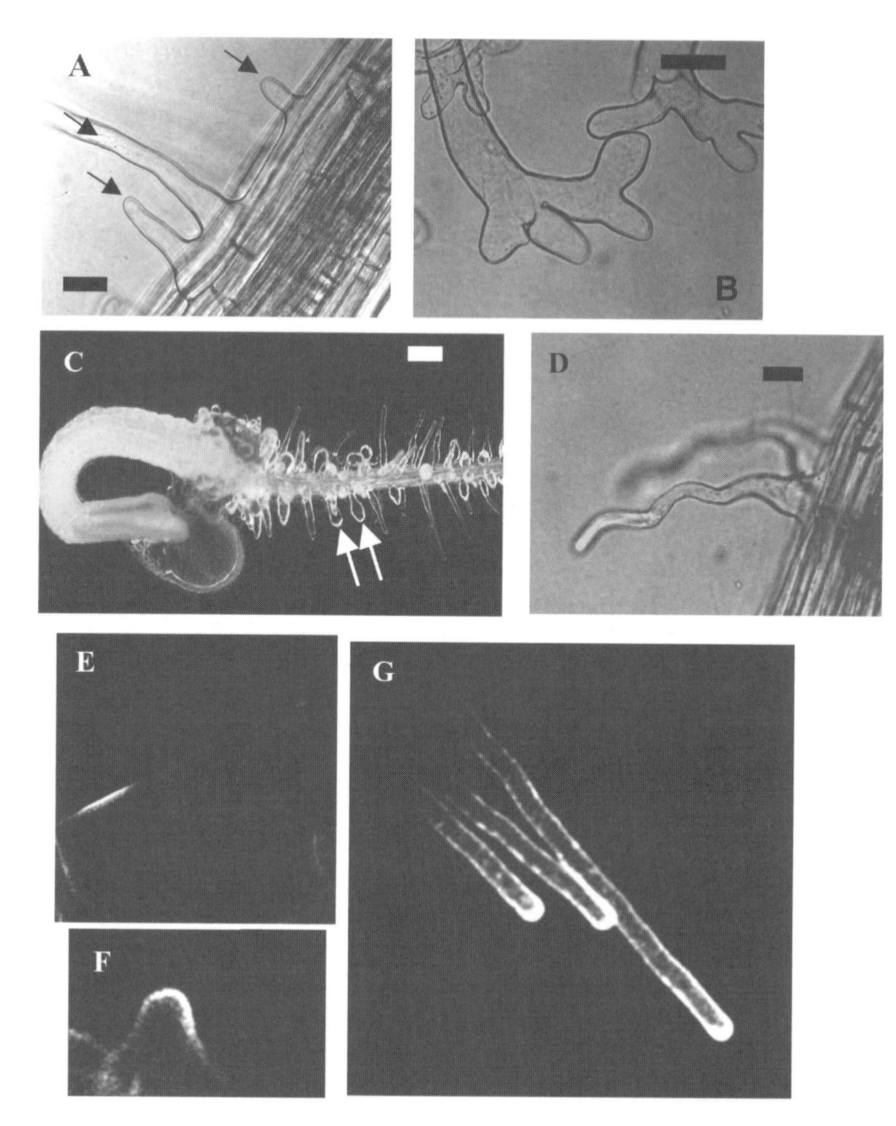

Fig. 7.5 The role of *Arabidopsis* ROP2 in root hair development. (A) Multiple hairs (arrows) on a single hair cell of a transgenic line expressing CArop2; (B) Branched tips at the ends of root hairs on plants overexpressing ROP2; (C) Depolarized root hairs (arrows) on a plant expressing CArop2; (D) Wavy root hair on a plant expressing DN-rop2; (E), (F), (G) GFP-ROP2 is located at the future site of root hair formation before growth begins (E), and remains at the site of growth throughout swelling formation (F) and tip growth (G).

Molendijk *et al.*, 2001) suggesting that actin is not required at the very beginning of the process. Further work is required to clarify whether actin is involved in specifying the root hair initiation site.

Samaj *et al.* (2002) showed that the stress-induced MAPK of alfalfa (SIMK) localized to the growth site very early in root hair initiation. Overexpressing SIMK affected root hair elongation in tobacco (see below), but the authors did not report any effects on root hair initiation. Any role for SIMK during root hair initiation remains to be elucidated.

7.3.3 Mechanisms regulating tip growth

Besides their effects on the number of root hairs produced, environmental signals can have strong effects on root hair length. Light signaling has strong effects on root hair length that are mediated, at least in part, by phytochromes. The signal from PHYB, and possibly other phytochromes, is mediated by HY5, which encodes a bZIP transcription factor that is localized in the nucleus of *Arabidopsis* cells (Oyama *et al.*, 1997; Schafer & Bowler, 2002), including growing root hairs (Simone *et al.*, unpublished). Regulation of root hair length by HY5 depends on *IRE*, a gene encoding a predicted serine/threonine kinase. *HY5* and *IRE* do not regulate elongation rate, but influence when root hair tip growth ceases. IRE is in the same group of protein kinases as MAST205, a microtubule-associated protein from mice, and it is possible that IRE regulates root hair length by acting on microtubules (Oyama *et al.*, 2002). It is not known whether other environmental signals apart from light act via HY5 or the IRE kinase.

Some environmental signals, such as low iron availability, act via auxin and/or ethylene signaling, but others (e.g. low phosphate) are independent of these pathways (Schmidt & Schikora, 2001). In general, increased auxin or ethylene signaling increases root hair length and decreased signaling decreases root hair length. The molecules linking auxin and ethylene signaling with root hair length have not been identified, but one possible mechanism might be through mitogen-activated protein (MAP) kinase cascades. MAP kinases are implicated in auxin and ethylene responses, and also in cell polarity and actin organization in plants (Jonak *et al.*, 2002). A stress induced MAP kinase (SIMK) has a clear role in root hair elongation. Samaj *et al.* (2002) showed that alfalfa SIMK colocalized with polymerized actin throughout tip growth. Treatment with latrunculin B depolymerized actin, and SIMK antibodies re-localized to the nucleus. Treatment with jasplakinolide produced thick actin cables that were decorated with SIMK. Treatment with the MAPK kinase inhibitor UO 126 disrupted SIMK localization, and distorted root hairs so that they lost their tip organization and became bulbous, suggesting that regulation of SIMK by an unidentified MAPK kinase is important for root hair growth. Crucially, overexpression of a gain-of-function SIMK in tobacco produced a shorter differentiation zone and longer root hairs, and could overcome UO 126 inhibition, confirming that regulation of SIMK by an MAPK kinase can control root hair length. Mechanisms that control the specificity of MAP kinase cascades are not understood, but it is tempting to speculate that there could be a link between auxin signaling, ethylene perception and the activity of MAP kinases such as SIMK. It is also not clear how SIMK contributes to root hair tip growth, although it is clearly associated with polymerized actin.

ROP small GTPases have a strong effect on the direction and amount of root hair tip growth. Plants expressing too much ROP2 have very long root hairs that repeatedly branch (Fig. 7.5 (B), Jones *et al.*, 2002), suggesting that the amount of ROP2 can be limiting for root hair growth. Plants overexpressing CArop have partly or completely (spherical) depolarized root hairs, suggesting that ROP must be able to cycle from the GTP-bound form to the GDP-bound form for the direction of tip growth to be controlled (Fig. 7.5 (C), Molendijk *et al.*, 2001; Jones *et al.*, 2002). Plants expressing DNrop2 have fewer, shorter root hairs that are sometimes wavy. The latter phenotype is reminiscent of root hairs grown in the presence of microtubule-disrupting drugs (Figs 7.4 (H), (I) and 7.5 (D), Bibikova *et al.*, 1999). Further work is required to establish whether ROP2 affects root hair microtubules, or produces this phenotype by a microtubule-independent mechanism (Jones *et al.*, 2002). There is evidence for a role of ROP2 in actin organization during tip growth. Jones *et al.* (2002) used GFP-mouse talin to visualize F-actin in living root hairs and showed that hairs expressing DNrop2 had less fine F-actin than wild type hairs, whereas hairs expressing CArop2 contained a dense network of actin. The mechanisms linking ROP2 to actin organization in root hairs are not known, but it is likely that actin-binding proteins are involved.

Profilin is a G-actin-binding protein (see Chapter 2). When G-actin is bound to profilin, nucleation and incorporation at the pointed end of F-actin is inhibited, whereas incorporation at the barbed ends continues at the normal rate (Pollard *et al.*, 2000). PFN-1 encodes one of four *Arabidopsis* profilins, and is expressed in root hairs. Transgenic plants overexpressing PFN-1 have root hairs that are twice as long as wild type hairs (Ramachandran *et al.*, 2000). Although some papers report an accumulation of profilin in root hair apices (Braun *et al.*, 1999; Baluska *et al.*, 2000), this localization is possibly caused by fixation artifacts, and profilin is likely to be distributed evenly through the cytoplasm in root hairs, as has been reported in pollen tubes (Vidali & Hepler, 1997; Hepler *et al.*, 2001).

ADF/cofilin binds both to G- and F-actin, it enhances depolymerization at the pointed end of F-actin (Carlier *et al.*, 1997) and has the ability to sever F-actin at high pH (8.0), whereas it binds F-actin at lower pH 6.0 (Gungabissoon *et al.*, 1998; see Chapter 2). ZmADF3 has been identified in root hairs (Jiang *et al.*, 1997) and its activity has been extensively characterized *in vitro*. ZmADF3 activity is inhibited by phosphorylation of Ser-6 (Smertenko *et al.*, 1998), and by binding to phosphatidylinositol 4,5-bisphosphate (PIP_2) or phosphatidylinositol 4-monophosphate (PIP) (Gungabissoon *et al.*, 1998). Phosphorylation of Ser-6 is regulated by a calmodulin-like domain protein kinase (CDPK) (Smertenko *et al.*, 1998; Allwood *et al.*, 2001). ZmADF3 enhances the actin-bundling activity of Zm elongation factor 1α *in vitro* (Gungabissoon *et al.*, 2001). The results described above are consistent with a model where ADF is deactivated in the apical area of a growing root hair, where the concentration of free cytoplasmic calcium is relatively high and CDPK is activated. In the alkaline subapical zone in root hairs, ADF would sever F-actin, which creates many free barbed ends, where polymerization can take place. The presence of profilin bound G-actin enables rapid incorporation of G-actin at the generated barbed ends. The regulation of actin dynamics by ADF and profilin in the subapical area may lead to the formation of fine F-actin (FB-actin). Further toward the root hair base, villin and/or ZmEF-1α mediated bundling of the actin filaments takes place, which creates the thick bundles of F-actin in the root hair tube.

7.3.4 Mechanisms acting at the end of tip growth

When root hairs stop growing, ROP protein (Molendijk *et al.*, 2001; Jones *et al.*, 2002), the calcium gradient (Wymer *et al.*, 1997), and calcium channel activity (Scheifelbein *et al.*, 1992; Very & Davies, 2000) are lost from the tip. The *Arabidopsis* genes *HY5* and *IRE* are involved in controlling the end of tip growth (see Section 7.3.3) by an unknown mechanism (Oyama *et al.*, 2002).

7.4 The genetic network controlling root hair morphogenesis in *Arabidopsis*

This section summarizes the large number of *Arabidopsis* genes that affect root hair production, but whose molecular functions are not yet clear, or have not been linked to the cytoskeleton. Tables 7.1, 7.2, and 7.3 list *Arabidopsis* genes with known roles in root hair development, and Fig. 7.6 illustrates the phenotypes of mutations affecting root hair morphogenesis. Fig. 7.7 summarizes the genetic pathways that control root hair patterning, initiation and tip growth. These genes are discussed in the order in which they contribute to hair production.

7.4.1 Genes involved in root hair patterning

Apart from *SCZ*, *WER*, *CPC*, *GL2*, *ERH3*, and *RHD6*, whose functions are discussed in Section 7.3.1, several other genes listed in Table 7.1 also affect the fate of epidermal cells, but no links between these genes and the cytoskeleton have been reported. In current models of root hair initiation, positional signals act through *GL2* and *RHD6*. Environmental cues, and information from other parts of the plant, act after *RHD6* to fine tune root hair initiation to the prevailing conditions. Some of this information is transmitted via auxin and/or ethylene signaling (Fig. 7.7 (A)).

7.4.2 Genes affecting initiation

Numerous genes affect the position, number, or size of the swelling that forms during root hair initiation. As swelling formation involves actin accumulation (see

Table 7.1 *Arabidopsis* genes controlling root epidermal cell fate

Locus	Gene product	Chromosomal location	Mutant phenotype	References
SCHIZORHIZA (*SCZ*)	Unknown	I	Ground tissue produces root hairs	Mylona *et al.*, 2002
CAPRICE (*CPC*)	MYB protein	II	Reduced number of root hairs	Wada *et al.*, 1997, 2002; Lee & Schiefelbein, 2002
ECTOPIC ROOT HAIR1 (*ERH1*)	Unknown	V	Ectopic root hairs	Schneider *et al.*, 1997
ECTOPIC ROOT HAIR3 (*ERH3*)	Katanin p60	I	Ectopic root hairs	Schneider *et al.*, 1997 Webb *et al.*, 2002
GLABRA2 (*GL2*)	Homeodomain protein	I	Ectopic root hairs	DiCristina *et al.*, 1996; Masucci *et al.*, 1996
POMPOM (*POM1*)	Unknown	I	Ectopic root hairs	Schneider *et al.*, 1997
ROOTHAIRLESS1 (*RHL1*)	Novel nuclear protein	I	Reduced number of root hairs	Schneider *et al.*, 1998
ROOTHAIRLESS2 (*RHL2*)	Unknown	V	Reduced number of root hairs	Schneider *et al.*, 1997
ROOTHAIRLESS3 (*RHL3*)	Unknown	III	Reduced number of root hairs	Schneider *et al.*, 1997
TRANSPARENT TESTA GLABRA (*TTG*)	WD repeat protein	V	Ectopic root hairs	Galway *et al.*, 1994; Walker *et al.*, 1999
WEREWOLF(*WER*)	MYB protein	V	Ectopic root hairs	Lee & Schiefelbein, 1999

Section 7.2.2.1) these genes must directly or indirectly affect actin organization in hair cells. *RHD6* affects the frequency and location of root hair initiation (Masucci & Schiefelbein, 1994); hair cells on roots mutated in the *RHD6* gene are usually hairless, but when hairs do form they are in the wrong place on the cell. Mutants in the *TINY ROOT HAIR1* (*TRH1*) gene sometimes have extra swellings (Rigas *et al.*, 2001), suggesting that *TRH1* prevents multiple swellings from forming on wild type hair cells. *TRH1* encodes a predicted potassium transporter (Rigas *et al.*, 2001). Further research is required to discover how this affects swelling number. Combining loss-of-function mutations in the *TIP1* and *CENTIPEDE2* (*CEN2*), *CEN2* and *SHV3*, or *SUPERCENTIPEDE1* (*SCN1*) and *RHD3* gene pairs prevents root hair formation (Parker *et al.*, 2000). The molecular functions of most of these genes are not known, but SCN1 has recently been identified as a predicted regulator of ROP activity (O'Sullivan *et al.*, unpublished), and work is in progress to test whether root hair initiation involves the regulation of ROP2 by SCN1.

Table 7.2 *Arabidopsis* genes controlling root hair initiation and tip growth

Locus	Gene product	Chromosomal location	Mutant phenotype	References
BRISTLED1 (BST1)	Unknown	V	Short hairs, sometimes branched	Parker *et al.*, 2000
CAN OF WORMS1 (COW1)	Unknown	IV	Short, wide hairs, sometimes branched at base	Grierson *et al.*, 1997; Parker *et al.*, 2000
CENTIPEDE1 (CEN1)	Unknown	I	Short, wide hairs, sometimes curled	Parker *et al.*, 2000
CENTIPEDE2 (CEN2)	Unknown	V	Short, wide hairs, sometimes branched and/or curled	Parker *et al.*, 2000
CENTIPEDE3 (CEN3)	Unknown	III	Short, wide hairs, some with wide bases, some hairs curled and/or branched	Parker *et al.*, 2000
KEULE	Sec1 protein	I	Hairs absent or stunted and swollen	Assaad *et al.*, 2001; Sollner *et al.*, 2002
CLUB	Unknown	V	Hairs stunted and swollen	Assaad *et al.*, 2001; Sollner *et al.*, 2002
KOJAK (KJK/AtCSLD3)	Cell wall synthase	III	Hairs burst after swellings form	Favery *et al.*, 2001
LRX1	Leucine-rich repeat/extensin	I	Hairs short, swollen, or branched	Baumberger *et al.*, 2001
PFN1	Profilin	–	Long hairs	Ramachandran *et al.*, 2000
PHYA	Phytochrome A	I	Altered hair length	De Simone *et al.*, 2000
PHYB	Phytochrome B	II	Altered hair length	Reed *et al.*, 1993; De Simone *et al.*, 2000
ROOT HAIR DEFECTIVE1 (RHD1)	Unknown	I	Wide swellings	Schiefelbein & Somerville, 1990
ROOT HAIR DEFECTIVE2 (RHD2)	Unknown	V	Hairs stop growing after swellings form	Schiefelbein & Somerville, 1990
ROOT HAIR DEFECTIVE3 (RHD3)	GTP-binding protein	III	Short, wavy hairs, sometimes branched	Schiefelbein & Somerville, 1990; Galway, *et al.*, 1997; Wang *et al.*, 1997
ROOT HAIR DEFECTIVE4 (RHD4)	Unknown	III	Short hairs with bulges and constrictions, sometimes branched	Schiefelbein & Somerville, 1990; Galway, *et al.*, 1999
ROOT HAIR DEFECTIVE6 (RHD6)	Unknown	I	Reduced number of root hairs; site of hair formation closer to basal end of cell; some cells with multiple hairs	Masucci & Schiefelbein, 1994

Table 7.2 (continued)

Locus	Gene product	Chromosomal location	Mutant phenotype	References
SHAVEN1 (*SHV1*)	Unknown	III	Hairs stop growing after swellings form	Parker *et al*., 2000
SHAVEN2 (*SHV2*)	Unknown	V	Hairs stop growing after swellings form	Parker *et al*., 2000
SHAVEN3 (*SHV3*)	Unknown	IV	Hairs stop growing after swellings form	Parker *et al*., 2000
SUPERCENTIPEDE (*SCN1*)	Unknown	III	Short, wide hairs, sometimes branched	Parker *et al*., 2000
TINY ROOT HAIR 1 (*TRH1*)	Potassium transporter	IV	Hairs stop growing after swellings form; some cells with multiple swellings	Rigas *et al*., 2001
TIP1	Unknown	V	Wide swellings, short, wide hairs, sometimes branched at base	Schiefelbein *et al*., 1993; Ryan *et al*., 1998
SOS4	Pyridoxal kinase	V	Reduced number of root hairs, impaired tip growth	Shi & Zhu, 2002

Table 7.3 *Arabidopsis* genes with roles in auxin or ethylene signaling that affect root hair development

Locus	Gene product	Chromosomal location	Mutant phenotype	References
AUX1	Auxin influx carrier	II	Site of hair formation closer to basal end of cell, hairs short, sometimes branched at base	Cernac *et al.*, 1997; Pitts *et al.*, 1998; Grebe *et al.*, 2002
AUXIN RESISTANT 1 (AXR1)	Subunit of RUB 1-activating enzyme	I	Short hairs, sometimes branched at base; some hairs stop growing after swellings form	Cernac *et al.*, 1997; Pitts *et al.*, 1998
AUXIN RESISTANT 2 (AXR2/IAA7)	AUX/IAA protein	III	Reduced number of root hairs; site of hair formation closer to basal end of cell	Masucci & Schiefelbein, 1994; Bates & Lynch, 1996; Nagpal *et al.*, 2000
AUXIN RESISTANT 3 (AXR3/IAA17)	AUX/IAA protein	I	Gain-of-function mutants have a reduced number of root hairs	Leyser *et al.*, 1996; Knox *et al.* (in press)
CONSTITUTIVE TRIPLE RESPONSE 1 (CTR1)	Raf-like protein kinase	V	Dominant mutants have ectopic root hairs	Kieber *et al.*, 1993
SHY2/IAA3	AUX/IAA protein	I	Gain-of-function mutants have longer root hairs that start to develop nearer the root tip; loss-of-function mutants have fewer root hairs that are slightly short	Knox *et al.* (in press)
ETHYLENE INSENSITIVE 2 (EIN2)	Unknown	V	Short hairs	Masucci & Schiefelbein, 1994; Pitts *et al.*, 1998
ETHYLENE OVERPRODUCER 1 (ETO1)	Unknown	V	Long hairs; site of hair formation closer to apical end of cell	Pitts *et al.*, 1998
ETHYLENE RESPONSE 1 (ETR1/EIN1)	Histidine kinase	I	Short hairs; site of hair formation closer to basal end of cell	Masucci & Schiefelbein, 1994; Pitts *et al.*, 1998
ETHYLENE RESPONSE 1 (ETR1/EIN1)	Putative ethylene receptor	I	Short hairs; site of hair formation closer to basal end of cell	Masucci & Schiefelbein, 1994; Pitts *et al.*, 1998

Table 7.3 (continued)

Locus	Gene product	Chromosomal location	Mutant phenotype	References
ETHYLENE RESPONSE SENSOR 1 (ERS1)	Putative ethylene receptor	II	Short hairs	Cho & Cosgrove, 2002
ETHYLENE RESPONSE SENSOR 2 (ERS2)	Putative ethylene receptor	I	Short hairs	Cho & Cosgrove, 2002
ETHYLENE RESPONSE 2 (ETR2)	Putative ethylene receptor	III	Short hairs	Cho & Cosgrove, 2002
SUPPRESSOR OF AUXIN RESISTANCE 1 (SAR1)	Synaptobrevin-related protein	II	Hairs have swollen ends	Cernac *et al.*, 1997; Pitts *et al.*, 1998

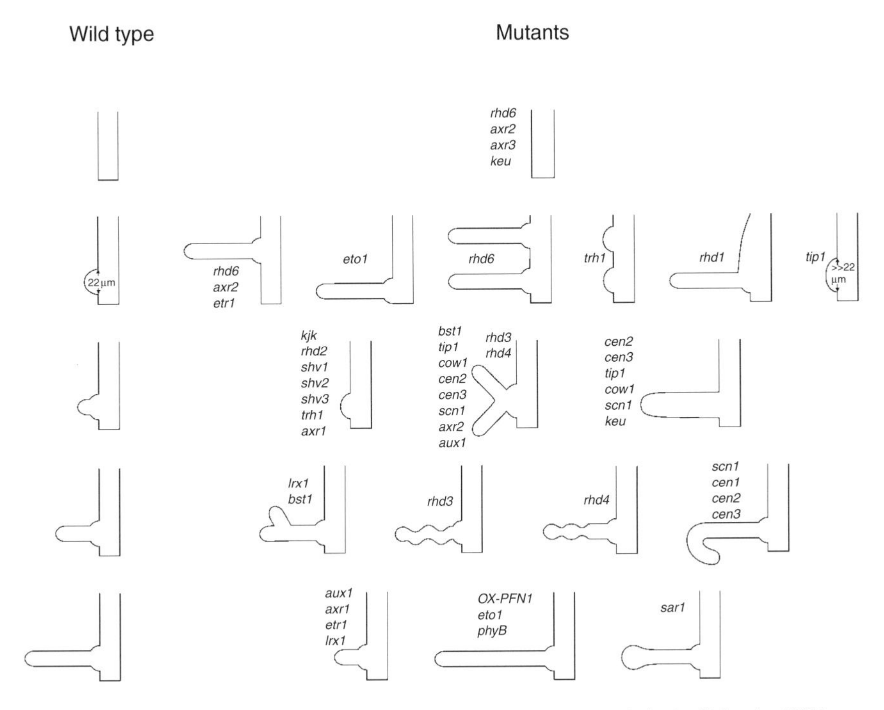

Fig. 7.6 Phenotypes of genes affecting root hair initiation and tip growth in *Arabidopsis*. Wild type development is shown on the left, with mutants affecting each stage of development on the right. Mutants are drawn more than once when they affect multiple stages of development. Not to scale.

In wild type *Arabidopsis* the amount of cell wall loosening during swelling formation is highly reproducible (Parker *et al.*, 2000). *tip1* and *root hair defective1* (*rhd1*) mutant plants have large swellings. In *tip1* mutants, swelling diameter is increased by about a third. *rhd1* mutants have huge swellings that take up most of the outer surface of the hair cell (Fig. 7.6). As *tip1* and *rhd1* are both loss-of-function mutants, these results suggest that the *RHD1* and *TIP1* genes restrict swelling size. They presumably do this by restricting the area of wall that is loosened (Schiefelbein & Somerville, 1990; Ryan *et al.*,1998; Parker *et al.*, 2000). RHD1 encodes an isoform of UDP-D-glucose 4-epimerase, an enzyme that acts in the formation of UDP-D-galactose. The RHD1 isoform is specifically required for the galactosylation of xyloglucan and type II arabinogalactan (AGII). Xyloglucan endotransglycosylases (XETs) are active during root hair initiation, but no link between RHD1 and XET action has been reported. *tip1 rhd1* double mutants have similar swelling dimensions to *rhd1* single mutants, suggesting that the *TIP1* gene cannot affect swelling size unless the *RHD1* gene product is present (Parker *et al.*, 2000).

lrx1 loss-of-function mutants have root hairs with swollen bases, suggesting that LRX1 is required for proper root hair initiation (Baumberger *et al.*, 2001).

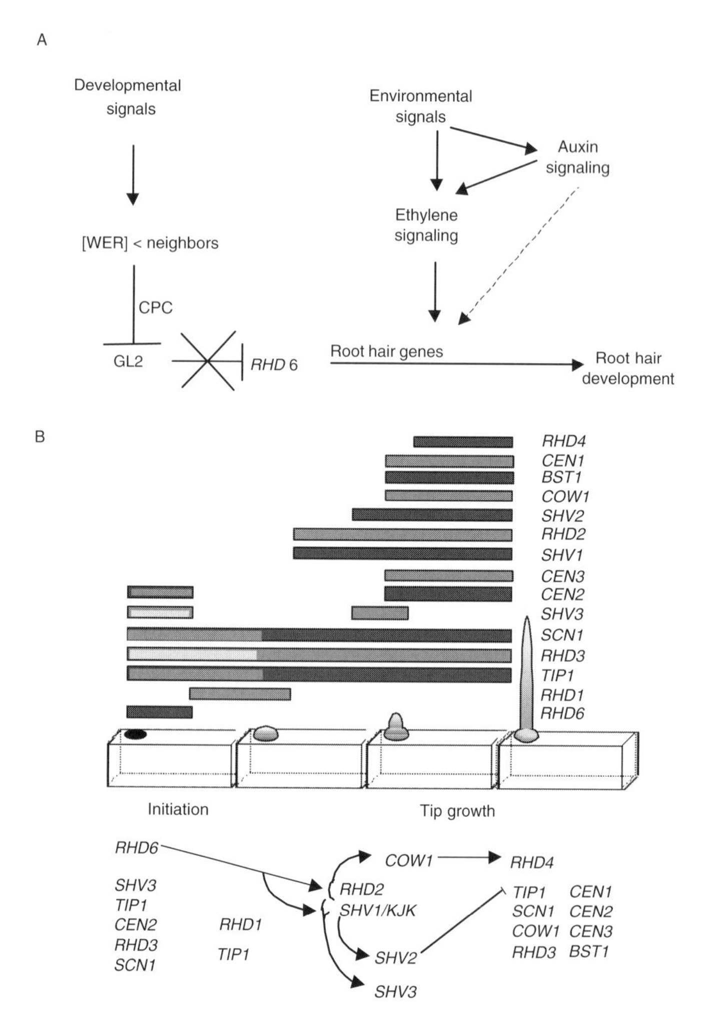

Fig. 7.7 Overview of genetic mechanisms controlling root hair formation. (A) Mechanisms controlling root hair patterning and initiation. Arrows show the sequence of events, with blunt-ended lines to represent repressive mechanisms. Root epidermal cells produce different amounts of WEREWOLF (WER) transcription factor in response to positional signals. In cells with less WER than their neighbors, CAPRICE (CPC) inhibits the expression of the *GLABRA 2* (*GL2*) gene. GL2 is a repressor of root hair formation, so CPC de-represses root hair development in these cells. Environmental signals, some acting through auxin and ethylene signaling, tailor the final number of root hairs to suit the prevailing growth conditions. (B) Genes involved in root hair growth. Stages of root hair development are shown with horizontal bars above to show the stages of root hair development when each gene contributes. Striped regions indicate stages that are not affected in single mutants, but are defective in some double mutant combinations (Parker *et al.*, 2000). At the bottom of the figure is a summary of the genetic pathway controlling root hair growth. Genes appear more than once if they affect more than one stage of growth. Arrows indicate epistatic interactions, with the gene before an arrow epistatic to the gene after the arrow. *SHV2* is a hypothetical negative regulator of *TIP1*, and this is shown by a blunt-ended line. Auxin and ethylene signaling affects the final length of root hairs, but this is not shown in this figure.

LRX1 encodes a protein with leucine-rich repeats and homology to extensins, but the mechanism of LRX1 action has not yet been determined.

7.4.3 Genes required for tip growth to be established

Tip growth is established by the time *Arabidopsis* root hairs reach 40 μm long (Dolan *et al.*, 1994). Root hairs without functional *ROOTHAIRDEFECTIVE2* (*RHD2*), *SHAVEN1/KOJAK* (*SHV1/KJK*), *SHV2*, *SHV3*, or *TRH1* genes usually stop growing before this stage (Fig. 7.6, Schiefelbein & Somerville, 1990; Parker *et al.*, 2000; Favery *et al.*, 2001; Rigas *et al.*, 2001). Mutations that damage the *CEN1*, *CEN2*, *CEN3*, *BRISTLED1*(*BST1*), *TIP1*, and *SCN1* genes can also stop hair growth before this stage, but only in certain double mutant combinations (Parker *et al.*, 2000). These results suggest that all of these genes are important for tip growth to be successfully established. As tip growth involves both the microtubular and actin cytoskeletons (see Sections 7.2.1 and 7.2.2), at least some of these genes are likely to affect cytoskeletal function.

There are at least two ways that growth can fail during the transition from swelling formation to tip growth. In the first case, wall loosening may not stop, or the balance between wall deposition and protoplast growth may fail so that the hair bursts. For example, the hairs of *kjk* mutants burst after swelling formation, killing the cells (Favery *et al.*, 2001; Wang *et al.*, 2001). *KJK* encodes an enzyme that synthesizes cell wall components for export to the growing tip. KJK resembles a cellulose synthase, but cellulose is formed at the plasma membrane whereas KJK is found in the endoplasmic reticulum. KJK probably contributes to the synthesis of polysaccharides such as beta-xylans, mannans or xyloglucan (Favery *et al.*, 2001). *kjk* mutant hairs have weak cell walls that cannot contain the growing protoplast and burst. The second way that growth can fail is if a crucial part of the machinery for tip growth is not functioning. For example, *trh1* mutants do not burst, but are unable to grow (Liam Dolan, personal communication). TRH1 is a potassium transporter that is specifically required for tip growth. *trh1* mutant roots have other functional potassium transporters and supplementing *trh1* mutant roots with high levels of potassium does not restore tip growth. These results suggest that tip growth depends on potassium transport that is precisely localized within the cell or closely coordinated with other events. This specialized potassium transport is carried out by the TRH1 protein and without it growth stops after swelling formation (Rigas *et al.*, 2001).

LRX1 localizes to the cell wall at the apex of tip-growing hairs. *lrx1* loss-of-function mutants have stunted and branched root hairs showing that LRX1 affects the amount and location of tip growth (Baumberger *et al.*, 2001). The cytoskeleton presumably plays a role in targeting LRX1 to the root hair tip by targeted secretion, but this has not yet been investigated.

SCN1 is unique among root hair genes in that *scn1* loss-of-function mutations interact strongly with mutations in every other root hair gene tested (Table 7.4,

Table 7.4 Genetic interactions between SCN1 and other root hair morphogenesis genes (Parker *et al.*, 2000)

Phenotype partly suppressed by *scn1*	Phenotype enhances *scn1*	Epistatic to *scn1*	Simply additive with *scn1* (i.e. genetically independent)
shv1, *shv2*	*rhd3*, *cen3*, *cow1*, *tip1*, *shv3*, *bst1*, *cen1*, *cen2*	*rhd6*	None

Parker *et al.*, 2000). This suggests that *SCN1* is at the heart of mechanisms controlling root hair growth. *SCN1* encodes a predicted regulator of ROP activity (O'Sullivan *et al.*, unpublished), and there are similarities between the *scn1* phenotype and the phenotypes of ROP2 transgenic lines, consistent with the idea that SCN1 regulates ROP2 throughout root hair growth (Jones & Grierson, unpublished). The molecular identities of genes that interact strongly with *SCN1* are being elucidated. For example, *SHV1*, whose loss-of-function phenotype is partly suppressed by *scn1-1*, is allelic with *KJK* (O'Sullivan *et al.*, unpublished). As more genes involved in root hair development are identified, the mechanisms linking and coordinating different aspects of root hair tip growth, including ROP, the cytoskeleton, and cell wall synthesis, should be revealed.

7.4.4 Genes required to sustain and direct tip growth

Plants carrying loss-of-function mutations in the *RHD2*, *SHV1/KJK*, *SHV2*, and *SHV3* genes occasionally make tip-growing hairs. In all of these cases, the hairs are short and deformed showing that all of these genes are required for normal tip growth (Parker *et al.*, 2000; Favery *et al.*, 2001). *bst1*, *cen1*, *cen2*, *cen3*, *cow1*, *rhd3*, *scn1*, *tip1*, and *rhd4*, mutants all have short root hairs so all of these genes are required for root hairs to achieve their usual length (Parker *et al.*, 2000).

Several genes affect the shape of hairs in a way that suggests that they control the number or location of vesicles that fuse at the growing tip, and hence influence the functions of the actin and/or microtubule cytoskeletons. *scn1*, *cen1*, *cen2*, and *cen3* hairs are often curved, showing that these genes are required to keep the elongating tube straight. In the case of *cen2*, hairs are curved even when grown through a semi-solid medium (Jones & Grierson, unpublished), confirming that they curve as they grow rather than bending after growth has ceased. These results suggest that the *CEN2* gene is required to repeatedly target vesicles accurately to the center of the growing tip, and that in *cen2* mutants, targeting becomes misplaced repeatedly to one side so that the hair curls as it grows. Therefore CEN2 is likely to be a molecular link between polarity and root hair growth. *rhd2-2*, *scn1*, *tip1* and *cow1* all have wide hairs (Parker *et al.*, 2000), suggesting that the *RHD2*, *SCN1*, *TIP1* and *COW1* genes somehow restrict the area at the tip of the hair where vesicles can fuse. These genes are therefore likely to encode proteins that regulate the actin cytoskeleton, or are involved in secretion that is targeted by the actin cytoskeleton. As *SCN1* encodes a predicted regulator of ROP activity (O'Sullivan *et al.*, unpublished), work is in progress to test how SCN1, ROP2, and actin might interact during root hair tip growth.

rhd3 mutant hairs are corkscrew shaped because vesicle fusion occurs at a point that rotates around the edges of the growing tip rather than being focused in the center (Galway *et al.*, 1997). This wavy phenotype is reminiscent of the effects of mutants and drug treatments affecting microtubules (Fig. 7.4), and of DNrop2 (Fig. 7.5D), so RHD3 is a candidate for a protein affecting microtubule structure, linking microtubule structure to targeted secretion, or affecting ROP function. *RHD3* encodes a protein with GTP-binding domains whose molecular function is unknown (Wang *et al.*, 1997). *rhd4* mutant hairs have inconsistent diameters and patches of thick cell wall, suggesting that the amount of material that is deposited varies as the tube grows, producing local constrictions and expansions along the length of the hair (Galway *et al.*, 1999). The *rhd4* phenotype is strongly reminiscent of the effects of actin disruption on root hair growth (Fig. 7.4, Ketelaar *et al.*, 2002), suggesting that *RHD4* could affect actin dynamics. Mutations in the *COW1* gene reduce the elongation rate of root hairs (Grierson *et al.*, 1997). The molecular identity of *COW1* has not been reported.

The Sec1 protein KEULE is required for normal root hair development. Loss-of-function *keule* mutants have absent or stunted, radially swollen root hairs. It is not clear whether keule effects root hair initiation, tip growth, or both. Sec1 proteins in other cell types are involved in mechanisms controlling vesicle targeting and vesicle fusion, so it is likely that KEULE contributes to root hair development by facilitating targeted vesicle fusion. *CLUB* is another gene with a similar set of mutant phenotypes to *KEULE*. *club* mutants also have defective root hairs, and it is likely that this is also due to problems with targeted secretion (Assaad *et al.*, 2001; Sollner *et al.*, 2002).

7.4.5 Genes involved in nuclear movement

The position of nuclei in *cen3* mutants is randomized (Grierson *et al.*, unpublished), suggesting that the function of actin in positioning the nucleus requires the CEN3 protein.

7.4.6 Genes with roles at the end of tip growth

One mutant affects coordination of events at the end of tip growth; *sar1* hairs that have stopped growing have fat tips. SAR1 acts downstream of *AXR1* in the auxin response, so this phenotype suggests that auxin signaling plays a coordinating role at the end of root hair growth (Cernac *et al.*, 1997). *SAR1* encodes a protein of unknown function that has sequence similarity to synaptobrevin (see http://www.*arabidopsis*.org/).

7.5 Concluding remarks

Root hair development relies heavily on the microtubule and actin cytoskeletons, from cell-fate determination, through to the end of tip growth. There are still many important questions to address. Is the cytoskeleton involved in mechanisms that set up, maintain, or interpret root and cell polarity? How do auxin and/or ethylene signaling impinge on the root hair cytoskeleton, for example to affect root hair length? Does all locally targeted growth in root hair cells involve actin, or is root hair initiation actin-independent? During tip growth, which is relatively well understood, we still need to find out how microtubules impinge on actin dynamics to focus root hair tip growth, and how the actin (and in some species the microtubule) cytoskeleton(s) position nuclei. Many aspects of ROP function remain to be elucidated. For example, molecules linking ROP activity to changes in the actin cytoskeleton need to be identified.

The combination of genetics and cell biology in *Arabidopsis* root hairs promises rapid progress in root hair research. We are beginning to identify and characterize genes that regulate the root hair cytoskeleton, and coordinate cytoskeletal function with other aspects of root hair differentiation and development. Research in legumes is also revealing how root hair growth can be manipulated by other organisms. It is an exciting time to be a root hair cell biologist.

Acknowledgements

CG would like to thank Mark Jones, who contributed images for Fig. 7.5, Alison Kemp and Piers Hemsley, who commented on the manuscript, John Schiefelbein, who contributed to some of the tables, and past and present members of CGs research team who contributed research that is described here. We would also like to thank The Royal Society (CG), the BBSRC (CG and TK), and the University of Bristol (CG) for financial support.

References

Allwood, E.G., Smertenko, A.P. & Hussey, P.J. (2001) Phosphorylation of plant actin-depolymerising factor by calmodulin-like domain protein kinase. *FEBS Lett.*, **499**, 97–100.

An, Y.Q., McDowell, J.M., Huang, S.R., McKinney, E.C., Chambliss, S. & Meagher, R.B. (1996) Strong, constitutive expression of the *Arabidopsis* ACT2/ACT8 actin subclass in vegetative tissues. *Plant J.*, **10**, 107–121.

Assaad, F.F., Huet, Y., Mayer, U. & Jurgens, G. (2001) The cytokinesis gene *KEULE* encodes a Sec1 protein that binds the syntaxin KNOLLE. *J. Cell Biol.*, **152**, 531–543.

Baluska, F., Salaj, J., Mathur, J., *et al.* (2000) Root hair formation: F-actin-dependent tip growth is initiated by local assembly of profilin-supported F-actin meshworks accumulated within expansin-enriched bulges. *Dev. Biol.*, **227**, 618–632.

Bao, Y.Q., Kost, B. & Chua, N.H. (2001) Reduced expression of α-tubulin genes in *Arabidopsis thaliana* specifically affects root growth and morphology, root hair development and root gravitropism. *Plant J.*, **28**, 145–157.

Bates, T.R. & Lynch, J.P. (1996) Stimulation of root hair elongation in *Arabidopsis thaliana* by low phosphorus availability. *Plant Cell Environ.*, **19**, 529–538. *Plant J.*, **28**, 145–157.

Bates, T.R. & Lynch, J.P. (2000a) Plant growth and phosphorus accumulation of wild type and two root hair mutants of *Arabidopsis thaliana* (Brassicaceae). *Am. J. Bot.*, **87**, 958–963.

Bates, T.R. & Lynch, J.P. (2000b) The efficiency of *Arabidopsis thaliana* (Brassicaceae) root hairs in phosphorus acquisition. *Am. J. Bot.*, **87**, 964–970.

Baumberger, N., Ringli, C. & Keller, B. (2001) The chimeric leucine-rich repeat/extensin cell wall protein LRX1 is required for root hair morphogenesis in *Arabidopsis thaliana*. *Genes Dev.*, **15**, 1128–1139.

Bibikova, T.N., Blancaflor, E.B. & Gilroy, S. (1999) Microtubules regulate tip growth and orientation in root hairs of *Arabidopsis thaliana*. *Plant J.*, **17**, 657–665.

Bibikova, T.N., Jacob, T., Dahse, I. & Gilroy, S. (1998) Localized changes in apoplastic and cytoplasmic pH are associated with root hair development in *Arabidopsis thaliana*. *Development*, **125**, 2925–2934.

Braun, M., Baluska, F., Von Witsch, M. & Menzel, D. (1999) Redistribution of actin, profilin and phosphatidylinositol-4,5-bisphosphate in growing and maturing root hairs. *Planta*, **209**, 435–443.

Cao, X.F., Linstead, P., Berger, F., Kieber, J. & Dolan, L. (1999) Differential ethylene sensitivity of epidermal cells is involved in the establishment of cell pattern in the *Arabidopsis* root. *Physiol. Plant.*, **106**, 311–317.

Cárdenas, L., Vidali, L., Domínguez, J., *et al.* (1998) Rearrangement of actin microfilaments in plant root hairs responding to Rhizobium etli nodulation signals. *Plant Physiol.*, **116**, 871–877.

Carlier, M.F., Laurent, V., Santolini, J., *et al.* (1997) Actin depolymerizing factor (ADF/cofilin) enhances the rate of filament turnover: Implication in actin-based motility. *J. Cell Biol.*, **136**, 1307–1322.

Cernac, A., Lincoln, C., Lammer, D. & Estelle, M. (1997) The *SAR1* gene of *Arabidopsis* acts downstream of the AXR1 gene in auxin response. *Development*, **124**, 1583–1591.

Chant, T. (1999) Cell polarity in yeast. *Ann. Rev. Cell Dev. Biol.*, **15**, 365–391.

Cho, H.T. & Cosgrove, D.J. (2002) Regulation of root hair initiation and expansin gene expression in *Arabidopsis*. *Plant Cell*, **14**(12), 3237–3253.

De Ruijter, N.C.A., Rook, M.B., Bisseling, T. & Emons, A.M.C. (1998) Lipochito-oligosaccharides reinitiate root hair tip growth in *Vicia sativa* with high calcium and spectrin-like antigen at the tip. *Plant J.*, **13**, 341–350.

De Ruijter, N.C.A. & Emons, A.M.C. (1999) Actin-binding proteins in plant cells. *Plant Biol.*, **1**, 26–35.

De Ruijter, N.C.A., Bisseling T. & Emons A.M.C. (1999) *Rhizobium* nod factors induce an increase in sub-apical fine bundles of actin filaments in *Vicia sativa* root hairs within minutes. *Mol. Plant Microbe Interactions*, **12**, 829–832.

De Simone, S., Oka, Y. & Inoue, Y. (2000) Effect of light on root hair formation in *Arabidopsis thaliana* phytochrome-deficient mutants. *J. Plant Res.*, **113**, 63–69.

Derksen, J. & Emons, A.M.C. (1990) Microtubules in tip growth systems, in *Tip Growth in Plant and Fungal Cells* (ed. I.B. Heath), San Diego, Academic Press, pp. 147–181.

DiCristina, M.D., Sessa, G., Dolan, L., *et al.* (1996) The *Arabidopsis* Athb-10 (GLABRA2) is an HD-Zip protein required for regulation of root hair development. *Plant J.*, **10**, 393–402.

Dolan, L., Duckett, C., Grierson, C., *et al.* (1994) Clonal relations and patterning in the root epidermis of *Arabidopsis*. *Development*, **120**, 2465–2474.

Eaton, S., Wepf, R. & Simons, K. (1996) Roles for Rac1 and Cdc42 in planar polarization and hair outgrowth in the wing of *Drosophila*. *J. Cell Biol.*, **135**, 1277–1289.

Emons, A.M.C. (1987) The cytoskeleton and secretory vesicles in root hairs of *Equisetum* and *Limnobium* and cytoplasmic streaming in root hairs of *Equisetum*. *Annals of Botany*, **60**, 625–632.

Emons, A.M.C. (1989) Helicoidal microfibril deposition in a tip-growing cell and microtubule alignment during tip morphogenesis: a dry-cleaving and freeze-substitution study. *Canadian J. Botany*, **67**, 2401–2408.

Esseling, J., De Ruijter, N. & Emons, A.M.C. (2000) The root hair cytoskeleton as backbone, highway, morphogenetic instrument and target for signalling, in *Root Hairs. Cell and Molecular Biology* (eds R.W. Ridge & A.M.C. Emons). Tokyo Berlin Heidelberg, New York: Springer-Verlag, pp. 29–52.

Favery, B., Ryan, E., Foreman J., *et al.* (2001) *KOJAK* encodes a cellulose synthase-like protein required for root hair cell morphogenesis in *Arabidopsis*. *Genes Dev.*, **15**, 79–89.

Galway, M.E., Masucci, J.D., Lloyd, A.M., Walbot, V., Davis, R.W. & Schiefelbein, J.W. (1994) The TTG gene is required to specify epidermal cell fate and cell patterning in the *Arabidopsis* root. *Dev. Biol.*, **166**, 740–754.

Galway, M.E., Heckman, J.W. & Schiefelbein, J.W. (1997) Growth and ultrastructure of *Arabidopsis* root hairs: the *rhd3* mutation alters vacuole enlargement and tip growth. *Planta*, **201**, 209–218.

Galway, M.E., Lane, D.C. & Schiefelbein, J.W. (1999) Defective control of growth rate and cell diameter in tip-growing root hairs of the *rhd4* mutant of *Arabidopsis thaliana*. *Can. J. Bot.*, **77**, 494–507.

Gilliland, L.U., Kandasamy, M.K., Pawloski, L.C. & Meagher, R.B. (2002) Both vegetative and reproductive actin isovariants complement the stunted root hair phenotype of the *Arabidopsis act2-1* mutation. *Plant Physiol.*, **130**, 2199–2209.

Grebe, M., Friml, J., Swarup, R., *et al.* (2002) Cell polarity signaling in *Arabidopsis* involves a BFA-sensitive auxin influx pathway. *Curr. Biol.*, **12**, 329–334.

Grierson, C.S., Roberts, K., Feldmann, K.A. & Dolan, L. (1997) The COW1 locus of *Arabidopsis* acts after RHD2, and in parallel with RHD3 and TIP1, to determine the shape, rate of elongation, and number of root hairs produced from each site of hair formation. *Plant Physiol.*, **115**, 981–990.

Gungabissoon, R.A., Jiang, C.-J., Drobak, B.J., Maciver, S.K. & Hussey, P.J. (1998) Interaction of maize actin-depolymerizing factor with F-actin and phosphoinositides and its inhibition of plant phospholipase C. *Plant J.*, **16**, 689–696.

Gungabissoon, R.A., Khan, S., Hussey, P.J. & Maciver, S.K. (2001) Interaction of elongation factor 1α from Zea mays (ZmEF-1α) with F-actin and interplay with the maize actin severing protein, ZmADF3. *Cell Motil. Cytoskeleton*, **49**, 104–111.

Grierson, C.S., Roberts, K., Feldmann, K.A. & Dolan, L. (1997) The COW1 locus of *Arabidopsis* acts after RHD2, and in parallel with RHD3 and TIP1, to determine the shape, rate of elongation, and number of root hairs produced from each site of hair formation. *Plant Physiol.*, **115**, 981–990.

Heath, I.B. (1974) A unified hypothesis for the role of membrane bound enzyme complexes and microtubules in plant cell wall synthesis. *J. Theoretical Biol.*, **48**, 445–449.

Heidstra, R., Geurts, R., Franssen, H., Spaink, H.P., Vankammen, A. & Bisseling, T. (1994) Root hair deformation activity of nodulation factors and their fate on *Vicia-Sativa*. *Plant Physiol.*, **105**, 787–797.

Hepler, P.K., Vidali, L. & Cheung, A.Y. (2001) Polarized cell growth in higher plants. *Ann. Rev. Cell Dev. Biol.*, **17**, 159–187.

Jiang, C.J., Weeds, A.G. & Hussey, P.J. (1997) The maize actin depolymerizing factor, ZmADF3, redistributes to the growing tip of elongating root hairs and can be induced to translocate into the nucleus with actin. *Plant J.*, **12**, 1035–1043.

Jonak, C., Okresz, L., Bogre, L. & Hirt, H. (2002) Complexity, cross talk and integration of plant MAP kinase signaling. *Curr. Opin. Plant Biol.*, **5**, 415–424.

Jones, M.A., Shen, J.J., Fu, Y., Li, H., Yang, Z.B. & Grierson, C.S. (2002) The *Arabidopsis* Rop2 GTPase is a positive regulator of both root hair initiation and tip growth. *Plant Cell*, **14**, 763–776.

Ketelaar, T. & Emons, A.M.C. (2000) The role of microtubules in root hair growth and cellulose microfibril deposition, in *Root Hairs Cell and Molecular Biology* (eds R.W. Ridge & A.M.C. Emons), Tokyo Berlin Heidelberg, New York: Springer-Verlag, pp. 17–28.

Ketelaar, T., Faivre-Moskalenko, C. & Esseling, J.J. (2002) Positioning of nuclei in *Arabidopsis* root hairs: An actin-regulated process of tip growth. *Plant Cell*, **14**(11), 2941–2955.

Ketelaar, T., de Ruijter, N.C.A. & Emons, A.M.C. (2003) Unstable F-actin specifies the area and microtubule direction of cell expansion in *Arabidopsis* root hairs. *Plant Cell*, **15**, 285–292.

Kieber, J.J., Rothenberg, M., Roman, G., Feldman, K.A. & Ecker, J.R. (1993) CTR1, a negative regulator of the ethylene response pathway in *Arabidopsis*, encodes a member of the Raf family of protein kinases. *Cell*, **72**, 427–441.

Knox, K., Grierson, C.S. & Leyser, O. AXR3 and SHY2 interact to regulate root hair development. *Development*. In press.

Lee, M.M. & Schiefelbein, J. (1999) WEREWOLF, a MYB-related protein in *Arabidopsis*, is a position-dependent regulator of epidermal cell patterning. *Cell*, **99**, 473–483.

Lee, M.M. & Schiefelbein, J. (2002) Cell pattern in the *Arabidopsis* root epidermis determined by lateral inhibition with feedback. *Plant Cell*, **14**, 611–618.

Lew, R.R. (1991) Electrogenic transport-properties of growing *Arabidopsis* root hairs – the plasma-membrane proton pump and potassium channels. *Plant Physiol.*, **97**, 1527–1534.

Lew, R.R. (1996) Pressure regulation of the electrical properties of growing *Arabidopsis thaliana* L. root hairs. *Plant Physiol.*, **112**, 1089–1100.

Lew, R.R. (1998) Immediate and steady state extracellular ionic fluxes of growing *Arabidopsis thaliana* root hairs under hyperosmotic and hypoosmotic conditions. *Plant Physiol.*, **104**, 397–404.

Lew, R.R. (2000) Electrobiology of root hairs, in *Root Hairs: Cell and Molecular Biology* (eds R.W. Ridge & A.M.C. Emons) Tokyo: Springer-Verlag, pp. 115–139.

Leyser, H.M.O., Pickett, F.B., Dharmasiri, S. & Estelle, M. (1996) Mutations in the AXR3 gene of *Arabidopsis* result in altered auxin response including ectopic expression from the SAUR-AC1 promoter. *Plant J.*, **10**, 403–413.

Lhuissier, F.G.P., De Ruijter, N.C.A., Sieberer, B.J., Esseling, J.J. & Emons, A.M.C. (2001) Time course of cell biological events evoked in legume root hairs by *Rhizobium* Nod factors: State of the art. *Ann. Botany*, **87**(3), 289–302.

Lloyd, C.W., Pearce, K.J., Rawlins, D.J., Ridge, R.W. & Shaw, P.J. (1987) Endoplasmic microtubules connect the advancing nucleus to the tip of legume root hairs, but F-actin is involved in basipetal migration. *Cell Motil. Cytoskeleton*, **8**, 27–36.

Masucci, J.D. & Schiefelbein, J.W. (1994) The RHD6 mutation of *Arabidopsis thaliana* alters root-hair initiation through an auxin- and ethylene-associated process. *Plant Physiol.*, **106**, 1335–1346.

Masucci, J.D. & Schiefelbein, J.W. (1996) Hormones act downstream of TTG and GL2 to promote root hair outgrowth during epidermis development in the *Arabidopsis* root. *Plant Cell*, **8**, 1505–1517.

Masucci, J.D., Rerie, W.G., Foreman, D.R., *et al.* (1996) The homeobox gene GLABRA2 is required for position-dependent cell differentiation in the root epidermis of *Arabidopsis thaliana. Development*, **122**, 1253–1260.

McDowell, J.M., An, Y.Q., Huang, S.R., McKinney, E.C. & Meagher, R.B. (1996) The *Arabidopsis* ACT7 actin gene is expressed in rapidly developing tissues and responds to several external stimuli. *Plant Physiol.*, **111**, 699–711.

Miller, D.D., De Ruijter, N.C.A., Bisseling, T. & Emons, A.M.C. (1999) The role of actin in root hair morphogenesis: Studies with lipochito-oligosaccharide as a growth stimulator and cytochalasin as an actin perturbing drug. *Plant J.*, **17**, 141–154.

Molendijk, A.J., Bischoff, F., Rajendrakumar, C.S.V., *et al.* (2001) *Arabidopsis thaliana* Rop GTPases are localized to tips of root hairs and control polar growth. *EMBO J.*, **20**, 2779–2788.

Mylona, P., Linstead, P., Martienssen, R. & Dolan, L. (2002) *SCHIZORIZA* controls an asymmetric cell division and restricts epidermal identity in the *Arabidopsis* root. *Development*, **129**, 4327–4334.

Nagpal, P., Walker, L.M., Young, J.C., *et al.* (2000) *AXR2* encodes a member of the Aux/IAA protein family. *Plant Physiol.*, **123**, 563–573.

Nebenfuhr, A., Ritzenthaler, C., & Robinson, D.G. (2002) Brefeldin A: Deciphering an enigmatic inhibitor of secretion, *Plant Physiol.*, **130**, 1102–1108.

Okada, K. & Shimura, Y. (1994) Modulation of root hair growth by physical stimuli, in *Arabidopsis* (eds E.M. Meyerowitz & C.R. Somerville) Cold Spring Harbor, NY: Cold Spring Harbor Laboratory Press, pp. 665–684.

Ovecka, M., Nadubinska, M., Volkmann, D. & Baluska, F. (2000) Actomyosin and exocytosis inhibitors alter root hair morphology in *Poa annua*. *Biologia.*, **55**, 105–114.

Oyama, T., Shimura, Y. & Okada, K. (1997) The *Arabidopsis HY5* gene encodes a bZIP protein that regulates stimulus-induced development of root and hypocotyl. *Genes Dev.*, **11**, 2983–2995.

Oyama, T., Shimura, Y. & Okada, K. (2002) The *IRE* gene encodes a protein kinase homologue and modulates root hair growth in *Arabidopsis*. *Plant J.*, **30**, 289–299.

Parker, J.S., Cavell, A.C., Dolan, L., Roberts, K. & Grierson, C.S. (2000) Genetic interactions during root hair morphogenesis in *Arabidopsis*. *Plant Cell*, **12**, 1961–1974.

Pitts, R.J., Cernac, A. & Estelle, M. (1998) Auxin and ethylene promote root hair elongation in *Arabidopsis*. *Plant J.*, **16**, 553–560.

Pollard, T.D., Blanchoin, L. & Mullins, R.D. (2000) Molecular mechanisms controlling actin filament dynamics in nonmuscle cells. *Annu. Rev. Bioph. Biom.*, **29**, 545–576.

Ramachandran, S., Christensen, H.E.M., Ishimaru, Y., *et al.* (2000) Profilin plays a role in cell elongation, cell shape maintenance, and flowering in *Arabidopsis*. *Plant Physiol.*, **124**, 1637–1647.

Reed, J.W., Nagpal, P., Poole, D.S., Furuya, M. & Chory, J. (1993) Mutations in the gene for the red far-red light receptor phytochrome-B alter cell elongation and physiological responses throughout *Arabidopsis* development. *Plant Cell*, **5**, 147–157.

Ridge, R.W., Uozumi, Y., Plazinski, J., Hurley, U.A. & Williamson, R.E. (1999) Developmental transitions and dynamics of the cortical ER of *Arabidopsis* cells seen with green fluorescent protein. *Plant Cell Physiol.*, **40**, 1253–1261.

Rigas, S., Debrosses, G., Haralampidis, K., *et al.* (2001) TRH1 encodes a potassium transporter required for tip growth in *Arabidopsis* root hairs. *Plant Cell*, **13**, 139–151.

Ringli, C., Baumberger, N., Diet, A., Frey, B. & Keller, B. (2002) ACTIN2 is essential for bulge site selection and tip growth during root hair development of *Arabidopsis*. *Plant Physiol.*, **129**, 1464–1472.

Romagnoli, S., Cai, G. & Cresti, M. (2003) *In vitro* assays demonstrate that pollen tube organelles use kinesin-related motor proteins to move along microtubules. *Plant Cell*, **15**, 251–269.

Ryan, E., Grierson, C.S., Cavell, A., Steer, M. & Dolan, L. (1998) *TIP1* is required for both tip growth and non-tip growth in *Arabidopsis*. *New Phytol.*, **138**, 49–58.

Ryan, E., Steer, M. & Dolan, L. (2001) Cell biology and genetics of root hair formation in *Arabidopsis thaliana*. *Protoplasma*, **215**, 140–149.

Samaj, J., Ovecka, M., Hlavacka, A., *et al.* (2002) Involvement of the mitogen-activated protein kinase SIMK in regulation of root hair tip growth. *EMBO J.*, **21**, 3296–3306.

Schafer, E. & Bowler, C. (2002) Phytochrome-mediated photoperception and signal transduction in higher plants. *EMBO Reports*, **3**, 1042–1048.

Schiefelbein, J.W. & Somerville, C. (1990) Genetic control of root hair development in *Arabidopsis thaliana*. *Plant Cell*, **2**, 235–243.

Schiefelbein, J.W., Shipley, A. & Rowse, P. (1992) Calcium influx at the tip of growing root-hair cells of *Arabidopsis thaliana*. *Planta*, **187**, 455–459.

Schiefelbein, J., Galway, M., Masucci, J. & Ford, S. (1993) Pollen tube and root-hair tip growth is disrupted in a mutant of *Arabidopsis thaliana*. *Plant Physiol.*, **103**, 979–985.

Schikora, A. & Schmidt, W. (2001) Acclimative changes in root epidermal cell fate in response to Fe and P deficiency: a specific role for auxin? *Protoplasma*, **218** (1–2), 67–75.

Schmidt, W. & Schikora, A. (2001) Different pathways are involved in phosphate and iron stress-induced alterations of root epidermal cell development. *Plant Physiol.*, **125**, 2078–2084.

Schneider, K., Wells, B., Dolan, L. & Roberts, K. (1997) Structural and genetic analysis of epidermal cell differentiation in *Arabidopsis* primary roots. *Development*, **124**, 1789–1798.

Schneider, K., Mathur, J., Boudonck, K., Wells, B., Dolan, L. & Roberts, K. (1998) The *ROOT HAIRLESS1* gene encodes a nuclear protein required for root hair initiation in *Arabidopsis*. *Genes Dev.*, **12**, 2013–2021.

Shaw, S.L., Dumais, J. & Long, S.R. (2000) Cell surface expansion in polarly growing root hairs of *Medicago truncatula*. *Plant Physiol.*, **123**, 959–969.

Shi, H.Z. & Zhu, J.K. (2002) *SOS4*, a pyridoxal kinase gene, is required for root hair development in *Arabidopsis*. *Plant Physiol.*, **129**, 585–593.

Sieberer, B. & Emons, A.M.C. (2000) Cytoarchitecture and pattern of cytoplasmic streaming in root hairs of *Medicago truncatula* during development and deformation by nodulation factors. *Protoplasma*, **214**, 118–127.

Sieberer, B.J., Timmers, A.C.J., Lhuissier, F.G.P. & Emons, A.M.C. (2002) Endoplasmic microtubules configure the subapical cytoplasm and are required for fast growth of *Medicago truncatula* root hairs. *Plant Physiol.*, **130**(2), 977–988.

Smertenko, A.P., Jiang, C.J., Simmons, N.J., Weeds, A.G., Davies, D.R. & Hussey, P.J. (1998) Ser6 in the maize actin-depolymerizing factor, ZmADF3, is phosphorylated by a calcium-stimulated protein kinase and is essential for the control of functional activity. *Plant J.*, **14**, 187–193.

Sollner, R., Glasser, G., Wanner, G., Somerville, C.R., Jurgens, G. & Assaad, F.F. (2002) Cytokinesis-defective mutants of *Arabidopsis*. *Plant Physiol.*, **129**, 678–690.

Tominaga, M., Morita, K., Sonobe, S., Yokota, E. & Shimmen, T. (1997) Microtubules regulate the organization of actin filaments at the cortical region in root hair cells of *Hydrocharis*. *Protoplasma*, **199**, 83–92.

Tominaga, M., Yokota, E., Vidali, L., Sonobe, S., Helper, P.K. & Shimmen, T. (2000) The role of plant villin in the organization of the actin cytoskeleton, cytoplasmic streaming and the architecture of the transvacuolar strand in root hair cells of Hydrocharis. *Planta*, **210**, 836–843.

Valster, A.H., Pierson, E.S., Valenta, R., Hepler, P.K. & Emons, A.M.C. (1997) Probing the plant actin cytoskeleton during cytokinesis and interphase by profilin microinjection. *Plant Cell*, **9**, 1815–1824.

Very, A.A. & Davies, J.M. (2000) Hyperpolarization-activated calcium channels at the tip of *Arabidopsis* root hairs. *Proc. Natl. Acad. Sci. USA*, **97**, 9801–9806.

Vidali, L. & Hepler, P.K. (1997) Characterization and localization of profilin in pollen grains and tubes of *Lilium longiflorum*. *Cell Motil. Cytoskeleton*, **36**, 323–338.

Vissenberg, K., Fry, S.C. & Verbelen, J.P. (2001) Root hair initiation is coupled to a highly localized increase of xyloglucan endotransglycosylase action in *Arabidopsis* roots. *Plant Physiol.*, **127**, 1125–1135.

Wada, T., Tachibana, T., Shimura, Y. & Okada, K. (1997) Epidermal cell differentiation in *Arabidopsis* determined by a Myb homolog, *CPC*. *Science*, **277**, 1113–1116.

Wada, T., Kurata, T., Tominaga, R., *et al.* (2002) Role of a positive regulator of root hair development, *CAPRICE*, in *Arabidopsis* root epidermal cell differentiation. *Development*, **129**, 5409–5419.

Walker, A.R., Davison, P.A., Bolognesi-Winfield, *et al.* (1999) The *TRANSPARENT TESTA GLABRA1* locus, which regulates trichome differentiation and anthocyanin biosynthesis in *Arabidopsis*, encodes a WD40 repeat protein. *Plant Cell*, **11**, 1337–1349.

Wang, H.-Y., Lockwood, S.K., Hoeltzel, M.F. & Schiefelbein, J.W. (1997) The *ROOT HAIR DEFECTIVE3* gene encodes an evolutionarily conserved protein with GTP-binding motifs and is required for regulated cell enlargement in *Arabidopsis*. *Genes Dev.*, **11**, 799–811.

Wang, X., Cnops, G., Vanderhaeghen, R., *et al.* (2001) AtCSLD3, a cellulose synthase-like gene important for root hair growth in *Arabidopsis*. *Plant Physiol.*, **126**, 575–586.

Webb, M., Jouannic, S., Foreman, J., Linstead, P. & Dolan, L. (2002) Cell specification in the *Arabidopsis* root epidermis requires the activity of ECTOPIC ROOT HAIR 3 – a katanin-p60 protein. *Development*, **129**, 123–131.

Whittington, A.T., Vugrek, O., Wei, K.J., *et al.* (2001) *MOR1* is essential for organizing cortical microtubules in plants. *Nature*, **411**, 610–613.

Wymer, C.L., Bibikova, T.N. & Gilroy, S. (1997) Cytoplasmic free calcium distributions during the development of root hairs of *Arabidopsis thaliana*. *Plant J.*, **12**, 427–439.

8 Signaling the cytoskeleton in pollen tube germination and growth

Rui Malhó and Luísa Camacho

8.1 Introduction

It seems likely that those who investigate the details of tissue-specific gene expression, the production of form, metabolism and cell growth will find increasingly that signaling constituents play a crucial role in the process. Signaling is an integral component in the establishment and maintenance of cellular identity, and it is clear that *Arabidopsis* devotes a significant percentage of its genome to cell signaling (~10%; Chory & Wu, 2002) – perhaps a consequence of the large number of environmental signals that need to be integrated with intrinsic developmental programs. These are rapidly changing biotic and abiotic signals that are perceived in different parts of the plant and which must be integrated to give a fine-tuned and appropriate growth response.

Tip-growing cells represent an ideal system in which to investigate signal transduction mechanisms and pollen tubes (PTs) offer one of the favourite models. These cells grow at extremely fast rates ($>1\,\mu m\,min^{-1}$ in some species), have an intense cytoplasmic streaming and respond quickly to intra- and extracellular cues. With the technical advances in molecular constructs and imaging detectors, PTs became excellent models for studying the role of signal perception and transduction in effecting morphogenetic changes. In PTs, the actin cytoskeleton plays an essential role during germination and tip growth, which involves targeted vesicle fusion, Ca^{2+}, protein kinases, cAMP, GTPases, etc. (Steer & Steer, 1989; Vidali & Hepler, 2001; Camacho & Malhó, 2002; Fig. 8.1) – a large web of signaling networks that crosstalk and generate a signal-specific pattern that is converted into a signal-specific response.

This chapter highlights some of the most recent advances made in PT signal transduction cascades. The emerging picture is one of great complexity with a great number of similarities to other tip growing systems (fungal hyphae, root hairs) but also to other less obvious cell types such as neurones. Branched informational networks, signal amplification loops, multiple signaling pathways, which characterise any complex biological system, are present and act in a concerted form to respond to the different signal inputs.

8.2 Different signaling pathways converge in the cytoskeleton

The principal signaling agents demonstrated to initiate changes within the actin network of animal cells are Ca^{2+} and lipids (Ridley & Hall, 1992; Janmey, 1994).

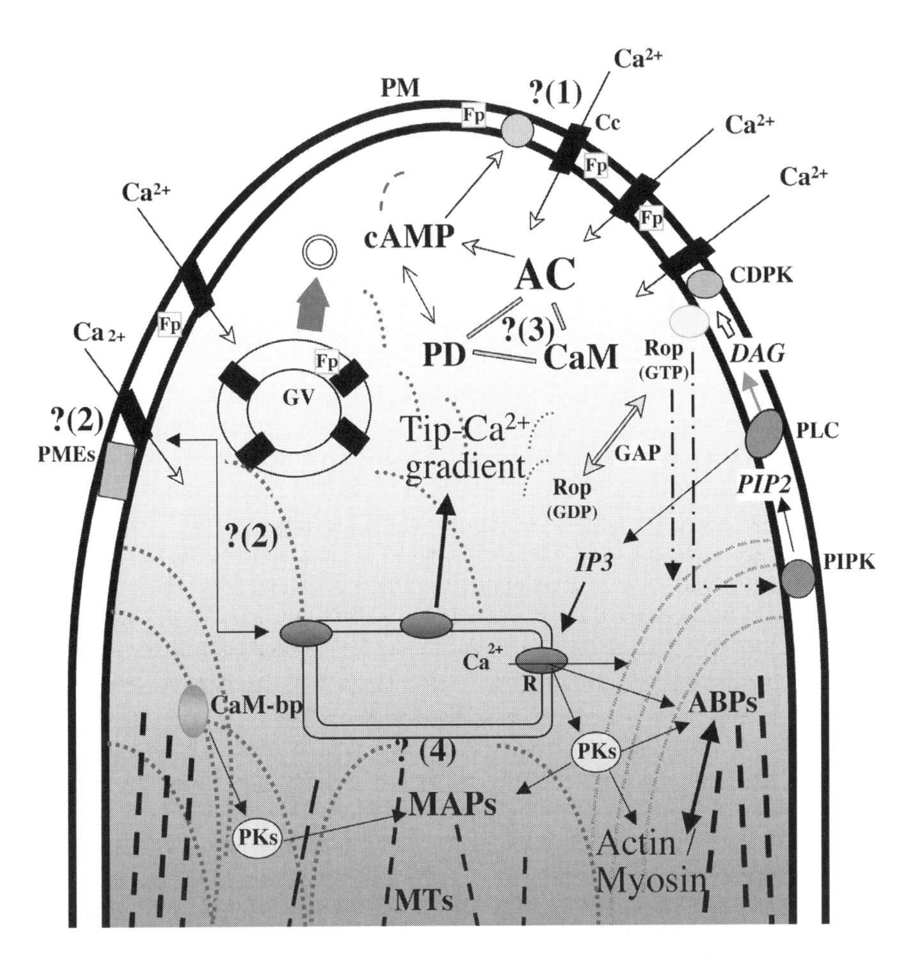

Fig. 8.1 Apical region of a growing pollen tube illustrating the main signal transduction pathways and their interactions. Question marks indicate areas that lack confirmation: (1) What are the targets of cAMP? (2) Ca^{2+} influx is regulated by internal stores or cell wall? (3) How is adenylyl cyclase and phosphodiesterase regulated? (4) The position of Ca^{2+} stores is determined by MFs, MTs or both? ABPs, actin-binding proteins; AC, Adenylyl cyclase; CaM-bp, Calmodulin-binding protein; Cc, Ca^{2+} channels; DAG, diacylglycerol; Fp, Fusogenic protein (SNAREs and/or anexins); GAP, Rop GTPase activating protein; GV, Golgi vesicle; IP_3, Inositol 1,4,5 triphosphate; PD, Phosphodiesterase; PIP_2, phosphatidylinositol-(4,5)-bis phosphate; PIPK, phosphatidylinositol kinase; PLC, phospholipase C; PM, plasma membrane; PMEs, pectin-methyl-esterases; R, IP_3 receptor.

These second messengers can trigger structural changes through interactions with actin-binding proteins, e.g. profilin (Staiger *et al.*, 1994), or through alterations in phosphorylation mediated by calmodulin (CaM) and protein kinases (Goeckeler & Wysolmerski, 1995) and phosphatases (Kimura *et al.*, 1996). Modulation of the integrity of the actin network through the regulation of F-actin assembly, the amount and type of actin-binding proteins (ABPs), and myosin binding and filament formation can, therefore, provide regulatory points for signal-mediated reorganisations of the actin network within specific domains of the cytoplasm. Such reorganisations may then promote topological changes in the transport of ions and metabolites across the plasma membrane within those regions (Derksen *et al.*, 1995). Numerous studies have provided evidence for the role of plant hormones and growth factors in modifying the tension within the actin network. These modifications in tension are proposed to result from the hormone-induced formation of Ca^{2+} gradients and lipid-signaling molecules, similar to the effectors of the actin network described in animal cells (Brownlee & Wood, 1986). Additional evidence for the involvement of CaM, Ca^{2+}-dependent protein kinases (CDPK and/or CaMK), phosphatases (calcineurin or calcineurin-type) as well as cyclic nucleotides as transducers of actin tension has been obtained, and suggest that myosin and/or actin binding proteins may be the target of these regulatory molecules (see Chapter 2).

8.3 The actin cytoskeleton is the major motor driving force in pollen tube growth

PT microfilaments (MFs), together with myosin, regulate the cytoplasmic streaming that translocates organelles (Taylor & Hepler, 1997). MFs are mainly longitudinally orientated and occur both in the cortical and central cytoplasm of the PTs. Although this system was believed to be more dense in the extreme apical region, a study of actin distribution in living PTs (Miller *et al.*, 1996) showed a different picture. While numerous MFs arranged axially were present in the sub-apical region, the tip itself appeared essentially devoid of these elements. Single MFs occur next to the plasma membrane within 5 μm from the tip, whereas MF bundles first appear 10–20 μm from the tip. The absence of MFs at the tip of the growing tube was suggested to be due to the tip-focused Ca^{2+} gradient indispensable for the growth of the PT (Malhó *et al.*, 1995; Pierson *et al.*, 1996; Messerli *et al.*, 1999). These data were updated using GFP-talin as a probe for actin (Kost *et al.*, 1998; Fu *et al.*, 2001). It was found that, although thick bundles are only visualised in the sub-apical region, thinner and transient MFs could be observed in the apex. However, PT growth after GFP-talin expression is strongly perturbed raising questions about the validity of this methodology also.

Cytoplasmic streaming with a reverse-fountain pattern is a typical feature of growing PTs. In this pattern, streaming is orientated towards the tip in the cortical region and towards the base in the subcortical part of the tube (Taylor & Hepler, 1997). Cytoplasmic streaming is involved in vesicle transport towards the tip but the function of the reverse streaming is unclear. Proteins homologous to those that regulate the structure and function of the actin cytoskeleton in animal and yeast cells have been identified in PTs (Vidali & Hepler, 2001). Antibodies raised against myosin I localised to the plasma membrane, to the surface of the nuclei, and was seen as spots in the PT cytoplasm (Miller *et al.*, 1995). Antibodies against myosin II subfragment (S1) and light meromyosin (LMM) visualised myosin II in a pattern that seemed to follow the MF tracks of the PT. Similarly, the spots visualised with the antibody against myosin V could be thought to follow the MF tracks (Miller *et al.*, 1995). Recently, analysis of the myosins in the *Arabidopsis* genome sequence has revealed, however, that plant myosins fall into two groups – class VIII and class XII (Reddy & Day, 2001). Myosins XI are closely related to class V.

Several genes encoding the small ABPs cofilin and profilin have been characterised from pollen (Staiger *et al.*, 1993; Mittermann *et al.*, 1995; Lopez *et al.*, 1996; Gibbon *et al.*, 1998; see Chapter 2). Cofilin and profilin are small polypeptides, suggested to have similar and opposite effects on MF dynamics (Bamburg, 1999). Cofilin belongs to the ADF (actin depolymerising factor) protein family, the members of which interact with actin monomers and filaments in a pH-sensitive manner. When ADF/cofilin binds to F-actin it induces a change in the helical twist and fragmentation, and accelerates the dissociation of subunits from the pointed ends of MFs (Bamburg, 1999; Cooper & Schafer, 2000). The properties of plant cofilin have been analysed by using recombinant proteins from the maize pollen specific cofilin gene *ZmADF1* (Hussey *et al.*, 1998) and from *ZmADF3*, which is suppressed in pollen but expressed in all other maize tissues. *ZmADF3* has the ability to bind monomeric (G) actin and F-actin and to decrease the viscosity of polymerised actin solutions indicating an ability to depolymerise actin filaments (Lopez *et al.*, 1996). *ZmADF3* is phosphorylated on Ser6 by a CDPK in plant extracts (Smertenko *et al.*, 1998). This suggests that phosphorylation and dephosphorylation could regulate actin-binding ability of cofilin and thus affect actin cytoskeleton stability.

Profilin is a G-actin binding protein that is known to interact with proline-rich motifs of other proteins and with phosphoinositides. The interaction of profilin with G-actin provides a mechanism to sequester actin monomers and promote actin depolymerisation. Profilin may also be involved in promoting actin polymerisation. This might take place by binding to proline-rich motifs in proteins that convey intra- or extracellular signals to reorganisation of actin cytoskeleton (Mullins, 2000). The presence of several pollen-specific profilin isoforms in maize has raised the question of whether they are functionally unique or redundant (Gibbon *et al.*, 1997, 1998). Microinjection of birch pollen-specific profilin into stamen hair cells caused a rapid irreversible change in the cellular organisation and cytoplasmic streaming due to depolymerisation of MFs (Staiger *et al.*, 1994), and these results have been confirmed by the use of bacterially expressed maize pollen profilin isoforms in similar experiments (Gibbon *et al.*, 1997, 1998). These results support the model that pollen profilins act as actin-sequestering proteins that promote actin depolymerisation (Staiger *et al.*, 1994). Recently it has been shown that one of the profilins expressed in PTs shows higher proline-rich motif binding capacity than others (Gibbon *et al.*, 1998), and that the actin-sequestering activity of profilins is dependent on Ca^{2+}concentration (Staiger, 2000).

Plants also have several genes with high homology to animal villin (Klahre *et al.*, 2000). The first plant villin was isolated from *L. longiflorum* PTs as a 135 kDa actin-bundling protein (Yokota *et al.*, 1998; Yokota & Shimmen, 1999). Its identity as a villin-gelsolin family member (Vidali *et al.*, 1999) was confirmed by partial amino acid sequencing and by isolating the corresponding cDNA from a pollen grain cDNA expression library. Immunodetection of villin revealed its co-localisation with MF bundles in PTs. Due to the gelsolin-like headpiece, villin may also act as an actin-severing protein, although this activity has not yet been demonstrated for plant villins. Ca^{2+} and CaM together inhibit the actin-bundling activity of PT villin (Yokota *et al.*, 2000), adding villin to the group of Ca^{2+}-regulated ABPs. The pollen villin also includes a phosphoinositide-binding domain (Vidali *et al.*, 1999) thus making this protein a putative target of the phosphoinositide signaling pathway.

8.4 Microtubules and microtubule-associated proteins in pollen tube growth

The role of microtubules (MTs) in PT growth is almost unknown, when compared to all the data on MFs. We do not know, for example, where MTs are organised, the extent of MT dynamics, and the polarity of MTs in the PT. Roles of MTs and related motors in organelle trafficking are not clear. There are several reasons for this but perhaps the most relevant is colchicine. This drug, known to disrupt MTs, when applied to *in vitro* PTs was virtually ineffective at inhibiting growth. This led to the assumption that MTs were not important for PT growth and so researchers neglected this area for many years. In 1994, Joos and co-workers (1994) showed that propham was about 100 times more effective than colchicine. This drug caused disorganisation of the cytoplasm, vacuolisation of the tip and to a deranged position of both the generative cell and vegetative nucleus (Joos *et al.*, 1994). Even in this case, pollen MTs proved to be extremely stable. The MT stability is probably due to cross-linking of the MTs by MT-associated proteins (MAPs) (Raudaskoski *et al.*, 1987; Palevitz & Tiezzi, 1992; Tiezzi *et al.*, 1992). We now have indications that MTs play a role in controlling the apical transport of secretory vesicles (Cai *et al.*, 1993), in the positioning of organelles (Joos *et al.*, 1994), and in pulsed growth (Geitmann *et al.*, 1995). Furthermore, evidence that MTs and actin filaments cooperate functionally to move vesicles and organelles in different types of cells (Goode *et al.*, 2000) suggests that the role of PT MTs in organelle transport could be complementary to actin-based transport. However, the only evidence of interactions between MFs and MTs in the PT is coalignment (Pierson *et al.*, 1986; Lancelle & Hepler, 1991); there is no indication that the two systems cooperate functionally during vesicle and organelle transport.

In PTs, the MTs are relatively thin but become thicker near the nuclear region. Helical orientation and bundling of MTs increases behind the nuclei close to the plasma membrane, whereas axial MTs occur in the central part of the tube and in the cytoplasm surrounding the nuclei (Raudaskoski *et al.*, 1987; Joos *et al.*, 1994). A common feature to all PTs is the absence of MTs from the apex and the short and randomly oriented MTs in the subapical region. Thick longitudinally oriented MT bundles between the plasma membrane and the generative nucleus are frequently observed (Raudaskoski *et al.*, 1987). The MT bundles appear to be attached to the plasma membrane at each end of the generative cell and it is possible that they control the shape of the nuclei through their movement in the PT. When the MTs are depolymerised, the spindle shape of the generative cell is lost and the cell dissociates from the vegetative nuclei (Åström *et al.*, 1995). These observations suggest that the nuclei shape (and movement) is dependent on MTs. The depolymerisation of the MT bundles in isolated generative cells (Åström *et al.*, 1995) further suggests that the formation and maintenance of the MT bundles in the generative cell requires an interaction with the cytoplasm surrounding it.

Although the structure and localisation of MTs in the PT is known (Derksen *et al.*, 1985; Pierson *et al.*, 1986; Del Casino *et al.*, 1993), their involvement in the translocation of organelles along the tube remains obscure. Regardless of scarce information, MT-based motors of both the kinesin and dynein families have been identified in the PT. Most of these MT motors have also been found in association with membrane-bounded organelles, which suggest that these proteins could translocate organelles or vesicles. A pollen kinesin homolog (PKH) (Tiezzi *et al.*, 1992) and dynein-related polypeptides (Moscatelli *et al.*, 1995) have been identified in the vegetative cytoplasm of PTs and localised in association with membrane structures (Liu *et al.*, 1994; Moscatelli *et al.*, 1998). The PKH is presumably involved in the process of tip growth, whereas the dynein-related polypeptide or polypeptides probably take part in the translocation of membrane-bounded organelles. Although specific antibodies have identified these proteins, their functional properties is unknown. They have a molecular weight typical of kinesins and dyneins, show MT-stimulated ATPase activity, and the kinesins have been shown to induce MTs to glide in motility assays *in vitro*. These proteins appear to be located in the PT cytoplasm but not in the generative cell. The PKH label occurred mainly in the apex (Tiezzi *et al.*, 1992; Cai *et al.*, 1993) whereas the 90kDa ATP-MAP-labelled organelles follow the MT tracks in the basal part of the tube (Cai *et al.*, 2000). The two 400kDa dynein-related polypeptides localised to membrane structures in PTs (Moscatelli *et al.*, 1998). In tobacco, two cDNAs for kinesin-like proteins have been isolated (Wang *et al.*, 1996). The product of one of these cDNAs, TCK1, binds CaM in a Ca^{2+}-dependent manner and kinesins with comparable structure are known in *Arabidopsis* (Wang *et al.*, 1996). This indicates a possible crosstalk between MT-based movement and the Ca^{2+}-CaM signaling pathway.

The stability of the MT cytoskeleton is also dependent on Ca^{2+} concentration in a CaM-dependent manner (Bartolo & Carter, 1992; Fisher *et al.*, 1996). The regulation by Ca^{2+} could explain the absence of MTs from the apex, the appearance of short MTs in the subapical region, and prominent MT tracks and spirals in the basal region of the tube behind the nuclei. After conveying the vegetative nucleus and generative cell from the grain into the tube (Åström *et al.*, 1995), MTs probably are involved in keeping them in the proximity of the tube tip. Short MTs in the subapical region (Raudaskoski *et al.*, 1987) are perhaps cooperating with the surrounding MFs, using motor molecules from both cytoskeletal systems as in the bud tip of yeast or at the edge of a migrating fibroblast (Goode *et al.*, 2000). This interaction could keep the nuclei and generative cell near the tip.

8.5 Ca^{2+}, modulator of the cytoskeleton

Our present understanding implicates cytosolic free calcium ($[Ca^{2+}]_c$) in the regulation of PT elongation and guidance. A tip-focused $[Ca^{2+}]_c$ gradient has been imaged with a high 1–3 μM Ca^{2+} concentration in the tip region and the low 0.2–0.3 μM Ca^{2+} concentration in the subapical and basal part of the tube (Malhó *et al.*, 1995; Pierson *et al.*, 1996; Messerli & Robinson, 1997; Malhó 1999). This gradient is absent in non-growing PTs and disruption of the gradient inhibits tube growth. It is pertinent that both the intracellular $[Ca^{2+}]_c$ gradient and extracellular Ca^{2+} influx suffer sinusoidal oscillations which are accompanied by oscillations in growth rate (Holdaway-Clarke *et al.*, 1997; Messerli *et al.*, 2000). The phase relationships differ however, with the gradient being delayed from growth ~4 s while the influx is delayed by ~11–13 s.

A substantial body of evidence reveals that cytoplasmic streaming is controlled by Ca^{2+}; at basal levels (100–200 nm) streaming occurs normally, however, when the concentration is elevated to 1 μM or higher, streaming is rapidly but reversibly inhibited (Taylor & Hepler, 1997). These results apply well to PTs, which also show a marked inhibition of streaming when the intracellular $[Ca^{2+}]_c$ has been elevated, and a recovery when the concentration returns to basal levels (Yokota *et al.*, 1999). In addition, high Ca^{2+} fragments F-actin in PTs (Yokota *et al.*, 1998) which explains why thick MFs are not visible in the tube apex. These inhibitory and fragmentary activities are certainly brought about by specific ABPs. $[Ca^{2+}]_c$ at the tube apex is also sufficiently high to inhibit myosin motor activity and thus to inhibit streaming. Apical $[Ca^{2+}]_c$may then have an important role slowing or stopping motion, and as a consequence regulate vesicle docking and fusion with the plasma membrane (see Section 8.11). These events are intimately regulated with PT reorientation and it has been shown that modification of $[Ca^{2+}]_c$ at the tube apex induced bending of the growth axis towards the zone of higher $[Ca^{2+}]_c$ (Malhó & Trewavas, 1996); the reorientation also seems to depend on Ca^{2+}-induced Ca^{2+}-release (CICR) and on the asymmetric activity of CDPKs (see Sections 8.6 and 8.8).

PT growth depends on polarised exocytosis at the growing tip and Ca^{2+} was suggested to be involved in this process (Malhó *et al.*, 2000). However, the quantity of membrane delivered by exocytosis is in excess for the PT growth rate, suggesting an underlying recycling process (Steer & Steer, 1989). Measurements of endo/exocytosis in growing PTs with the fluorescent probe FM 1–43 showed a fluorescence hotspot at the apex (Camacho & Malhó, 2003) and, when the tube reorients, the side of the dome to which the cell bent became more fluorescent. This suggests vesicle relocation and an asymmetric distribution of the fusion events. The cargo, which is released upon vesicle fusion, provides the material for the construction of the new cell wall needed for growth (not simply elongation of existing wall).

These results strongly resemble those obtained when $[Ca^{2+}]_c$ was mapped during this process (Malhó & Trewavas, 1996). This suggests that $[Ca^{2+}]_c$ is a regulator of the coupling between growth and endo/exocytosis. Indirect support for this statement can be found in older literature. Reiss and Herth (1978) found that chlortetracycline disturbed the apical cytoplasmic gradient in PTs, and induced the formation of abnormal wall thickenings composed of aggregates of the dictyosome-derived wall precursors bodies. These authors attributed the effect to the chelating of Ca^{2+} and the consequent disruption of exocytosis. But the same authors also found that A23187, a calcium ionophore, induced thickening of the wall at the tip by the accretion of the precursors bodies. However, the results from Camacho and Malhó (2003) also suggest that cell growth is not strictly dependent on a Ca^{2+}-mediated stimulation of exocytosis. This is in agreement with the results of Roy *et al.* (1999) using the Yariv reagent thus suggesting that a Ca^{2+}-dependent exocytosis serves mainly to secrete cell wall components. This explains why rises in apical Ca^{2+} alone led to augmented secretion, reorientation of the growth axis but not increase in growth rates.

The mechanisms which regulate $[Ca^{2+}]_c$ at the PT apex are still controversial. Holdaway-Clarke *et al.* (1997) found an apparent discrepancy between internal Ca^{2+} measurements and external Ca^{2+} fluxes, with the fluxes an order of magnitude greater than the amount needed to support the observed intracellular apical Ca^{2+} gradient. Based on these findings it was suggested that the cell wall could act as a buffer for Ca^{2+}. Pectin-methyl-esterases responsible for cross-linking unesterified pectins were proposed to control the availability of cell wall binding sites for Ca^{2+}. However, aequorin measurements of apical $[Ca^{2+}]_c$ (Messerli & Robinson, 1997) suggest that peak values for $[Ca^{2+}]_c$ in the apical region are about one order of magnitude higher than previous estimates using Ca^{2+}-sensitive dyes thus challenging this hypothesis. An alternative explanation is that $[Ca^{2+}]_c$ influx is regulated by capacitative entry. Trewavas and Malhó (1997) suggested that the ion channels in tip growing cells might have their activity regulated by the degree of emptiness of intracellular Ca^{2+} stores; when these stores are full, plasma membrane channels would close and vice versa. The two hypotheses are not mutually exclusive and it is quite possible that intracellular communication between internal stores, plasma membrane and cell wall may all participate in the regulation of Ca^{2+}-permeable channels. This could be achieved through the association of proteins in *transducon-like* entities (see Section 8.12).

8.6 Signaling the cytoskeleton through phosphoinositides

Phosphoinositides are a vast family of molecules playing a major role in signaling but in plants the information is still scarce (Drøbak, 1992). Ins(1,4,5)P_3 is the most investigated member of this family because it functions as a cytosolic second messenger in animal cells by stimulating Ca^{2+} release from intracellular stores (Berridge, 1995). In plants, evidence has been gathered for a similar role either through photolysis of caged Ins(1,4,5)P_3 (Franklin-Tong *et al.*, 1996; Malhó, 1998) or addition to membrane vesicles (Muir & Sanders, 1997). However, unequivocal measurements of Ins(1,4,5)P_3 concentration in living cells are still missing.

In PTs, Ins(1,4,5)P_3 binding does not seem to be required for the activation of Ca^{2+} entry, but instead to transduce signals to further regions of the cell (Malhó, 1998). It has also been shown in animal cells that injection of heparin, an inhibitor of Ins(1,4,5)P_3-dependent Ca^{2+} release, completely blocked responses to photolysis of caged Ins(1,4,5)P_3 but had no apparent effect on stimulus-evoked Ca^{2+} entry (Petersen & Berridge, 1994). Following initial Ca^{2+} influx, the signal transduction pathway probably includes further Ca^{2+} release from Ins(1,4,5)P_3-dependent stores, a form of signal amplification. These observations in PTs (Malhó, 1998) indicate that the putative Ins(1,4,5)P_3 receptors may have an asymmetric activity depending on their spatial localisation. This is in agreement with the kinetics of Ins(1,4,5)P_3 receptors being profoundly affected by Ca^{2+} ions (Dawson, 1997): in the nM range, increasing $[Ca^{2+}]_c$ potentiates Ca^{2+} release by Ins(1,4,5)P_3 to the extent that Ca^{2+} and Ins(1,4,5)P_3 can be regarded as co-agonists for Ca^{2+} release; in the μM range, however, Ca^{2+} is found to be inhibitory and the receptor undergoes an intrinsic inactivation when Ins(1,4,5)P_3 is bound.

Among the possible targets for the phosphoinositide pathway, actin MFs and ABPs are particularly important. Profilin can bind phosphatidyl 4,5-bisphosphate (PIP_2) (Drøbak *et al.*, 1994), effectively decreasing the profilin concentration which in turn would change the physical state of the cytoskeletal network (see Chapter 2). Profilin was shown to distribute evenly in PTs (Vidali & Hepler, 1997) but subsequent work raised the possibility that at least some profilin isoforms could be located at distinct subcellular domains (von Witsch *et al.*, 1998). Profilin interacts with soluble components resulting in dramatic alterations in the phosphorylation of several proteins (Clarke *et al.*, 1998) which stresses the need to map protein activity and not only its distribution. Profilin also binds to phospholipase C, inhibiting the function of the enzyme, which could be a way of regulating the levels of PIP_2 (Drøbak *et al.*, 1994). The Ca^{2+} released by Ins(1,4,5)P_3 can activate gelsolin, a protein that severs actin. In animal cells, Ins(1,4,5)P_3 receptor-like proteins were also shown to be linked to actin filaments (Fujimoto *et al.*, 1995) and it is likely that part of the self-incompatibility mechanism described by Franklin-Tong *et al.* (1996) involves Ins(1,4,5)P_3-induced changes in the cytoskeleton (Clarke *et al.* 1998). Indeed, it has been shown that increases in Ca^{2+} induced by Ins(1,4,5)P_3 can mimick the incompatibility response (Franklin-Tong *et al.*, 1996) and that Ca^{2+}-induced actin depolymerisation is a key event in this process (Geitmann *et al.*, 2000; Snowman *et al.*, 2002). These authors reported dramatic and rapid changes to the actin cytoskeleton in incompatible PTs of *P. rhoeas* undergoing the SI reaction which included marginalisation of actin to the plasma membrane and fragmented F-actin. It was also observed that a release of higher concentrations of Ins(1,4,5)P_3 in the apical region resulted frequently in tip bursting (Malhó, 1998). Whether this is an experimental artefact or a physiological effect can not be presently assessed. We can only speculate that an Ins(1,4,5)P_3 increase might cause two independent, but not mutually exclusive scenarios: (1) an increase in Ca^{2+} crossing a *recovery* threshold or (2) an inhibition of phospholipase C with the consequent increase in PIP_2 and modulation of profilin and/or ADF activity (Gungabissoon *et al.*, 1998; Chen *et al.*, 2002).

Another target of the phosphoinositide pathway in PTs is anion fluxes across the plasma membrane. Using an ion-specific vibrating probe, Zonia and co-workers (Zonia *et al.*, 2002) identified a new circuit that involves oscillatory efflux of the anion Cl^- at the apex and steady influx along the tube starting at 12 μm distal to the tip. This spatial coupling of influx and efflux sites predicts that a vectorial flux of Cl^- ion traverses the apical region. Inositol 3,4,5,6-tetra*kis*phosphate [Ins(3,4,5,6)P_4] was found to induce cell volume increases and disrupted Cl^- efflux. These effects were specific for Ins(3,4,5,6)P_4 and were not mimicked by either Ins(1,3,4,5)P_4 or Ins(1,3,4,5,6)P_5.

8.7 Calmodulin, a primary Ca^{2+} sensor

Ca^{2+}-binding proteins play a key role in decoding Ca^{2+} signatures and transducing signals by activating specific targets and pathways. CaM is a Ca^{2+} sensor known to modulate the activity of many proteins. And yet, surprisingly little is known about its action mechanism and targets in plants (Snedden & Fromm, 2001). One of the reasons for this is that many targets appear to be unique to plants and the extended family of CaM-related proteins in plants consists of evolutionary divergent members (see Luan *et al.*, 2002 for further details on CaM structure, mode of action and targets). The concentration of CaM in plant cells has been estimated to be between 1 and 5 μM (Braam & Davis, 1990), although differences can occur depending on cell type and physiological state. Examples include aleurone cells – 1.3 μM (Schuurink *et al.*, 1996) and carrot cell lines – 4 μM (Kurosaki & Kaburaki, 1994).

In PTs, mapping of CaM with fluorescent phenothiazines (Haußer *et al.*, 1984) suggested the existence of a tip-to-base gradient, similar to the membrane-associated Ca^{2+} gradient visualised with chlortetracycline (Reiss & Herth, 1978). However, this technology disrupts tube growth and is no longer accepted as reliable. Experiments with microinjected fluorescently labelled CaM in living PTs suggest instead that the protein distribute evenly (Moutinho *et al.*, 1998). Similar observations have been made during mitosis (Vos & Hepler, 1998) thus stressing the need for a fundamental reassessment of early-acquired data. However, these data do not preclude an important role for CaM. Moutinho *et al.* (1998) also observed that the binding capacity of labelled CaM appears to be more intense in the sub-apical region, suggesting a higher concentration of CaM-target molecules, possibly cytoskeletal elements. It can be argued that the distribution observed is the result of a high concentration of exogenous microinjected CaM, leading to the formation of a pattern that is otherwise absent. However, CaM functions are necessarily dependent on the localisation of its targets. Romoser *et al.* (1997) concluded that the CaM concentration in the cell is limiting, and that essentially all the $(Ca^{2+})_4$-CaM present in the cell must be bound to its targets.

The MF distribution observed in the sub-apical region of both freeze-substituted (Lancelle *et al.*, 1997) and living PTs (Miller *et al.*, 1996) resembles the V-shaped collar reported for CaM binding (Moutinho *et al.*, 1998). A direct interaction between CaM and actin has therefore been hypothesised. The interaction of CaM with proteins of the actin cytoskeleton is very well documented in animal cells, specially in the case of myosin light-chain kinase (James *et al.*, 1995). In yeast, Myop2, an unconventional myosin required for polarised growth, is a CaM binding protein targeted to sites of cell growth (Brockerhoff *et al.*, 1994). The involvement of CaM and CaM-binding peptides in polarised actin-dependent processes has already been suggested (Kobayashi & Fukuda, 1994; Capelli *et al.*, 1997). In *S. cerevisiae* it has been shown that CaM controls organisation of the actin cytoskeleton via regulation of phosphatidylinositol (4,5)-bisphosphate synthesis (Desrivières *et al.*, 2002), thus linking CaM to the phosphoinositide signaling pathway.

Another possible function for CaM is the regulation of Ca^{2+} stores. In PTs, $[Ca^{2+}]_c$ drops within a few microns from about 1–3 μM in the apex to 200–220 nm in the sub apical region (Pierson *et al.*, 1996; Messerli & Robinson, 1997). Thus, a tight control between Ca^{2+} channels and pumps in this latter region must exist. CaM regulation of Ca^{2+} stores and Ins(1,4,5)P_3 receptors (Ainger *et al.*, 1993; Patel *et al.*, 1997) suggests that CaM may allow both feedback control of membrane receptors and integration of inputs from other signaling pathways. In plants modulation of ion channel and pump activity by CaM has been reported (Snedden & Fromm, 2001).

An even distribution of CaM does not necessarily mean an even functionality. CaM activity is dependent on binding of Ca^{2+} ions, thus on $[Ca^{2+}]_c$. High levels of $[Ca^{2+}]_c$, such as those found in the PT apex, may result in localised activation of CaM. Gough and Taylor (1993) showed that while the protein itself is uniformly distributed, CaM binding is elevated in the tails of highly polarised and migrating fibroblasts. The presence of a high local concentration of a CaM target and sufficient Ca^{2+} to induce some interactions, could result in an increased local activation of one CaM target relative to another. Such a positive feedback mechanism would allow the targeting of the response to particular sub-sets of the Ca^{2+}-CaM binding sites in particular regions of the cell (Gough & Taylor, 1993).

In animal cells, CaM is known to act both intra- and extracellularly (Hoeflich & Ikura, 2002). In plants, although studies point to a mainly intracellular one, CaM has also been found in extracellular areas (Murdoch & Hardham, 1998). In vitro studies have revealed extracellular roles of CaM in cell division and protoplast wall regeneration (Crocker *et al.*, 1988; Sun *et al.*, 1995) as well as pollen germination and tube growth (Ligeng & Sun, 1997). However the function of this protein in the extracellular matrix remains unexplained. In PTs, Ma and co-workers (Ma *et al.*, 1999) found that CaM, when added directly to the medium of plasma membrane vesicles, significantly activated GTPase activity and promoted tube growth. These results suggest the existence of an extracellular CaM-sensor in the plasma membrane somehow linked to a G-protein signaling cascade. This hypothesis clearly requires additional testing. CaM inhibitors such as calmidazolium, W-7 and W-12, when added extracellularly, were found to inhibit tube growth (Estruch *et al.*, 1994; Moutinho *et al.*, 1998). However, when diffusing from a microelectrode placed near the tip of growing tubes, they were unable to promote reorientation of the tube (Moutinho *et al.*, 1998). Furthermore, in *A. umbellatus* PTs, extracellular CaM itself (0.5 $\mu g\,\mu l^{-1}$) was found to perturb growth (Moutinho *et al.*, 1998). Clearly, additional experimentation is required to firmly establish a putative role for extracellular CaM in PT growth and germination.

8.8 Protein kinases and phosphatases

Phosphorylation cascades regulated by protein kinases and phosphatases represent primary, downstream transduction routes. In plants, many of the kinases identified so far are Ca^{2+}-dependent (Harmon *et al.*, 2000). Four types of protein kinases constitute the Ca^{2+}-dependent protein kinase or CaM-like protein kinase superfamily (CDPK). These kinases differ in whether they are regulated by binding Ca^{2+} (CDPKs), Ca^{2+}/CaM (CaM kinases), a combination of both (CCaMKs), or neither CDPK-related protein kinases (CRKs). CDPKs are the most abundant and were suggested to be involved in the regulation of apical $[Ca^{2+}]_c$ levels, and reorientation of PTs (Moutinho *et al.*, 1998a). CCaMKs are rarer than CDPKs and might be expressed in a few plant tissues only; they have been found in anthers (Poovaiah *et al.*, 1999), so it is tempting to speculate that their presence is associated with intense signaling cells such as PTs.

It has been proposed that transduction of Ca^{2+} signals can occur via the coupled action of Ca^{2+}-activated phosphatases and Ca^{2+}-independent kinases (Goldbeter *et al.*, 1990). De Koninck and Schulman (1998) have shown that CaM kinase II can decode the frequency of Ca^{2+} spikes into distinct amounts of kinase activity. Indirect evidence for a similar mechanism in plant cells has been obtained; within a time window, tobacco cells seem to *remember* previous hypo-osmotic shocks (Takahashi *et al.*, 1997) which cause protein kinases-controlled $[Ca^{2+}]_c$ transients.

Using a cell optical displacement assay in soybean cultured cells, Grabski *et al.* (1998) measured changes in actin tension upon incubation with several agonists and inhibitors of protein kinases. Based on their observation, the authors suggested that CDPKs and phosphatases are involved in the regulation of tension. Changes in the organisation of the actin network also occur upon Ca^{2+} transients through the stimulation of proteins containing CaM-like domains or Ca^{2+}/CaM-dependent regulatory proteins. Myosin and/or actin cross-linking proteins are likely to be the principal regulator(s) of tension within the actin network, and the principal targets of CDPKs and phosphatases. Results from experiments utilising CaM inhibitors suggest that the effect of Ca^{2+} on the tension within the actin network is CaM-dependent and involves the coordinate regulation of a CDPK and/or CaMK and a CaM-dependent phosphatase. The results with the CaM inhibitors proved more complex than those observed with kinase and phosphatase inhibitors in that there was a concentration-dependent effect on tension (Grabski *et al.*, 1998). These dose-dependent results may be interpreted to suggest that there is a greater sensitivity to CaM inhibitors of CaM or CaM-like domains associated with MLCK and/or CDPK than the CaM associated with the CaM-dependent phosphatase. Results using kinase inhibitors provided evidence that the principal target of kinase activity may be either myosin or an ABP capable of bundling or cross-linking F-actin filaments.

Moutinho *et al.* (1998a) found that PTs contain a CDPK showing functional homology to PKC. This CDPK is involved in PT growth and orientation and its activity is higher in the apical region of growing cells. When apical Ca^{2+} values were modulated (using caged-probes), the activity of the CDPK changed accordingly and induced reorientation of the growth axis. A previous study in pollen found a CDPK to co-localise with F-actin (Putnam-Evans *et al.*, 1989). That would suggest that the protein kinase activity measured by Moutinho and co-workers correspond to a different member of this family. Ritchie and Gilroy (1998) have found a member of the CDPK family in barley aleurone which does not cross-react with an antibody against the CDPK detected in pollen. However, Putnam-Evans *et al.* (1989) did not show data of how the antibody distributes in the apical region. Therefore, a direct comparison between the different results cannot be established and more definitive conclusions require further results.

Protein kinases are also involved in the self-incompatibility response of *P. rhoeas* (Rudd *et al.*, 1996); two soluble pollen proteins (p26 and p68) have been identified that showed rapid and S-specific increases in phosphorylation that was SI-specific. Another class of protein kinase, a MAPK, involved in the SI response has been identified (Rudd *et al.*, 2001). This p52-MAPK has low basal levels in growing PTs, and is activated in a SI-specific manner. The targets of this protein are not known, so it is premature to speculate on a crosstalk with the Ca^{2+}-response that leads to F-actin depolymerisation (Snowman *et al.*, 2002). Many other protein kinases have been identified in pollen but mainly associated with plasma membrane receptors or to cell cycle regulation. These will not be discussed here and we refer the reader to other references (Muschietti *et al.*, 1998; Heberle-Bors *et al.*, 2001; Kachroo *et al.*, 2002).

8.9 14-3-3 proteins

The 14-3-3 proteins are a new group of proteins whose function seems to be to join together other transduction proteins. A role in regulation and docking of protein targets has been established (Chung, 1999). 14-3-3 proteins bind to several CDPKs and also to sites in proteins that have been phosphorylated by CDPK (Chung, 1999). One family of 14-3-3 proteins, the so-called fusicoccin receptor, controls plasma membrane ATPase activity through a specific Ca^{2+}-dependent and fatty acid-dependent protein kinase (Van der Hoeven *et al.*, 1996). In pollen, 14-3-3 proteins have been recently identified and found to affect ATPase activity (Pertl *et al.*, 2001). 14-3-3 proteins are known to activate Raf, protein kinases, and phosphatases (Trewavas & Malhó, 1997) and in *Pseudomonas aeruginosa* to cause an indirect disruption of the actin cytoskeleton (Henriksson *et al.*, 2000). Further studies in apical growing cells to explore additional 14-3-3 targets are thus required.

8.10 The role of cyclic nucleotides

The role of cyclic nucleotides (CNs) in plant cell signaling has been clouded by much controversy. The presence of CNs in plant tissues has been established chemically (Newton *et al.*, 1999); so synthetic enzymes have to be present. CNs are synthesised by adenylyl/guanylyl cyclases and degraded by specific phosphodiesterases. Phosphodiesterases are usually regulated by CaM; thus an intimate relation with Ca^{2+} signaling is to be expected and this is slowly being uncovered (Malhó *et al.*, 2000). So far, there is only one work reporting the identification of a plant adenylyl cyclase (AC) (Moutinho *et al.*, 2001). The purification of plasma membrane-bound AC from animal cells has been managed only recently but there is no obvious equivalent to these large 120 kDa enzymes in the *Arabidopsis* database. Recently a new form of mammalian AC was detected in rat with a catalytic domain very similar to AC from cyanobacteria and myxobacteria (Buck *et al.*, 1999; Chen *et al.*, 2000), thus indicating great diversity in this family. Soluble ACs that show sequence differences from the classic mammalian enzymes have been reported in algae and fungi (Yashiro *et al.*, 1996) and it is likely that other cyclases might emerge.

In pollen, cloning of a putative AC revealed common motifs with its fungal counterpart (Moutinho *et al.*, 2001) but also with proteins involved in disease responses. cAMP is believed to be involved in such responses (Cooke *et al.*, 1994), and parallels between PT growth in the style and fungal hyphal infection are frequently drawn. This cDNA, although not full length, caused accumulations of cAMP when expressed in *E. coli* and complemented a catabolic defect in carbohydrate fermentation in an *E. coli cya*A mutant (Moutinho *et al.*, 2001). Transformation with antisense oligos directed against this pollen cDNA or treatment with antagonists of AC caused disruption of PT growth suggesting a requirement for continued cAMP synthesis. This was supported by the imaging of cAMP in growing PTs where it was observed that forskolin, an AC activator, transiently increased cAMP whilst dideoxyadenosine, an inhibitor of ACs, caused a temporary decline (Moutinho *et al.*, 2001). Transient changes in cAMP also indicated that degrading enzymes such as phosphodiesterases must be present in PTs. It is thus important that other libraries are screened in the *cya*A bacterial line to aid identification of other AC in plants.

The targets of the cAMP signaling pathway are still largely unknown. Recently two independent reports (Ehsan *et al.*, 1998; Volotovski *et al.*, 1998) suggested a role for cAMP in the regulation of Ca^{2+} levels. A similar situation has been reported for PTs (Malhó *et al.*, 2000). When growing PTs were submitted to different treatments that putatively affect the levels of cAMP (photolysis of caged cAMP; external addition of membrane-permeant cAMP; external addition of forskolin), $[Ca^{2+}]_c$ transients were observed. After caged release of an estimated ~1–2 μM cAMP, growth rates temporally declined and reorientation of the tube growth axis occurred. This was accompanied by a transient $[Ca^{2+}]_c$ elevation in the apical region but not in the sub-apical. In contrast, the addition of external cAMP led to a complete growth arrest and a $[Ca^{2+}]_c$ increase in the apical and sub-apical region. When compared to the release of a small concentration of cAMP by flash photolysis, this suggests a toxic effect and strong perturbation of ion fluxes. This suggests that a cAMP pathway, acting together with Ca^{2+}, is involved in PT reorientation. Interestingly, experiments with a membrane-permeable version of cGMP diffusing from a microneedle placed near the tip of growing PTs, failed to cause a visible response (Moutinho *et al.*, 2001). Further experimentation is thus required to evaluate the role of other CNs in PT growth and reorientation.

8.11 GTPases, the signaling switches

Small GTP-binding proteins (GTPases) are versatile signaling switches responsible for an extensive cross-talk and cooperation between signal transduction pathways (Bar-Sagi & Hall, 2000). They all belong to the Ras superfamily which includes five families: Ras, Rab, Arf, Ran and Rho. Despite several common features, they exhibit a remarkable diversity in both structure and function (Yang, 2002). They act as molecular switch regulators, cycling between a GDP-bound (inactive) state and a GTP-bound (active) state. Conversion is regulated by a guanine nucleotide exchange factor (GEF) but the GTP form exhibits a weak intrinsic GTPase activity for GTP hydrolysis, requiring a GTPase-activating protein (GAP) for efficient deactivation. In addition, most small GTPases cycle between membrane-bound and cytosolic forms. Because only membrane-associated GTPases can be activated by GEF, their removal by a cytosolic factor called guanine nucleotide dissociation inhibitor (GDI) negatively regulates these GTPases (Yang, 2002).

Rab is the largest family of small GTPases and control transport and docking of specific vesicles. A Rab2 homologue, necessary for membrane traffic between ER and Golgi in mammalian cells (Tisdale *et al.*, 1992), is present in *Arabidopsis* pollen grains (Moore *et al.*, 1997) and was recently shown to be important for PT growth (Cheung *et al.*, 2002). The authors found that a pollen-predominant Rab2 functions in the secretory pathway between the endoplasmic reticulum and the Golgi in elongating PTs. GFP–NtRab2 fusion protein localised to Golgi bodies and dominant-negative mutations in NtRab2 proteins inhibited this localisation, blocked the delivery of Golgi-resident as well as plasmalemma and secreted proteins to their normal locations, and inhibited PT growth. These observations indicate that NtRab2 is important for trafficking between the endoplasmic reticulum and the Golgi bodies in PTs and may be specialised to optimally support the high secretory demands in these tip growth cells.

Rho GTPases were initially shown to regulate the organisation of the actin cytoskeleton and cell polarity development in eukaryotes (Yang, 2002) but Rho signaling controls many diverse processes, including gene expression, cell wall synthesis, H_2O_2 production, endocytosis, exocytosis, cytokinesis, cell cycle progression, and cell differentiation in various eukaryotic organisms (Yang, 2002). The Rho family is composed of conserved subfamilies (Cdc42, Rac and Rho) but plants do not contain orthologs for any of the fungal and animal Rho family GTPases. However, a large number of Rho-related GTPases have been found in plants, all belonging to a unique subfamily named ROP (for Rho-related GTPase from plants) (Yang, 2002). Interestingly, many Rop interactors are novel or unique to plants, consistent with the hypothesis that Rops belong to a plant-specific branch of the Rho family of small GTPases (Yang, 2002). The naming of Rop GTPases is confusing in the literature and here we adopted the terminology of Yang (2002).

Rops have been shown to play a significant role in PT elongation in pea, *Arabidopsis*, and tobacco (Yang & Watson, 1993; Lin & Yang, 1997; Kost *et al.*, 1999). Both in *Arabidopsis* and *N. tabacum* the small GTPases were found to accumulate at the tube tip and similar observations were made in root hairs (Molendjik *et al.*, 2001). In pollen, the constitutively active, GTP-bound form led to swollen tip growth whereas the dominant-negative, GDP-bound form led to cessation of tube growth, which proves that the protein has a central role in the regulation of tube extension. At the PT tip, Rops seem to activate phosphatidylinositol kinase leading to the formation of phosphatidylinositol 4,5-phosphate, which could function in release of Ca^{2+} from intracellular stores or activate entry from the extracellular space (see Section 8.5). Transient expression of the mutant GTP- and GDP-bound proteins in tobacco PTs led to formation of extensive and reduced actin cables, respectively, which shows that Rops play a role in PT actin organisation (Kost *et al.*, 1999).

GTP-binding proteins have also been claimed to regulate endo-exocytosis (Lin *et al.*, 1996; Li *et al.*, 1999), and the relationships between exocytosis, endocytosis, and the actin cytoskeleton in the growing PT tip make an interesting question to which no clear answer can yet be provided. The non-hydrolysable analogues of guanine nucleotide interfere in biological processes and it was observed that GDPβS decreases growth rate, while GTPγS has the opposite effect (Ma *et al.*, 1999; Camacho & Malhó, 2003) thus confirming the importance of GTPases for apical growth. The release of GTPγS was also found to promote exocytosis, a process which seems to be more dependent on the active state of GTPases rather then on the cycling between GTP-GDP bound state (Zheng & Yang, 2000). Selective exposure to elevated Ca^{2+} alone is not sufficient to explain the selectivity of membrane retrieval (Smith *et al.*, 2000). It is therefore possible that GTPases play an active role in endo/exocytosis by coupling the actin cytoskeleton to the sequential steps underlying membrane trafficking at the site of exocytosis. This suggestion finds support in the results of Fu *et al.* (2001) who reported that F-actin dynamics in the tip of PTs is GTPase-dependent. The results obtained in PTs (Camacho &

Malhó, 2003) favour the existence of a rapid endocytosis mechanism. This type of endocytosis is a Ca^{2+}-dependent process, coupled to exocytosis that requires GTP hydrolysis and dynamin but not clathrin. Dynamin is a GTPase involved in the pulling and detachment of the endocytic vesicle from the plasma membrane. This protein, already identified in plant cells (Jin *et al.*, 2001), is thought to be involved both in constitutive and rapid endocytosis and to require the hydrolysis of GTP to accomplish its function. Along with this rapid endocytosis, a second mechanism where endo- and exocytosis are uncoupled should be present. This mode is favoured if Ca^{2+} is lowered and/or GTP increases. According to this hypothesis, secretion is modulated by Ca^{2+} and GTPase activity to attain optimal conditions for endo/exocytosis.

Which signals activate the ROP signaling pathway and confine its activity to the tube apex is not clear. A Rop-like protein was found to associate with a receptor-like kinase (Trotochaud *et al.*, 1999) and kinase activity is higher in the PT apex (Moutinho *et al.*, 1998a). Using microinjection of specific antibodies, Li *et al.* (1999) reported an interaction between Ca^{2+} and the expression of a Rop1 protein. The authors demonstrated that inhibition of Rop1At proteins modifies Ca^{2+} influx but could not determine inhibition of the Ca^{2+} influx altogether (which would require electrophysiological methods). It was actually found that growth inhibition was reversed by higher extracellular Ca^{2+}, which means that influx was still possible. Using antisense oligonucleotides to reduce Rop expression, Camacho and Malhó (2003) made similar observations. These experiments revealed (1) that perturbation of polar growth can be accomplished by reducing the expression of Rop proteins; (2) that provided growth is not totally disrupted, a $[Ca^{2+}]_c$ gradient, thus Ca^{2+} influx, can be maintained; (3) that Rop inhibition results in a substantial decrease in the rate of endocytosis but not on its overall pattern. The data support a model in which Ca^{2+} and Rop GTPases act differentially but in a concerted form in the sequential regulation of PT secretion and membrane retrieval. However, inhibition (or knock-out) of any protein crucial for tip growth will certainly affect the Ca^{2+} gradient either directly or through a signaling loop. Although there may be several different interpretations to these data, we suggest that Ca^{2+} plays a major role in the secretion of cell wall components while Rop GTPases appear to play a key role in fusion of docked vesicles and endocytosis. This hypothesis is supported by recent identification of two structurally distinct putative ROP1 targets in the control of polar growth in PTs, RIC3 and RIC4 (Wu *et al.*, 2001). It is clear, though, that the exact position of these elements in the signaling cascade controlling PT growth and orientation is still dubious and requires further investigation.

8.12 Transducons – the unity for signaling

The problems of specificity of signal transduction are still outstanding. The same set of molecules seem to be used to transduce a whole variety of different signals. In this case it is thought that particular spatial distributions of transduction molecules and machinery are critically coupled with temporal aspects of the second messengers and protein kinases. Cytosolic Ca^{2+} exemplifies both properties because it has (1) a highly constrained mobility in the cytoplasm and (2) differential sensitivity of proteins to Ca^{2+} ensuring that transients of different sizes and temporal properties induce specific transduction sequences.

Other molecules do not suffer the same diffusion constraints as Ca^{2+} (e.g. H^+, CNs) which does not imply that highly localised changes do not occur. Rich and co-workers (Rich *et al.*, 2000, 2001) observed that, within the same cell, there are different compartments for CNs that behave in response to the same signal with different kinetics. One of the compartments exhibited an extremely short transient signal as a result of AC clustering within the vicinity of receptors and transduction through tethered protein kinases. Diffusion of cAMP was severely retarded by high levels of phosphodiesterase activity and thus cAMP can behave like Ca^{2+}. A similar situation may happen in PTs. Moutinho *et al.* (2001) observed that during reorientation transient elevations in cAMP occur but restricted to the apical region of the cell. It is thus clear that spatial inhomogeneity of signaling is the base for specificity of cellular responses to different initial signals. Indeed, mechanisms that anchor kinases and phosphatases to the cytoskeleton through particular protein subunits are being uncovered rapidly (Faux & Scott, 1996). These *aggregates* of receptors, channels, bound CaM, protein kinases, etc. were designated by *Transducons* (Trewavas & Malhó, 1997). They may be specifically positioned within a cell and operate as on/off switches in the metabolic circuitry. Such interactions would help to improve the rate and accuracy of information transfer and processing. Transducons could also act as integration and information storage facilities for the cell. Using this analogy, the phenotype of a plant represents the indissoluble linkage between genes, and the signaling networks. The keys in this programming machine are punched by environmental and developmental cues.

8.13 Concluding remarks

Apical growing cells, namely PTs, are excellent models to study signal transduction mechanisms given their extreme dynamics and polarisation of sub-cellular structures. We know that this is largely due to cytoskeletal elements but surprisingly this is an area where our knowledge is still obscure. Despite the well-known characteristics of molecular self-assembly and patterning of tubulin and actin, *in vivo* research has focused mainly on positional information cues (e.g. which signals lead to a specific arrangement). How the cytoskeleton reorganises to produce that arrangement i.e 'what type of assembly dynamics does it exhibit' and 'how does it integrate multiple signalling cues?' are questions that remain largely unanswered. It is thus crucial to design experiments where imaging of cytoskeleton dynamics is coupled to second messenger signaling. Thanks to molecular constructs this is now largely possible. However, if we aim to complete this puzzle, we must not forget that signaling cascades are tightly interconnected and act in a concerted way producing, not a *second messenger signature* but instead a *cell signature* response.

Acknowledgements

Research in R.M. lab is supported by Centro de Biotecnologia Vegetal and Fundação Ciência e Tecnologia, Portugal (Grants no BCI/32605/99, BCI/37555/2001 and BD/19642/99).

References

Ainger, K., Avossa, D., Morgan, F., Hill, S.J., Barry, C., Barbarese, E. & Carson, J.H. (1993) Transport and localisation of exogenous myelin basic protein mRNA microinjected into oligodendrocytes, *J. Cell Biol.*, **123**, 431–441.

Åström, H., Sorri, O. & Raudaskoski, M. (1995) Role of the microtubules in the movement of the vegetative nucleus and generative cell in tobacco pollen tubes, *Sex. Plant Reprod.*, **8**, 61–69.

Bamburg, J.R. (1999) Proteins of the ADF/cofilin family: essential regulators of actin dynamics, *Ann. Rev. Cell Dev. Biol.*, **15**, 185–230.

Bar-Sagi, D. & Hall, A. (2000) Ras and Rho GTPases: a family reunion, *Cell*, **103**, 227–238.

Bartolo, M.E. & Carter, J.V. (1992) Lithium decreases cold-induced microtubule depolymerization in mesophyll cells of spinach, *Plant Physiol.*, **99**, 1716–1718.

Berridge, M.J. (1995) Capacitative calcium entry, *Biochem. J.*, **312**, 1–11.

Braam, J. & Davis, R.W. (1990) Rain-, wind-, and touch-induced expression of calmodulin and calmodulin-related genes in *Arabidopsis*, *Cell*, **60**, 357–364.

Brockerhoff, S.E., Stevens, R.C. & Davis, T. (1994) The unconventional myosin Myop2 is a calmodulin target at sites of cell growth in *Saccharomyces cerevisiae*, *J. Cell Biol.*, **124**, 315–323.

Brownlee, C. & Wood, J.W. (1986) A gradient of cytoplasmic free calcium in growing rhizoid cells of *Fucus serratus*, *Nature*, **320**, 624–626.

Buck, J., Sinclair, M.L., Schapal, L., Cann, M.J. & Levin, L.R. (1999) Cytosolic adenylyl cyclase defines a unique signaling molecule in mammals, *Proc. Nat. Acad. Sci. USA*, **96**, 79–84.

Cai, G., Bartalesi, A., Del Casino, C., Moscatelli, A., Tiezzi, A. & Cresti, M. (1993) The kinesin-immunoreactive homologue from *Nicotiana tabacum* pollen tube: biochemical properties and subcellular localization, *Planta*, **191**, 496–506.

Cai, G., Romagnoli, S., Moscatelli, A., *et al.* (2000) Identification and characterization of novel microtubule-based motor associated with membranous organelles in tobacco pollen tubes, *Plant Cell*, **12**, 1719–1736.

Camacho, L. & Malhó, R. (2003) Endo-exocytosis in the pollen tube apex is diferentially regulated by Ca^{2+} and GTPases, *J. Exptl. Bot.*, **54**, 83–92.

Capelli, N., Barja, F., Van Tuinen, D., Monnat, J., Turian, G. & Ortega Perez, R. (1997) Purification of a 47kDa calmodulin-binding polypeptide as an actin-binding protein from *Neurospora crassa*. *FEMS Microb. Lett.*, **147**, 215–220.

Chen, C.Y., Wong, E.I., Vidali, L., *et al.* (2002) The regulation of actin organization by actin-depolymerizing factor in elongating pollen tubes, *Plant Cell*, **14**, 2175–2190.

Chen, Y., Cann, M.J., Litvin, T.N., *et al.* (2000) Soluble adenylyl cyclase as an evolutionarily conserved bicarbonate sensor, *Science*, **289**, 625–628.

Cheung, A.Y., Chen, C.Y.-H., *et al.* (2002) Rab2 GTPase regulates vesicle trafficking between the endoplasmic reticulum and Golgi bodies and is important to pollen tube growth, *Plant Cell*, **14**, 945–962.

Chory, J. & Wu, D. (2003) Weaving the complex web of signal transduction, *Plant Physiol.*, **125**, 77–80.

Chung, H.J. (1999) The 14-3-3 proteins: cellular regulators of plant metabolism, *Trends Plant Sci.*, **4**, 367–371.

Clarke, S.R., Staiger, C.J., Gibbon, B.C. & Franklin-Tong, V.E. (1998) A potential signaling role for profilin in pollen of *Papaver rhoeas*, *Plant Cell*, **10**, 967–979.

Cooke, C.J., Smith, C.J., Walton, T.J. & Newton, R.P. (1994) Evidence that cyclic AMP is involved in the hypersensitive response of *Medicago sativa* to a fungal elicitor, *Phytochem.*, **35**, 889–895.

Cooper, J.A. & Schafer, D.A. (2000) Control of actin assembly and disassembly at filament ends, *Curr. Opin. Cell Biol.*, **12**, 97–103.

Crocker, C., Dawson, R.A., Barton, C.H. & MacNeil, S. (1988) An extracellular role for calmodulin-like activity in cell proliferation, *Biochem. J.*, **25**, 877–844.

Dawson, A.P. (1997) Calcium signaling How do IP_3 receptors work?, *Curr. Biol.*, **7**, R544–RR547.

De Koninck, P. & Schulman, H. (1998) Sensitivity of CaM Kinase II to the frequency of Ca^{2+} oscillations, *Science*, **279**, 227–230.

Del Casino, C., Li, Y., Moscatelli, A., Scali, M., Tiezzi, A. & Cresti, M. (1993) Distribution of microtubules during the growth of tobacco pollen tubes, *Biol. Cell*, **79**, 125–132.

Derksen, J., Pierson, E.S. & Traas, J.A. (1985) Microtubules in vegetative and generative cells of pollen tubes, *Eur. J. Cell Biol.*, **38**, 142–148.

Derksen, J., Rutten, T., van Amstel, T., de Win, A., Doris, F. & Steer, M. (1995) Regulation of pollen tube growth, *Acta Bot. Neerl.*, **44**, 93–119.

Desrivières, S., Cooke, F.T., Morales-Johansson, H., Parker, P.J. & Hall, M.N. (2002) Calmodulin controls organization of the actin cytoskeleton via regulation of phosphatidylinositol (4,5)-bisphosphate synthesis in *Saccharomyces cerevisiae*, *Biochem. J.*, **366**, 945–951.

Drøbak, B.K. (1992) The plant phosphoinositide system, *Biochem. J.*, **288**, 697–712.

Drøbak, B.K., Watkins, P.A.C., Valenta, R., Dove, S.K., Lloyd, C.W. & Staiger, C.J. (1994) Inhibition of plant plasma membrane phosphoinositide phospholipase C by the actin-binding protein, profilin, *Plant J.*, **6**, 389–400.

Ehsan, H., Reichheld, J.P., Roef, L., *et al.* (1998) Effect of indomethacin on cell cycle dependent cyclic AMP fluxes in tobacco BY-2 cells, *FEBS Lett.*, **422**, 165–169.

Estruch, J.J., Kadwell, S., Merlin, E. & Crossland, L. (1994) Cloning and characterisation of a maize pollen-specific calcium-dependent calmodulin-independent protein kinase, *Proc. Nat. Acad. Sci. USA*, **91**, 8837–8841.

Faux, M.C. & Scott, J.D. (1996) More on target with protein phosphorylation conferring specificity by location, *Trends Biochem. Sci.*, **21**, 312–315.

Fisher, D.D., Gilroy, S. & Cyr, J.R. (1996) Evidence for opposing effects of calmodulin on cortical microtubules, *Plant Physiol.*, **112**, 1079–1087.

Franklin-Tong, V.E., Drøbak, B.K., Allan, A.C., Watkins, P.A.C. & Trewavas, A.J. (1996) Growth of pollen tubes of *Papaver rhoeas* is regulated by a slow moving calcium wave propagated by inositol triphosphate, *Plant Cell*, **8**, 1305–1321.

Fu, Y., Wu, G. & Yang, Z. (2001) Rop GTPase-dependent dynamics of tip-localized F-actin controls tip growth in pollen tubes, *J. Cell Biol.*, **152**, 1019–1032.

Fujimoto, T., Miyawaki, A. & Mikoshiba, K. (1995) Inositol 1,4,5-trisphosphate receptor-like protein in plasmalemmal caveolae is linked to actin filaments, *J. Cell Sci.*, **108**, 7–15.

Geitmann, A., Li, Y.-Q. & Cresti, M. (1995) The role of cytoskeleton and dictyosome activity in the pulsatory growth of *Nicotiana tabacum* and *Petunia hybrida* pollen tubes, *Bot. Acta*, **109**, 102–109.

Geitmann, A., Snowman, B.N., Emons, A.M.C. & Franklin-Tong, V.E. (2000) Alterations in the actin cytoskeleton of pollen tubes are induced by the self-incompatibility reaction in *Papaver rhoeas*, *Plant Cell*, **12**, 1239–1251.

Gibbon, B.C., Ren, H. & Staiger, C.J. (1997) Characterization of maize (*Zea mays*) pollen profilin function *in vitro* and in live cells, *Biochem J.*, **327**, 909–915.

Gibbon, B.C., Zonia, L.E., Kovar, D.R., Hussey, P.J. & Staiger, C.J. (1998) Pollen profilin function depends on interaction with proline-rich motifs, *Plant Cell*, **10**, 981–993.

Goeckeler, Z.M. & Wysolmerski, R.B. (1995) Myosin light chain kinase regulated endothelial cell contraction: the relationship between isometric tension, actin polymerization, and myosin phosphorylation, *J. Cell Biol.*, **130**, 613–627.

Goldbeter, A., Dupont, G. & Berridge, M.J. (1990) Minimal model for signal-induced Ca^{2+} oscillations and for their frequency encoding through protein phosphorylation, *Proc. Nat. Acad. Sci. USA*, **87**, 1461–1465.

Goode, B.L., Drubin, D.G. & Barnes, G. (2000) Functional cooperation between the microtubule and actin cytoskeletons, *Curr. Opin. Cell Biol.*, **12**, 63–71.

Gough, A.H. & Taylor, D.L. (1993) Fluorescence anisotropy imaging microscopy maps calmodulin binding during cellular contraction and locomotion, *J. Cell Biol.*, **121**, 1095–1107.

Grabski, S., Arnoys, E., Busch, B. & Schindler, M. (1998) Regulation of actin tension in plant cells by kinases and phosphatases, *Plant Physiol.*, **116**, 279–290.

Gungabissoon, R.A., Jiang, C.-J., Drobak, B.K., Maciver, S.K. & Hussey, P.J. (1998) Interaction of maize actin-depolymerising factor with actin and phosphoinositides and its inhibition of plant phospholipase C, *Plant J.*, **16**, 689–696.

Harmon, A.C., Gribskov, M. & Harper, J.F. (2000) CDPKs – a kinase for every Ca^{2+} signal?, *Trends Plant Sci.*, **5**, 154–159.

Haußer, I., Herth, W. & Reiss, H.-D. (1984) Calmodulin in tip-growing plant cells, visualized by fluorescing calmodulin-binding phenothiazines, *Planta*, **162**, 33–39.

Heberle-Bors, E., Voronin, V., Touraev, A., Testillano, P.S., Risueño, M.C. & Wilson, C. (2001) MAP kinase signaling during pollen development, *Sex. Plant Reprod.*, **14**, 15–19.

Henriksson, M.L., Trollér, U. & Hallberg. B. (2000) 14-3-3 proteins are required for inhibition of Ras by Exoenzyme S, *Biochem. J.*, **349**, 697–701.

Hoeflich, K.P. & Ikura, M. (2002) Calmodulin in action: diversity in target recognition and activation mechanisms, *Cell*, **108**, 739–742.

Holdaway-Clarke, T., Feijó, J.A., Hackett, G.R., Kunkel, J.G. & Hepler, P.K. (1997) Pollen tube growth and the intracellular cytosolic calcium gradient oscillates in phase while extracellular calcium influx is delayed, *Plant Cell*, **9**, 1999–2010.

Hussey, P.J., Yuan, M., Calder, G., Khan, S. & Lloyd, C.W. (1998) Microinjection of pollen-specific actin-depolymerizing factor, ZmADF1, reorientates F-actin strands in *Tradescantia* stamen hair cells, *Plant J.*, **14**, 353–357.

James, P., Vorherr, T. & Carafoli, E. (1995) Calmodulin-binding domains: just two faced or multi-faceted?, *Trends Biochem. Sci.*, **20**, 38–42.

Janmey, P.A. (1994) Phosphoinositides and calcium as regulators of cellular actin assembly and disassembly, *Ann. Rev. Physiol.*, **56**, 169–191.

Jin, J.B., Kim, Y.A., Kim, S.J., *et al.* (2001) A new dynamin-like protein, ADL6, is involved in trafficking from the trans-Golgi network to the central vacuole in *Arabidopsis*, *Plant Cell*, **13**, 1511–1526.

Joos, U., van Aken, J. & Kristen, U. (1994) Microtubules are involved in maintaining the cellular polarity in pollen tubes of *Nicotiana sylvetris*, *Protoplasma*, **179**, 5–15.

Kachroo, A., Nasrallah, M.E. & Nasrallah, J.B. (2002) Self-incompatibility, receptor-ligand signaling, and cell–cell communication in plant reproduction, *Plant Cell*, **14**, S227–S238.

Kimura, K., Ito, M., Amano, M., *et al.* (1996) Regulation of myosin phosphatase by rho and rho-associated kinase (rhokinase), *Science*, **273**, 245–248.

Klahre, U., Friederich, E., Kost, B., Louvard, D. & Chua, N.H. (2000) Villin-like actin-binding proteins are expressed ubiquitously in *Arabidopsis*, *Plant Physiol.*, **122**, 35–47.

Kobayashi, H. & Fukuda, H. (1994) Involvement of calmodulin and calmodulin-binding peptides in the differentiation of tracheary elements in *Zinnia cells*, *Planta*, **194**, 388–394.

Kost, B., Spielhofer, P. & Chua, N.-H. (1998) A GFP-mouse talin fusion protein labels plant actin filaments *in vivo* and visualizes the actin cytoskeleton in growing pollen tubes, *Plant J.*, **16**, 393–401.

Kost, B., Lemichez, E., Spielhofer, P., *et al.* (1999) Rac homologues and compartmentalized phosphatidilinositol 4,4-biphosphate act in a common pathway to regulate polar pollen tube growth, *J. Cell Biol.*, **145**, 317–330.

Kurosaki, F. & Kaburaki, H. (1994) Calmodulin-dependency of a Ca^{2+}-pump at the plasma membrane of cultured carrot cells, *Plant Sci.*, **104**, 23–30.

Lancelle, S.A., Cresti, M. & Hepler, P.K. (1997) Growth inhibition and recovery in freeze-substituted *Lilium longiflorum* pollen tubes: structural effects of caffeine, *Protoplasma*, **196**, 21–33.

Lancelle, S.A. & Hepler, P.K. (1991) Association of actin with cortical microtubules revealed by immunogold localization in Nicotiana pollen tubes, *Protoplasma*, **165**, 167–172.

Li, H., Lin, Y., Heath, R.M., Zhu, M.X. & Yang, Z. (1999) Control of pollen tube tip growth by a Rop GTPase-dependent pathway that leads to tip-localized calcium influx, *Plant Cell*, **11**, 1731–1742.

Ligeng, M. & Sun, D.Y. (1997) The effects of extracellular calmodulin on initiation of *Hippeastrum rutilum* pollen germination and tube growth, *Planta*, **202**, 336–340.

Lin, Y. & Yang, Z. (1997) Inhibition of pollen tube elongation by microinjected anti-Rop1Ps antibodies suggests a crucial role for Rho-type GTPases in the control of tip growth, *Plant Cell*, **9**, 1647–1659.

Lin, Y., Wang, Y., Zhu, J.-K. & Yang, Z. (1996) Localization of a Rho GTPase implies a role in tip growth and movement of the generative cell in pollen tubes, *Plant Cell*, **8**, 293–303.

Liu, G.Q., Cai, G., Del Casino, C., Tiezzi, A. & Cresti, M. (1994) Kinesin-related polypeptide is associated with vesicles from *Corylus avellana* pollen, *Cell Motil. Cytoskeleton*, **29**, 155–166.

Lopez, I., Anthony, R.G., Maciver, S.K., *et al.* (1996) Pollen specific expression of maize genes encoding actin depolymerizing factor-like proteins, *Proc. Nat. Acad. Sci. USA*, **93**, 7415–7420.

Luan, S., Kudla, J., Rodriguez-Concepcion, M., Yalovsky, S. & Gruissem, W. (2002) Calmodulins and calcineurin-B-like proteins: calcium sensors for specific signal response coupling in plants, *Plant Cell*, **14**, S389–S400.

Ma, L., Xu, X., Cui, S. & Sun, D. (1999) The presence of a heterotrimeric G protein and its role in signal transduction of extracellular calmodulin in pollen germination and tube growth, *Plant Cell*, **11**, 1351–1364.

Malhó, R. (1998) The role of inositol(1,4,5)triphosphate in pollen tube growth and orientation, *Sex. Plant Reprod.*, **11**, 231–235.

Malhó, R. (1999) Coding information in plant cells. The multiple roles of Ca^{2+} as a second messenger, *Plant Biol.*, **1**, 487–494.

Malhó, R., Camacho, L. & Moutinho, A. (2000) Signalling pathways in pollen tube growth and reorientation, *Ann. Bot.*, **85**(Suppl. A), 59–68.

Malhó, R., Read, N.D., Trewavas, A.J. & Pais, M.S. (1995) Calcium channel activity during pollen tube growth and reorientation, *Plant Cell*, **7**, 1173–1184.

Malhó, R. & Trewavas, A.J. (1996) Localized apical increases of cytosolic free calcium control pollen tube orientation, *Plant Cell*, **8**, 1935–1949.

Messerli, M.A., Créton, R., Jaffe, L.F. & Robinson, K.R. (2000) Periodic increases in elongation rate precede increases in cytosolic Ca^{2+} during pollen tube growth, *Dev. Biol.*, **222**, 84–98.

Messerli, M.A., Danuser, G. & Robinson, K.R. (1999) Pulsatile influxes of H^+, K^+ and Ca^{2+} lag growth pulses of *Lilium longiflorum* pollen tubes, *J. Cell Sci.*, **112**, 1497–1509.

Messerli, M. & Robinson, K.R. (1997) Tip localized Ca^{2+} pulses are coincident with peak pulsatile growth rates in pollen tubes of *Lilium longiflorum*, *J. Cell Sci.*, **110**, 1269–1278.

Miller, D.D., Lancelle, S.A. & Hepler, P.K. (1996) Actin filaments do not form a dense meshwork in *Lilium longiflorum* pollen tube tips, *Protoplasma*, **195**, 123–132.

Miller, D.D., Scordilis, S.P. & Hepler, P.K. (1995) Identification and localization of three classes of myosins in pollen tubes of *Lilium longiflorum* and *Nicotiana alata*, *J. Cell Sci.*, **108**, 2549–2563.

Mittermann, I., Swoboda, I., Pierson, E., *et al.* (1995) Molecular cloning and characterization of profilin from tobacco (*Nicotiana tabacum*): increased profilin expression during pollen maturation, *Plant Mol. Biol.*, **27**, 137–146.

Molendjik, A.J., Bischoff, F. & Rajendrakumar, C.S.V. (2001) *Arabidopsis thaliana* Rop GTPases are localized to tips of root hairs and control polar growth, *Embo J.*, **20**, 2779–2788.

Moore, I., Diefenthal, T., Zarsky, V., Shell, J. & Palme, K. (1997) A homolog of the mammalian GTPase Rab2 is present in *Arabidopsis* and is expressed predominantly in pollen grains and seedlings, *Proc. Nat. Acad. Sci. USA*, **94**, 762–767.

Moscatelli, A., Del Casino, C., Lozzi, L., *et al.* (1995) High molecular weight polypeptides related to dynein heavy chains in *Nicotiana tabacum* pollen tubes, *J. Cell Sci.*, **108**, 1117–1125.

Moscatelli, A., Cai, G., Ciampolini, F. & Cresti, M. (1998) Dynein heavy chain-related polypeptides are associated with organelles in pollen tubes of *Nicotiana tabacum*, *Sex. Plant Reprod.*, **11**, 31–40.

Moutinho, A., Love, J., Trewavas, A.J. & Malhó, R. (1998) Distribution of calmodulin protein and mRNA in growing pollen tubes, *Sex. Plant Reprod.*, **11**, 131–139.

Moutinho, A., Hussey, P.J., Trewavas, A.J. & Malhó, R. (2001) cAMP acts as a second messenger in pollen tube growth and reorientation, *Proc. Nat. Acad. Sci. USA*, **98**, 10481–10486.

Moutinho, A., Trewavas, A.J. & Malhó, R. (1998a) Relocation of a Ca^{2+}-dependent protein kinase activity during pollen tube reorientation, *Plant Cell*, **10**, 1499–1510.

Muir, S.R. & Sanders, D. (1997) Inositol 1,4,5-trisphosphate-sensitive Ca^{2+} release across non vacuolar membranes in cauliflower, *Plant Physiol.*, **114**, 1511–1521.

Mullins, R.D. (2000) How WASP-family proteins and the Arp2/3 complex convert intracellular signals into cytoskeletal structures, *Curr. Opin. Cell Biol.*, **12**, 91–96.

Murdoch, L.J. & Hardham, A.R. (1998) Components in the haustorial wall of the flax rust fungus, *Melampsora lini*, are labelled by three anti-calmodulin monoclonal antibodies, *Protoplasma*, **201**, 180–193.

Muschietti, J., Eyal, Y. & McCormick, S. (1998) Pollen tube localization implies a role in pollen-pistil interactions for the tomato receptor-like protein kinases LePRK1 and LePRK2, *Plant Cell*, **10**, 319–330.

Newton, R.P., Roef, L., Witters, E. & Van Onckelen, H. (1999) Cyclic nucleotides in higher plants: the enduring paradox, *New Phytol.*, **143**, 427–455.

Palevitz, B.A. & Tiezzi, A. (1992) Organization, composition, and function of the generative cell and sperm cytoskeleton, *Int. Rev. Cytol.*, **140**, 149–185.

Patel, S., Morris, S.A., Adkins, C.E., O'Beirne, G. & Taylor, C.W. (1997) Ca^{2+}-independent inhibition of inositol trisphosphate receptors by calmodulin: redistribution of calmodulin as a possible means of regulating Ca^{2+} mobilization, *Proc. Nat. Acad. Sci. USA*, **94**, 11627–11632.

Pertl, H., Himly, M., Gehwolf, R., *et al.* (2001) Molecular and physiological characterisation of a 14-3-3 protein from lily pollen grains regulating the activity of the plasma membrane H^+ ATPase during pollen grain germination and tube growth, *Planta*, **213**, 132–141.

Petersen, C.C.H. & Berridge, M.J. (1994) The regulation of capacitative calcium entry by calcium and protein kinase C in *Xenopus* oocytes, *J. Biol. Chem.*, **269**, 32246–32253.

Pierson, E.S., Derksen, J. & Traas, J.A. (1986) Organization of microfilaments and microtubules in pollen tubes grown *in vitro* or *in vivo* in various angiosperms, *Eur. J. Cell Biol.*, **41**, 14–18.

Pierson, E.S., Miller, D.D., Callaham, D.A., van Aken, J., Hackett, G. & Hepler, P.K. (1996) Tip-localized calcium entry fluctuates during pollen tube growth, *Dev. Biol.*, **174**, 160–173.

Poovaiah, B.W., Xia, M., Liu, Z., *et al.* (1999) Developmental regulation of the gene for chimeric calcium/calmodulin-dependent protein kinase in anthers, *Planta*, **209**, 161–171.

Putnam-Evans, C., Harmon, A.C., Palevitz, B.A., Fechheimer, M. & Cormier, M.J. (1989) Calcium-dependent protein kinase is localized with F-actin in plant cells, *Cell Mot. Cytoskeleton*, **12**, 12–22.

Raudaskoski, M., Åström, H., Perttila, K., Virtanen, I. & Louhelainen, J. (1987) Role of microtubule cytoskeleton in pollen tubes: An immunochemical and ultrastructural approach, *Biol. Cell*, **61**, 177–188.

Reddy, A.S.N. & Day, I.S. (2001) Analysis of the myosins enconded in the recently completed *Arabidopsis thaliana* genome sequence, *Gen. Biol.*, **2**(7), research0024.1–0024.17.

Reiss, H.-D. & Herth, W. (1978) Visualization of the Ca^{2+}-gradient in growing pollen tubes of *Lilium longiflorum* with chlorotetracycline fluorescence, *Protoplasma*, **97**, 373–377.

Rich, T.C., Fagan, K.A., Tse, T.E., Schaak, J., Cooper, D.M.F. & Karpen, J.W. (2001) A uniform extracellular stimulus triggers distinct cAMP signals in different compartments of a simple cell, *Proc. Nat. Acad. Sci. USA*, **98**, 13049–13054.

Rich, T.C., Gagan, K.A., Nakata, H., Schaak, J., Cooper, D.M.F. & Karpen, J.W. (2000) Cyclic nucleotide gated channels co-localise with adenylyl cyclase in regions of restricted cAMP diffusion, *Proc. Nat. Acad. Sci. USA*, **116**, 147–162.

Ridley, A.J. & Hall, A. (1992) The small GTP-binding protein rho regulates the assembly of focal adhesions and actin stress fibers in response to growth factors, *Cell*, **70**, 389–399.

Ritchie, S. & Gilroy, S. (1998) Calcium-dependent protein phosphorylation may mediate the giberellic acid response in barley aleurone, *Plant Physiol.*, **116**, 765–776.

Romoser, V.A., Hinkle, P.M. & Persechini, A. (1997) Detection in living cells of Ca^{2+}-dependent changes in the fluorescence emission of an indicator composed of 2 GFP variants linked by a CaM-binding sequence, *J. Biol. Chem.*, **272**, 13270–13274.

Roy, S.J., Holdaway-Clarke, T.L., Hackett, G.R., Kunkel, J.G., Lord, E.M. & Hepler, P.K. (1999) Uncoupling secretion and tip growth in lily pollen tubes: evidence for the role of calcium in exocytosis, *Plant J.*, **19**, 379–386.

Rudd, J.J., Franklin, F.C.H., Lord, J.M. & Franklin-Tong, V.E. (1996) Increased phosphorylation of a 26-kD pollen protein is induced by the self-incompatibility response in *Papaver rhoeas*, *Plant Cell*, **8**, 713–724.

Rudd, J.J. & Franklin-Tong, V.E. (2001) Unravelling response-specificity in Ca^{2+} signalling pathways in plant cells, *New Phytol.*, **151**, 7–33.

Schuurink, R.C., Chan, P.V. & Jones, R.L. (1996) Modulation of calmodulin mRNA and protein levels in barley aleurone, *Plant Physiol.*, **111**, 371–380.

Snedden, W.A. & Fromm, H. (2001) Calmodulin as a versatile calcium signal transducer in plants, *New Phytol.*, **151**, 35–66.

Snowman, B.N., Kovar, D.R., Schevchenko, G., Franklin-Tong, V.E. & Staiger, C.J. (2002) Signal-mediated depolymerization of actin in pollen during the self-incompatibility response, *Plant Cell*, **14**, 2613–2626.

Smertenko, A.P., Jiang, C.-J., Simmons, N.J., Weeds, A.G., Davies, D.R. & Hussey, P.J. (1998) Ser6 in the maize actin-depolymerizing factor, ZmADF3, is phosphorylated by a calcium-stimulated protein kinase and is essential for the control of functional activity, *Plant J.*, **14**, 187–193.

Smith, R.M., Baibakov, B., Ikebuchi, Y., *et al.* (2000) Exocytotic insertion of calcium channels constrains compensatory endocytosis to sites of exocytosis, *J. Cell Biol.*, **148**, 755–767.

Staiger, C.J. (2000) Signaling to the actin cytoskeleton in plants, *Ann. Rev. Plant Physiol. Plant Mol. Biol.*, **51**, 257–288.

Staiger, C.J., Goodbody, K.C., Hussey, P.J., Valenta, R., Drøbak, B.K. & Lloyd, C.W. (1993) The profilin multigene family of maize: differential expression of three isoforms, *Plant J.*, **4**, 631–641.

Staiger, C.J., Yuan, M., Valenta, R., Shaw, P.J., Warn, R.M. & Lloyd, C.W. (1994) Microinjected profilin affects cytoplasmic streaming in plant cells by rapidly depolymerizing actin microfilaments, *Curr. Biol.*, **4**, 215–219.

Steer, M.W. & Steer, J.L. (1989) Pollen tube tip growth, *New Phytol.*, **111**, 323–358.

Sun, D.Y., Bian, Y.Q., Zhao, B.H., Zhao, L.Y., Yu, X.M. & Shengjun, D. (1995) The effects of extracellular calmodulin on cell wall regeneration of protoplast and cell division, *Plant Cell Physiol.*, **36**, 133–138.

Takahashi, K., Isobe, M. & Muto, S. (1997) An increase in cytosolic calcium ion concentration precedes hypoosmotic shock-induced activation of protein kinases in tobacco suspension culture cells, *FEBS Lett.*, **401**, 202–206.

Taylor, L.P. & Hepler, P.K. (1997) Pollen germination and tube growth, *Ann. Rev. Plant Physiol. Plant Mol. Biol.*, **48**, 461–491.

Tiezzi, A., Moscatelli, A., Cai, G., Bartalesi, A. & Cresti, M. (1992) An immunoreactive homolog of mammalian kinesin in *Nicotiana tabacum* pollen tubes, *Cell Motil. Cytoskeleton*, **21**, 132–137.

Tisdale, E.J., Bourne, J.R., Khosravi-Far, R., Der, C.J. & Balch, W.E. (1992) GTP-binding mutants of rab1 and rab2 are potent inhibitors of vesicular transport from the endoplasmic reticulum to the Golgi complex, *J. Cell Biol.*, **119**, 749–761.

Trewavas, A.J. & Malhó, R. (1997) Signal perception and transduction the origin of the phenotype, *Plant Cell*, **7**, 1181–1195.

Trotochaud, A.E., Hao, T., Wu, G., Yang, Z. & Clark, S.E. (1999) The CLAVATA1 receptor-like kinase requires CLAVATA3 for its assembly into a signaling complex that includes KAPP and a Rho-related protein, *Plant Cell*, **11**, 393–405.

Van der Hoeven, P.C.J., Siderius, M., Korthout, H.A.A.J., Drabkin, A.V. & De Boer, A.H. (1996) Calcium and free fatty acid-modulated protein kinase as putative effector of the fusicoccin 14-3-3 receptor, *Plant Physiol.*, **111**, 857–865.

Vidali, L. & Hepler, P.K. (1997) Characterization and localization of profilin in pollen grains and tubes of *Lilium longiflorum*, *Cell Mot. Cytoskeleton*, **36**, 323–338.
Vidali, L. & Hepler, P.K. (2001) Actin and the pollen tube growth, *Protoplasma*, **215**, 64–76.
Vidali, L., Yokota, E., Cheung, A.Y., Shimmen, T. & Hepler, P.K. (1999) The 135 kDa actin-bundling protein from *Lilium longiflorum* pollen is the plant homologue of villin, *Protoplasma*, **209**, 283–291.
Volotovski, I.D., Sokolovsky, S.G., Molchan, O.V. & Knight, M.R. (1998) Second messengers mediate increases in cytosolic calcium in tobacco protoplasts, *Plant Physiol.*, **117**, 1023–1030.
von Witsch, M., Baluska, F., Staiger, C.J. & Volkmann, D. (1998) Profilin is associated with the plasma membrane in microspores and pollen, *Eur. J. Cell Biol.*, **77,** 303–312.
Vos, J.W. & Hepler, P.K. (1998) Calmodulin is uniformly distributed during cell division in living stamen hair cells of *Tradescantia virginiana*, *Protoplasma*, **20**, 158–171.
Wang, W., Takezawa, D., Narasimhulu, S.B., Reddy, A.S.N. & Poovaiah, B.W. (1996) A novel kinesin-like protein with a calmodulin-binding domain, *Plant Mol. Biol.*, **31**, 87–100.
Wu, G., Gu, Y., Li, S. & Yang, Z. (2001) A genome-wide analysis of *Arabidopsis* Rop-interactive CRIB motif-containing proteins that act as Rop GTPase targets, *Plant Cell*, **13**, 2841–2856.
Yang, Z. (2002) Small GTPases: versatile signaling switches in plants, *Plant Cell*, **14**, S375–388.
Yang, Z. & Watson, J.C. (1993) Molecular cloning and characterization of rho, a ras-related small GTP-binding protein from the garden pea, *Proc. Nat. Acad. Sci. USA*, **90**, 8732–8736.
Yashiro, K., Sakamoto, T. & Ohmori, M. (1996) Molecular characterization of an adenylate cyclase gene of the cyanobacterium *Spirulina platensis*, *Plant Mol. Biol.*, **31**, 175–181.
Yokota, E. & Shimmen, T. (1999) The 135-kDa actin-bundling protein from lily pollen tubes arranges F-actin into bundles with uniform polarity, *Planta*, **209**, 264–266.
Yokota, E., Muto, S. & Shimmen, T. (1999) Inhibitory regulation of higher-plant myosin by Ca^{2+} ions, *Plant Physiol.*, **119**, 231–239.
Yokota, E., Muto, S. & Shimmen, T. (2000) Calcium-calmodulin suppresses the filamentous actin-binding activity of a 135-kilodalton actin-bundling protein isolated from lily pollen tubes, *Plant Physiol.*, **123**, 645–654.
Yokota, E., Takahara, K. & Shimmen, T. (1998) Actin-bundling protein isolated from pollen tubes of lily: biochemical and immunocytochemical characterization, *Plant Physiol.*, **116**, 1421–1429.
Zheng, Z.-L. & Yang, Z. (2000) The Rop GTPase switch turns on polar growth in pollen, *Trends Plant Sci.*, **5**, 298–303.
Zonia, L., Cordeiro, S., Tupy, J. & Feijó, J.A. (2002) Oscillatory chloride efflux at the pollen tube apex has a role in growth and cell volume regulation and is targeted by Inositol 3,4,5,6-tetrakisphosphate, *Plant Cell*, **14**, 2233–2249.

9 Cytoskeletal requirements during *Arabidopsis* trichome development

Mark Beilstein and Dan Szymanski

9.1 Introduction

Arabidopsis leaf trichomes are unicellular hair-like structures that are found on most aerial organs of the plant. Trichome morphogenesis employs gene functions that are used widely during plant development. A genetic analysis of this process is an efficient approach to increase our understanding of plant cell growth mechanisms. Developing trichomes execute a highly reproducible morphogenetic program that is under well-defined genetic and cytoskeletal control. Many trichome-shape mutants clearly fall within either microtubule or actin filament-dependent morphogenesis pathways, and several of the affected genes have been cloned. This chapter integrates the cell biology of trichome development with recent advances using molecular genetics.

9.2 Trichome morphogenesis

9.2.1 Arabidopsis

Trichomes are unicellular, hair-like projections that extend from the epidermis of shoot-derived organs. They are widespread among terrestrial plants and their shapes are extremely diverse, ranging from an unbranched spike to ornate umbrella-shaped cells with multiple branches (Johnson, 1975). *Arabidopsis* trichome shape is respectably complex, and includes multiple branches and cell walls that are decorated with papillae or tubercles (Fig. 9.1C). *Arabidopsis* trichome development has been used as a model process to better understand the regulation of cell fate and morphogenesis. For example, molecular genetic studies identified a hierarchy of transcriptional regulation that controls trichome initiation in the leaf epidermis. Trichome precursor cells subsequently execute a highly reproducible morphogenetic program: cytoplasmic reorganization, endoreduplication (DNA replication in the absence of cell division), and cell expansion are tightly integrated. Cortical sites for polarized growth are repeatedly established during the initial outgrowth of the stalk and branches. The mature trichome is highly polarized, and reaches a final length of ~500 μm. Not surprisingly cytoskeleton-dependent activities are highly constrained throughout trichome development, and this highly specialized cell type is sensitized to mutations that affect the cell growth machinery.

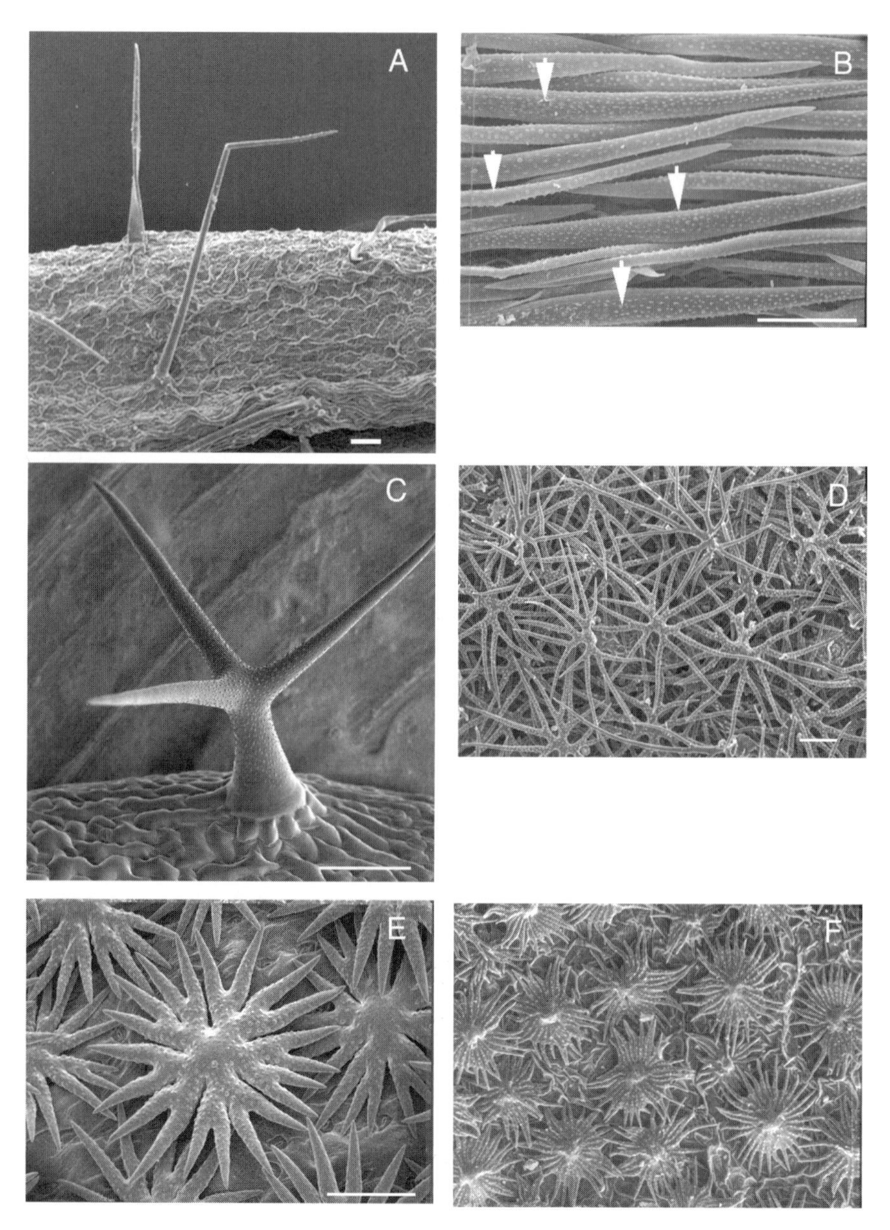

Fig. 9.1 Scanning electron micrographs of leaf trichomes from selected members of the Brassicacae. Leaves from herbarium samples – A, C–F, or freshly frozen material – B, were used for electron microscopy. (A) *Sisymbrium* leaf trichomes. (B) *Farsetia* leaf trichomes. (C) *Arabidopsis* leaf trichomes. (D) *Alyssum* leaf trichomes. (E) *Physaria goodrichii* leaf trichomes. (F) *Physaria genetea* leaf trichomes. Arrows – location of branch position; bars = 100 μm.

Because trichomes are both non-essential for plant survival and are physically accessible, a genetic analysis of trichome development is a powerful approach to identify genes that control polarized growth. This chapter summarizes the current molecular genetic, biochemical, and cytological knowledge about trichome development and its relationship to plant cell growth in general.

9.2.2 *Members of the Brassicaceae*

Before delving into *Arabidopsis* trichome development, we would like to place *Arabidopsis* trichomes within the broader context of non-secretory trichome morphologies in the family Brassicaceae. The family contains roughly 3500 species in over 350 genera (Al-Shehbaz, 1984). In addition to *Arabidopsis*, economically important species of the family include *Brassica oleracea* (the cultivars broccoli, cabbage, and Brussels sprouts) and *Brassica napus* (the seeds of which are the source of Canola oil). Most family members harbor trichomes on the leaf surface. Studies of the adaptive significance of trichomes suggest a function in plant defense against insect herbivory (Mauricio & Rausher, 1997; Mauricio *et al.*, 1997; Mauricio, 1998). The role of trichomes in preventing water loss in arid environments is experimentally unexplored; however, species in *Physaria* occur in primarily alpine habitats in North and South America with different levels of precipitation and these levels may be correlated to the extent of webbing between trichome rays (Rollins & Banerjee, 1975, 1976). In addition to a role in water loss, several Brassicaceae genera exhibit specialized barbed trichomes on fruits that ostensibly facilitate animal dispersal. For example, the leaf trichomes of *Asperuginoides axillaris* have a simple pointed tip, while fruit trichomes terminate in a 3-pronged barb.

Historically, trichomes played an important role in classifications of the family. The Brassicaceae leaf epidermis ranges from completely glabrous to densely hirsute. Trichome shape variation was used in efforts to delineate both genera and species (Al-Shehbaz, 1989, 1994a,b; Mulligan, 1995). The earliest diverging lineage in the Brassicaceae, *Aethionema saxatile* (Koch *et al.*, 2001), lacks trichomes entirely. Despite this fact, simple unbranched trichomes, like those in the genera *Sisymbrium* (Fig. 9.1A) and *Barbarea*, are found throughout the family and in the closest sister lineage to the Brassicaceae, Cleomoideae:Capparaceae (Hall *et al.*, 2002). It is likely, therefore, that trichomes were either absent or simple in ancestral Brassicaceae. If ancestral Brassicaceae exhibited simple trichomes, trichomes with more than one branch may have evolved independently in several lineages within the family. Molecular phylogenies of the Brassicaceae indicate a complex scenario of trichome evolution with several distinct lineages sharing similar trichome morphologies.

The simplest branching pattern in the family is one in which the growth axis bifurcates near the base at an early stage of development. Sessile trichomes (those lacking a stalk) occur in two closely related genera (*Farsetia* and *Lobularia*: Beilstein, unpublished results) and this character may unite the pair. A dense mat of aligned sessile trichomes covers the surface of *Farsetia* leaves (Fig. 9.1B). These trichomes resemble those of *zwi* and an *Arabidopsis* mutants. *Arabidopsis* (Fig. 9.1C) and related genera (*Capsella* and *Boechera*) (O'Kane *et al.*, 1996; Koch *et al.*, 2001) contain trichomes with three or four branches that radiate from the stalk. The branches occasionally give rise to small secondary branches of greatly reduced stature. Unlike true stellate trichomes, the branches are not parallel to the leaf surface, but grow away from the leaf surface at a slight angle. In *Arabidopsis*, the branch–stalk junction contains intersecting helical networks of microtubules (Mathur & Chua, 2000). Layers of cellulose microfibrils with a similar arrangement could impart structural stability to the elongated branch. In contrast, the primary branches in truly stellate trichomes originate from a central stalk domain and grow parallel to the epidermal surface. The dendritic branches of *Alyssum* stellate trichomes retain their ability to generate additional sites of secondary branch initiation (Fig. 9.1D). The most ornate trichomes in the family are found in *Physaria* (the genus formerly known as *Lesquerella*) (Rollins & Banerjee, 1975, 1976). *Physaria goodrichii* contains stellate dendritic trichomes (Fig. 9.1E), but the cells are more swollen than in related species, and the branch–stalk interface zone is fused. Apparently, the balance between tip-directed and diffuse growth is altered in this species. *Physaria gentea* trichomes are stellate, but webbed (Fig. 9.1F). Tubercles, which appear to require direct secretion from the cytoplasm, are absent in the webbed regions of the cell. The extent of webbing varies between species, and in some cases extends to branch tips.

9.3 *Arabidopsis* trichome development

9.3.1 Initiation and leaf development

Among the Brassicaceae, trichome development is best studied in *Arabidopsis*. Trichome initiation is tightly integrated with leaf development. Leaf primordia do not initiate trichomes until they reach a length of ~100 μm (Larkin *et al.*, 1996). The interval of leaf development that supports trichome initiation differs between ecotypes. In the *Landsberg erecta* ecotype, the first leaf makes about ten trichomes when the leaf is between ~100 and ~400 μm in length (Larkin *et al.*, 1996). In the *Columbia* background, trichome initiation continues until the first leaf is about 1 mm in length and the average number of trichomes is ~30 (Larkin *et al.*, 1996; Szymanski & Marks, 1998). The precise characteristics of the trichome-competent population of cells are not known, but dependence on cell-cycle status is likely. Trichome initiation is restricted to regions of the leaf that contain fields of dividing cells and occurs among groups of cells that are at or near the transition from the mitotic to the endomitotic cell cycle (Lloyd *et al.*, 1994; Larkin *et al.*, 1996). Once a trichome precursor cell is defined it grows isodiametrically. The enlarged spherical cell eventually undergoes additional rounds of DNA replication in the absence of cell and nuclear division (endoreduplication) and expands perpendicular to the plane of the leaf in a highly polarized manner. The transition from mitotic to endomitotic cycling occurs systemically in the leaf, but occurs to a greater extent (Melaragno *et al.*, 1993; Szymanski & Marks, 1998) and perhaps earlier (Hülskamp *et al.*, 1994) in trichomes. During the later stages of leaf development trichome initiation stops, and the existing trichomes expand in close association with a ring of socket cells that provide physical support.

9.3.2 Genetics of initiation

Arabidopsis leaf trichome development was originally used to study the question of how cell fate and pattern formation are regulated during leaf development (reviewed in Larkin *et al.*, 1997; Marks, 1997). Mutant screens were conducted to identify genes that are required for normal trichome initiation. *GLABROUS1* (*GL1*) and *TRANSPARENT TESTA GLABRA1* (TTG1) were discovered based on reduced trichome number and branching phenotypes. The regulatory triad of a myb (*GL1*), bHLH (*GL3*), and a WD-40 repeat (*TTG1*) that positively regulates trichome initiation constitutes an evolutionarily conserved, transcription control module that is also employed during anthocyanin biosynthesis and root hair patterning (Lee & Schiefelbein, 1999; Schiefelbein, 2000). Genetic and yeast two-hybrid experiments suggest that GL1, TTG, and GL3 physically interact and function to regulate expression of downstream genes that are involved in trichome morphogenesis and endoreduplication (Szymanski *et al.*, 1998a; Larkin *et al.*, 1999; Payne *et al.*, 2000). The superimposition of developmental status and the composition or levels of the trichome-promoting complex of transcription factors may contribute to the patterning of trichomes on the leaf surface (de Vetten *et al.*, 1997).

Trichome initiation is also subject to negative regulation (reviewed in Szymanski *et al.*, 2000). The activity of the trichome-promoting complex is antagonized by at least two additional MYB-class transcription factors. *CAPRICE* (*CPC*) and *TRIPTYCHON* (*TRY*) each encode truncated MYB-like proteins that contain only one MYB-repeat and no transcription activation domain (Wada *et al.*, 1997; Schellmann *et al.*, 2002). Although *cpc* and *try* single mutants have subtle defects in trichome initiation and spacing, the *cpc try* double mutants display a striking increase in the clustering of trichomes around a single initiation site (Schellmann *et al.*, 2002). Like the trichome-promoting factors, *CPC* and *TRY* are expressed diffusely in fields of dividing epidermal cells, and it appears that local fluctuations in positively and negatively acting factors control trichome initiation and spacing. The genetic tug of war that regulates initiation must be resolved at a very early stage of development since clusters of spherical trichome precursor cells are extremely rare (Larkin *et al.*, 1996).

The genes that function antagonistically during trichome initiation also regulate endoreduplication in developing trichomes. *GL1*, *TTG*, and *GL3* promote endoreduplication in epidermal pavement cells and trichomes, while *TRY* negatively regulates DNA replication (Hülskamp *et al.*, 1994; Szymanski *et al.*, 1998a,b; Szymanski & Marks, unpublished results). Although there are exceptions, the level of endoreduplication usually correlates with cell size and the number of trichome branches. *GL1*, *TTG*, and *GL3* also regulate additional downstream transcription factors that direct gene expression patterns that are needed for normal morphogenesis. For example, the homeodomain-containing transcription factor *GLABRA2* (*GL2*) and the WRKY transcription factor *TTG2* are required for normal polarized growth (Rerie *et al.*, 1994; Johnson *et al.*, 2002). In addition, the *AtMYB5* transcription factor is expressed following branch initiation (Johnson *et al.*, 2002; Szymanski, unpublished results). Presumably, the function of these transcription factors is to regulate the expression of the large number of genes that are needed to construct a multi-branched, 500 μm long cell. However, the target genes that directly control morphogenesis are not known.

9.4 *Arabidopsis* trichome morphogenesis

Developing trichomes undergo dramatic cytoplasmic rearrangements during development (Fig. 9.2). Because of the position of the trichome on the surface of the leaf it is not possible to use transmitted light to visualize living cells. Therefore, confocal reflected light and fluorescence microscopy can be used to observe the overall cellular organization and the localization of a specific compartment respectively. The images in Fig. 9.2 were obtained from intact transgenic plants expressing a GFP marker for nucleus localization (a GFP fusion to an ankrin-repeat protein, N7) (Cutler *et al.*, 2000). The first visual hallmark of trichome formation is the isodiametric expansion of a precursor cell in a field of otherwise normally shaped cells (Fig. 9.2A). Although the DNA content of trichomes at this stage of development has not been measured, trichome precursor cells clearly contain an enlarged nucleus. Based on increased nucleus size, Hülskamp originally proposed that endoreduplication is functionally linked to trichome initiation (Hülskamp *et al.*, 1994). After the trichome precursor reaches a diameter of ~15 μm, the developing stalk emerges from the apical surface of the cell. At the cell surface, the initiating stalk first appears as a rounded bulge with a diameter of ~10 μm (Szymanski *et al.*, 1999). Intrinsic apical/basal polarity of the epidermis and/or physical contact with neighboring socket cells may influence the positioning of the stalk bulge. Mutations in the *GL2* gene cause trichomes to expand in amorphous manner in the plane of the epidermis. Outgrowth of the stalk is microtubule-dependent, but the mechanism of cortical site selection is far from clear (see below). After stalk initiation occurs the cell elongates, but the diameter does not substantially increase and the nucleus remains near the base of the stalk (Fig. 9.2B). Unbranched stalk cells that reach a length of ~30 μm display a polarized cytoplasm with a central cylindrical membrane compartment that extends toward the cell apex (Fig. 9.2B).

Once the stalk reaches ~30 μm, branch initiation occurs (Szymanski *et al.*, 1999). Branch buds are distinguished from stalk buds only by their position on the

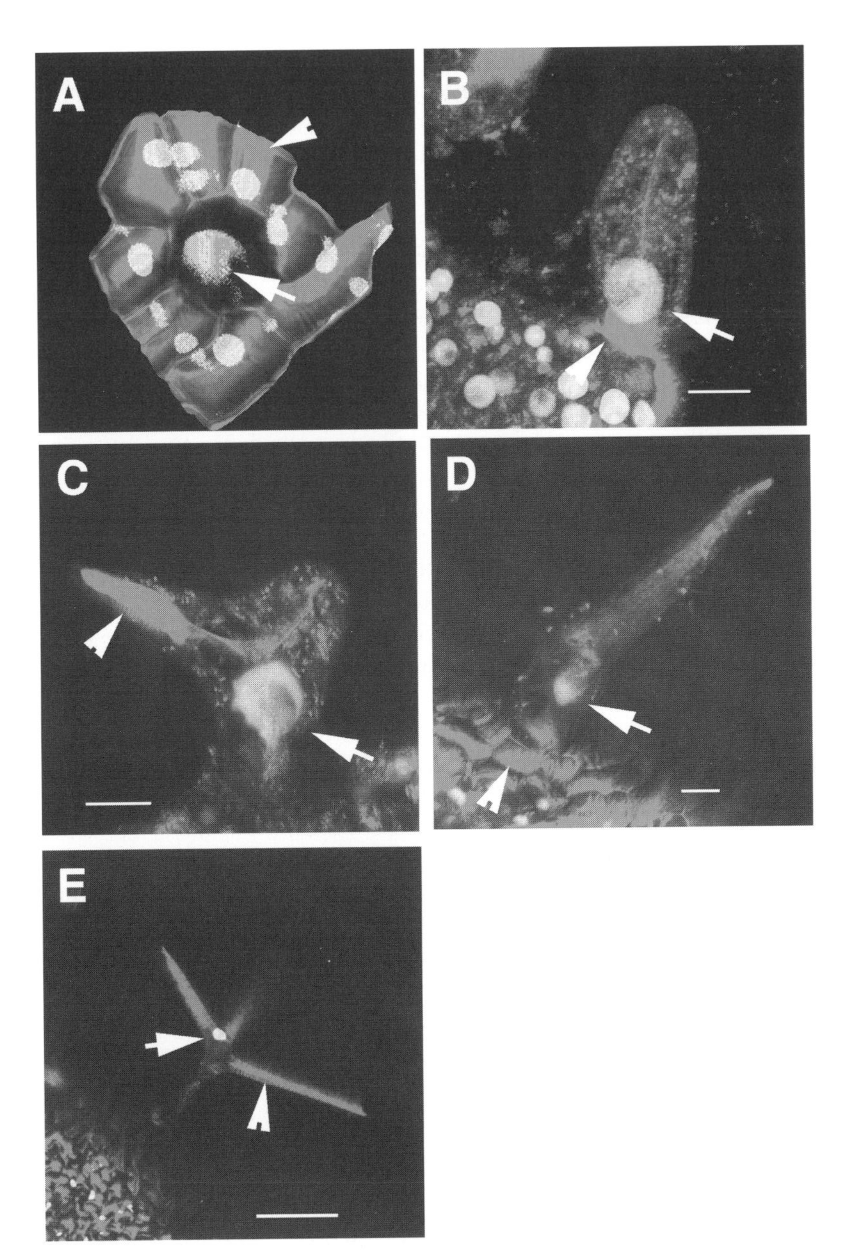

Fig. 9.2 Nuclear position and cellular organization of developing *Arabidopsis* trichomes development. Confocal images of trichomes obtained using GFP fluorescence from a nucleus localized marker protein (arrows) and reflected light from living cells. (A) Spherical trichome precursor cell encircled by epidermal socket cells. (B) Elongated trichome stalk prior to branch initiation. (C) Developing trichome containing both blunt and pointed branch tips. (D) Stalk and branch of a trichome in the cell expansion phase. (E) Elongated branches of a mature trichome. Arrows – trichome nuclei; arrowheads – cell surface reflections; (A–D) bar = 10 μm; (E) bar = 100 μm.

lateral face of the expanded stalk. The first branch bud is often aligned with the long axis of the leaf, and subsequent buds are formed at adjacent sites near the stalk apex. The source of the information that controls branch bud positioning is not known. Prior to and during branch formation, the nucleus resides in a basal position relative to the developing branches. Cylindrical membrane compartments radiate toward the growing branch tips from a central expanded membrane-bound domain (Fig. 9.2C). The identity of this membrane compartment is not known, but it is not marked with GFP probes that accumulate in the endoplasmic reticulum, golgi, or vacuole compartments (Szymanski, unpublished results). The appearance and expansion of this polarized membrane system coincides with the transition to the actin-dependent phase of trichome morphogenesis, and its positioning may define the general pattern of vesicle production and membrane recycling during branch and stalk growth. Once the branches reach a length of ~15–20 μm, the tips acquire a more pointed morphology and the nucleus is maintained in its basal position (Fig. 9.2D). A vast majority of the cell volume is generated during this stage of development (Szymanski *et al.*, 1999). In addition to a central radial membrane compartment, extremely fine membrane tubules populate the cortical region of the developing branches; a subset of these structures is aligned with the long axis of the branch. Perhaps these structures are important for localized vesicle transport and/or membrane recycling. Mature trichomes are highly vacuolated and the nucleus is usually positioned near the base of one of the branches. The surface features and a typical nucleus position of a mature trichome are shown in Fig. 9.2E.

9.4.1 Cytoskeletal inhibitors

Inhibitor experiments demonstrated the non-overlapping requirements for the microtubule and actin filament cytoskeletons during trichome development. This was clearly shown using a dexamethasone-regulated form of the maize *R* gene (Lloyd *et al.*, 1994) to induce synchronous trichome formation in the presence or absence of cytoskeleton-disrupting drugs (Szymanski *et al.*, 1999). In this genetic background, trichome initiation is dexamethasone-dependent, the induced trichomes are highly polarized and contain either two or three branches. Exposure of the developing leaves to high concentrations of oryzalin prior to dexamethasone-induced initiation causes isotropic cell expansion without branch formation. However, at lower oryzalin concentrations branch initiation is observed, but branch number is reduced (Mathur *et al.*, 1999). Precise actin organization is not required for polarity establishment: elongation of the trichome stalk and branch initiation are not affected noticeably by F-actin disrupting agents (Szymanski *et al.*, 1999). The apparent unimportance of F-actin during growth pattern establishment is not due to an absence of F-actin during these stages of development. Using conventional fixation techniques coupled with the freeze shattering permeabilization approach (Wasteneys *et al.*, 1997), anti-actin antibodies detected intricate networks of F-actin at all stages of trichome development (Szymanski *et al.*, 1999). Perhaps prior to branch formation, F-actin plays a minor role in facilitating the apical transport of organelles and/or vesicles, but in its absence, bulk flow or microtubule-dependent function is sufficient. Branch initiation and the early phase of branch elongation appear to be a reiteration of the stalk initiation and growth process. The formation of this bulge is sensitive to microtubule-disrupting agents, but the microtubule organization that is functionally linked to this process is not known. In this regard, the pharmacological sensitivities of the early stages of trichome growth are similar to those of cotton fibers (Tiwari & Wilkins, 1995), but distinct from the actin-dependent tip growth that is observed in pollen tubes, root hairs, and fungal hyphae (Picton & Steer, 1981; Jackson & Heath, 1993; Miller *et al.*, 1999). At this time the mode of trichome expansion during stalk and branch outgrowth is not clear, but based on available microtubule localization data may involve a novel type of microtubule-dependent tip-directed growth (see below).

F-actin-disrupting drugs have no noticeable effect in earlier stages of development, but following branch formation drug-treated stalks and branches are twisted and swollen. The transition to actin-dependent growth roughly coincides with the refinement of the branch tip to a more pointed morphology and a dramatic cell expansion phase. Cell expansion at this stage includes diffuse growth of the stalk and branches as well as potential tip-directed growth of the elongating branches. Precise organelle positioning, vesicle delivery, and membrane recycling are likely actin-dependent components. Unlike microtubule-disrupting drugs, cytochalasin D and latrunculin B do not block cell expansion or branch initiation. Even at high drug concentrations, enlarged trichomes resembling those of the distorted trichome mutants develop over a period of days when either drug is applied topically. Microtubule function remains important throughout trichome development (Szymanski, 2001). Branched trichomes treated with microtubule depolymerizing drugs display uniform swelling and little indication of persistent polarized growth (Szymanski, 2001). The microtubule cytoskeleton may define cytoplasmic boundaries, within which actin filaments refine growth patterns.

9.4.2 Cytoskeletal organization in developing trichomes

9.4.2.1 Microtubules

Analyses of the cytoskeletal organization in developing trichomes provide functional clues and important assays for altered organization in mutants. Traditional immunological detection and live cell, green fluorescent protein (GFP)-based methods for cytoskeletal organization have been used. However, in the case of microtubules, antibody-based localization methods have been difficult to establish. The only published images of trichome microtubules employ a MAP4 microtubule-binding domain fused to GFP (GFP-MBD) (Marc *et al.*, 1998). GFP-MBD in mature trichome stalks and branches revealed a dense population of microtubules with a net alignment with the long axis of the cell (Mathur & Chua, 2000). A direct fusion of GFP to α-TUBULIN6 (TUA6-GFP) (Ueda *et al.*, 1999) detects a similar organization (Szymanski, unpublished results). The function of these microtubules is not known, as microtubule-dependent transport activity and cellulose microfibril orientation have not been examined.

A key unanswered question is centered on which microtubule arrays affect stalk and branch initiation. It has been proposed that microtubule reorientation regulates the establishment of polarity during branch formation (Mathur & Chua, 2000). Microtubule-stabilizing drugs can cause the formation of aster-like microtubule structures and increase branching in some genetic backgrounds. In addition, discordant or unaligned microtubules are observed at abortive branch points in mature mutant trichomes (Mathur & Chua, 2000). However, high-resolution images of TUA6-GFP and GFP-MBD in developing branches suggest that cortical microtubules do not reorient from a transverse to longitudinal orientation until well after branch formation (Fig. 9.3C,D and Folkers *et al.*, 2002). In fact unbranched stalks and early stage branches display a similar microtubule organization: a wide band of circumferential, transverse microtubules that occupies the apical region of stalks and branches (Fig. 9.3). The microtubule organization at the extreme apex is not obvious. Using TUA6-GFP as a reporter, the apex is depleted of micotubules relative to the adjacent cortex (Fig. 9.3A). The GFP-MBD reporter detects a more restricted microtubule-depleted zone at the apex, that often contains randomly oriented, short microtubules (Fig. 9.3B,D). These differences at the apex may be due to the microtubule-stabilizing activity of the GFP-MBD probe (Olson *et al.*, 1995; Marc *et al.*, 1998).

It is not clear which cytoplasmic region or population of microtubules contains the information for polarity establishment. Cells at very early stages of branch formation contain a transverse cortical array, an apical domain with short, randomly oriented microtubules, and a dense cytoplasmic network of microtubules distal to the branch bud (Fig. 9.3D). It is difficult to imagine how the transverse cortical array could self-organize to initiate the site of polarized growth. Perhaps membrane trafficking and/or a localized calcium gradient lies upstream from cytoskeletal reorganization. This appears to be the case in root hairs. In root hair-containing cell files ROP GTPases affect polarized growth, and the polarized localization of ROP predicts the site of bulge formation (see Chapter 7). Polar localization of ROP is insensitive to cytoskeleton-disrupting drugs such as latrunculin B and cyctochalasin D. However, brefeldin A, which inhibits vesicle production from the golgi, disrupts polar localization of ROP protein (Molendijk *et al.*, 2001). Although the pharmacological sensitivity of trichomes differs from those of root hairs (Bibikova *et al.*, 1999; Baluska *et al.*, 2000), similar cellular activities may control polarity establishment.

9.4.2.2 Actin filaments

The organization of the actin cytoskeleton has also been examined in trichomes at each developmental stage (Szymanski *et al.*, 1999). Spherical trichome precursor cells, unbranched stalks, and cells at the branch initiation stage contain dense networks of actin filaments that are detectable with antibodies. However, in regions of the cell in which growth patterns are being established, such as branch buds and the apical regions of stage 2 stalks, the actin signal is punctate and diffuse, with little evidence of actin filaments. Slightly more advanced branches in

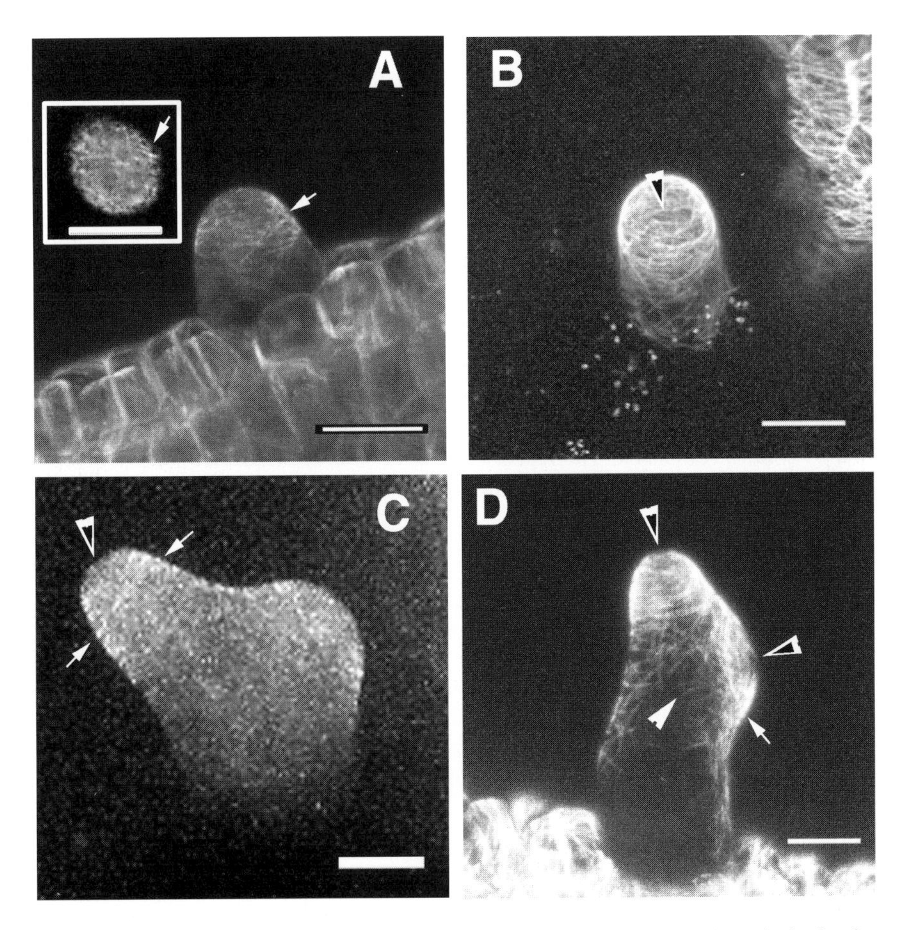

Fig. 9.3 Microtubule organization of developing trichome stalks and branches obtained using TUA6-GFP (A, C) and GFP-MBD (B, D). Confocal images are a maximum projection of individual images. (A) Lateral cortex of an elongated, unbranched trichome stalk; inset, the apical surface of a trichome stalk. (B, C) Cortex of a recently branched trichome (stage 4). Arrows – regions of increased cortical microtubule density; white arrowhead – cytoplasmic microtubules; white and black arrowhead – apical zone with reduced microtubules; bar = 10 μm.

the same cell contain an extensive network of actin filaments that are loosely aligned with the main growth axis. This population of filaments terminates near the apical plasma membrane of the growing blunt branch tip. Perhaps actin filaments in the apical cortex of branch buds terminate within the apical microtubule-depleted zone and act as a scaffold for membrane attachment and/or tracks for vesicle delivery. As the branch bud transitions to a more refined tip structure, clearly aligned actin filaments or bundles populate the cytoplasm. In addition, a dense cortical meshwork is observed in growing trichomes following branch formation. Directed and continuous delivery of raw materials into the growing branch and stalk is actin-dependent and involves the coordination of multiple cellular activities: organelle positioning and biogenesis, membrane trafficking, cytoskeletal dynamics. The observed actin arrays are likely to have multiple functions in the context of persistent polarized growth and cell shape control. The daunting task of gaining mechanistic insight into its form and function remains.

9.5 Genetics of trichome morphogenesis

A genetic approach to cell shape control is an efficient strategy to identify genes of functional significance, and ultimately gain insight into mechanisms. Trichomes are an ideal cell type because microtubule and actin-based defects are easily distinguished and can guide morphology-based mutant screens toward genes that directly control cell shape. The recent cloning of several cytoskeleton-related genes required for normal trichome branching confirms the utility of this approach and also provides some surprises (Table 9.1).

9.5.1 Reduced branching mutants: microtubule-based functions

9.5.1.1 ZWICHEL (ZWI)

The *ZWI* gene encodes a minus end-directed kinesin-like motor protein that is required for normal branch formation and stalk elongation. The reduced branching and swollen stalk phenotype of the *zwi* trichomes is very similar to the shape defects of cells that are exposed to microtubule-disrupting drugs (Mathur *et al.*, 1999).

Table 9.1 Genes that participate in *Arabidopsis* trichome morphogenesis

Gene (alias)	Protein	Proposed function	Trichome phenotype	References
ZWI (*KCBP*)	Kinesin	Microtubule motor protein; end directed	Reduced stalk elongation, reduced trichome branching	Oppenheimer *et al.*, 1997
ATKSS (*BOT*, *ERH3*, *FRC2*)	p60 katanin	Microtubule severing	Reduced trichome branching	Burk *et al.*, 2001
AN	CTBP/BARS	Transcriptional repressor; Golgi vesicle budding	Reduced branching	Folkers *et al.*, 2002; Kim *et al.*, 2002
POR[C]	Tubulin folding cofactor-C	Heterodimer release from the folding complex	Reduced branching, uniform swelling	Kirik *et al.*, 2002b
KIS[A]	Tubulin folding cofactor-A	β-tubulin folding	Reduced branching, uniform swelling	Kirik *et al.*, 2002a
SPK1	Putative ROP-GEF	Signaling to the cytoskeleton	Reduced branching, variable swelling	Qiu *et al.*, 2002

Cloning of the *ZWI* gene provided the first evidence for microtubule function during trichome development (Oppenheimer *et al.*, 1997). The *ZWI* gene was first isolated in a screen for calmodulin-binding proteins, and was shown to encode a kinesin-like calmodulin-binding protein (KCBP) (Reddy *et al.*, 1996a,b). Biochemical tests revealed ZWI/KCP has minus end-directed microtubule motor activity (Song *et al.*, 1997). Both the C-terminal domain, which contains the motor domain, and an N-terminal domain are capable of bundling microtubules *in vitro* (Kao *et al.*, 2000). The microtubule-bundling and binding activities are inhibited by Ca^{2+}-calmodulin (Narasimhulu & Reddy, 1998; Kao *et al.*, 2000). Presumably Ca^{2+}-calmodulin binds the C-terminal calmodulin-binding domain of KCBP/ZWI and blocks the microtubule-binding activity of the motor domain. In terms of *in vivo* function, current data suggest that KCBP/ZWI affects the organization of microtubules during discrete stages of the cell cycle, particularly during the transition from metaphase to anaphase (Vos *et al.*, 2000). Localization of this motor protein to the mitotic spindle and the phragmoplast is consistent with its potential role in cell division (Smirnova *et al.*, 1998).

The stalk elongation and branch initiation defect of *zwi* trichomes demonstrates its importance during trichome cell growth, but the cause of the cell shape defects in mutant cells is not known. ZWI may transport organelles, vesicles, and/or microtubules to the developing branch. Because ZWI contains both a C-terminal minus end-directed motor and an N-terminal microtubule-binding domain, ZWI may control microtubule organization. Microtubule motors with a similar domain organization can generate parallel or converged microtubule arrays by dragging microtubules in a directed manner on existing microtubule tracks (reviewed in Sharp *et al.*, 2000). ZWI may facilitate alignment of microtubules and/or the stabilization of minus ends. The apical region of developing stalks and branches contain roughly parallel arrays of transverse microtubules and a microtubule-depleted zone at the extreme apex (Fig. 9.3). Perhaps there is a tip-directed gradient of Ca^{2+}-calmodulin that negatively regulates ZWI activity at the apex. It will be important to determine the localization of ZWI at early developmental stages and its relationship to microtubule organization. Additional functional information may come from analysis of extragenic suppressors of the *zwi* phenotype. Three distinct loci partially suppress the branching defect of *zwi-3* mutants and may participate directly in ZWI-dependent processes (Krishnakumar & Oppenheimer, 1999).

9.5.1.2 Tubulin folding cofactors (TFCs)

The TFC complex determines the amount of cytoplasmic tubulin heterodimer in the cell (reviewed in Lewis *et al.*, 1997). Like actin and γ-tubulin, α- and β-tubulin polypeptides transiently interact with the cytoplasmic chaperonin complex in order to acquire a natively folded state. α- and β-tubulins bind tightly to the cytosolic chaperonin. α- and β-tubulin must bind to TFC-B and -A respectively in order to be released from the chaperonin complex (Tian *et al.*, 1996, 1997). Three additional TFCs are required to form polymerization-competent subunits. TFC-D binds tightly to β-tubulin and displaces TFC-A. The subsequent association of the β-tubulin-TFC-D complex with TFCs-C and -E, leads to the release of polymerization-competent β-tubulin (Tian *et al.*, 1996). α-tubulin is recruited to the TFC complex via sequential interactions with TFC-B and TFC-E (Tian *et al.*, 1997). The α- and β-tubulin folding pathways converge during the formation of a multimeric complex consisting of TFC-D/β, TFC-E/α, and TFC-C; subunit association and dimer release follows. Forward genetic screens in *Arabidopsis* first demonstrated the importance of TFCs in the development of multicellular organisms (McElver *et al.*, 2001; Steinborn *et al.*, 2002; Tzafrir *et al.*, 2002; Kirik *et al.*, 2002a,b).

Yeast TFC-like genes were discovered in genetic screens for mutants with altered microtubule-based functions. In *S. cerevisciae*, TFC genes are not essential; however TFC mutations cause an increased frequency of chromosome loss, genetic interactions with tubulin mutants, and hypersensitivity to microtubule-disrupting drugs (Archer *et al.*, 1995; Hoyt *et al.*, 1997). Consistent with the notion that TFCs function as a complex, double and triple mutant combinations among TFCs did not cause more severe phenotypes (Archer *et al.*, 1995). *S. pombe* relies more heavily on microtubule-based functions during morphogenesis, and this may explain the lethality of TFC mutations. There is a clear functional hierarchy among yeast TFCs. For example, over-expression of *S. pombe* TFC-D and -E could bypass the requirement for TFC-B. Between TFC-D and -E, only over-expression of TFC-D could functionally substitute for TFC-E (Radcliffe *et al.*, 1999). Similar results were obtained with budding yeast (Hoyt *et al.*, 1997). Therefore, TFC-D appears to have the most specialized function. TFC-D is the only cofactor that has been colocalized with microtubules and is negatively regulated by the small TFC GTPase ARL2 (Hoyt *et al.*, 1997; Hirata *et al.*, 1998; Bhamidipati *et al.*, 2000).

Although TFC biochemical activity was originally discovered using vertebrate proteins, the extent to which TFCs participate in multicellular development was not known. Two different morphology-based mutant screens demonstrated the importance of TFCs as essential players in plant morphogenesis. In the first approach, embryo lethal mutant screens identified *Arabidopsis* homologs of TFC-A, -C, -D, -E, and the GTPase ARL2 (McElver *et al.*, 2001; Steinborn *et al.*, 2002). The *TITAN/PILZ* group mutants display abortive embryos with a severely reduced cell number, defective cell plates, and enlarged nuclei. With the exception of the TFC-A-like gene *KIESEL*^{A} (*KIS*^{A}), *TITAN/PILZ* group TFCs were also required for normal endosperm development. Perhaps *KIS*^{A} gene function is more important during pre-prophase band formation, which does not occur during endosperm development (reviewed in Otegui & Staehelin, 2000a,b). Surprisingly, the haploid phase of the plant was not affected, as the TFC alleles were efficiently transmitted through both male and female gametophytes. Developing *porcino*^{C} (*por*^{C}) embryos have reduced levels of α-tubulin. The effects seem to be specific to the microtubule cytoskeleton, as the organization of the actin cytoskeleton was not dramatically altered.

Plant TFCs are also important during cell growth in non-dividing cells, and were discovered in screens for mutants with reduced trichome branching

(Kirik *et al.*, 2002a,b). Weak alleles of *KIS*^{A} and *POR*^{C} were found based on reduced trichome branching and cell swelling. *kis*^{A} and *por*^{C} plants also displayed dwarfism, sporophytic sterility, and hypocotyl cell shape defects in elongating dark-grown plants. Over-expression of either the endogenous gene or a vertebrate homolog was sufficient to rescue *por*^{C} and *kis*^{A} phenotypes. Given that TFC-A promotes β-tubulin folding, it is surprising that over-expression of α-tubulin, but not β-tubulin, is sufficient to bypass the requirement for *KIS*^{A} (Kirik *et al.*, 2002a). Perhaps the *kis*^{A} phenotype is due to the presence of toxic levels of free β-tubulin (Weinstein & Solomon, 1990), and excess α-tubulin relieves the toxicity via dimer formation. In growing trichome branches, the microtubule cytoskeleton eventually re-orients from a transverse to parallel orientation with respect to the long axis of the cell. *por*^{C} and *kis*^{A} trichome branches display an extensive microtubule cytoskeleton, but the orientation remains fixed in the transverse direction. The relationship between the altered microtubule cytoskeleton and compromised TFC function is not obvious. It will be important to determine the effects of TFC mutations on steady-state levels of tubulin and on microtubule stability. Plant microtubules display a higher degree of dynamic instability compared to animal microtubules, which may depend on isotype composition (Moore *et al.*, 1997). *Arabidopsis* contains 6 α- and 9 β-tubulin genes (Kopczak *et al.*, 1992; Snustad *et al.*, 1992). It may be that TFCs regulate the supply, dimerization specificity, as well as the turnover of existing dimers and microtubules. Microtubule dynamics is dramatically affected by the concentration of free dimers. A TFC-dependent, spatially regulated supply of dimers would be an efficient method to regulate microtubule polymerization and stability in large eukaryotic cells.

9.5.1.3 Arabidopsis *katanin small subunit (AtKSS)*

The catalytic subunit (p60) of the microtubule-severing protein KATANIN (AtKSS) is required for normal morphogenesis in many different cell types. Mutations in the *AtKSS* affect hypocotyl elongation, the physical strength of the stem, root hair patterning, and trichome development. Consequently the gene was identified independently by several laboratories. *BOTERO1*, *FRAGILE FIBER2* (*FRA2*), and *ECTOPIC ROOT HAIR 3* (ERH3) all correspond to the gene that encodes the p60 subunit of KATANIN in *Arabidopsis* (*AtKSS*) (Bichet *et al.*, 2001; Burk *et al.*, 2001; Webb *et al.*, 2002). *AtKSS* is the only AAA (ATPase associated with various cellular activities) family member that shares a high degree of amino acid identity (43%) with sea urchin KATANIN, and displays microtubule severing activity *in vitro* (Stoppin-Mellet *et al.*, 2002).

Sea urchin KATANIN was isolated from egg extracts as a stable complex of two proteins, p60 and p80. The KATANIN complex was capable of severing microtubules in an ATP-dependent manner (McNally & Vale, 1993). The p60 subunit is an AAA ATPase family member, and the p80 subunit is a WD-40 repeat-containing protein. Using recombinant proteins, it was shown that p60 alone has microtubule severing activity, but the severing activity is stimulated upon addition of p80 (Hartman *et al.*, 1998). Endogenous sea urchin KATANIN is localized to centrosomes via interaction with the p80 subunit (McNally *et al.*, 1996; Hartman *et al.*, 1998; Srayko *et al.*, 2000). The WD-40 repeat domain of p80 alone is sufficient for centrosomal localization, while the conserved C-terminal domain binds to p60 (Hartman *et al.*, 1998). Centrosomal KATANIN may sever and depolymerize microtubules at their minus ends during anaphase (McNally *et al.*, 1996). Katanin function and microtubule turnover in the context of meiosis and deflagellation has been reviewed (Quarmby, 2000).

Genetic data have clearly established the importance of *AtKSS* in non-dividing cells. Cell shape and organ architecture are affected throughout the root and shoot (Burk *et al.*, 2001; Webb *et al.*, 2002). In the root epidermis, which displays a clear developmental gradient from zones of cell division to one of rapid cell expansion, the mutant phenotype is restricted to non-dividing cells. Reduced anisotropic growth was correlated with randomization of cortical microtubules and cellulose microfibrils in the elongation zone (Bichet *et al.*, 2001; Burk *et al.*, 2001; Burk & Ye, 2002). Burk *et al.* (2001) closely examined microtubule organization dividing cells in 5-day-old *fra2* and wild-type root tips. In *fra2* cells, a delay in the redistribution of nuclear membrane-associated microtubules to the cortex as cells exit the cell cycle was reported. Localization of AtKSS to the nuclear membrane is consistent with the phenotype (McClinton *et al.*, 2001), and the idea that AtKSS at the nuclear membrane clips microtubules and allows them to recycle into the cortical array is attractive. However, microtubule localization in whole mounted *fra2* roots failed to detect a persistent nuclear membrane-associated population (Burk & Ye, 2002).

The pleiotropic phenotypes in *AtKSS* mutants may have more to do with secondary effects on the organization of the cell wall. Cellulose microfibrils are randomized in the primary wall, and the normal stratification of primary and secondary walls is not seen in *fra2* roots (Burk & Ye, 2002). Coordination of cell growth not only within cells but also between tissues may require *AtKSS* function (Webb *et al.*, 2002). Although the primary defect of mutations in *AtKSS* appears to be restricted to non-dividing cells, a role for *AtKSS* in mitosis and meiosis cannot be ruled out. A null allele has yet to be identified (Webb *et al.*, 2002), and microtubule organization in mitotic cells was examined using in *fra2* allele, which is unlikely to be a null (Burk *et al.*, 2001). *AtKSS* mutants also display reduced trichome branching and lobe-formation in leaf epidermal pavement cells (Burk *et al.*, 2001). *FURCA2* is one of many loci that may positively regulate branch formation during trichome development (Luo & Oppenheimer, 1999), and is in the same complementation group as *FRA2* (Szymanski, unpublished results). Trichome development may prove to be a useful cell type to dissect AtKSS function.

In *C. elegans*, meiosis-specific KATANIN p60 is regulated by p80 and ME1, and mutations in each of the above genes cause a similar phenotype (Srayko *et al.*, 2000). There are four *Arabidopsis* genes that share a high degree of amino acid identity with the WD-40 repeat and C-terminal p60-binding domain of KATANIN p80. Perhaps this gene family regulates the localization and activity of *AtKSS* in different developmental contexts. If the p80-like genes, like those of animals, are required for the *in vivo* function of *AtKSS*, they may also be identified in screens for trichome branching mutants.

9.5.1.4 *ANGUSTIFOLIA (AN)*

Cotyledons and leaves of *angustifolia* (*an*) plants are more narrow than those of the wild type and in most cases the trichomes contain only two branches instead of the usual three (Tsuge *et al.*, 1996). Consistent with the pleiotropic phenotypes, AN is expressed in all major organs (Folkers *et al.*, 2002; Kim *et al.*, 2002). The deduced *AN* gene product shares amino acid identity with the *CTBP/BARS* family members. CTBP/BARS is a dual function protein that was identified using completely different strategies. C-terminal binding protein (CTBP) was fished out of a two-hybrid screen using a C-terminal fragment of the adenovirus E1A protein (Schaeper *et al.*, 1995). Genetic and molecular data from several experimental systems indicate that CTBPs function as transcriptional co-repressors (reviewed in Chinnadurai, 2002). Brefeldin A-ADP-ribosylated substrate (BARS) was isolated biochemically from rat brain cytosol as one of two major proteins that are ADP-ribosylated following brefeldin A treatment (Spanfò *et al.*, 1999). Subsequent characterization of BARS showed that it had enzymatic activity, and was capable of acylating lysophoshatidic acid using purified components (Weigert *et al.*, 1999). When incubated with an appropriate acyl lipid donor and purified golgi membranes, BARS promotes fission of golgi membranes, presumably via regulation of the physical properties of the bilayer. AN shares ~30% identity with animal CTBP/BARS proteins with the highest similarity found within a putative NAD-binding site. AN also contains a C-terminal extension of 155 amino acids that is not present in the animal proteins, but is found in a rice ortholog (genebank accession AAM19105).

Plant CTBP/BARS proteins may have dual functions in the nucleus and the cytoplasm. Overexpression of functional, GFP-tagged forms of AN in onion epidermal cells and in trichomes demonstrated that most of the AN-GFP signal accumulated in the nucleus (Folkers *et al.*, 2002; Kim *et al.*, 2002). Kim *et al.* (2002) conducted a microarray experiment using wild-type and *an-1* RNA isolated from two-week-old shoots. Eight genes with a 3-fold or greater increase in mRNA levels in the *an-1* background compared to the wild-type were detected (consistent with its role as a transcriptional repressor, only up-regulated genes were detected in the mutant). Future experiments will focus on determining if any of the eight candidate genes are directly related to *AN*-dependent morphogenesis.

The organization of the microtubule cytoskeleton was examined in the *an* leaves and trichomes. The leaf epidermal pavement cells in *an* plants display reduced polarized growth compared to the highly lobed morphology of the wild type (Tsuge *et al.*, 1996). In the elongated regions of the mutant and wild-type pavement cells, the cortical microtubules reflect cell shape, and adopt an overall transverse orientation. Localized regions of bundled microtubules on the anticlinal walls may predict sites of asymmetric growth in *Arabidopsis* pavement cells, and could be a useful assay for *AN* function at early developmental stages

(Qiu *et al.*, 2002). Kirik *et al.* (2002a,b) used the GFP-MBD reporter to examine microtubule organization in developing trichomes. The transverse organization of the cortical microtubules in unbranched and recently branched cells was similar in the wild type and the mutant. However, mutant trichomes lacked the apical/basal stratification of cortical microtubules that was observed in the wild type (Fig. 9.2A,B). For example, a quantitative analysis of microtubule density along the long axis of the stalk demonstrated that cortical microtubules in unbranched *an* trichomes were more or less uniformly distributed. Perhaps the first branching event is linked to the presence of an apical circumferential microtubules, and subsequent branching events require apical/basal stratification of cortical microtubules. What is the relationship between AN function and microtubule organization in trichomes? Is there a BARS side to AN function that regulates membrane trafficking from the golgi? Is there a cytoplasmic or golgi-associated pool of AN during the early stages of trichome development? Is there a relationship between vesicle trafficking and the establishment or maintenance of apical/basal cell polarity? Trichome development is an ideal context to dissect the possible dual functions of AN in transcription and morphogenesis.

9.5.1.5 SPIKE1 (SPK1)

The *SPK1* gene was identified in a screen for seedling-lethal mutants that also display trichome morphology defects. *spk1* plants have reduced trichome branching, narrow leaf and cotyledon, conditional seedling lethality, and sporophytic sterility (Qiu *et al.*, 2002). Polarized growth is disrupted during lobe formation in epidermal pavement cells. Cytoskeletal organization was examined in wild-type and *spk1* cotyledon pavement cells during lobe initiation and growth. In wild-type cells, laterally clustered microtubules are present along the anticlinal walls prior to lobe formation. As the pavement cells develop, positions of increased microtubule clustering correlate with the presence of cell indentations where neighboring pavement cell lobes protrude. Pavement cell protrusions tend to be populated with actin filaments (Fu *et al.*, 2002; Qiu *et al.*, 2002). The simplest explanation is that sites of microtubule clustering define specialized cortical and cell wall domains that yield to neighboring cell protrusions. If distinct microtubule and actin filament-dependent functions in neighboring cells coordinate cell morphogenesis and tissue development, it is not known which cellular activity initiates the process. Lobes are completely absent in *spk1*. An extensive microtubule cytoskeleton is present, but the extensive lateral clustering observed in the wild type is absent. Actin filaments are present in *spk1* pavement cells, but do not display an alternating cortical pattern with microtubules. Although the reduced trichome branching and lateral microtubule clustering in epidemal pavement cells suggest *spk1* function impinges on the microtubule cytoskeleton, a functional role in coordinating either or both cytoskeletal arrays cannot be ruled out.

The degree to which *SPK1* directly affects the cytoskeleton is not known, however *SPK1* shares amino acid identity with the *DOCK* family of adaptor proteins that transmit signal to the cytoskeleton (Cote & Vuori, 2002). Although endogenous

DOCK protein has not been localized in any cells type, over-expression experiments suggest DOCK proteins accumulate at the plasma membrane in response to spatially defined signals. DOCK recruitment may lead to localized activation of RAC GTPases and cytoskeletal reorganization. Human DOCK180, the founding member of the family, was isolated based on its interaction with the SH3-domain-containing protein CRK (Hasegawa *et al.*, 1996). DOCK family members from flies and worms were identified genetically based on cell migration, cell fusion, and phagocytosis defects (Erickson *et al.*, 1997; Wu & Horvitz, 1998). A wealth of genetic and co-immunoprecipitation experiments implicate RAC as a potential downstream target of DOCK signaling (Kiyokawa *et al.*, 1998; Nolan *et al.*, 1998; Reddien & Horvitz, 2000); however, direct evidence for a physical interaction was only recently demonstrated. A C-terminal-conserved domain that is present in all DOCK family members directly binds to RHO family GTPases. A DOCK-A family member, DOCK180, preferentially binds the nucleotide-free form of RAC (Brugnera *et al.*, 2002) and displays guanine nucleotide exchange factor (GEF) activity (Cote & Vuori, 2002). The C-terminal domain of a DOCK-D family member preferentially binds to CDC42 (Meller *et al.*, 2002), and displays GEF activity (Cote & Vuori, 2002). The effectors of DOCK/RAC-mediated signaling, and the cytoskeletal consequences of signaling through the pathway, are not known. However, based on the mutant defects in actin-dependent processes such as phagocytosis, and the accumulation of over-expressed DOCK180 at focal adhesions, an actin-based defect is hypothesized.

SPK1 is an excellent candidate as a regulator of RHO-family GTPase in plants. Plants contain a unique family of RHO GTPases termed ROPs (Li & Yang, 2000; see Chapter 8). ROPs, like all other small GTPases, exist in the cell in an inactive (GDP-bound) or active (GTP-bound) state. Several ROP GTPase-activating proteins that promote GTP hydrolysis and negative regulation of ROP activity are present in the *Arabidopsis* genome (Wu *et al.*, 2000). Animal RHO family GTPases are positively regulated by DBL homology-containing GEFs, but no DBL-encoding gene is present in the *Arabidopsis* genome sequence. Proteins that have GEF can actively adopt very diverse tertiary structures, and it may be that the conserved C-terminal domain of *SPK1* is a novel ROP-GEF. The *spk1*-like phenotypes of plants that overexpress mutant forms of ROP2 provide some circumstantial evidence that the two genes function in a similar pathway (Fu *et al.*, 2002). It is possible that *SPK1* regulates the activity of multiple ROPs in several developmental contexts. The localization, ROP-binding, and potential GEF activity of *SPK1* is currently being examined.

9.5.2 The distorted trichome shape mutants: actin-based functions

The stage-specific effects of actin-filament-disrupting drugs phenocopy the distorted group of trichome mutants. For example, both drug-treated, *gnarled (grl)* and *crooked* (*crk*) distorted trichomes exhibit stage-specific stalk swelling and reduced branch elongation (Szymanski *et al.*, 1999, 2000). Distorted mutants do not display either the uniform swelling or branch number reduction that is characteristic of microtubule-based defects (Szymanski, 2001). Fully developed distorted mutant trichomes often contain abortive or shortened branches and elongated stalks that are swollen and twisted. Based on their similarities to latrunculin- and cytochalasin-treated cells, the distorted mutants are excellent candidate regulators of actin-dependent growth in plant cells.

The cellular reorganization during the transition to actin-dependent branch and stalk expansion likely requires multiple cellular activities. The presence of at least ten independent distorted group loci is consistent with this notion. The *distorted1* (*dis1*) and *distorted2* trichome mutants were historically used as visual markers for classical mapping experiments (Feenstra, 1978). Eight additional distorted mutants were identified in screens for trichome morphology mutants (Hülskamp *et al.*, 1994; Szymanski, unpublished results). The trichome phenotypes of the ten different distorted mutants are very similar. However, the *dis1* mutant displays reduced pavement cell lobing and both *dis1* and *dis4* plants have reduced fresh weight accumulation compared to the wild type (Szymanski, unpublished results). Without having null alleles for each of the distorted group genes, it is not known if the different phenotypes reflect gene function or differences in the severity of the mutant alleles. It is also clear that relying on cell shape as the terminal phenotype will be of limited use. If the transition to actin-dependent growth requires the sequential activity of ten different genes, all single and double mutant combinations will have the same cell shape defect. However, if the positioning and morphology of multiple cell compartments are analysed in the mutants, unique gene functions could be resolved. A genetic pathway of gene functions would be very useful in understanding the sequence of events that occur during the transition to actin-dependent growth.

Live cell and immunological detection of actin is one useful assay for cellular defects in distorted group mutants. Because distorted group mutations have only mild effects on the plant growth in general, it is not surprising that none of the mutants examined thus far display a striking reduction in filamentous actin (Mathur *et al.*, 1999; Szymanski *et al.*, 1999). In most cases, the severity of the defects in the actin cytoskeleton correlates with the extent to which cell shape is altered in a particular cell. A careful immunological examination of the actin cytoskeleton at the onset of the mutant phenotype (following branch initiation) is a more informative, but ultimately incomplete exercise. Sensitive and non-toxic live-cell probes that decorate the actin cytoskeleton in living plants are needed to visualize the actin cytoskeleton over time.

9.6 Concluding remarks

Arabidopsis trichomes appear to employ a combination of diffuse and tip-directed growth during their development. Pharmacological, genetic, and localization data point to a novel function for microtubules in defining apical domains of cell expansion. Elongated cells such as cotton fibers or stigmatic papillae cells may share similar growth mechanisms. Actin filaments also play a prominent role during trichome development, and are likely to control organelle positioning, vesicle motility, and membrane recycling. It is apparent, based on pleiotropic phenotypes, that many proteins that function during trichome development are also important in other cell types and developmental contexts. Therefore, new knowledge about the generation of cell asymmetry, polarized growth, and cell size control obtained using the trichome model will be generally applicable to plant cells. Progress will come from the systematic localization of important proteins, their binding partners, and a better understanding of cell dynamics in developing trichomes.

Acknowledgements

Thanks to Elizabeth Kellogg for supporting Mark Beilstein and to both Elizabeth and Ihshan Al-Shehbaz for helpful comments on the manuscript. This work was supported by National Science Foundation grant 0110817-IBN and Department of Energy grant DE-FG02-02ER15357 to DBS.

References

Al-Shehbaz, I. (1984) The tribes of the Cruciferae (Brassicaceae) in the southeastern United States. *J. Arnold Arboretum*, **65**, 343–373.

Al-Shehbaz, I. (1989) *Dactylocardamum* (Brassicaceae), a remarkable new genus from Peru. *J. Arnold Arboretum*, **70**, 515–521.

Al-Shehbaz, I. (1994a) *Petroravenia* (Brassicaceae), a new genus from Argentina. *Novon*, **4**, 191–196.

Al-Shehbaz, I. (1994b) Three new South American species of *Draba* (Brassicaceae). *Novon*, **4**, 197–202.

Archer, J.E., Vega, L.R. & Solomon, F. (1995) Rbl2p, a yeast protein that binds to β-tubulin and participates in microtubule function *in vivo*. *Cell*, **82**, 425–434.

Baluska, F., Salaj, J., Mathur, J., *et al.* (2000) Root hair formation: F-actin-dependent tip growth is initiated by local assembly of profilin-supported F-actin meshworks accumulated within expansion-enriched bulges. *Dev. Biol.*, **227**, 618–632.

Bhamidipati, A., Lewis, S.A. & Cowan, N.J. (2000) ADP ribosylation factor-like protein 2 (Arl2) regulates the interaction of tubulin-folding cofactor D with native tubulin. *J. Cell Biol.*, **149**, 1087–1096.

Bibikova, T.N., Blancaflor, E.B. & Gilroy, S. (1999) Microtubules regulate tip growth and orientation in root hairs of *Arabidopsis thaliana*. *Plant J.*, **17**, 657–665.

Bichet, A., Desnos, T., Turner, S., Grandjean, O. & Hofte, H. (2001) *BOTERO1* is required for normal orientation of cortical microtubules and anisotropic cell expansion in *Arabidopsis*. *Plant J.*, **25**, 137–148.

Brugnera, E., Haney, L., Grimsley, C., *et al.* (2002) Unconventional Rac-GEF activity is mediated through the Dock180-ELMO complex. *Nat. Cell Biol.*, **4**, 574–582.

Burk, D.H., Liu, B., Zhong, R., Morrison, W.H. & Ye, Z.H. (2001) A katanin-like protein regulates normal cell wall biosynthesis and cell elongation. *Plant Cell*, **13**, 807–827.

Burk, D.H. & Ye, Z. (2002) Alteration of oriented deposition of cellulose microfibrils by mutation of a katanin-like microtubule-severing protein. *Plant Cell*, **14**, 2145–2160.

Chinnadurai, G. (2002) CtBP, an unconventional transcriptional corepressor in development and oncogenesis. *Mol. Cell*, **9**, 213–224.

Cote, J.-F. & Vuori, K. (2002) Identification of an evolutionarily conserved superfamily of DOCK180-related proteins with guanine nucleotide exchange activity. *J. Cell Sci.*, **115**, 4901–4913.

Cutler, S.R., Ehrhardt, D.W., Griffitts, J.S. & Somerville, C.R. (2000) Random GFP::cDNA fusions enable visualization of subcellular structures in cells of *Arabidopsis* at a high frequency. *Proc. Natl. Acad. Sci. USA*, **97**, 3718–3723.

de Vetten, N., Quattrocchio, F., Mol, J. & Koes, R. (1997) The *an11* locus controlling flower pigmentation in petunia encodes a novel WD-repeat protein conserved in yeast, plants, and animals. *Genes Dev.*, **11**, 1422–1434.

Erickson, M.R.S., Galletta, B.S. & Abmayr, S.M. (1997) *Drosophila myoblast city* encodes a conserved protein that is essential for myoblast fusion, dorsal closure, and cytoskeletal organization. *J. Cell Biol.*, **138**, 589–603.

Feenstra, W.J. (1978) Contiguity of linkage groups I and IV as revealed by linkage relationship of two newly isolated markers *dis-1* and *dis-2*. *Arab. Inf. Serv.*, **15**, 35–38.

Folkers, U., Kirik, V., Schöbinger, U., *et al.* (2002) The cell morphogenesis gene *ANGUSTIFOLIA* encodes a CtBP/BARS-like protein and is involved in the control of the microtubule cytoskeleton. *EMBO J.*, **21**, 1280–1288.

Fu, Y., Li, H. & Yang, Z. (2002) The ROP2 GTPase controls the formation of cortical fine F-actin and the early phase of directional cell expansion during *Arabidopsis* organogenesis. *Plant Cell*, **14**, 777–794.

Hall, J.C., Systsma, K.J. & Ilstis, H.H. (2002) Phylogeny of Capparaceae and Brassicaceae based on chloroplast sequence data. *Am. J. Bot.*, **89**, 1826.

Hartman, J.J., Mahr, J., McNally, K., *et al.* (1998) Katanin, a microtubule-severing protein, is a novel AAA ATPase that targets to the centrosome using a WD40-containing subunit. *Cell*, **93**, 277–287.

Hasegawa, H., Kiyokawa, E., Tanaka, S., *et al.* (1996) DOCK180, a major CRK-binding protein, alters cell morphology upon translocation to the cell membrane. *Mol. Cell Biol.*, **16**, 1770–1776.

Hirata, D., Masuda, H., Eddison, M. & Toda, T. (1998) Essential role of tubulin-folding cofactor D in microtubule assembly and its association with microtubules in fission yeast. *EMBO J.*, **17**, 658–666.

Hoyt, M.A., Macke, J.P., Roberts, B.T. & Geiser, J.R. (1997) *Saccharomyces cerevisiae* PAC2 functions with CIN1, 2 and 4 in a pathway leading to normal microtubule stability. *Genetics*, **146**, 849–857.

Hülskamp, M., Misra, S. & Jürgens, G. (1994) Genetic dissection of trichome cell development in *Arabidopsis*. *Cell*, **76**, 555–566.

Jackson, S.L. & Heath, I.B. (1993) The dynamic behavior of cytoplasmic F-actin in growing hyphae. *Protoplasma*, **173**, 23–34.

Johnson, H.B. (1975) Plant pubescence: an ecological perspective. *Bot. Rev.*, **41**, 233–258.

Johnson, C.S., Kolevski, B. & Smyth, D.R. (2002) *TRANSPARENT TESTA GLABRA2*, a trichome and seed coat development gene of *Arabidopsis*, encodes a WRKY transcription factor. *Plant Cell*, **14**, 1359–1375.

Kao, Y.-L., Deavours, B.E., Phelps, K.K., Walker, R.A. & Reddy, A.S.N. (2000) Bundling of microtubules by motor and tail domains of a kinesin-like calmodulin-binding protein from *Arabidopsis*: regulation by Ca^{2+}/Calmodulin. *Biochem. Biophys. Res. Commun.*, **267**, 201–207.

Kim, G., Shoda, K., Tsuge, T., *et al.* (2002) The *ANGUSTIFOLIA* gene of *Arabidopsis*, a plant CtBP gene, regulates leaf-cell expansion, the arrangement of cortical microtubules in leaf cells and expression of a gene involved in cell-wall formation. *EMBO J.*, **21**, 1267–1279.

Kirik, V., Grini, P.E., Mathur, J., *et al.* (2002a) The *Arabidopsis* TUBULIN-FOLDING COFACTOR A gene is involved in the control of the α-/β-tubulin monomer balance. *Plant Cell*, **14**, 2265–2276.

Kirik, V., Mathur, J., Grini, P.E., *et al.* (2002b) Functional analysis of the tubulin folding cofactor C in *Arabidopsis thaliana*. *Curr. Biol.*, **12**, 1519–1523.

Kiyokawa, E., *et al.* (1998) Evidence that DOCK180 up-regulates signals from the crkII-p130(cas) complex. *J. Biol. Chem.*, **273**, 24479–24484.

Koch, M., Haubold, B. & Mitchell-Olds, T. (2001) Molecular systematics of the Brassicaceae: evidence from coding plastidic matK and nuclear Chs sequences. *Am. J. Bot.*, **88**, 534–544.

Kopczak, S., *et al.* (1992) The small genome of *Arabidopsis* contains at least six expressed α-tubulin genes. *Plant Cell*, **4**, 539–547.

Krishnakumar, S. & Oppenheimer, D.G. (1999) Extragenic suppressors of the *Arabidopsis zwi-3* mutation identify new genes that function in trichome branch formation and pollen tube growth. *Development*, **126**, 3079–3088.

Larkin, J.C., Young, N., Prigge, M. & Marks, M.D. (1996) The control of trichome spacing and number in *Arabidopsis*. *Development*, **122**, 997–1005.

Larkin, J.C., Marks, M.D., Nadeau, J. & Sack, F. (1997) Epidermal cell fate and patterning in leaves. *Plant Cell*, **9**, 1109–1120.

Larkin, J.C., Walker, J.D., Bolognesi-Winfield, A.C., Gray, J.C. & Walker, A.R. (1999) Allele-specific interactions between *ttg* and *gl1* during trichome development in *Arabidopsis thaliana*. *Genetics*, **151**, 1591–1604.

Lee, M.M. & Schiefelbein, J. (1999) WEREWOLF, a MYB-related protein in *Arabidopsis*, is a position-dependent regulator of epidermal cell patterning. *Cell*, **99**, 473–483.

Lewis, S.A., Tian, G. & Cowan, N.J. (1997) The α- and β-tubulin folding pathways. *Trends Cell Biol.*, **7**, 479–484.

Li, H. & Yang, Z. (2000) Rho GTPases and the actin cytoskeleton. *Actin: a dynamic framework for multiple plant cell functions*, 301–322.

Lloyd, A.M., Schena, M., Walbot, V. & Davis, R.W. (1994) Epidermal cell fate determination in *Arabidopsis*: patterns defined by a steroid-inducible regulator. *Science*, **266**, 436–439.

Luo, D. & Oppenheimer, D.G. (1999) Genetic control of trichome branch number in *Arabidopsis*: the roles of the *FURCA* loci. *Development*, **126**, 5547–5557.

Marc, J., Granger, C.L., Brincat, J., *et al.* (1998) A GFP-MAP4 reporter gene for visualizing cortical microtubule rearrangements in living epidermal cells. *Plant Cell*, **10**, 1927–1940.

Marks, M.D. (1997) Molecular genetic analysis of trichome development in *Arabidopsis*. *Annu. Rev. Plant Physiol. Plant Mol. Biol.*, **48**, 137–163.

Mathur, J., Spielhofer, P., Kost, B. & Chua, N. (1999) The actin cytoskeleton is required to elaborate and maintain spatial patterning during trichome cell morphogenesis in *Arabidopsis thaliana*. *Development*, **126**, 5559–5568.

Mathur, J. & Chua, N.-H. (2000) Microtubule stabilization leads to growth reorientation in *Arabidopsis* trichomes. *Plant Cell*, **12**, 465–477.

Mauricio, R. & Rausher, M.D. (1997) Experimental manipulation of putative selective agents provides evidence for the role of natural enemies in the evolution of plant defense. *Evolution*, **51**, 1435–1444.

Mauricio, R., Rausher, M.D. & Burdick, D.S. (1997) Variation in the defense strategies of plants: are resistance and tolerance mutually exclusive? *Ecology*, **78**, 1301–1311.

Mauricio, R. (1998) Costs of resistance to natural enemies in field populations of the annual plant *Arabidopsis thaliana*. *Am. Nat.*, **151**, 20–28.

McClinton, R.S., Chandler, J.S. & Callis, J. (2001) cDNA isolation, characterization, and protein intracellular localization of a katanin-like p60 subunit from *Arabidopsis thaliana*. *Protoplasma*, **216**, 181–190.

McElver, J., Tzafrir, I., Aux, G., *et al.* (2001) Insertional mutagenesis of genes required for seed development in *Arabidopsis thaliana*. *Genetics*, **159**, 1751–1763.

McNally, F.J. & Vale, R.D. (1993) Identification of Katanin, an ATPase that severs and disassembles stable microtubules. *Cell*, **75**, 419–429.

McNally, F.J., Okawa, K., Iwamatsu, A. & Vale, R. (1996) Katanin, the microtubule-severing ATPase, is concentrated at centrosomes. *J. Cell Sci.*, **109**, 561–567.

Melaragno, J.E., Mehrota, B. & Coleman, A.W. (1993) Relationship between endoploidy and cell size in epidermal tissue of *Arabidopsis*. *Plant Cell*, **5**, 1661–1668.

Meller, N., Irani-Tehrani, M., Kiosses, W.B., Del Pozo, M.A. & Schwartz, M.A. (2002) Zizimin1, a novel Cdc42 activator, reveals a new GEF domain for Rho proteins. *Nat. Cell Biol.*, **4**, 639–647.

Miller, D.D., de Ruijter, N.C.A., Bisseling, T. & Emons, A.M.C. (1999) The role of actin in root hair morphogenesis: studies with lipochito-oligosaccharide as a growth stimulator and cytochalasin as an actin perturbing drug. *Plant J.*, **17**, 141–154.

Molendijk, A.J., Bischoff, F., Rajendrakumar, C.S., *et al.* (2001) *Arabidopsis thaliana* Rop GTPases are localized to tips of root hairs and control polar growth. *EMBO J.*, **20**, 2779–2788.

Moore, R., Zhang, M., Cassimeris, L. & Cyr, R. (1997) In vitro assembled plant microtubules exhibit a high state of dynamic instability. *Cell Motil. Cytoskel.*, **38**, 278–286.

Mulligan, G.A. (1995) Synopsis of the genus *Arabis* (Brassicaceae) in Canada, Alaska and Greenland. *Rhodora*, **97**, 109–163.

Narasimhulu, S.B. & Reddy, A.S.N. (1998) Characterization of microtubule binding domains in the *Arabidopsis* kinesin-like calmodulin binding protein. *Plant Cell*, **10**, 957–965.

Nolan, K.M., *et al.* (1998) Myoblast city, the *Drosophila* homolog of DOCK180/CED-5, is required in a rac signaling pathway utilized for multiple developmental processes. *Genes Dev.*, **12**, 3337–3342.

O'Kane, S.L.J., Schaal, B.A. & Al-Shehbaz, I.A. (1996) The origins of *Arabidopsis suecica* (Brassicaceae) as indicated by nuclear rDNA sequences. *Syst. Bot.*, **21**(4), 559–566.

Olson, K.R., McIntosh, J.R. & Olmstead, J.B. (1995) Analysis of MAP 4 function in living cells using green fluorescent protein (GFP) chimeras. *J. Cell Biol.*, **130**, 639–650.

Oppenheimer, D.G., Pollock, M.A., Vacik, J., *et al.* (1997) Essential role of a kinesin-like protein in *Arabidopsis* trichome morphogenesis. *Proc. Natl. Acad. Sci. USA*, **94**, 6261–6266.

Otegui, M. & Staehelin, L. (2000a) Cytokinesis in flowering plants: more than one way to divide a cell. *Cell Biol.*, **3**, 493–502.

Otegui, M. & Staehelin, L.A. (2000b) Synctial-type cell plates: a novel kind of cell plate involved in endosperm cellularization of *Arabidopsis*. *Plant Cell*, **12**, 933–947.

Payne, C.T., Zhang, F. & Lloyd, A.M. (2000) *GL3* is a bHLH protein that regulates trichome development in *Arabidopsis* through interaction with GL1 and TTG1. *Genetics*, **156**, 1349–1362.

Picton, J.M. & Steer, M.W. (1981) Determination of secretory vesicle production rates by dictyosomes in pollen tubes of *Tradescantia* using cytochalasin D. *J. Cell Sci.*, **49**, 261–272.

Qiu, J.L., Jilk, R., Marks, M.D. & Szymanski, D.B. (2002) The *Arabidopsis SPIKE1* gene is required for normal cell shape control and tissue development. *Plant Cell*, **14**, 101–118.

Quarmby, L. (2000) Cellular samurai: katanin and the severing of microtubules. *J. Cell Sci.*, **113**, 2821–2827.

Radcliffe, P.A., Hirata, D., Vardy, L. & Toda, T. (1999) Functional dissection and hierarchy of tubulin-folding cofactor homologues in fission yeast. *Mol. Biol. Bell*, **10**, 2987–3001.

Reddien, P.W. & Horvitz, H.R. (2000) CED-2/crkII and CED-10/rac control phagocytosis and cell migration in *Caenorhabditis elegans*. *Nat. Cell Biol.*, **2**, 131–136.

Reddy, A., Narasimhulu, S., Safadi, F. & Golovkin, M. (1996a) A plant kinesin heavy chain-like protein is a calmodulin-binding protein. *Plant J.*, **10**, 9–21.

Reddy, A., Safadi, F., Narasimhulu, S., Golovkin, M. & Hu, X. (1996b) A novel plant calmodulin-binding protein with a kinesin heavy chain motor domain. *J. Biol. Chem.*, **271**, 7052–7060.

Rerie, W.G., Feldmann, K.A. & Marks, M.D. (1994) The GLABRA2 gene encodes a homeo domain protein required for normal trichome development in *Arabidopsis*. *Genes Dev.*, **8**, 1388–1399.

Rollins, R.C. & Banerjee, U.C. (1975) *Atlas of the Trichomes of Lesquerella (Cruciferae)*, Bussey Institute, Harvard University.

Rollins, R.C. & Banerjee, U.C. (1976) *Trichomes in Studies of the Cruciferae*, Academic Press Inc., London.

Schaeper, U., Boyd, J.M., Verma, S., Uhlmann, E., Subramanian, T. & Chinnadurai, G. (1995) Molecular cloning and characterization of a cellular phoshoprotein that interacts with a conserved C-terminal domain of adenovirus E1A involved in negative modulation of oncogenic transformation. *Proc. Natl. Acad. Sci. USA*, **92**, 10467–10471.

Schellmann, S., Schnittger, A., Kirik, V., *et al.* (2002) TRIPTYCHON and CAPRICE mediate lateral inhibition during trichome and root hair patterning in *Arabidopsis*. *EMBO J.*, **21**, 5036–5046.

Schiefelbein, J.W. (2000) Constructing a plant cell. The genetic control of root hair development. *Plant Physiol.*, **124**, 1525–1531.

Sharp, D.J., Rogers, G.C. & Scholey, J.M. (2000) Microtubule motors in mitosis. *Nature*, **407**, 41–47.

Smirnova, E., Reddy, A., Bowser, J. & Bajer, A. (1998) Minus end-directed kinesin-like motor protein, Kcbp, localizes to anaphase spindle poles in *Haemanthus* endosperm. *Cell Motil. Cytoskel.*, **41**, 271–280.

Snustad, D., Haas, N., Kopczak, S. & Silflow, C. (1992) The small genome of *Arabidopsis* contains at least nine expressed β-tubulin genes. *Plant Cell*, **4**, 549–556.

Song, H., Golovkin, M., Reddy, A.S. & Endow, S.A. (1997) *In vitro* motility of ATKCBP, a calmodulin-binding kinesin protein of *Arabidopsis*. *Proc. Natl. Acad. Sci. USA*, **94**, 322–327.

Spanfò, S., Silletta, M.G., Colanzi, A., *et al.* (1999) Molecular cloning and functional characterization of brefeldin A-ADP-ribosylated substrate. *J. Biol. Chem.*, **274**, 17705–17710.

Srayko, M., Buster, D.W., Bazirgan, O.A., McNally, F.J. & Mains, P.E. (2000) MEI-1/MEI-2 katanin-like microtubule severing activity is required for *Caenorhabditis elegans* meiosis. *Gene Dev.*, **14**, 1072–1084.

Steinborn, K., Maulbetsch, C., Priester, B., *et al.* (2002) The *Arabidopsis PILZ* group genes encode tubulin-folding cofactor orthologs required for cell division but not cell growth. *Gene Dev.*, **16**, 959–971.

Stoppin-Mellet, V., Gaillard, J. & Vantard, M. (2002) Functional evidence for *in vitro* microtubule severing by the plant katanin homologue. *Biochem. J.*, **365**, 337–342.

Szymanski, D.B. & Marks, M.D. (1998) *GLABROUS1* overexpression and *TRIPTYCHON* alter the cell cycle and trichome cell fate in *Arabidopsis*. *Plant Cell*, **10**, 2047–2062.

Szymanski, D.B., Jilk, R.A., Pollock, S.M. & Marks, M.D. (1998a) Control of GL2 expression in *Arabidopsis* leaves and trichomes. *Development*, **125**, 1161–1171.

Szymanski, D.B., Klis, D.A., Larkin, J.C. & Marks, M.D. (1998b) *cot1*: A regulator of *Arabidopsis* trichome initiation. *Genetics*, **149**, 565–577.

Szymanski, D.B., Marks, M.D. & Wick, S.M. (1999) Organized F-actin is essential for normal trichome morphogenesis in *Arabidopsis*. *Plant Cell*, **11**, 2331–2347.

Szymanski, D.B., Lloyd, A.M. & Marks, M.D. (2000) Progress in the molecular genetic analysis of trichome initiation and morphogenesis in *Arabidopsis*. *Trends Plant Sci.*, **5**, 214–219.

Szymanski, D.B. (2001) *Arabidopsis* trichome morphogenesis: a genetic approach to studying cytoskeletal function. *J. Plant Growth Regul.*, **20**, 131–140.

Tian, G., Huang, Y., Rommelaere, H., Vandekerckhove, J., Ampe, C. & Cowan, N.J. (1996) Pathway leading to correctly folded β-tubulin. *Cell*, **86**, 287–296.

Tian, G., Lewis, S.A., Feierbach, B., *et al.* (1997) Tubulin subunits exist in an activated conformational state generated and maintained by protein cofactors. *J. Cell Biol.*, **138**, 821–832.

Tiwari, S.C. & Wilkins, T.A. (1995) Cotton (*Gossypium hirsutum*) seed trichomes expand via diffuse growing mechanism. *Can. J. Bot.*, **73**, 746–757.

Tsuge, T., Tsukaya, H. & Uchimiya, H. (1996) Two independent and polarized processes of cell elongation regulate leaf blade expansion in *Arabidopsis thaliana* (L.) Heynh. *Development*, **122**, 1589–1600.

Tzafrir, I., McElver, J.A., Liu, C.M., *et al.* (2002) Diversity of *TITAN* functions in *Arabidopsis* seed development. *Plant Physiol.*, **128**, 38–51.

Ueda, K., Matsuyama, T. & Hashimoto, T. (1999) Visualization of microtubules in living cells of transgeneic *Arabidopsis thaliana*. *Protoplasma*, **206**, 201–206.

Vos, J., Safadi, F., Reddy, A. & Hepler, P. (2000) The kinesin-like calmodulin binding protein is differentially involved in cell division. *Plant Cell*, **12**, 979–990.

Wada, T., Tachibana, T., Shimura, Y. & Okada, K. (1997) Epidermal cell differentiation in *Arabidopsis* determined by a *Myb* homolog, *CPC*. *Science*, **277**, 1113–1116.

Wasteneys, G.O., Willingale-Theune, J. & Menzel, D. (1997) Freeze shattering: A simple and effective method for permeabilizing higher plant cell walls. *J. Microsc.*, **188**, 51–61.

Webb, M., Jouannic, S., Foreman, J., Linstead, P. & Dolan, L. (2002) Cell specification in the *Arabidopsis* root epidermis requires the activity of ECTOPIC ROOT HAIR 3 a katanin-p60 protein. *Development*, **129**, 123–131.

Weigert, R., Silletta, M.G., Spano, S., *et al.* (1999) CtBP/BARS induces fusion of Golgi membranes by acylating lysophatidic acid. *Nature*, **402**, 429–433.

Weinstein, B. & Solomon, F. (1990) Phenotypic consequences of tubulin overproduction in *Saccharomyces cerevisiae*: differences between α-tubulin and β-tubulin. *Mol. Cell Biol.*, **10**, 5295–5304.

Wu, Y.C. & Horvitz, H.R. (1998) *C. elegans* phagocytosis and cell-migration protein CED-5 is similar to human DOCK180. *Nature*, **392**, 501–504.

Wu, G., Li, H. & Yang, Z. (2000) *Arabidopsis* RopGAPs are a novel family of rho GTPase-activating proteins that require the Cdc42/Rac-interactive binding motif for rop-specific GTPase stimulation. *Plant Physiol.*, **124**, 1625–1636.

10 Signaling and the cytoskeleton in guard cells

Paula Duque, Juan-Pablo Sánchez and Nam-Hai Chua

10.1 Introduction

Stomata are pore-forming structures found on the surface of the leaves and stems of most higher plants. Two guard cells – specialized kidney-shaped epidermal cells whose changes in volume control the pore size – comprise each stoma (Fig. 10.1). The regulation of stomatal aperture is a central mechanism by which terrestrial plants respond to changing environmental conditions and balance the needs for two of the most important processes to the vegetative plant: photosynthesis and transpiration.

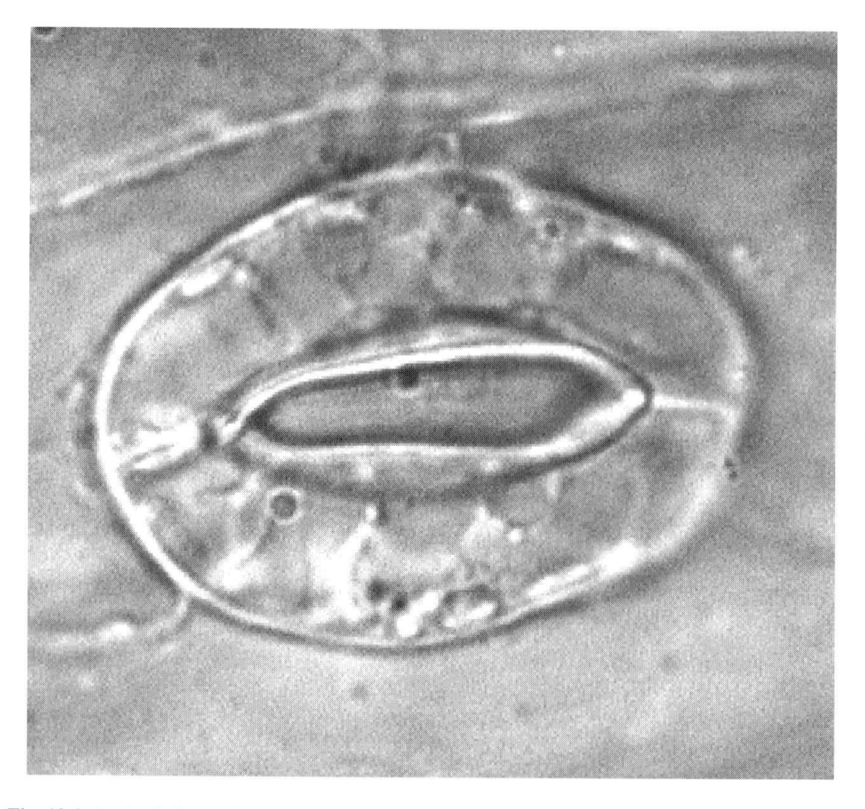

Fig. 10.1 An *Arabidopsis thaliana* stoma visualized under light transmission microscopy.

Under conditions of adequate water status, guard cells are fully turgid and constraints on their inner walls maintain the pore open, allowing CO_2 fixation for photosynthesis. By contrast, release of guard cell turgor pressure closes the stomata to prevent water vapor loss, an essential response under stress conditions such as drought. These turgor and volume changes of guard cells occur in response to various environmental and endogenous stimuli – including light, CO_2, temperature, humidity, plant pathogen elicitors, air pollutants and hormones such as abscisic acid (ABA) – working through different signal transduction pathways.

Owing to their ability to integrate and respond rapidly at the single cell level to multiple physiological and abiotic stimuli, guard cells have become a popular and highly developed higher-plant cell model for exploring early signal transduction in plants. Guard cell swelling (i.e. stomatal opening) and turgor loss (i.e. stomatal closing) are driven primarily, although not exclusively, by potassium and anion influx and efflux, respectively. Indeed, the central role of ion fluxes across the plasma membrane and the tonoplast in stomatal movements is undisputed, but the exact upstream events involved remain unclear. Many signaling components, such as calcium ions, cytosolic pH, phospholipases, cyclic ADP-ribose and protein kinases and phosphatases, have been implicated. Furthermore, several studies indicate that cytoskeleton elements may also act as intermediates in guard cell signaling.

In the present chapter, we briefly review the main components involved in guard cell signaling, before focusing on the accumulating experimental evidence that, since the mid-1990s, has ascribed cytoskeletal reorganization (particularly of the actin cytoskeleton) a role in the regulation of stomatal movements. Several valuable reviews on different aspects of guard cell signal transduction have been published in recent years (Blatt, 2000; Assmann & Wang, 2001; Hetherington, 2001; Schroeder *et al.*, 2001a,b).

10.2 Guard cell signaling

The receptors for the vast majority of the signals that elicit guard cell responses remain to be identified. Moreover, the precise signaling events leading to the corresponding stomatal response are far from being fully understood, despite the fact that recent developments have added significantly to the understanding of cellular signaling in guard cells. It is nevertheless unchallenged that the pathways involved must converge on guard cell volume control, largely through the coordination of ion channel function.

Auxins (Lohse & Hedrich, 1992), blue light (Assmann *et al.*, 1985; Shimazaki *et al.*, 1986) and red light (Serrano *et al.*, 1988) induce stomatal opening by activating plasma membrane H^+-ATPases. Extrusion of H^+ causes plasma membrane hyperpolarization, which drives K^+ uptake via inward-rectifying K^+ channels (Schroeder *et al.*, 1987; Thiel *et al.*, 1992). Charge balance for the K^+ flux is maintained by parallel anion (Cl^-) transport, thought to occur via H^+/anion symporters or anion/OH^- antiporters in the plasma membrane (Schroeder *et al.*, 2001a). In addition, sucrose levels rise (Talbot & Zeiger, 1998). Consequently, water enters the guard cells, and the stomata open.

Other stimuli induce stomatal closure, which requires ion efflux from guard cells. Channel-mediated anion efflux from guard cells causes membrane depolarization, which in turn leads to the inhibition of inward-rectifying K^+ channels and activation of outward-rectifying K^+ channels (Schroeder *et al.*, 1987). Concomitantly, sucrose is removed from guard cells (Talbot & Zeiger, 1998). The result is loss of guard cell turgor and consequent closing of the stomata. As the vacuole of mature guard cells comprises 80–90% of the total cell volume (Blatt, 2000), most solutes released across the plasma membrane need first to traverse the tonoplast in a coordinated manner (MacRobbie, 1999). Stomatal closure may be induced by high CO_2 concentrations in the intercellular spaces of the leaf mesophyll during respiration in the dark (Assmann, 1999). Similarly, elevated atmospheric CO_2 levels are thought to affect gas exchange by reducing stomatal apertures (Drake *et al.*, 1997). Plant pathogen elicitors have also been reported to induce closing of the stomata – and thus restrict invasion of pathogens through the stomatal pores – via the production of reactive oxygen species (ROS) in guard cells (Lee *et al.*, 1999). Moreover, high concentrations of ozone have been shown to promote stomatal closing by inhibiting inward-rectifying K^+ channels, thereby enabling plants to limit further damage by this pollutant (Torsethaugen *et al.*, 1999). Last but not least, biosynthesis of the phytohormone ABA in response to water deficit also elicits stomatal closure.

In fact, stomatal closure evoked by ABA has, in the past decade, yielded some of the major and most exciting advances in the understanding of the mechanisms underlying guard cell signal transduction. ABA induces elevations in the concentration of free Ca^{2+} in the cytosol of guard cells, which in turn signal the ionic currents leading to stomatal closure. Various lines of evidence suggest that early ABA signaling events involve lipid-derived second messengers and/or cyclic nucleotides such as cyclic ADP-ribose. Furthermore, phosphorylation events are known to be mediators in guard cell ABA signaling, although very few specific target proteins have been identified. More recently, other components – namely sphingosine-1-phosphate (S1P) and an mRNA cap-binding protein – have been added to the list of intermediates. ABA has also been shown to trigger cytoskeletal reorganization in guard cells. The latter line of work is the main focus of this chapter and will be dealt with separately, after a brief coverage of the key players in guard cell signal transduction, particularly of those acting downstream of ABA perception.

10.2.1 Cytosolic calcium

Free Ca^{2+} is a second messenger for many stimuli that affect stomatal aperture. Increases in cytosolic Ca^{2+} concentration $[Ca^{2+}]_{cyt}$, arising from both release from intracellular stores and influx across the plasma membrane via different

Ca^{2+} channel types, are the target for several pathways acting downstream of ABA (McAinsh *et al.*, 1990). The elevation in $[Ca^{2+}]_{cyt}$ then inhibits proton-extruding plasma membrane H^+-ATPases (Kinoshita *et al.*, 1995) and inward-rectifying K^+ channels and activates anion channels (Schroeder & Hagiwara, 1989), hence both inhibiting stomatal opening and promoting stomatal closure. Experimental elevation of Ca^{2+} induces stomatal closure while mimicking several effects of ABA on ion channels in guard cells (McAinsh *et al.*, 1995). Moreover, Allen *et al.* (1999) have provided genetic evidence for the role of $[Ca^{2+}]_{cyt}$ elevations in ABA guard cell signaling. In the ABA-insensitive *Arabidopsis thaliana* mutants *abi1* and *abi2*, ABA-induced $[Ca^{2+}]_{cyt}$ increases are impaired and, in addition, the stomatal response phenotype is suppressed by experimentally imposing a rise in $[Ca^{2+}]_{cyt}$ (Allen *et al.*, 1999). Recently, ROS were identified as an upstream regulator of Ca^{2+}-permeable channels in *Arabidopsis* guard cells (Pei *et al.*, 2000), providing the first indication on the mechanisms by which ABA activates plasma membrane Ca^{2+} channels and hence Ca^{2+} uptake by guard cells. In this study, Schroeder and coworkers found that ABA causes a rapid increase in ROS levels and activation of Ca^{2+} channels, and that these responses are disrupted in the ABA-insensitive mutant *gca2* (Pei *et al.*, 2000). Subsequent work showed that this ABA-dependent activation of Ca^{2+} channels also requires cytosolic NAD(P)H (Murata *et al.*, 2001).

In animal cells, a few studies have shown that a sequence of short repetitive increases in $[Ca^{2+}]_{cyt}$ might control the efficiency and specificity of cellular responses (DeKoninck & Schulman, 1998; Dolmetsch *et al.*, 1998). Such oscillations are produced through the integration of Ca^{2+} influx and efflux from the cell and Ca^{2+} sequestration and release from intracellular stores. Spatial (Taylor *et al.*, 1996) or temporal (Ehrhardt *et al.*, 1996; Bauer *et al.*, 1998) $[Ca^{2+}]_{cyt}$ oscillations in response to external stimuli have been identified in different plant cell types, but it remains unclear whether these waves are essential for triggering physiological responses. In guard cells, since the first findings indicating that specific oscillation patterns encode the information required to distinguish between different Ca^{2+}-mobilizing stimuli (McAinsh *et al.*, 1995), it has been proposed that $[Ca^{2+}]_{cyt}$ oscillations are necessary for stomatal closure. Indeed, several stomatal closing stimuli produce these oscillations in wild-type *Arabidopsis* guard cells (Grabov & Blatt, 1998; Staxén *et al.*, 1999), and two recent studies have provided genetic evidence that stimulus-specific $[Ca^{2+}]_{cyt}$ oscillations are required for stomatal closure. In guard cells of the *Arabidopsis* V-type H^+-ATPase mutant, *det3*, some stimuli that induce stomatal closure in wild-type plants elicited non-oscillatory $[Ca^{2+}]_{cyt}$ increases and did not lead to stomatal closure, whereas others elicited $[Ca^{2+}]_{cyt}$ oscillations and did not interfere with stomatal closure. Furthermore, experimentally imposing $[Ca^{2+}]_{cyt}$ oscillations in *det3* guard cells rescued stomatal closure (Allen *et al.*, 2000). In a subsequent study using the *gca2* mutant, in which the capacity to generate appropriate guard cell $[Ca^{2+}]_{cyt}$ oscillations is clearly affected, Allen *et al.* (2001) further demonstrated that the kinetics (frequency, number, duration and amplitude) of such oscillations can control steady-state stomatal aperture changes.

Whereas the above studies clearly indicate that ABA-induced stomatal closure is Ca^{2+}-dependent, it is likely that additional Ca^{2+}-independent components or pathways are required for the overall control of guard cell volume. *Commelina communis* plants grown at suboptimal temperatures fail to exhibit any ABA-induced increase in guard cell $[Ca^{2+}]_{cyt}$ despite stomatal closure (Allan *et al.*, 1994). In fact, $[Ca^{2+}]_{cyt}$ increases are not universally associated with ABA and related stimuli or even with stomatal closure (Blatt, 2000). On the other hand, not all the ion channels whose activity regulates guard cell turgor changes are regulated by increases in $[Ca^{2+}]_{cyt}$ (Hetherington, 2001).

10.2.2 Cytosolic pH

Studies mainly from Blatt's group have led to the proposal of a functional role for cytosolic pH (pH_{cyt}) in ABA guard cell signaling. In an early work using *Vicia faba* guard cells, Blatt (1992) observed that lowering pH_{cyt} inactivates outward-rectifying K^+ channels. A subsequent study showed that ABA treatment leads to a transient elevation in the pH_{cyt} of guard cells and that this rise is both necessary and sufficient to account for the activation of outward-rectifying K^+ channels through which K^+ efflux occurs during stomatal closure (Blatt & Armstrong, 1993). The pH_{cyt} dependence of these channels is consistent with the cooperative binding of at least two H^+ (Blatt & Armstrong, 1993; Grabov & Blatt, 1997). Further studies have suggested an extended role for pH_{cyt} in guard cells. These include the control of K^+ channels by auxin (Blatt & Thiel, 1994), the ABA-mediated control of K^+ channels via the ABI1 protein phosphatase (Armstrong *et al.*, 1995; Leube *et al.*, 1998) and the control of anion channels (Schulz-Lessdorf *et al.*, 1996).

Observations reported by Grabov and Blatt (1997) point to a distinction between the actions of pH_{cyt} and $[Ca^{2+}]_{cyt}$ and suggest a role for pH_{cyt} as a second messenger acting in parallel but independently of $[Ca^{2+}]_{cyt}$. In fact, the effect of pH_{cyt} on K^+ channels seems to be voltage-independent (contrasting with the effect of $[Ca^{2+}]_{cyt}$) and to occur without a measurable change in $[Ca^{2+}]_{cyt}$. Although these results do not rule out the titration of Ca^{2+}-binding sites by H^+, they indicate that the effects of these two ions on the K^+ channel are separable.

10.2.3 Cyclic ADP-ribose

In non-plant systems, the importance of cyclic ADP-ribose (cADPR) in the mobilization of intracellular Ca^{2+} stores has been well established in a variety of cell types. The molecule is synthesized via the cyclization of NAD^+ by ADP-ribosyl cyclase or its homolog CD38 (Aarhus *et al.*, 1995), and numerous studies indicate that it exerts its action by modulating the Ca^{2+} sensitivity of ryanodine receptor (RYR) channels (see Lee, 2001). In plants, cADPR also exerts its effects by way of intracellular Ca^{2+} release and has been identified as a signaling molecule in the ABA response (Wu *et al.*, 1997; Leckie *et al.*, 1998).

ABA stimulates the production of cADPR in *Arabidopsis* and, in tomato hypocotyl cells, microinjection of cADPR induces the expression of ABA-responsive genes, whereas microinjection of a structural analog, 8-NH_2-cADPR, blocks the induction of these genes by ABA (Wu *et al.*, 1997). In *Commelina communis* guard cells, microinjection of cADPR leads to a reduction in turgor which is preceded by increases in $[Ca^{2+}]_{cyt}$, while microinjection of 8-NH_2-cADPR reduces the rate of turgor loss in response to ABA (Leckie *et al.*, 1998). However, it should be noted that this ABA-induced stomatal closure is only partially inhibited by microinjection of the referred analog or of nicotinamide, an inhibitor of cADPR production (Leckie *et al.*, 1998; Jacob *et al.*, 1999), indicating that additional mechanisms of Ca^{2+} mobilization are required. In another study, blocking RYR-activated Ca^{2+} channels with ryanodine inhibited $[Ca^{2+}]_{cyt}$ elevations in *Vicia faba* guard cells (Grabov & Blatt, 1999), further substantiating the Ca^{2+} release role of cADPR in guard cells.

10.2.4 Inositol 1,4,5-trisphosphate and other lipid-derived second messengers

Inositol 1,4,5-trisphosphate ($InsP_3$) is a product of the hydrolysis of phosphatidylinositol 4,5-bisphosphate catalyzed by phospholipase C (PLC). In guard cells, $InsP_3$ turnover is activated in response to ABA (Lee *et al.*, 1996) and PLC seems to play a role in the generation of ABA-induced oscillations in $[Ca^{2+}]_{cyt}$ (Staxén *et al.*, 1999). Moreover, earlier experiments had shown that release of sequestered $InsP_3$ inhibits guard cell inward-rectifying K^+ channels (Blatt *et al.*, 1990) and stimulates $[Ca^{2+}]_{cyt}$ increases leading to stomatal closure (Gilroy *et al.*, 1990).

The cADPR and the PLC/$InsP_3$-mediated Ca^{2+}-mobilizing pathways constitute two recognized second messenger systems for release of Ca^{2+} from internal stores in guard cells. Simultaneous application of a PLC and a cADPR inhibitor impairs ABA-induced stomatal closing in *Vicia faba* (Jacob *et al.*, 1999) and K^+ efflux in *Commelina communis* (MacRobbie, 2000), while either of the inhibitors alone causes only partial inhibition of these responses. This suggests that the two pathways operate in parallel during ABA signaling. Furthermore, results obtained with *Commelina* guard cells suggest that strong activation of plasma membrane Ca^{2+} influx occurs only at high ABA concentrations (1 μM and above), whereas at lower ABA concentrations (0.1 or 0.2 μM) both pathways of internal release predominate and Ca^{2+} influx plays a minor role (MacRobbie, 2000).

A recent study (Lemtiri-Chlieh *et al.*, 2000) indicates that another inositol phosphate can also function as a second messenger in ABA signaling. In guard cells, inositol hexakisphosphate ($InsP_6$) levels were found to increase in response to ABA to a greater extent than $InsP_3$. In addition, experimentally elevating $InsP_6$ in guard cell protoplasts strongly inhibited inward-rectifying K^+ channels in a Ca^{2+}-dependent manner. The molecule was also significantly more potent in regulating inward K^+ currents than $InsP_3$ (Lemtiri-Chlieh *et al.*, 2000). Taken together, these results suggest that $InsP_6$ may play a major, if not dominant, role in the physiological guard cell response to ABA. It is still unknown whether $InsP_6$ leads to $[Ca^{2+}]_{cyt}$ elevations or acts via a parallel signaling pathway.

Following its implication in the initial steps of ABA signal transduction in aleurone cells (Ritchie & Gilroy, 1998), phospholipase D (PLD) has also been implicated in ABA guard cell signaling. In *Vicia faba* guard cell protoplasts, the activity of this phospholipase transiently increases after ABA application, and treatment with phosphatidic acid, a product of PLD hydrolytic activity, promotes partial stomatal closure, inhibits stomatal opening, and leads to inhibition of the activity of inward-rectifying K^+ channels (Jacob *et al.*, 1999). However, no increase in $[Ca^{2+}]_{cyt}$ was observed in response to phosphatidic acid, suggesting that PLD is either acting through a Ca^{2+}-independent pathway or downstream of $[Ca^{2+}]_{cyt}$ elevations. Whereas application of an inhibitor of PLD activity partially inhibited ABA-induced stomatal closure, this effect was enhanced by simultaneous application of nicotinamide, indicating that PLD operates in parallel with the cADPR-mediated pathway (Jacob *et al.*, 1999).

Two phosphatidylinositol monophosphates, phosphatidylinositol 3-phosphate (PI3P) and phosphatidylinositol 4-phosphate (PI4P), have very recently been implicated in the control of stomatal movements (Jung *et al.*, 2002). Treatment of *Commelina* guard cells with inhibitors of the kinases responsible for the synthesis of PI3P and PI4P promoted light-induced stomatal opening and inhibited ABA-induced stomatal closure. Consistent with this, overexpression of fusion proteins which specifically bind PI3P and PI4P also enhanced stomatal opening and partially inhibited ABA-induced closure. Moreover, inhibitors of PI3P and PI4P synthesis also inhibited ABA-induced $[Ca^{2+}]_{cyt}$ elevations, suggesting that the reduction of ABA-induced stomatal closure caused by reduced levels of PI3P and PI4P is due, at least in part, to impaired Ca^{2+} signaling (Jung *et al.*, 2002).

10.2.5 Protein kinases and phosphatases

Suppression of stomatal responses by inhibitors and mutations of protein kinases and phosphatases provide evidence that both protein phosphorylation and dephosphorylation play central roles in the signaling cascades involved in stomatal movements.

A guard-cell-specific, ABA-activated protein kinase (AAPK) showing Ca^{2+}-independent serine–threonine protein kinase activity has been identified and characterized in *Vicia faba* guard cells (Li & Assmann, 1996; Mori & Muto, 1997). The *AAPK* gene was subsequently cloned and expression of a dominant mutant allele shown to disrupt ABA-induced stomatal closure and impair ABA activation of plasma membrane anion channels (Li *et al.*, 2000). Recently, an RNA-binding protein named AKIP1 (AAPK-interacting protein 1) was identified as a substrate of AAPK (Li *et al.*, 2002). The latter study shows that phosphorylation of AKIP1 by AAPK is necessary for interaction of the former protein with the ABA-inducible dehydrin mRNA.

Other phosphorylation events have also been implicated in guard cell signal transduction. Two examples are a Ca^{2+}-dependent kinase that phosphorylates the inward-rectifying K^+ channel KAT1 (Li *et al.*, 1998) and the blue-light-dependent phosphorylation of the C-terminus of the H^+-ATPase by a serine-threonine protein kinase whose activity is modulated by a 14-3-3 protein (Kinoshita & Shimazaki, 1999). The latter functions as a scaffolding protein by linking the protein kinase to its target.

Two dominant *Arabidopsis* mutants, *abi1-1* and *abi2-1*, isolated from genetic screens for insensitivity to ABA inhibition of seed germination, are impaired in ABA-induced stomatal closure (Koornneef *et al.*, 1984). The corresponding genes, *ABI1* and *ABI2*, encode type 2C protein phosphatases (PP2C) and the mutations interfere with guard cell ABA signaling events that act upstream of $[Ca^{2+}]_{cyt}$. Indeed, PP2C activity is necessary for normal $[Ca^{2+}]_{cyt}$ elevations (Allen *et al.*, 1999), activation of anion channels (Pei *et al.*, 1997), and regulation of inward- and outward-rectifying K^+ channels (Armstrong *et al.*, 1995) in guard cells responding to ABA. As these are dominant mutations, it had remained unclear whether ABI1 and ABI2 were positive or negative regulators of ABA signaling. However, intragenic revertants of both mutants have recently been isolated, with the *abi1-1* revertants exhibiting hypersensitive responses to ABA (Gosti *et al.*, 1999; Merlot *et al.*, 2001). A double mutant of both revertants was more responsive to ABA than each of the parental single mutants, suggesting that ABI1 and ABI2 are negative regulators of ABA signaling with overlapping roles in controlling ABA action (Merlot *et al.*, 2001). Recently, Himmelbach *et al.* (2002) have identified a homeodomain protein, ATHB6, that interacts with ABI1. Transgenic plants overexpressing this transcriptional regulator are insensitive to a number of ABA responses, suggesting that ATHB6 is a negative regulator of ABA signaling (Himmelbach *et al.*, 2002). Furthermore, the ABI1 and ABI2 phosphatases have also been implicated in the activation of plasma membrane Ca^{2+} channels by ROS, with ABI1 and ABI2 probably acting up- and downstream of NAD(P)H-dependent ROS production, respectively (Murata *et al.*, 2001).

In addition, type 1 protein phosphatases (PP1) and type 2A protein phosphatases (PP2A) are active in guard cells. These phosphatases are sensitive to okadaic acid and calyculin A, with guard cell K^+ (Thiel & Blatt, 1994) and anion currents, as well as ABA-induced stomatal movements (Schmidt *et al.*, 1995; Pei *et al.*, 1997), being affected by these inhibitors. Various lines of evidence indicate that some PP1 or PP2A act as negative regulators, whereas others act as positive regulators of guard cell ABA signaling (see Schroeder *et al.*, 2001a).

10.2.6 Membrane trafficking

Stomatal movements require large and repetitive changes in the volume of guard cells that cannot be accommodated simply by expansion and compression of the membrane bilayer (Blatt, 2000). In fact, changes in stomatal aperture are accompanied by substantial changes in the total membrane surface area of guard cells, which are coupled to vesicle trafficking (Homann & Thiel, 1999; Kubitscheck *et al.*, 2000). Recent evidence indicates that in guard cell protoplasts, the number of both inward- and outward-rectifying K^+ channels also changes in parallel with the plasma membrane surface area (Homann & Thiel, 2002).

Syntaxins and related SNARE proteins play a central role in vesicle trafficking and membrane fusion. A tobacco syntaxin homolog, *Nt-SYR1*, was isolated by Blatt and colleagues in an expression cloning screen for an ABA receptor in *Xenopus laevis* oocytes (Leyman *et al.*, 1999). In tobacco guard cells, microinjection of non-functional *Nt-SYR1* or of an inhibitor of syntaxin-dependent vesicle trafficking impaired ABA regulation of inward- and outward-rectifying K^+ channels, as well as of anion channels (Leyman *et al.*, 1999). These results point to a role of *Nt-SYR1* as a positive regulator early in the ABA signaling pathway and suggest a link between ABA, membrane trafficking and ion channel control. A possible function for *Nt-SYR1* as a scaffolding protein or as a second messenger element has been proposed (Leyman *et al.*, 1999). However, in spite of the subsequent identification of protein-binding partners of *Nt-SYR1* (Kargul *et al.*, 2001), further evidence is needed in order to gain insight into the function of this syntaxin-related protein in guard cells.

10.2.7 New key intermediates

Sphingosine-1-phosphate (S1P) has recently been identified as a calcium-mobilizing molecule in plants (Ng *et al.*, 2001). In animal cells, S1P plays a role in Ca^{2+} mobilization from intracellular stores via an $InsP_3$-independent pathway (Mattie *et al.*, 1994), and is involved in mediating various cellular responses such as proliferation, survival, motility, differentiation and apoptosis (Pyne & Pyne, 2000). After confirming the presence of S1P in *Commelina communis* extracts, the Hetherington group (Ng *et al.*, 2001) found that its levels increase in whole leaves upon drought treatment. This prompted them to investigate the role of the molecule in stomatal movements. Incubation of epidermal strips with S1P promoted stomatal closure and induced $[Ca^{2+}]_{cyt}$ oscillations in guard cells. Moreover, the effect of ABA on stomatal closure was attenuated by treatment with a competitive inhibitor of sphingosine kinase, the enzyme that catalyzes the phosphorylation of sphingosine to yield S1P (Ng *et al.*, 2001). These results place S1P upstream of elevated $[Ca^{2+}]_{cyt}$, as a new intermediate in the ABA guard cell signal transduction pathway.

Another recent study has implicated an mRNA cap binding protein in the control of guard cell sensitivity to ABA (Hugouvieux *et al.*, 2001). While screening for ABA hypersensitivity in *Arabidopsis* seed germination, Schroeder and coworkers isolated a loss-of-function mutant, *abh1*, which turned out to also be hypersensitive to ABA-induced stomatal closure. The *abh1* gene encodes the *Arabidopsis* homolog of a nuclear mRNA cap-binding protein, which functions in the cap binding complex known to play important roles in the processing of mRNA. Microarray analysis revealed that only a few transcripts are downregulated in the *abh1* mutant (Hugouvieux *et al.*, 2001), several of which are known to be regulated by ABA. Interestingly, one of these genes encodes a PP2C previously proposed to function as a negative regulator of ABA signaling (Sheen, 1998). In addition, *abh1* guard cells display ABA hypersensitive $[Ca^{2+}]_{cyt}$ increases. These findings led Hugouvieux *et al.* (2001) to propose that ABH1 is a modulator of ABA guard cell signaling that directly controls the abundance of transcripts involved in early ABA signaling.

The characterization of two knockout alleles of *GPA1*, the *Arabidopsis* gene encoding the heterotrimeric GTP-binding (G) protein α subunit, has demonstrated the involvement of G proteins in the regulation of guard cell ion channels and in the ABA signaling pathway (Wang *et al.*, 2001). With these experiments, Assmann's group showed that both ABA inhibition of guard cell inward-rectifying K^+ channels and pH-independent ABA activation of anion channels require the presence of functional *GPA1*. In addition, both *GPA1* alleles were impaired in the inhibition of stomatal opening by ABA and exhibited greater water loss rates than wild-type plants (Wang *et al.*, 2001). In animal cells, heterotrimeric G proteins were already known to play key roles in the regulation of ion channels and to be among the most important components of signaling pathways. This study by Wang *et al.* (2001), together with other reports showing that the *GPA1* gene is also a regulator of cell division (Ullah *et al.*, 2001) and seed germination (Ullah *et al.*, 2002) in *Arabidopsis*, have shown that G proteins are also central regulators of plant growth and development.

Inward-rectifying K^+ channels have also been shown to be regulated by polyamines (Liu *et al.*, 2000). Polyamine levels increase in plant cells upon exposure to stress conditions such as drought, salinity and air pollutants (Evans & Malmberg, 1989), which are also known to elicit stomatal responses. Liu *et al.* (2000) investigated the role of all natural polyamines in stomatal movements and found that these compounds inhibit opening and induce closure of stomata. Whole-cell patch-clamp analysis showed that polyamines inhibit K^+ inward currents. However, this inhibition was not observed when isolated membrane patches were used, suggesting that regulation of K^+ channels by polyamines requires unknown cytoplasmic factors (Liu *et al.*, 2000). These findings link stress conditions to stomatal regulation through polyamine levels.

ATP-binding cassette (ABC) proteins comprise a large family of transporters that actively translocate solutes across membranes in a wide range of species. Recently, evidence is accumulating that these proteins may play a role in the regulation of guard cell ABA signaling. Pharmacological approaches first showed that inhibitors of ABC proteins block anion and outward-rectifying K^+ channels, and abolish stomatal closure triggered by ABA and Ca^{2+} in *Vicia* and *Commelina* (Leonhardt *et al.*, 1999). A subsequent study has identified a new ABC protein in *Arabidopsis*, *At*MRP5, that is involved in the control of stomatal movements (Gaedeke *et al.*, 2001). Results from this work indicate that *At*MRP5 is an anion transporter. In addition, experiments on epidermal strips showed that a specific ABC protein inhibitor which triggers stomatal opening in wild-type plants had no effect on stomatal opening in a T-DNA knockout mutant for *AtMRP5* (Gaedeke *et al.*, 2001). Taken together, these results demonstrate that this novel *Arabidopsis* ABC protein is involved in maintaining stomatal closure and raise the possibility that *At*MRP5 itself may be a component of, or act as, a guard cell anion channel.

10.3 The cytoskeleton in guard cell function

Several plant cell responses to physiological and external stimuli, such as pollen tube tip growth (see Chapter 8) or resistance to fungal invasion, involve dramatic rearrangements of the cytoskeleton (Staiger, 2000). Evidence is accumulating that cytoskeletal proteins may function as signal transducers in plant systems as has been well established to be the case in animal cells, where structural changes in these dynamic polymers are important in several signaling cascades and in cell shape changes.

Changes in the organization pattern of actin microfilaments also occur in guard cells during stimulus-mediated stomatal movements, with experimental data increasingly pointing to a participation of actin polymers in guard cell signaling via actin-binding proteins. These proteins modulate the organization and functional properties of the cytoskeleton by controlling actin (de)polymerization and stabilization. Actin-regulatory proteins are extremely sensitive to the cellular levels of signaling intermediates and second messengers, and can therefore act as efficient sensors of the intracellular environment (Staiger *et al.*, 2000; see Chapter 2).

On the other hand, microtubules are well known to be important for guard cell development, during which they guide the deposition of microfibrils in the cell wall (see Chapter 3), but their involvement in signal transduction for mature guard cell responses has been tested in several reports with conflicting results. The involvement of microtubules in stomatal function remains therefore controversial.

10.3.1 (Re)organization of actin filaments

The first link between guard cell actin filaments and stomatal responses was reported by Lee's group in a study using *Commelina communis*. Kim *et al.* (1995) observed that actin microfilaments are radially distributed in mature guard cells, fanning outward from the stomatal pore, and used a pharmacological approach to assess the involvement of actin filament polymerization/stabilization or depolymerization in stomatal movements. An actin-stabilizing agent, phalloidin, inhibited ABA-induced stomatal closure and light-induced stomatal opening, whereas an actin-depolymerizing compound, cytochalasin D, stimulated stomatal opening in both darkness and light (Kim *et al.*, 1995). Hwang *et al.* (2000) later reported that cytochalasin D also enhances ABA-induced stomatal closure. The early studies by Kim and colleagues (1995) indicated that dynamic changes in the actin-based cytoskeleton are involved in the regulation of stomatal aperture and led the authors to propose that actin filaments, presumably via actin-binding proteins, modulate the activity of ion channels in guard cells.

To test their hypothesis, Lee and coworkers next examined the possible regulation of inward-rectifying K^+ channels by changes in the state of actin polymerization. To this end, patch clamp analysis of ion channel activity in *Vicia faba* guard cells was performed in the presence of actin antagonists (Hwang *et al.*, 1997). In agreement with prior results (Kim *et al.*, 1995), phalloidin inhibited, whereas cytochalasin D promoted, light-induced stomatal opening. Furthermore, Hwang *et al.* (1997) showed that phalloidin inhibits inward-rectifying K^+ channel activity, whereas cytochalasin D potentiates inward K^+ currents. These findings suggest that cytochalasin D enhances stomatal opening by facilitating K^+ influx into guard cells, whereas phalloidin induces the opposite effect. Moreover, they provided the first evidence that plant cell microfilaments can regulate ion channel activity and thereby physiological responses.

The Lee group was also the first to report that guard cell actin filaments are differently organized in open and closed stomata (Eun & Lee, 1997). The distribution and number of *Commelina* guard cell microfilaments, visualized by immunofluorescence microscopy, were found to change in response to light and ABA. When stomata were open under white light, actin filaments were localized in the cortex as previously described (Kim *et al.*, 1995), arranged in a pattern that radiates from the stomatal pore. However, in stomata closed by darkness or ABA, the radial pattern of actin arrays was abolished and short, randomly oriented microfilaments were observed instead, indicating depolymerization of microfilaments. In addition, Eun and Lee (1997) observed that these actin rearrangements occur very rapidly after exposure to the physiological stimulus (as early as 3 min, in the case of ABA), thereby constituting one of the fastest stimulus-mediated cytoskeletal responses described to date (Staiger *et al.*, 2000).

Eun and Lee (2000) subsequently used fusicoccin, a potent stomatal opening agent that stimulates H^+-ATPases, to further validate the role of actin filaments as a signal mediator in guard cell responses. If actin arrays altered their organization as a consequence of changes in stomatal aperture, then all stomatal opening and closing stimuli should lead to actin polymerization and depolymerization, respectively, and fusicoccin should therefore induce radial arrangement of microfilaments. However, induction of stomatal opening by fusicoccin was accompanied by depolymerization of actin filaments, and treatment of illuminated guard cells with this compound further disintegrated preformed actin filaments (Eun & Lee, 2000). It is not clear whether this ability of fusicoccin to depolymerize actin is a direct effect or a consequence of the activation of H^+-ATPases. Furthermore, stabilization of microfilaments with phalloidin delayed fusicoccin-induced stomatal opening (Eun & Lee, 2000). These observations indicate that actin polymerization and depolymerization are not simply occurring as a consequence of changes in stomatal aperture and suggest that microfilament depolymerization may accelerate (and, conversely, intact actin filaments delay) the opening and closing processes.

In a later study, in which the actin cytoskeleton was visualized by staining with rhodamine-phalloidin instead of immunolocalized, an additional pattern of actin organization was reported (Hwang & Lee, 2001). After the depolymerization of cortical actin filaments during stomatal closure, and when stomata were stably closed in the presence of ABA, long microfilaments in a sparse and random distribution were observed. This finding led Hwang and Lee (2001) to hypothesize that this *closed actin pattern* might help maintain the closed state and that filament disintegration in response to opening stimuli might allow prompt opening of stomata and reorganization into the *open actin pattern*.

The visualization of actin filaments in fixed guard cells using immunolocalization or actin staining techniques (Kim *et al.*, 1995; Eun & Lee, 1997, 2000; Hwang *et al.*, 2000; Eun *et al.*, 2001; Hwang & Lee, 2001) or by microinjection of fluorescent phalloidin into living guard cells (Kim *et al.*, 1995) has its drawbacks. Fixation procedures in particular may alter actin cytoskeleton organization (Kost *et al.*, 1999) and, in the specific case of guard cells, release turgor pressure and hence lead to stomatal closure. Observation of the plant cytoskeleton has greatly benefited from the development of green fluorescent protein (GFP)-based markers, which provide a powerful tool for the non-invasive visualization of cytoskeletal elements in genetically transformed, living plant cells or tissues. Kost *et al.* (1998) showed that transgenic *Arabidopsis* lines expressing GFP fused to the F-actin binding domain of mouse talin allow the *in vivo* visualization of actin arrays in a range of different plant cells including guard cells.

Lemichez *et al.* (2001) made use of this *Arabidopsis* plant expressing a GFP fusion protein targeted to the actin cytoskeleton (Kost *et al.*, 1998) to visualize actin dynamics during ABA-induced stomatal movements. Their observations confirm and reinforce previous work from the Lee group in that they show ABA-induced disruption of the actin cytoskeleton in guard cells during stomatal closure. In guard cells of open stomata, thick radial actin cables bridge from the pore site to the dorsal side of the guard cell (Fig. 10.2A). On ABA addition, actin cables become rapidly disorganized and shortened (Fig.10.2B). The majority of the stomata were closed and guard cells contained substantially fewer actin cables 15 min after the hormone treatment (Lemichez *et al.*, 2001).

10.3.1.1 Cytosolic calcium

A well-established second messenger playing a central role in guard cell signaling, $[Ca^{2+}]_{cyt}$ is also known to regulate the activity of several actin-regulatory proteins in animal cells. Indeed, several proteins involved in the cross-linking, severing, or capping of actin filaments are Ca^{2+}-dependent (Puius *et al.*, 1998). Therefore, $[Ca^{2+}]_{cyt}$ is an obvious candidate for the mediation of actin changes during stomatal movements. To date, only one study has addressed its involvement in guard cell actin reorganization.

Results obtained by Hwang and Lee (2001) indicate that $[Ca^{2+}]_{cyt}$ is indeed involved in ABA-induced actin reorganization in *Commelina* guard cells. Actin reorganization observed after treatment with $CaCl_2$ closely resembled that induced by ABA. Furthermore, removal of extracellular Ca^{2+} with the chelator EGTA slowed down ABA-induced actin changes. Taken together, these results suggest that an increase in $[Ca^{2+}]_{cyt}$ acts as a signal mediator and is necessary for guard cell

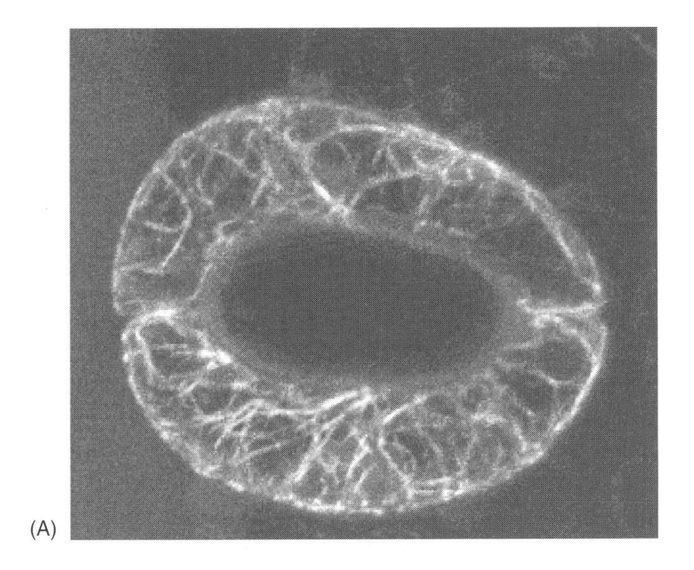

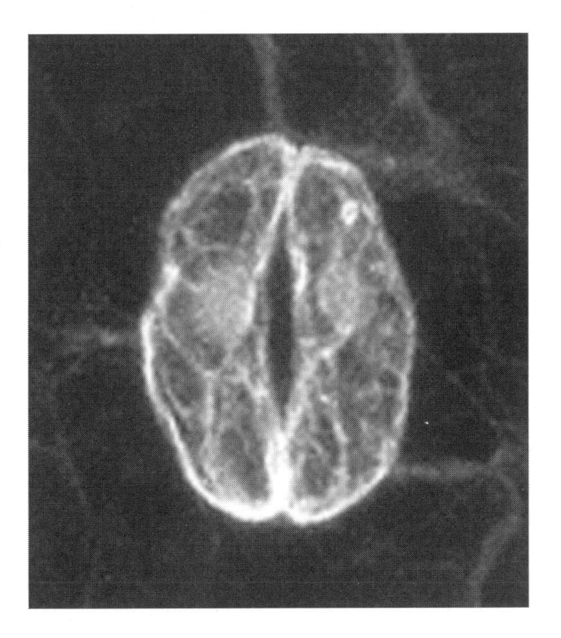

Fig. 10.2 ABA-induced actin cytoskeleton reorganization in *Arabidopsis thaliana* guard cells. Transgenic plants expressing GFP-mouse talin (Kost *et al.*, 1998), visualized using confocal microscopy, before (A) and 15 min after (B) treatment with 50 μM ABA.

actin reorganization in response to ABA. However, the fact that EGTA treatment did not totally abolish ABA-induced actin reorganization suggests that alternative, Ca^{2+}-independent pathways contribute to this process (Hwang & Lee, 2001).

10.3.1.2 Protein kinases and phosphatases

Protein (de)phosphorylation is known to play a central role in the regulation of actin-binding proteins in animal cells and has been found to control actin dynamics in a variety of plant cell types as well (Ressad *et al.*, 1998; Guillén *et al.*, 1999; Allwood *et al.*, 2001). As they are also known intermediates in guard cell signaling, protein kinases and phosphatases represent other potential mediators of actin changes in guard cells.

Hwang and Lee (2001) used a pharmacological approach to investigate the participation of protein kinases and phosphatases in actin reorganization during ABA-induced stomatal closure in *Commelina*. In open stomata, a broad-range protein kinase inhibitor, staurosporine, significantly increased the stomatal aperture, whereas PP1 and PP2A inhibitors such as calyculin A or okadaic acid had the opposite effect, inducing stomatal closure and depolymerization of actin filaments. Moreover, staurosporine increased the proportion of cells with long radial cortical microfilaments and inhibited actin reorganization in guard cells treated with ABA or $CaCl_2$. By contrast, calyculin A enhanced actin disintegration in control, ABA- or $CaCl_2$-treated guard cells (Hwang & Lee, 2001). These data suggest that the critical step for actin depolymerization involves activation of staurosporine-sensitive kinases, whereas the critical step for the formation of long actin filaments involves the activity of phosphatases sensitive to calyculin A (PP1 and PP2A).

In an attempt to test whether PP2C is an actin regulator in guard cells, Eun *et al.* (2001) compared ABA-induced actin reorganization in *Arabidopsis* guard cells of the wild type and of the *abi1-1* mutant, which is impaired in its ABA stomatal response. In contrast with the wild-type, no stomatal closure nor depolymerization of microfilaments was observed after ABA treatment in the *abi1-1* mutant, indicating that PP2C participates in ABA-dependent actin changes in guard cells. However, treatment of leaves with cytochalasin D (which induces actin depolymerization) prior to addition of ABA failed to alleviate the ABA-insensitive stomatal response of *abi1-1*, suggesting an effect of PP2C on ABA guard cell signaling components other than the actin cytoskeleton (Eun *et al.*, 2001).

10.3.1.3 Rho GTPases

The first piece of genetic evidence definitely showing that actin changes are actually necessary during ABA-induced stomatal closure came from the work of Lemichez *et al.* (2001). The study demonstrated that a Rho GTPase, *At*RAC1, down-regulates ABA-induced actin reorganization in *Arabidopsis* guard cells.

Members of the Rho family of small GTP-binding proteins – Rho, Rac and Cdc42 – are known key regulators of the actin cytoskeleton in yeast and animal cells (Hall, 1998). In these systems, they are thought to exert their effects through the activation of protein kinases, which ultimately alter the activity of multiple actin-binding proteins (Henson, 1999). Rho GTPases thereby transduce signals from cell surface receptors to reorganize actin architecture in the cytoplasm.

Lemichez *et al.* (2001) cloned *AtRAC1*, a Rho-related *Arabidopsis* GTPase, and found that, although ubiquitously expressed, its promoter activity is higher in guard cells than in the surrounding epithelial cells. This observation and the ascertained role of Rho GTPases in the regulation of actin dynamics in other systems prompted the authors to investigate whether AtRAC1 is involved in ABA-dependent actin reorganization in guard cells. In addition to disruption of the guard cell actin cytoskeleton during stomatal closure, ABA treatment caused AtRAC1 inactivation. However, neither AtRAC1 inactivation nor actin filament depolymerization were induced by addition of the hormone in the *abi1-1* mutant, which is impaired in ABA-induced stomatal closure. To further study the role of this gene, dominant-negative and dominant-positive *AtRAC1* mutants were generated (Lemichez *et al.*, 2001). Expression of the dominant-negative mutant led to stomatal closure in the absence of ABA, thereby mimicking the effects of the hormone. By contrast, ABA-induced actin reorganization and stomatal closure were blocked in the dominant-positive *AtRAC1* mutant. Moreover, the dominant-negative form of AtRAC1 rescued stomatal closure in the *abi1-1* mutant, demonstrating that the ABA-induced actin disruption in guard cells is dependent on the ABI1 protein, PP2C.

Taken together, these results show that AtRAC1 is a negative regulator of ABA-induced actin reorganization in guard cells, whose ABI1-dependent inactivation is required for stomatal closure.

10.3.1.4 Cell volume regulation

Numerous studies in yeast and animal model systems have suggested that the concentration of actin filaments in the submembranous cortex region of many cell types is involved in the regulation of cell volume. This connection is supported by several lines of evidence: Firstly, changes in actin organization are known to accompany cell volume changes and many works report F-actin-dependence of regulatory volume responses; Secondly, structural and functional links between the actin cytoskeleton and the membrane transporters known to be involved in cell volume homeostasis have been documented (Henson, 1999). In plant cells too, many cytoskeletal elements are located near the plasma membrane (Staiger, 2000) and two reports have indicated a relationship between actin dynamics and the shape and volume changes in guard cells, which regulate stomatal aperture.

Following the evidence provided by the work of Hwang *et al.* (1997) that guard cell microfilaments can regulate K^+ channel activity, Liu and Luan (1998) reported that both K^+ channels and actin filament arrays are also sensitive to extracellular osmolarity. The results from this study led the authors to propose that microfilaments serve as an osmosensor and target inward K^+ channels for turgor regulation in guard cells, accelerating changes in cell shape and volume. Patch-clamp analysis of *Vicia faba* guard cell protoplasts exposed to osmotic gradients showed that, under hypotonic (i.e. cell swelling) conditions, inward-rectifying K^+ channels are activated and outward-rectifying K^+ channels are inactivated; conversely, hypertonic (i.e. cell shrinking) conditions inactivate inward K^+ currents and activate outward K^+ currents (Liu & Luan, 1998). Moreover, Liu and Luan (1998) observed a network of microfilaments distributed throughout the body of guard cell protoplasts in hypertonic conditions and disruption of this actin network in hypotonic conditions or after cytochalasin D treatment, which also activated inward-rectifying K^+ channels in hypertonic conditions. This led the authors to postulate that hypotonic swelling leads to activation of inward-rectifying K^+ channels through actin depolymerization. It should be noted however that the patterns of actin organization visualized in this study with protoplasts do not necessarily coincide with those in walled guard cells under equivalent conditions. These results are yet to be confirmed in intact guard cells.

Another possible transduction pathway linking the actin cytoskeleton to cell volume regulation involves Rho GTPases (Henson, 1999). Although the connection is far from being established, studies in animal cells indicate a relationship between Rho GTPases and membrane transporters. A few studies show that a non-hydrolyzable GTP analog induces swelling-activated anion currents under isotonic conditions (see Okada, 1997). On the other hand, under hypotonic conditions, epithelial cells treated with a Rho inhibitor show reduced swelling-induced anion efflux and actin reorganization coincides with an increase in a Rho-dependent kinase, phosphatidylinositol-3-kinase, whose inhibition also reduces the activation of the osmosensitive ion efflux (Tilly *et al.*, 1996). In addition, a protein known to bind the anion channel plCln is homologous to the yeast Skb1 protein which interacts with Shk1, in turn a homolog of the mammalian Cdc42/Rac-activated kinases (Krapivinsky *et al.*, 1998). Cdc42/rac-dependent kinases regulate actin-mediated cell morphology and were shown to inhibit myosin light chain kinase activity (Sanders *et al.*, 1999), which has been linked to alterations in cell volume and membrane transport activity (see Henson, 1999). The involvement of Rho GTPases – namely of AtRAC1, a central component of ABA guard cell signaling (Lemichez *et al.*, 2001) – in the ion channel-mediated regulation of guard cell volume remains to be investigated.

10.3.1.5 Membrane trafficking

The substantial changes in plasma membrane surface area which accompany guard cell volume changes are also likely to be coupled to the actin cytoskeleton network along which vesicle traffic must occur during exocytic and endocytic events.

An interesting recent study addressed the issue of fusion of vesicular membranes with the plasma membrane in *Vicia* guard cell protoplasts (Bick *et al.*, 2001). Using patch clamp measurements of membrane capacitance, a parameter proportional to the surface area, Bick and coworkers (2001) found that applied hydrostatic pulses led to immediate increases in capacitance which could be provoked several times with successive pressure stimuli. However, the rate of capacitance rise decreased exponentially with time for subsequent pressure pulses. This was the result of a desensitization of the plasma membrane to mechanical stimulation rather than of shortage of membrane material required for incorporation into the membrane

(Bick *et al.*, 2001). The authors proposed that stretching of the membrane initiates a reactive process that fortifies and stabilizes the plasma membrane of guard cell protoplasts. Importantly, depolymerization of guard cell protoplast microfilaments with cytochalasin D practically abolished the desensitization process. Desensitization may therefore result from an increase in plasma membrane stability by actin filaments, achieved by an increase in actin polymerization and/or by increased crosslinking of existing microfilaments by actin-binding proteins (Bick *et al.*, 2001). The physiological significance of desensitization in intact, walled guard cells during stomatal movements requires further investigation.

The loss-of-function *era1 Arabidopsis* mutant is hypersensitive to ABA inhibition of seed germination and the corresponding gene encodes the β subunit of a protein farnesyltransferase (Cutler *et al.*, 1996). The *ERA1* gene is also a negative regulator of ABA signaling in guard cells, where it acts upstream of $[Ca^{2+}]_{cyt}$ increases (Allen *et al.*, 2002). The fact that farnesyltransferases are known to modify soluble proteins for membrane localization suggests that a negative regulator of ABA signaling might need to be farnesylated by ERA1 to function. Additionally, Blatt (2000) has proposed a connection between ERA1, membrane trafficking and, by implication, the actin cytoskeleton by pointing out that some Rac and Rho GTPases are commonly farnesylated to anchor and target GTPase activity for multiple cellular functions. The posterior identification of AtRAC1 as an important component of ABA-induced, actin-mediated stomatal closure (Lemichez *et al.*, 2001) sheds new light on Blatt's suggestion. Blatt (2000) drops another two clues indicating a potential link of recently identified intermediates in guard cell signaling to membrane trafficking and the actin cytoskeleton. In animal cells, PLD activity has been associated with the function of Rho GTPases in vesicle secretion, and SNARE proteins have been found to interact with microfilament-associated proteins (see Blatt, 2000).

10.3.1.6 Other hints of signaling to the guard cell actin cytoskeleton

Recent work by MacRobbie (2002) has provided evidence for the involvement of protein tyrosine phosphatases (PTPases) in the regulation of stomatal aperture. Tyrosine phosphorylation is an important mechanism for cellular regulation in animals, but until recently PTPases had not been identified in plants where their function remains largely unknown. To date, only two PTPases have been characterized in *Arabidopsis* (Gupta *et al.*, 1998; Xu *et al.*, 1998). However, around 20 genes encoding putative PTPases have been identified from the *Arabidopsis* genome, suggesting that these enzymes serve important functions in plants (Luan, 2002). MacRobbie (2002) shows that stomatal closure induced by four different stimuli – ABA, extracellular Ca^{2+}, darkness and H_2O_2 – is prevented by several specific PTPase inhibitors. Furthermore, flux measurements with $^{86}Rb^+$ identified efflux from the vacuole (but not efflux at the plasma membrane) as the target for the PTPase inhibitors, indicating that the PTPase component acts at or downstream of the $[Ca^{2+}]_{cyt}$ increases responsible for K^+ efflux from the vacuole (MacRobbie, 2002). These data show that PTPase activity is essential for stomatal closure and suggest that a protein related to tonoplast K^+ channels is regulated by tyrosine (de)phosphorylation. MacRobbie (2002) further suggests that actin is a good candidate for this regulation. In fact, several lines of evidence support this hypothesis. First, two studies have indicated that actin can regulate K^+ channel activity in guard cells (Hwang *et al.*, 1997; Liu & Luan, 1998). Second, an exciting recent report (Kameyama *et al.*, 2000) shows that the degree of actin tyrosine-phosphorylation regulates leaf movement in *Mimosa pudica* and that both these processes are inhibited by one of the specific PTPase inhibitor used in MacRobbie's study (2002). The exact mechanism by which actin and its tyrosine-phosphorylation status might regulate ion channel activity remains obscure.

It is well known that phosphoinositides, a family of inositol-containing phospholipids playing an increasingly important role as second messengers in guard cell signaling, may regulate cytoskeletal reorganization. Indeed, polyphosphoinositides – particularly phosphatidylinositol 4,5-bisphosphate – inhibit the activity of proteins of the gelsolin family, actin-regulatory proteins involved in the severing and capping of microfilaments (Puius *et al.*, 1998).

A calcium-mobilizing molecule recently identified as a guard cell signaling component by Ng *et al.* (2001) is another potential candidate for signaling to the actin cytoskeleton in guard cells. In mammalian cells, S1P inhibits cell motility by affecting actin filament reorganization and is implicated in the induction of other biological responses via the activation of Rho GTPases (Pyne & Pyne, 2000).

10.3.2 Involvement of microtubules

It is well established that microtubules are involved in guard cell development. During the differentiation of guard cells, cortical microtubules control the orientation and guide the asymmetric deposition of cellulose microfibrils in the cell wall, thereby playing an important role in determining cell shape (McDonald *et al.*, 1993). The fact that in guard cells the array of cortical microtubules remains well organized after differentiation is completed, has led to the suggestion that microtubules have an additional role in these cells beyond morphogenesis and are necessary for stomatal function (Marcus *et al.*, 2001). This hypothesis has been tested by several investigators, but the involvement of microtubules in stomatal movements remains a matter of debate in spite of a recent significant study.

Microtubule dynamics in *Vicia faba* guard cells have been observed during stomatal movements using immunofluorescence microscopy (Assmann & Baskin, 1998; Fukuda *et al.*, 1998) or by microinjecting fluorescent tubulin into living cells (Yu *et al.*, 2001). Fukuda *et al.* (1998) reported a change in the organization of microtubules during the diurnal cycle. At dawn, a radial array of cortical microtubules which increased in number during the morning was formed, accompanying an increase in stomatal aperture; towards night, destruction of microtubules which became localized near the nucleus was observed, following a decrease in stomatal aperture (Fukuda *et al.*, 1998). In a subsequent study, Fukuda *et al.* (2000) found that changes in the expression of α- and β-tubulin accompany these diurnal changes in microtubule organization. Furthermore, as treatment with mRNA and protein synthesis inhibitors in the early morning (but not at other times of the day) interfered with the radial organization of microtubules and stomatal opening, it was suggested that diurnal changes in microtubule organization and stomatal aperture may be regulated by *de novo* synthesis and degradation of tubulin in guard cells (Fukuda *et al.*, 2000). More recently, Yu *et al.* (2001) described microtubules in fully opened stomata as being transversely oriented from the stomatal pore to the dorsal wall. During dark-induced closure, microtubules became twisted and broke down into diffuse fragments, but the transversally oriented pattern was reestablished when stomata opened in response to light (Yu *et al.*, 2001). By contrast, Assmann and Baskin (1998) observed an array of microtubules radiating outwards from the stomatal pore in *Vicia* guard cells but no consistent difference in this pattern between dark-closed stomata and stomata open under light.

The findings of Fukuda *et al.* (1998) and Yu *et al.* (2001) are also in marked contrast with another two studies (Eun & Lee, 1997; Lemichez *et al.*, 2001), which focused on actin reorganization but also visualized microtubules to investigate whether they were equally redistributed in guard cells in response to stimulus-mediated stomatal movements. Using immunolocalization in fixed guard cells of *Commelina communis*, Eun and Lee (1997) observed that in open stomata the organization of cortical microtubules was similar to that of actin filaments, but that this organization was not affected during stomatal closure induced by darkness or ABA treatment. In addition, Lemichez *et al.* (2001) made use of an *Arabidopsis* plant expressing a GFP fusion protein targeted to tubulin and found that microtubule cables were organized in a fan-shaped pattern in guard cells of open stomata. However, in agreement with Eun and Lee's study (1997), no reorganization was observed when stomata closed in response to application of ABA (Lemichez *et al.*, 2001).

To determine whether microtubules control stomatal function, pharmacological assays have also been performed with conflicting results. The effect of treatment with anti-microtubule drugs which depolymerize the microtubule cytoskeleton, such as colchicine, propyzamide, oryzalin and trifluralin, or with microtubule-stabilizing agents, such as taxol, on stomatal movements has been assessed. In the study of Fukuda *et al.* (1998), early morning stomatal opening was inhibited by propyzamide but not by taxol, whereas evening stomatal closure was suppressed by taxol but not by propyzamide. However, in a previous study from the same laboratory (Jiang *et al.*, 1996), propyzamide had no effect on stomatal aperture for 1 h and taxol did not affect the ability of stomata to close rapidly in response to ABA. It was therefore suggested that changes in guard cell microtubule organization played a role only in relatively slow stomatal responses (Fukuda *et al.*, 1998). Assmann and Baskin (1998) found no effect of colchicine or taxol treatment on dark-closed *Vicia* stomata; in addition, neither opening induced by light or fusicoccin nor closing in response to Ca^{2+} or darkness were affected by any of the drugs. In yet another study, both oryzalin and taxol were reported to suppress light-induced opening and dark-induced stomatal closure in *Vicia* (Yu *et al.*, 2001). Thus, reported pharmacological data had not helped clarify the involvement of guard cell cortical microtubules in stomatal function.

A recent study has made a significant attempt to help resolve the debate of the role of microtubules in stomatal movements. Marcus *et al.* (2001) used an approach that involves the expression in *Vicia* guard cells of GFP fused to the microtubule-binding domain of the mammalian microtubule-associated protein, MAP4. The chimeric gene product binds specifically to microtubules affecting their function. Transformed guard cells failed to open in response to white, blue and red light, and low ambient CO_2. Moreover, under these stimuli, propyzamide, oryzalin and trifluralin – but not colchicine – also prevented stomatal opening. Marcus and colleagues (2001) argue that colchicine has a low binding affinity to plant tubulin and may need a longer exposure time, thereby explaining also the absence of effect reported previously with this drug (Assmann & Baskin, 1998). Interestingly, the proton pump activator, fusicoccin, induced opening of guard cells expressing the GFP fusion protein or treated with propyzamide, oryzalin and trifluralin; this indicates that microtubules are acting upstream of the regulation of ion fluxes (Marcus *et al.*, 2001). Finally, consistent with results presented earlier by Fukuda *et al.* (1998), guard cells in which microtubules were stabilized with taxol were able to open in response to light. This was not the case when microtubules were stabilized by expression of the microtubule-binding GFP fusion protein, high levels of which displace endogenous MAPs; the authors therefore postulated the involvement of an MAP in guard cell function (Marcus *et al.*, 2001).

In summary, owing to highly conflicting results that stem from a considerable number of studies, further work is required in order to assess the role of microtubules in stomatal function.

10.4 Conclusions and perspectives

Recent efforts have addressed the question of whether cytoskeleton components are involved in stomatal movements. Unlike microtubules whose participation in guard cell physiology remains controversial, it is now clear that actin filaments play an important role in stimulus-induced volume and shape changes of guard cells. In fact, pharmacological and genetic evidence demonstrates that actin reorganization is necessary for a change in stomatal aperture to occur, but results also suggest that the actin cytoskeleton does not determine the direction of this change. Instead, a role for actin rearrangement in promoting both the opening and closing processes appears more likely. The modulation of ion channels by microfilaments may contribute to the progress of the stomatal movement, but this effect is most probably overridden by the stimulus-specific, actin-independent regulation of ion fluxes, which ultimately determines opening or closure. Polymerization of actin filaments may on the other hand stabilize the stomatal aperture and control the magnitude of the change. In addition to the frequently reported radial organization of actin filaments in open stomata, a stabilized actin pattern in closed stomata has

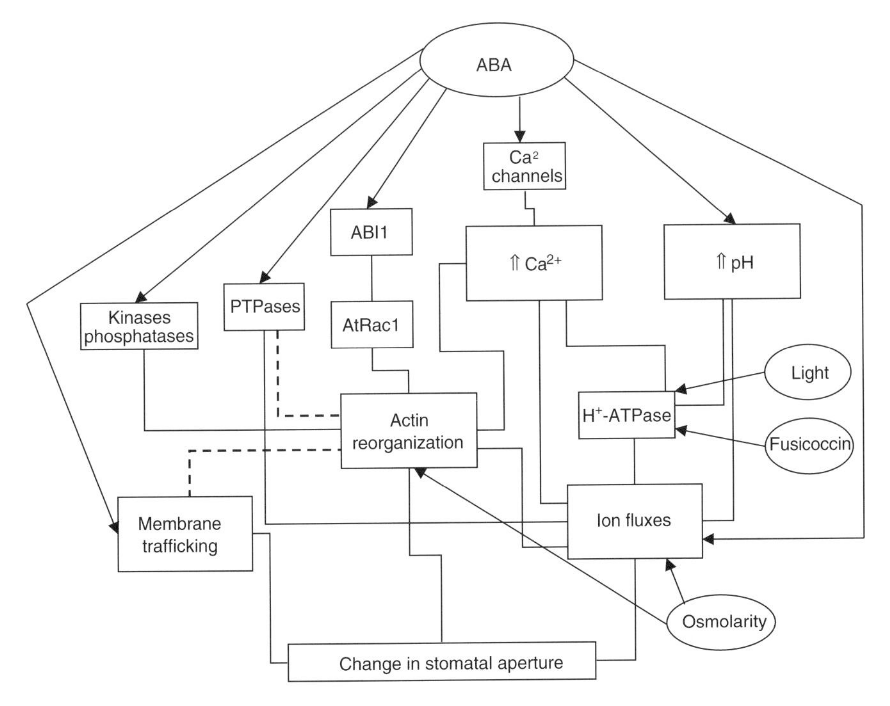

Fig. 10.3 A model for the integration of actin filament reorganization in the guard cell signaling network. For simplification purposes, only some of the known intermediates are shown. Dashed lines indicate hypothetical connections not validated experimentally or inferred from data obtained using guard cell protoplasts.

also recently been identified. Therefore, available data support a scenario in which actin polymerization helps maintain the aperture status, whereas rearrangement of microfilaments serves as a conformational switch required for a change in stomatal aperture.

A working model for the positioning of actin reorganization in the complex signaling network involved in stomatal movements is presented in Fig. 10.3. Consistent with a role for actin as a signaling intermediate acting upstream of stomatal movements, a number of the presently identified components in guard cell signaling, such as $[Ca^{2+}]_{cyt}$ elevations or phosphorylation events, have been found to regulate microfilaments during stomatal movements. Moreover, other such components have been structurally and/or functionally linked to the actin cytoskeleton in other eukaryotic cell systems. Indeed, virtually all known guard cell signaling intermediates can be included in the group of actin modulator candidates. The potential effect of the vast majority of these candidates on actin filaments during the regulation of stomatal aperture remains to be investigated. Furthermore, future research on the possible regulation by actin of channel components acting downstream in the guard cell signal transduction pathway also holds much promise for understanding the overall role of the actin cytoskeleton in stomatal function.

Acknowledgments

We thank Y. Wu and E. Lemichez for supplying the images used in Figs 10.1 and 10.2, and G. Jedd for critically reading the manuscript. P.D. was supported by a fellowship (PRAXISXXI/BPD/20190/99) from the Portuguese FCT (EU III framework program and MCES).

References

Aarhus, R., Graeff, R.M., Dickey, D.M., Walseth, T.F. & Lee, H.C. (1995) ADP-ribosyl cyclase and CD38 catalyze the synthesis of a calcium-mobilizing metabolite from NADP. *J. Biol. Chem.*, **270**, 30327–30333.

Allan, A.C., Fricker, M.D., Ward, J.L., Beale, M.H. & Trewavas, A.J. (1994) Two transduction pathways mediate rapid effects of abscisic acid in *Commelina* guard cells. *Plant Cell*, **6**, 1319–1328.

Allen, G.J., Kuchitsu, K., Chu, S.P., Murata, Y. & Scroeder, J.I. (1999) *Arabidopsis abi1-1* and *abi2-1* phosphatase mutations reduce abscisic acid-induced cytoplasmic calcium rises in guard cells. *Plant Cell*, **11**, 1785–1798.

Allen, G.J., Chu, S.P., Schumacher, K., *et al.* (2000) Alteration of stimulus-specific guard cell calcium oscillations and stomatal closing in *Arabidopsis det3* mutant. *Science*, **289**, 2338–2342.

Allen, G.J., Chu, S.P., Harrington, C.L., *et al.* (2001) A defined range of guard cell calcium oscillation parameters encodes stomatal movements. *Nature*, **411**, 1053–1057.

Allen, G.J., Murata, Y., Chu, S.P., Nafisi, M. & Schroeder, J.I. (2002) Hypersensitivity of abscisic acid-induced cytosolic calcium increases in the *Arabidopsis* farnesyltransferase mutant *era1-2*. *Plant Cell*, **14**, 1649–1662.

Allwood, E.G., Smertenko, A.P. & Hussey, P.J. (2001) Phosphorylation of plant actin-depolymerizing factor by calmodulin-like domain protein kinase. *FEBS Lett.*, **499**, 97–100.

Armstrong, F., Leung, J., Grabov, A., Brearley, J., Giraudat, J. & Blatt, M.R. (1995) Sensitivity to abscisic acid of guard cell K^+ channels is suppressed by *abi1-1*, a mutant *Arabidopsis* gene encoding a putative protein phosphatase. *Proc. Natl. Acad. Sci. USA*, **92**, 9520–9524.

Assmann, S.M. (1999) The cellular basis of guard cell sensing of rising CO_2. *Plant Cell Environ.*, **22**, 629–637.

Assmann, S.M. & Baskin, T.I. (1998) The function of guard cells does not require an intact array of cortical microtubules. *J. Exp. Bot.*, **49**, 163–170.

Assmann, S.M. & Wang, X.-Q. (2001) From milliseconds to millions of years: guard cells and environmental responses. *Curr. Opin. Plant Biol.*, **4**, 421–428.

Assmann, S.M., Simoncini, L. & Schroeder, J.I. (1985) Blue light activates electrogenic ion pumping in guard cell protoplasts of *Vicia faba*. *Nature*, **318**, 285–287.

Bauer, C.S., Plieth, C., Bethmann, B., *et al.* (1998) Strontium-induced repetitive calcium spikes in a unicellular green alga. *Plant Physiol.*, **117**, 545–557.

Bick, I., Thiel, G. & Homann, U. (2001) Cytochalasin D attenuates the desensitisation of pressure-stimulated vesicle fusion in guard cell protoplasts. *Eur. J. Cell Biol.*, **80**, 521–526.

Blatt, M.R. (1992) K^+ channels of stomatal guard cells: characteristics of the inward-rectifier and its control by pH. *J. Gen. Physiol.*, **99**, 615–644.

Blatt, M.R. (2000) Cellular signaling and volume control in stomatal movements in plants. *Annu. Rev. Cell Dev. Biol.*, **16**, 221–241.

Blatt, M.R. & Armstrong, F. (1993) K^+ channels of stomatal guard cells: abscisic acid-evoked control of the outward-rectifier mediated by cytoplasmic pH. *Planta*, **191**, 330–341.

Blatt, M.R. & Thiel, G. (1994) K^+ channels of stomatal guard cells: bimodal control of the K^+ inward-rectifier evoked by auxin. *Plant J.*, **5**, 55–68.

Blatt, M.R., Thiel, G. & Trentham, D.R. (1990) Reversible inactivation of K^+ channels of *Vicia* stomatal guard cells following the photolysis of caged inositol 1,4,5-trisphosphate. *Nature*, **346**, 766–769.

Cutler, S., Ghassemian, M., Bonetta, D., Cooney, S. & McCourt, P. (1996) A protein farnesyltransferase involved in abscisic acid signal transduction in *Arabidopsis*. *Science*, **273**, 1239–1241.

DeKoninck, P. & Schulman, H. (1998) Sensitivity of CaM kinase II to the frequency of Ca^{2+} oscillations. *Science*, **279**, 227–230.

Dolmetsch, R.E., Xu, K. & Lewis, R.S. (1998) Calcium oscillations increase the efficiency and specificity of gene expression. *Nature*, **392**, 933–936.

Drake, B.G., Gonzalez-Meler, M.A. & Long, S.P. (1997) More efficient plants: a consequence of rising atmospheric CO_2? *Annu. Rev. Plant Physiol. Plant Mol. Biol.*, **48**, 609–639.

Ehrhardt, D.W., Wais, R. & Long, S.R. (1996) Calcium spiking in plant-root hairs responding to *Rhizobium* nodulation signals. *Cell*, **85**, 673–681.

Eun, S.-O. & Lee, Y. (1997) Actin filaments of guard cells are reorganized in response to light and abscisic acid. *Plant Physiol.*, **115**, 1491–1498.

Eun, S.-O. & Lee, Y. (2000) Stomatal opening by fusicoccin is accompanied by depolymerization of actin filaments in guard cells. *Planta*, **210**, 1014–1017.

Eun, S.-O., Bae, S.-H. & Lee, Y. (2001) Cortical actin filaments in guard cells respond differently to abscisic acid in wild-type and *abi1-1* mutant *Arabidopsis*. *Planta*, **212**, 466–469.

Evans, P.T. & Malmberg, R.L. (1989) Do polyamines have roles in plant development? *Annu. Rev. Plant Physiol. Plant Mol. Biol.*, **40**, 235–269.

Fukuda, M., Hasezawa, S., Asai, N., Nakajima, N. & Kondo, N. (1998) Dynamic organization of microtubules in guard cells of *Vicia faba* L. with diurnal cycle. *Plant Cell Physiol.*, **39**, 80–86.

Fukuda, M., Hasezawa, S., Nakajima, N. & Kondo, N. (2000) Changes in tubulin protein expression in guard cells of *Vicia faba* L. accompanied with dynamic organization of microtubules during the diurnal cycle. *Plant Cell Physiol.*, **41**, 600–607.

Gaedeke, N., Klein, M., Kolukisaoglu, U., *et al.* (2001) The *Arabidopsis thaliana* ABC transporter *At*MRP5 controls root development and stomata movement. *EMBO J.*, **20**, 1875–1887.

Gilroy, S., Read, N.D. & Trewavas, A.J. (1990) Elevation of cytoplasmic Ca^{2+} by caged calcium or caged inositol triphosphate initiates stomatal closure. *Nature*, **346**, 769–771.

Gosti, F., Beaudoin, N., Serizet, C., Webb, A.A., Vartanian, N. & Giraudat, J. (1999) ABI1 protein phosphatase 2C is a negative regulator of abscisic acid signaling. *Plant Cell*, **11**, 1897–1910.

Grabov, A. & Blatt, M.R. (1997) Parallel control of the inward-rectifier K^+ channel by cytosolic free Ca^{2+} and pH in *Vicia* guard cells. *Planta*, **201**, 84–95.

Grabov, A. & Blatt, M.R. (1998) Membrane voltage initiates Ca^{2+} waves and potentiates Ca^{2+} increases with abscisic acid in stomatal guard cells. *Proc. Natl. Acad. Sci. USA*, **95**, 4778–4783.

Grabov, A. & Blatt, M.R. (1999) A steep dependence of inward-rectifying potassium channels on cytosolic free calcium concentration increase evoked by hyperpolarization in guard cells. *Plant Physiol.*, **119**, 277–287.

Guillén, G., Valdés-López, V., Noguez, R., *et al.* (1999) Profilin in *Phaseolus vulgaris* is encoded by two genes (only one expressed in root nodules) but multiple isoforms are generated *in vivo* by phosphorylation on tyrosine residues. *Plant J.*, **19**, 497–508.

Gupta, R., Huang, Y., Kieber, J.J. & Luan, S. (1998) Identification of a dual-specificity protein phosphatase that inactivates a MAP kinase from *Arabidopsis*. *Plant J.*, **16**, 581–590.

Hall, A. (1998) Rho GTPases and the actin cytoskeleton. *Science*, **279**, 509–514.

Henson, J.H. (1999) Relationships between the actin cytoskeleton and cell volume regulation. *Microsc. Res. Tech.*, **47**, 155–162.

Hetherington, A.M. (2001) Guard cell signaling. *Cell*, **107**, 711–714.

Himmelbach, A., Hoffmann, T., Leube, M., Höhener, B. & Grill, E. (2002) Homeodomain protein ATHB6 is a target of the protein phosphatase ABI1 and regulates hormone responses in *Arabidopsis*. *EMBO J.*, **21**, 3029–3038.

Homann, U. & Thiel, G. (1999) Unitary exocytotic and endocytotic events in guard-cell protoplasts during osmotically driven volume changes. *FEBS Lett.*, **460**, 495–499.

Homann, U. & Thiel, G. (2002) The number of K^+ channels in the plasma membrane of guard cell protoplasts changes in parallel with the surface area. *Proc. Natl. Acad. Sci. USA*, **99**, 10215–10220.

Hugouvieux, V., Kwak, J.M. & Schroeder, J.I. (2001) An mRNA cap binding protein, ABH1, modulates early abscisic acid signal transduction in *Arabidopsis*. *Cell*, **106**, 477–487.

Hwang, J.-U. & Lee, Y. (2001) Abscisic acid-induced actin reorganization in guard cells of dayflower is mediated by cytosolic calcium levels and by protein kinase and protein phosphatase activities. *Plant Physiol.*, **125**, 2120–2128.

Hwang, J.-U., Suh, S., Yi, H., Kim, J. & Lee, Y. (1997) Actin filaments modulate both stomatal opening and inward K^+-channel activities in guard cells of *Vicia faba* L. *Plant Physiol.*, **115**, 335–342.

Hwang, J.-U., Eun, S.-O. & Lee, Y. (2000) Structure and function of actin filaments in mature guard cells, in *Actin: A Dynamic Framework for Multiple Plant Cell Functions* (eds C.J. Staiger, F. Baluska, D. Volkmann & P.W. Barlow), Kluwer Academic Publishers, Dordrecht, The Netherlands, pp. 427–436.

Jacob, T., Ritchie, S., Assmann, S.M. & Gilroy, S. (1999) Abscisic acid signal transduction in guard cells is mediated by phospholipase D activity. *Proc. Natl. Acad. Sci. USA*, **96**, 12192–12197.

Jiang, C.-J., Nakajima, N. & Kondo, N. (1996) Disruption of microtubules by abscisic acid in guard cells of *Vicia faba* L. *Plant Cell Physiol.*, **37**, 697–701.

Jung, J.-Y., Kim, Y.-W., Kwak, J.M., *et al.* (2002) Phosphatidylinositol 3- and 4-phosphate are required for normal stomatal movements. *Plant Cell*, **14**, 2399–2412.

Kameyama, K., Kishi, Y., Yoshimura, M., Kanzawa, N., Sameshima, M. & Tsuchiya, T. (2000) Tyrosine phosphorylation in plant bending. *Nature*, **407**, 37.

Kargul, J., Gansel, X., Tyrrell, M., Sticher, L. & Blatt, M.R. (2001) Protein-binding partners of the tobacco syntaxin NtSyr1. *FEBS Lett.*, **508**, 253–258.

Kim, M., Hepler, P.K., Eun, S.-O., Ha, K.S. & Lee, Y. (1995) Actin filaments in mature guard cells are radially distributed and involved in stomatal movement. *Plant Physiol.*, **109**, 1077–1084.

Kinoshita, T. & Shimazaki, K. (1999) Blue light activates the plasma membrane H^+-ATPase by phosphorylation of the C-terminus in stomatal guard cells. *EMBO J.*, **18**, 5548–5558.

Kinoshita, T., Nishimura, M. & Shimazaki, K.-I. (1995) Cytosolic concentration of Ca^{2+} regulates the plasma membrane H^+-ATPase in guard cells of fava bean. *Plant Cell*, **7**, 1333–1342.

Koornneef, M., Reuling, G. & Karssen, C.M. (1984) The isolation of abscisic acid-insensitive mutants of *Arabidopsis thaliana*. *Plant Physiol.*, **61**, 377–383.

Kost, B., Spielhofer, P. & Chua, N.-H. (1998) A GFP-mouse talin fusion protein labels plant actin filaments *in vivo* and visualizes the actin cytoskeleton in growing pollen tubes. *Plant J.*, **16**, 393–401.

Kost, B., Mathur, J. & Chua, N.-H. (1999) Cytoskeleton in plant development. *Curr. Opin. Plant Biol.*, **2**, 462–470.

Krapivinsky, G., Pu, W., Wickman, K., Krapivinsky, L. & Clapham, D.E. (1998) PlCln binds to a mammalian homolog of a yeast protein involved in regulation of cell morphology. *J. Biol. Chem.*, **273**, 10811–10814.

Kubitscheck, U., Homann, U. & Thiel, G. (2000) Osmotically evoked shrinking of guard cell protoplasts causes vesicular retrieval of plasma membrane into the cytoplasm. *Planta*, **210**, 423–431.

Leckie, C.P., McAinsh, M.R., Allen, G.J., Sanders, D. & Hetherington, A.M. (1998) Abscisic acid-induced stomatal closure mediated by cyclic ADP-ribose. *Proc. Natl. Acad. Sci. USA*, **95**, 15837–15842.

Lee, H.C. (2001) Physiological functions of cyclic ADP-ribose and NAADP as calcium messengers. *Annu. Rev. Pharmacol. Toxicol.*, **41**, 317–345.

Lee, Y., Choi, Y.B., Suh, S., Lee, J. & Assmann, S.M. (1996) Abscisic acid-induced phosphinositide turn over in guard cell protoplasts of *Vicia faba*. *Plant Physiol.*, **110**, 987–996.

Lee, S., Choi, H., Suh, S., *et al.* (1999) Oligogalacturonic acid and chitosan reduce stomatal aperture by inducing the evolution of reactive oxygen species from guard cells of tomato and *Commelina communis*. *Plant Physiol.*, **121**, 147–152.

Lemichez, E., Wu, Y., Sánchez, J.P., Mettouchi, A., Mathur, J. & Chua, N.-H. (2001) Inactivation of AtRac1 by abscisic acid is essential for stomatal closure. *Genes Dev.*, **15**, 1808–1816.

Lemtiri-Chlieh, F., MacRobbie, E.A.C. & Brearley, C.A. (2000) Inositol hexakisphosphate is a physiological signal regulating the K^+-inward rectifying conductance in guard cells. *Proc. Natl. Acad. Sci. USA*, **97**, 8687–8692.

Leonhardt, H.N., Vavasseur, A. & Forestier, C. (1999) ATP binding cassette modulators control abscisic acid-regulated slow anion channels in guard cells. *Plant Cell*, **11**, 1141–1152.

Leube, M.P., Grill, E. & Amrhein, N. (1998) ABI1 of *Arabidopsis* is a protein serine/threonine phosphatase highly regulated by the proton and magnesium ion concentration. *FEBS Lett.*, **424**, 100–104.

Leyman, B., Geelen, D., Quintero, F.J. & Blatt, M.R. (1999) A tobacco syntaxin with a role in hormonal control of guard cell ion channels. *Science*, **283**, 537–540.

Li, J. & Assmann, S.M. (1996) An abscisic acid-activated and calcium-independent protein kinase from guard cells of fava bean. *Plant Cell*, **8**, 2359–2368.

Liu, K. & Luan, S. (1998) Voltage-dependent K^+ channels as targets of osmosensing in guard cells. *Plant Cell.*, **10**, 1957–1970.

Li, J., Lee, Y.R. & Assmann, S.M. (1998) Guard cells possess a calcium-dependent protein kinase that phosphorylates the KAT1 potassium channel. *Plant Physiol.*, **116**, 785–795.

Liu, K., Fu, H., Bei, Q. & Luan, S. (2000) Inward potassium channel in guard cells as a target for polyamine regulation of stomatal movements. *Plant Physiol.*, **124**, 1315–1326.

Li, J., Wang, X.Q., Watson, M.B. & Assmann, S.M. (2000) Regulation of abscisic acid-induced stomatal closure and anion channels by guard cell AAPK kinase. *Science*, **287**, 300–303.

Li, J., Kinoshita, T., Pandey, S., *et al.* (2002) Modulation of an RNA-binding protein by abscisic-acid-activated protein kinase. *Nature*, **418**, 793–797.

Lohse, G. & Hedrich, R. (1992) Characterization of the plasma membrane H^+ ATPase from *Vicia faba* guard cells; modulation by extracellular factors and seasonal changes. *Planta*, **188**, 206–214.

Luan, S. (2002) Tyrosine phosphorylation in plant cell signaling. *Proc. Natl. Acad. Sci. USA*, **99**, 11567–11569.

MacRobbie, E.A.C. (1999) Vesicle trafficking: a role in *trans*-tonoplast ion movements? *J. Exp. Bot.*, **50**, 925–934.

MacRobbie, E.A.C. (2000) ABA activates multiple Ca^{2+} fluxes in stomatal guard cells, triggering vacuolar $K^+(Rb^+)$ release. *Proc. Natl. Acad. Sci. USA*, **97**, 12361–12368.

MacRobbie, E.A.C. (2002) Evidence for a role for protein tyrosine phosphatase in the control of ion release from the guard cell vacuole in stomatal closure. *Proc. Natl. Acad. Sci. USA*, **99**, 11963–11968.

Marcus, A.I., Moore, R.C. & Cyr, R.J. (2001) The role of microtubules in guard cell function. *Plant Physiol.*, **125**, 387–395.

Mattie, M., Brooker, G. & Spiegel, S. (1994) Sphingosine-1-phosphate, a putative second messenger, mobilizes calcium from internal stores via an inositol trisphosphate-independent pathway. *J. Biol. Chem.*, **269**, 3181–3188.

McAinsh, M.R., Brownlee, C. & Hetherington, A.M. (1990) Abscisic acid-induced elevation of guard cell cytosolic Ca^{2+} precedes stomatal closure. *Nature*, **343**, 186–188.

McAinsh, M.R., Webb, A.A.R., Taylor, J.E. & Hetherington, A.M. (1995) Stimulus-induced oscillations in guard cell cytosolic free calcium. *Plant Cell*, **7**, 1207–1219.

McDonald, A.R., Liu, B., Joshi, H.C. & Palevitz, B.A. (1993) Gamma-tubulin is associated with a cortical-microtubule organizing zone in the developing guard cells of *Allium cepa* L. *Planta*, **191**, 357–361.

Merlot, S., Gosti, F., Guerrier, D., Vavasseur, A. & Giraudat, J. (2001) The ABI1 and ABI2 protein phosphatases 2C act in a negative feedback regulatory loop of the abscisic acid signalling pathway. *Plant J.*, **25**, 295–303.

Mori, I.C. & Muto, S. (1997) Abscisic acid activates a 48-kilodalton protein kinase in guard cell protoplasts. *Plant Physiol.*, **113**, 833–839.

Murata, Y., Pei, Z.-M., Mori, I.C. & Schroeder, J. (2001) Abscisic acid activation of plasma membrane Ca^{2+} channels in guard cells requires cytosolic NAD(P)H and is differentially disrupted upstream and downstream of reactive oxygen species production in *abi1-1* and *abi2-1* protein phosphatase 2C mutants. *Plant Cell*, **13**, 2513–2523.

Ng, C.K.-Y., Carr, K., McAinsh, M.R., Powell, B. & Hetherington, A.M. (2001) Drought-induced guard cell signal transduction involves sphingosine-1-phosphate. *Nature*, **410**, 596–599.

Okada, Y. (1997) Volume expansion-sensing outward-rectifier Cl^- channel: fresh start to the molecular identity and volume sensor. *Am. J. Physiol.*, **273**, C755–C789.

Pei, Z.-M., Kuchitsu, K., Ward, J.M., Schwarz, M. & Schroeder, J.I. (1997) Differential abscisic acid regulation of guard cell slow anion channels in *Arabidopsis* wild-type and *abi1* and *abi2* mutants. *Plant Cell*, **9**, 409–423.

Pei, Z.-M., Murata, Y., Benning, G., *et al.* (2000) Calcium channels activated by hydrogen peroxide mediate abscisic acid signalling in guard cells. *Nature*, **406**, 731–734.

Puius, Y.A., Mahoney, N.M. & Almo, S.C. (1998) The modular structure of actin-regulatory proteins. *Curr. Opin. Cell Biol.*, **10**, 23–34.

Pyne, S. & Pyne, N.J. (2000) Sphingosine 1-phosphate signalling in mammalian cells. *Biochem. J.*, **349**, 385–402.

Ressad, F., Didry, D., Xia, G.X., *et al.* (1998) Kinetic analysis of the interaction of actin-depolymerizing factor (ADF)/cofilin with G- and F-actins. Comparison of plant and human ADFs and effect of phosphorylation. *J. Biol. Chem.*, **273**, 20894–20902.

Ritchie, S. & Gilroy, S. (1998) Abscisic acid signal transduction in the barley aleurone is mediated by phospholipase D activity. *Proc. Natl. Acad. Sci. USA*, **95**, 2697–2702.

Sanders, L.C., Matsumura, F., Bokoch, G.M. & Lanerolle, P. (1999) Inhibition of myosin light chain kinase by p21-activated kinase. *Science*, **283**, 2083–2085.

Schmidt, C., Schelle, I., Liao, Y.-J. & Schroeder, J.I. (1995) Strong regulation of slow anion channels and abscisic acid signaling in guard cells by phosphorylation and dephosphorylation events. *Proc. Natl. Acad. Sci. USA*, **92**, 9535–9539.

Schroeder, J.I. & Hagiwara, S. (1989) Cytosolic calcium regulates ion channels in the plasma membrane of *Vicia faba* guard cells. *Nature*, **338**, 427–430.

Schroeder, J.I., Raschke, K. & Neher, E. (1987) Voltage dependence of K^+ channels in guard cell protoplasts. *Proc. Natl. Acad. Sci. USA*, **84**, 4108–4112.

Schroeder, J.I., Allen, G.J., Hugouvieux, V., Kwak, J.M. & Waner, D. (2001a) Guard cell signal transduction. *Annu. Rev. Plant Physiol. Plant Mol. Biol.*, **52**, 627–658.

Schroeder, J.I., Kwak, J.M. & Allen, G.J. (2001b) Guard cell abscisic acid signalling and engineering drought hardiness in plants. *Nature*, **410**, 327–330.

Schulz-Lessdorf, B., Lohse, G. & Hedrich, R. (1996) GCAC1 recognizes the pH gradient across the plasma membrane: a pH-sensitive and ATP-dependent anion channel links guard cell membrane potential to acid and energy metabolism. *Plant J.*, **10**, 993–1004.

Serrano, E.E., Zeiger, E. & Hagiwara, S. (1988) Red light stimulates an electrogenic proton pump in *Vicia* guard cell protoplasts. *Proc. Natl. Acad. Sci. USA*, **85**, 436–440.

Sheen, J. (1998) Mutational analysis of protein phosphatase 2C involved in abscisic acid signal transduction in higher plants. *Proc. Natl. Acad. Sci. USA*, **95**, 975–980.

Shimazaki, K.-I., Iino, M. & Zeiger, E. (1986) Blue light-dependent proton extrusion by guard cell protoplasts of *Vicia faba*. *Nature*, **319**, 324–327.

Staiger, C.J. (2000) Signaling to the actin cytoskeleton in plants. *Annu. Rev. Plant Physiol. Plant Mol. Biol.*, **51**, 257–288.

Staxén, I., Pical, C., Montgomery, L.T., Gray, J.E., Hetherington, A.M. & McAinsh, M.R. (1999) Abscisic acid induces oscillations in guard-cell cytosolic free calcium that involve phosphoinositide-specific phospholipase C. *Proc. Natl. Acad. Sci. USA*, **96**, 1779–1784.

Talbot, L.D. & Zeiger, E. (1998) The role of sucrose in guard cell osmoregulation. *J. Exp. Bot.*, **49**, 329–337.

Taylor, A.R., Manison, N.F.H., Fernandez, C., Wood, J. & Brownlee, C. (1996) Spatial organization of calcium signaling involved in cell volume control in the *Fucus* rhizoid. *Plant Cell*, **8**, 2015–2031.

Thiel, G. & Blatt, M.R. (1994) Phosphatase antagonist okadaic acid inhibits steady-state K^+ currents in guard cells of *Vicia faba*. *Plant J.*, **5**, 727–733.

Thiel, G., MacRobbie, E.A.C. & Blatt, M.R. (1992) Membrane transport in stomatal guard cells: the importance of voltage control. *J. Memb. Biol.*, **126**, 1–18.

Tilly, B.C., Edixhoven, M.J., Tertoolen, L.G., *et al.* (1996) Activation of the osmo-sensitive chloride conductance involves P21RHO and is accompanied by a transient reorganization of the F-actin cytoskeleton. *Mol. Biol. Cell*, **7**, 1419–1427.

Torsethaugen, G., Pell, E.J. & Assmann, S.M. (1999) Ozone inhibits guard cell K^+ channels implicated in stomatal opening. *Proc. Natl. Acad. Sci. USA*, **96**, 13577–13582.

Ullah, H., Chen, J.G., Young, J.C., Im, K.H., Sussman, M.R. & Jones, A.M. (2001) Modulation of cell proliferation by heterotrimeric G protein in *Arabidopsis*. *Science*, **292**, 2066–2069.

Ullah, H., Chen, J.G., Wang, S. & Jones, A.M. (2002) Role of a heterotrimeric G protein in regulation of *Arabidopsis* seed germination. *Plant Physiol.*, **129**, 897–907.

Wang, X.-Q., Ullah, H., Jones, A.M. & Assman, S.M. (2001) G protein regulation of ion channels and abscisic acid signaling in *Arabidopsis* guard cells. *Science*, **292**, 2070–2072.

Wu, Y., Kuzma, J., Maréchal, E., *et al.* (1997) Abscisic acid signaling through cyclic ADP-ribose in plants. *Science*, **278**, 2126–2130.

Xu, Q., Fu, H., Gupta, R. & Luan, S. (1998) Molecular characterization of a tyrosine-specific protein phosphatase encoded by a stress-responsive gene in *Arabidopsis*. *Plant Cell*, **10**, 849–857.

Yu, R., Huang, R.-F., Wang, X.-C. & Yuan, M. (2001) Microtubule dynamics are involved in stomatal movement of *Vicia faba* L. *Protoplasma*, **216**, 113–118.

Index